Elementary Functions

ELEMENTARY FUNCTIONS

Elvis H. Sullivan
San Antonio College

HOLBROOK PRESS, INC. **BOSTON**

Printed in the United States of America.

Library of Congress Cataloging in Publication Data

Sullivan, Elvis H
Elementary functions.

Includes index.
1. Functions. I. Title.
QA331.3.S92 515 74-28480
ISBN 0-205-04478-6

CONTENTS

PREFACE

This textbook is designed for a one semester course, to be taken by those students who do not have the necessary background for a first-year calculus course. The CUPM outline for the two-year college version of MATH 0 served as a guide in writing this text, though I do not make the claim that this outline was followed verbatim. It can be claimed, however, that the elementary functions are the primary concern of this text.

Each section of the text represents what I believe to be a reasonable amount of material for one class meeting. The problem sets are built on the basis of this belief. Each problem set contains enough problems so that the instructor will be able to make a selection. In most problem sets there are enough problems for a second "go" at the same topic when this becomes necessary. In the early chapters, however, there are some problems that I feel should always be assigned. These are marked with square brackets [ ].

It is assumed that the student has completed two years of high school algebra and could therefore begin this course with no need for the review sections of this text. My own teaching experience indicates that this is not always a realistic appraisal of the situation. But it is equally unrealistic to completely review high school algebra. For these reasons, then, I have felt it would be wise to provide, at the beginning of the text, a review which contains those axioms and their consequences with which the student is presumed to be familiar. This does not mean that the student has or can prove the theorems that appear in the review. They are placed there simply because there might exist a student who is curious about how they can be established. One should keep in mind the possibility that, at this time, the theorems in the review are so obvious that they do not require proof. In addition, I have included in the review my own selection of those ideas and manipulative skills that seem to give students trouble.

In the final analysis, though, the instructor is the only one who can and should decide how much review (and of what) is required by his class. It may even be a good idea to wait until it becomes clear that there is a need to review and then to use that review section that is pertinent to the need.

The definition of the elementary function L is not the traditional one but is as simple as the traditional definition and is the definition he will encourter in his calculus course. It is, therefore, a definition he will not have to "unlearn" in his calculus course.

Finally, for any portion of this text which has some merit, I am indebted to my teachers. For any blunders or shortcomings, I am unable to find anyone but myself to blame.

E. H. Sullivan
San Antonio College

Though not intended for such a course, this text can be, and has been, used in a one-semester course of college algebra for those students who intend to major in either mathematics, engineering, or the physical sciences. It is assumed that such a course would be a 3-semester-hour course and could use the following portions of this textbook:

Chapter 1	7 hours
Chapter 2, sections 1–3 only	3 hours
Chapter 3	6 hours
Chapter 4	3 hours
Chapter 5	4 hours
Chapter 6	6 hours
Chapter 7	4 hours
Chapter 9	6 hours
Chapter 10	4 hours
Total 43 hours	

Review

We shall begin by assuming the existence of objects called *real numbers*, some of which are called *counting numbers* or *natural numbers*, and we take for granted certain statements about numbers called *axioms*. Since we shall confine our attention to the real numbers exclusively, the statement that "x is a number" shall mean that x is a real number.

Section 1—Axioms The first few axioms (their traditional names are also given) are as follows:

AXIOM A1. (*closure for addition*). *If each of x and y is a number, then $x + y$ (read "x plus y") is a number called the sum of x and y.*

AXIOM A2. (*associative property of addition*). *If each of x, y, and z is a number, then $x + (y + z)$ is $(x + y) + z$.*

AXIOM A3. (*commutative property of addition*). *If each of x and y is a number, then $x + y$ is $y + x$.*

AXIOM A4. (*identity for addition*). 0 *is a number such that, if x is a number,* $0 + x$ *is x.*

AXIOM A5. (*inverse for addition*). *If x is a number, then $-x$ is a number such that $x + (-x)$ is* 0.

A study of these first few axioms can suggest a question which, when we answer it, leads us to discover a consequence of these axioms—that is, a theorem. It may even happen that one question will lead to another. For example, if we consider axiom A4, we may pose the question, Is 0 the only number such that, if x is a number, $0 + x$ is x? We shall now try to show that this question is answered affirmatively by some of the axioms, and we have our first theorem.

THEOREM A1. *If z is a number such that, for some number x, $z + x$ is x,* then z is 0.

Proof. $z + x = x$ given

Since x is a number, then $-x$ is a number by axiom A5, hence

$[z + x] + (-x) = x + (-x)$ by axiom A1

$z + [x + (-x)] = x + (-x)$ by axiom A2

$z + 0 = 0$ by axiom A5

$0 + z = 0$ by axiom A3

hence $z = 0$ by axiom A4

In a similar manner, after studying axiom A5, we might pose the question, If the sum of two numbers is 0, is one of the numbers the additive inverse of the other? Again we can show that the answer is yes, and so we have another theorem.

THEOREM A2. *If x is a number and z is a number such that $x + z$ is* 0, *then z is* $(-x)$.

The proof of this statement is so similiar to the one above that it is left for any student who desires to try his hand at discovering how to prove theorem A2.

Remark. Call the number $-x$ *the negative of* x.

If $-x$ is a number, then axiom A5 assures us that $-(-x)$ is a number and that $(-x) + [-(-x)]$ is 0. If we think about this for a little bit, we should get the idea that the following statement is a theorem.

THEOREM A3. *If x is a number, then $-(-x)$ is x.*

Proof. $x + (-x) = 0$ by axiom A5 thus
$(-x) + x = 0$ by axiom A3 hence
$x = -(-x)$ by theorem A2.

There is a small lesson to be learned here, namely, once a theorem has been shown to be true (i.e., once a theorem is proven), then that theorem as well as the axioms can be used to establish any later statement or theorem.

The preceding proofs were offered as models of how we can prove our theorems—that is, as sample proofs. A further study of the axioms can suggest other questions and theorems which would certainly include the following:

THEOREM A4. -0 *is* 0.

THEOREM A5. *If each of x, y, and z is a number and $x + y$ is z, then y is $z + (-x)$.*

Remark. The number $z + (-x)$ may be denoted by $z - x$ and called *the difference z minus x* or simply *z minus x.*

The student is encouraged to try and prove the above theorems; for mathematics is not a spectator sport and one can learn only by trying. To help you with the proof of theorem A4 may we make the following suggestions: First explain why $0 + 0$ is 0 and then try to see that theorem A2 can be of some use since it has already been established. The next few axioms are as follows:

AXIOM M1. (*closure for multiplication*). *If each of x and y is a number then $x \cdot y$ (read "x times y") is a number called the product of x and y.*

AXIOM M2. (*associative property of multiplication*). *If each of x, y, and z is a number, then $x \cdot (y \cdot z)$ is $(x \cdot y) \cdot z$.*

AXIOM M3. (*commutative property of multiplication*). *If x and y are numbers, then $x \cdot y$ is $y \cdot x$.*

AXIOM M4. (*identity for multiplication*). 1 *is a number such that, if x is a number,* $1 \cdot x$ *is x.*

AXIOM M5. (*inverse for multiplication*). *If x is a number and x is not* 0, *then $1/x$ is a number such that $x \cdot 1/x$ is* 1.

It should be noted that the last five axioms are similar to the first five with one important difference: in axiom M5 there is the restriction x is not 0, there is no counterpart to this in axiom A5. Thus 0 has no multiplicative inverse and hence division by 0 is not defined. If 0 had a multiplicative inverse, call it w, then $0 \cdot w$ is 1. This would contradict theorem M3 (below) and hence would be inconsistent with the axioms upon which the proof of theorem M3 is based. It is in this sense, then, that 1/0 is not defined. We continue our list of axioms with the following:

DISTRIBUTIVE AXIOM. *If each of x, y, and z is a number, then $x \cdot (y + z)$ is $(x \cdot y) + (x \cdot z)$.*

A study of these last axioms can, of course, generate many questions which in turn would lead to additional consequences of our axioms—that is, theorems. As before, I shall simply list some of the useful theorems to be proven.

THEOREM M1. *If y is a number such that, for some number x distinct from* 0, *$y \cdot x$ is x, then y is* 1.

THEOREM M2. *If x is a number different from* 0 *and y is a number such that $x \cdot y$ is* 1, *then y is $1/x$.*

Remark. If x is a number different from 0, call the number $1/x$ the *reciprocal of* x. The number $y \cdot 1/x$ may be denoted by y/x and called the *quotient* y *over* x or simply y *over* x.

THEOREM M3. *If x is a number, then $0 \cdot x$ is 0.*

THEOREM M4. *If x is a number, then $-1 \cdot x$ is $-x$.*

THEOREM M5. *If each of x and y is a number, then $(-x) \cdot y$ is $-(x \cdot y)$ and $(-x) \cdot (-y)$ is $x \cdot y$.*

THEOREM M6. *If each of x and y is a number and $x \cdot y$ is 0, then x is 0 or y is 0 (i.e., if x is not 0, then y is 0).*

It is convenient to use the symbol $=$ to mean "is." Thus, if each of x and y is a number, $x = y$ means that x is y. The denial of this may be expressed by $x \neq y$, x is not y. Further, if each of x and y is a number, the product $x \cdot y$ may be written as xy omitting the dot. Also, the product xx is denoted by x^2, xx^2 by x^3, xx^3 by x^4, and so forth. The number x^2 is called x *square*, x^3 is called x *cube*, x^4 is called x *to the fourth*. The use of these notations is illustrated in the following:

THEOREM P. *If each of x and y is a number, then $(x-y)(x+y) = x^2-y^2$ and $(x+y)^2 = x^2 + 2xy + y^2$.*

THEOREM N. *If each of x and y is a number, then $-(x+y) = -x-y$ and $-(x-y) = -x+y$.*

When the preliminary barrage has lifted, that is, when this brief review has been completed, we shall start our assault on the order of the real numbers. This assault will begin with a list of axioms called *order axioms*. One of these axioms tells the distinguishing things about the *counting numbers*, mentioned earlier. The student will surely notice that in the review, and in the remainder of this text for that matter, we shall assume (without specifically stating these assumptions) a great deal of knowledge about the sums and products of counting numbers. For example, if each of x and y is a counting number, then $x + y$ is a counting number and $x \cdot y$ is a counting number.

Section 2—Number Sets The primary objective in this section is to present enough about set theory to enable us to use the language of sets to describe some mathematical concepts. By a "set" we mean a nonempty collection of objects, that is, a collection which contains at least one object. The individual objects are called *elements* of the set. For example, we could talk about a set of poker chips or a set of tools or a set of good-looking females, but we will not. However, we will talk a great deal about a set, each element of which is a number. Such a set will be called a *number set*.

In this text, we shall use capital letters to denote number sets. In particular, R will be used to denote the set of all real numbers and N the set of counting numbers. We will denote the fact that the number b is an element of the set B by $b \in B$. The notation $b \in B$ is read "b is an element of B," "b belongs to B," or "b is in B." A number set is said to be defined if, for some number x, it can be determined that either $x \in B$ or $x \notin B$ (read "x is not an element of B").

There are two commonly used methods of describing a number set. They are called the *roster method* and the *rule method.* To describe a number set by the roster method, we list the elements of the set and enclose them with braces. For example, if A is the number set whose only elements are 1, 3, and 5, we write

$$A = \{1, 3, 5\}.$$

The roster method is very convenient for sets with only a few elements but for sets that have many elements (even an infinite number) this method is, to say the least, awkward. To denote a number set by the rule method, we write a statement or rule that tells us which numbers belong to the set and which do not and enclose this rule within braces. Thus the notation $B = \{x \mid x = 2n - 1, n \in N\}$ should be read "B is the set of all x such that x is $2n - 1$ for some counting number n." For a second example of the use of the rule method:

$$C = \left\{x \,\middle|\, x = \frac{n-1}{n}, n \in N\right\}$$

(Read this "C is the set of all x such that x is $\frac{n-1}{n}$ for some counting number n.") Though we shall usually prefer to use the rule method, it may be instructive to give the sets B and C by the roster method.

$$B = \{1, 3, 5, 7, \ldots\} \qquad C = \{0, 1/2, 2/3, 3/4, \ldots\}.$$

We read the three dots as "and so on." For another example, consider a number set Q defined by the notation

$$Q = \{x \mid x^2 - 3x - 10 = 0, x \in R\}.$$

We read the above notation as "Q is the set of all x such that $x^2 - 3x - 10 = 0$ for some number x." It should be easy to see that $Q = \{-5, 2\}$.

DEFINITION. *If each of A and B is a set, then the statement that "A is a subset of B," denoted by $A \subset B$, means that each element of A is also an element of B.*

$A \subset B$ is often read "A is contained in B."

Example. If $P = \{1, 3, 5\}$, $M = \{0, 1, 2, 3, 4, 5\}$, and $U = \{0, 1, 2, 5\}$ then $P \subset M$ and $U \subset M$ but $P \not\subset U$.

Remark. The definition above implies that a set is a subset of itself. Thus $A \subset A$ is always true.

DEFINITION. *Two sets are equal if and only if they are identical, that is, if and only if they are composed of exactly the same elements.*

Thus $A = B$ if and only if $A \subset B$ and $B \subset A$.

In addition to these relations between sets, there are two operations on number sets which we shall use.

DEFINITION. *If each of A and B is a number set, then the union of A and B, denoted by $A \cup B$, is the number set whose elements are the numbers that are in A or in B.*

Thus $x \in A \cup B$ if and only if $x \in A$ or $x \in B$.

Remark. The word "or" as used in mathematics has the meaning of "and/or" in colloquial language.

DEFINITION. *If each of A and B is a number set, then the intersection of A and B, if it exists, is denoted by $A \cap B$, and is the number set whose elements are the numbers that are both A and B.*

Thus, $x \in A \cap B$ if and only if $x \in A$ and $x \in B$.

Example. Suppose $A = \{0, 1, 2\}$, $B = \{0, 2, 4)$, and $C = \{1, 3, 5\}$. $A \cup B = \{0, 1, 2, 4\}$, $A \cap B = \{0, 2\}$, and $B \cup C = \{0, 1, 2, 3, 4, 5\}$. But $B \cap C$ is not a number set. This explains the reason for the phrase "if it exists" in the definition above.

For Student Eyes Only Each of the last three definitions is followed by a statement that begins with the word "thus." In each case, this statement is a compactly written equivalent of the definition which it follows.

Also, in each case, this statement includes the phrase "if and only if." My main purpose in writing these statements is to goad you into either asking your instructor what this phrase means or to read the following: If S_1 and S_2 are two statements, then the sentence "S_1 if and only if S_2" means that S_1 implies S_2 and S_2 implies S_1. For example, in order that the sentence "$x \cdot y = 0$ if and only if $x = 0$ or $y = 0$" be true, one must show that each of the following is true:

1. If $x \cdot y = 0$, then $x = 0$ or $y = 0$.
2. If $x = 0$ or $y = 0$, then $x \cdot y = 0$.

PROBLEMS FOR REVIEW

Group I: Sets

Indicate each of the following by the roster method.

1. $\{x \mid x = 2n + 1, n \in N\}$ **2.** $\{x \mid x = n^2 - 1, n \in N\}$

3. $\{x \mid x = 1/n, n \in N\}$ **4.** $\{x \mid x = \dfrac{n+1}{n}, n \in N\}$

5. $\{x \mid x = 3n - 2, n \in N\}$ **6.** $\{x \mid x = \dfrac{1}{n(n+1)}, n \in N\}$

Indicate each of the following by the rule method.

7. $\{1, 4, 9, 16, \ldots\}$ **8.** $\{1/3, 1/5, 1/7, 1/9, \ldots\}$

9. $\{2, 4, 6, 8, \ldots\}$ **10.** $\{7, 5, 3, 1, \ldots\}$

If $K = \{-2, 0, 2, 4\}$, $L = \{0, 1, 2, 3, 4, 5\}$, and $M = \{1, 4, 7\}$, give the following number sets (if they exist):

11. $K \cup L$ **12.** $K \cap L$ **13.** $L \cup M$ **14.** $L \cap M$ **15.** $K \cap M$

16. Give the intersection of the number sets given in problems 1 and 10 above.

17. Give the intersection of the number sets given in problems 3 and 8 above.

Section 3—Special Products The intent of this section is to review some consequences of the distributive axiom which, because they occur so frequently in mathematics, warrant special attention. Although these special products are primarily applications of the distributive axiom, we shall not hesitate to use the exponent notation mentioned just prior to theorem P, for compactness. For example, if x is a number and $m \in N$, then $x^m x$ shall be denoted by x^{m+1}.

It is admitted that the results shown on these pages do not exhaust the topic. Rather, we select only those products that occur with great regularity, so let us begin.

I If each of p, q, and x is a number, then $(x + p)(x + q)$ is $(x + p)x + (x + p)q$ by the distributive axiom, but $(x + p)x + (x + p)q$ is $x(x + p) + q(x + p)$ by axiom M3, and $x(x + p) + q(x + p)$ is $x^2 + px + qx + pq$, hence

$$(x + p)(x + q) \text{ is } x^2 + (p + q)x + pq.$$

Attention is invited to the idea that, given p and q, part of the problem is to find their sum $p + q$ and their product pq. This is particularly useful when we are trying to express the sum $x^2 + (p + q)x + pq$ as a product—that is, factoring.

Examples. To express the sum $x^2 + 3x + 2$ as a product, we try to find two numbers whose sum is 3 and whose product is 2. It is easy to see that 1 and 2 is such a pair, thus

$$x^2 + 3x + 2 = (x + 2)(x + 1).$$

To factor $x^2 - 4x + 21$, we find a pair of numbers whose sum is -4 and whose product is 21. Hence

$$x^2 - 4x + 21 = (x - 7)(x + 3).$$

Examples. These are examples of the product $(x + p)(x + q)$.

$$(x + 3)(x + 6) = x^2 + (3 + 6)x + 3 \cdot 6 = x^2 + 9x + 18.$$
$$(x - 5)(x - 7) = x^2 - (5 + 7)x + (-5)(-7)$$
$$= x^2 - 12x + 35.$$
$$(x + 8)(x - 1) = x^2 + 7x - 8.$$
$$(2t + 3)(2t + 5) = (2t)^2 + (3 + 5)2t + 15$$
$$= 4t^2 + 16t + 15.$$
$$(3s + 1)(3s - 4) = 9s^2 - 9s - 4.$$
$$(z^m + 8)(z^m + 3) = (z^m)^2 + 11z^m + 24 = z^{2m} + 11z^m + 24.$$

Try to see one reason why the product $(2w + 1)(w + 4)$ is not an example of this special product.

If we let $q = p$ or $q = -p$ in the statement

$$(x + p)(x + q) \text{ is } x^2 + (p + q)x + pq$$

we can obtain some useful variations of this statement.

If $q = p$, then we have $(x + p)^2 = x^2 + 2px + p^2$ which is a variation that is the basis of a process called *completing the square*, a process which we will examine later.

If q is $-p$, we have that $(x + p)(x - p) = x^2 - p^2$. This variation occurs so often that it has been labeled the *difference of two squares.*

Examples.

$$r^2 - 1 = (r - 1)(r + 1).$$
$$(3w + 2)(3w - 2) = 9w^2 - 4.$$
$$(x^m - 5)(x^m + 5) = (x^m)^2 - 25 = x^{2m} - 25.$$
$$(x + 9)^2 = x^2 + 18x + 81.$$
$$(z - 7)^2 = z^2 - 14z + 49.$$
$$(5y + 3)^2 = 25y^2 + 2 \cdot 15y + 9 = 25y^2 + 30y + 9.$$
$$(x - \sqrt{3})(x + \sqrt{3}) = x^2 - 3.$$
$$(x - \sqrt{2})^2 = x^2 - 2\sqrt{2} \cdot x + 2.$$

II If each of k and m is a number different from 0, and if each of p, q, and x is a number, then

$$(kx + p)(mx + q) = (kx + p)mx + (kx + p)q \quad \text{or}$$
$$(kx + p)(mx + q) = kmx^2 + (pm + kq)x + pq.$$

Admittedly I is a special case of II, but it is one that is simple and occurs so often that it deserves being treated separately.

Examples.

$$\begin{aligned}(2w + 3)(w + 4) &= 2w^2 + (3 \cdot 1 + 2 \cdot 4)w + 12 \\ &= 2w^2 + 11w + 12.\end{aligned}$$
$$(3t - 2)(2t + 7) = 6t^2 + 17t - 14.$$
$$(2y - 1)(3y + 2) = 6y^2 + y - 2.$$

The ability to write certain sums as a product (this is called factoring) is based almost entirely upon knowing products of the sort already indicated. In the examples which follow, we will examine only the more difficult or sophisticated applications of our knowledge of special products.

The first collection of examples illustrate the use made of the distributive axiom. This is sometimes labeled *common factoring.*

Examples.

$$\begin{aligned}x^2 + 3x + ax + 3a &= x(x + 3) + a(x + 3) \\ &= (x + 3)(x + a).\end{aligned}$$
$$x^{2m+1} + x^{2m} = x^{2m} \cdot x + x^{2m} = x^{2m}(x + 1).$$
$$\begin{aligned}(x - 3)^2 + 2(x - 3) &= (x - 3)[(x - 3) + 2] \\ &= (x - 3)(x - 1).\end{aligned}$$

Examples. Difference of two squares.

$$\begin{aligned}(2x + 7)^2 - 4 &= (2x + 7)^2 - 2^2\\ &= [(2x + 7) - 2][(2x + 7) + 2]\\ &= (2x + 5)(2x + 9).\end{aligned}$$

$$t^{2m} - 25 = (t^m - 5)(t^m + 5).$$

Note that use is made of the fact that $t^{2m} = t^m \cdot t^m = (t^m)^2$.

Example. To factor $(3t - 1)^2 - 4(3t - 1) - 5$, it helps to let $z = (3t - 1)$. Thus $(3t - 1)^2 - 4(3t - 1) - 5 = z^2 - 4z - 5$ which is easily factored as $(z - 5)(z + 1)$. Replacing z, we learn that

$$\begin{aligned}(3t - 1)^2 - 4(3t - 1) - 5 &= [(3t - 1) - 5][(3t - 1) + 1]\\ &= 9t(t - 2).\end{aligned}$$

Our final example of factoring will be based upon the idea that we can show that $(z - p)(z^2 + pz + p^2)$ is $z^3 - p^3$, the *difference of two cubes.*

Example. Factor $64m^3 - 125 = (4m)^3 - 5^3$
$= (4m - 5)(16m^2 + 20m + 25).$

Our final effort in this section of the review is an examination of the technique called *completing the square.* This technique is useful and is based upon the fact that

$$(x + p)^2 = x^2 + 2px + p^2.$$

The following examples will help you to remember the details of this technique.

Example. Suppose $y = x^2 + 8x - 3$, then

$$y + 3 = x^2 + 8x$$

and now we think $2p = 8$, so p is 4, and so p^2 is 16. Thus

$$y + 19 = x^2 + 8x + 16 = (x + 4)^2$$

from which we can deduce that

$$\begin{aligned}&y = (x + 4)^2 - 19, \text{ that is,}\\ &x^2 + 8x - 3 = (x + 4)^2 - 19.\end{aligned}$$

For some purposes, it is useful to write this as a difference of two squares. Thus

$$x^2 + 8x - 3 = (x + 4)^2 - (\sqrt{19})^2$$

for a difference of two squares is always easily factored.

Example. Suppose that $y = 2x^2 + 6x + 1$. Since the coefficient of x^2 is not 1, we have a small delay in which we write

$$y/2 = x^2 + 3x + 1/2$$

and then resume our normal pattern with

$$y/2 - 1/2 = x^2 + 3x$$

and, after thinking $2p = 3$ or $p^2 = 9/4$, we write

$$y/2 - 1/2 + 9/4 = x^2 + 3x + 9/4$$

or $y/2 + 7/4 = (x + 3/2)^2$, so that

$y/2$ is $(x + 3/2)^2 - 7/4$, therefore,
y is $2(x + 3/2)^2 - 7/2$

and if we wish to write this as a difference of two squares we can do so by observing that

$$2(x + 3/2)^2 - 7/2 = 2[(x + 3/2)^2 - 7/4]$$

hence $y = 2[(x + 3/2)^2 - (\sqrt{7}/2)^2]$.

Example. If $y = 2x^2 - 2x + 5$, then $y/2 = x^2 - x + 5/2$ so that

$y/2 - 5/2 = x^2 - x$ (here $2p = -1$ so $p^2 = 1/4$)
$y/2 - 5/2 + 1/4 = x^2 - x + 1/4$
$y/2 - 9/4 = (x - 1/2)^2$
$y/2 = (x - 1/2)^2 + 9/4$
$y = 2(x - 1/2)^2 + 9/2$.

This is not a difference of two squares.

Example. If $y = 1 - 10x - x^2$, then $-y = x^2 + 10x - 1$ so

$1 - y = x^2 + 10x$ (1) note $2p$ is 10 (1)
$26 - y = x^2 + 10x + 25 = (x + 5)^2$
$-y = (x + 5)^2 - 26$ and so $y = 26 - (x + 5)^2$.

PROBLEMS FOR REVIEW

Group II: Special Products

A Find the following products.

1. a. $(x + r)(x + s)$
b. $(x + s)^2$
c. $(x + p)(x^2 - px + p^2)$
d. $(x + r)(x - r)(x^2 + r^2)$

2. a. $(z - \sqrt{m})(z + \sqrt{m})$
b. $(kx + r)(x + s)$
c. $(x - p)(x^2 + px + p^2)$
d. $(u^m - a)(u^m + a)$

3. **a.** $(x + 2)(x - 3)$
b. $(x + 5)(x - 2)$
c. $(z - 7)(z - 8)$
d. $(t + 4)(t + 9)$

4. **a.** $(x + 2)(2x - 3)$
b. $(w - 1)(3w + 2)$
c. $(2s + 1)(s + 1)$
d. $(4x - 3)(x + 2)$

5. **a.** $(m + 5)(m + 4)$
b. $(x - 1)(x + 8)$
c. $(2t + 3)(2t - 1)$
d. $(s + 8)(s - 9)$

6. **a.** $(2t + 3)(2t + 1)$
b. $(t - 5)(2t + 3)$
c. $(x + 3)(3x - 5)$
d. $(3n + 1)(2n - 1)$

7. **a.** $(x + 1/2)(x + 2)$
b. $(x + 1/3)(x + 3)$
c. $(x - 3/2)(x + 2)$
d. $(x + 3/4)(x - 1/3)$

8. **a.** $(y - \sqrt{3})(y + \sqrt{3})$
b. $(2x + 1)(2x - 1)$
c. $(x + \sqrt{5})(x - \sqrt{5})$
d. $(3s + 2t)(3s - 2t)$

9. **a.** $x^{2n-1}(x + 1)$
b. $p^m(p - 1)$
c. $z^m(z + 1)$
d. $z^{m-1}(z - 1)$

10. **a.** $[(x+1) - \sqrt{2}][(x+1) + \sqrt{2}]$
b. $[(s+1) - \sqrt{3}][(s+1) + \sqrt{3}]$
c. $[(t-3) + \sqrt{5}][(t-3) - \sqrt{5}]$
d. $[(z+2) + \sqrt{7}][(z+2) - \sqrt{7}]$

11. **a.** $(m - 3)(m^2 + 3m + 9)$
b. $(t - 2)(t^2 + 2t + 4)$
c. $(z - \sqrt{2})(z^2 + \sqrt{2}\cdot z + 2)$
d. $(2x - 1)(4x^2 + 2x + 1)$

12. **a.** $(t + 1)(t^2 - t + 1)$
b. $(x + 3)(x^2 - 3x + 9)$
c. $(2m + 3)(4m^2 - 6m + 9)$
d. $(s + \sqrt{3})(s^2 - \sqrt{3}\cdot s + 3)$

13. **a.** $(x + 12)^2$
b. $(x - 8)^2$
c. $(5 - s)^2$
d. $(2t + 7)^2$

14. If each of r, s, and x is a number, verify that

$$[x + r/2]^2 - \frac{r^2 - 4s}{4} \text{ is } x^2 + rx + s.$$

15. Verify that $\dfrac{r^2 + 4s}{4} - [x - r/2]^2$ is $s + rx - x^2$.

B Express each of the following sums as a product.

Common Factor

16. **a.** $3x - 3a + x^2 - ax$
b. $x^2 - tx + 3x - 3t$
c. $x^2 - a^2 + 2x - 2a$
d. $m^2 - t^2 - 7m + 7t$

17. **a.** $x^2 - ax - 3x + 3a$
b. $4x - 4a - x^2 + ax$
c. $x^2 - a^2 - 5x + 5a$
d. $x^2 - x - mx + m$

18. a. $z^m + z^{m-1}$
b. $t^{m+1} - t^m$
c. $x^m - x$
d. $x^m - x^{m-1}$

$x^2 + (p + q)x + pq$ is $(x + p)(x + q)$

19. a. $x^2 + 6x + 8$
b. $y^2 + 11y + 30$
c. $s^2 + 11s + 28$
d. $x^2 + 13x + 42$

20. a. $w^2 - 13w + 40$
b. $x^2 - 10x + 24$
c. $m^2 - 5m + 6$
d. $x^2 - 8x + 12$

21. a. $x^2 - 14x + 49$
b. $x^2 - 2x + 1$
c. $w^2 - 10w + 25$
d. $t^2 - 8t + 16$

22. a. $m^2 - 3m - 28$
b. $x^2 + 2x - 3$
c. $x^2 + 5x - 14$
d. $z^2 - z - 56$

23. a. $3 - 2x - x^2$
b. $6 - x - x^2$
c. $5 + 4t - t^2$
d. $12 + t - t^2$

24. a. $t^2 + 12t + 36$
b. $m^2 + 18m + 81$
c. $x^2 + 6x + 9$
d. $x^2 + 2x + 1$

Difference of Two Squares

25. a. $4x^2 - a^2$
b. $w^2 - 36$
c. $x^2 - 64$
d. $9m^2 - 4$

26. a. $z^{2n} - 4$
b. $x^{2m} - 1$
c. $x^{2m} - 144$
d. $z^{2n} - p^2$

27. a. $(x + 2)^2 - 9$
b. $(s - 1)^2 - 4$
c. $(m + 3)^2 - 1$
d. $(s + 1)^2 - 16$

$(kx + p)(mx + q)$ is $kmx^2 + (mp + kq)x + pq$

28. a. $2x^2 + 7x + 3$
b. $3x^2 - 7x + 2$
c. $4w^2 + w - 5$
d. $3z^2 - 11z - 4$

29. a. $3x^2 + 2x - 1$
b. $2x^2 - 9x + 9$
c. $4m^2 - 9m + 2$
d. $5s^2 + 11s + 2$

30. a. $4t^2 + 4t - 15$
b. $6x^2 + 5x + 1$
c. $6s^2 - s - 1$
d. $8x^2 + 10x - 3$

Sum or Difference of Two Cubes
(See Problems 1c and 2c)

31. a. $m^3 - 1$
b. $x^3 - 8$
c. $w^3 - 1$
d. $x^3 - 27$

32. a. $x^3 + 8$
b. $m^3 + 1$
c. $w^3 + 64$
d. $z^3 + 1$

33. a. $8z^3 - 27$
b. $27x^3 - 1$
c. $x^3 - 8a^3$
d. $x^3 + 8t^3$

Miscellaneous

34. a. $(x - a)^2 + 2(x - a)$
b. $(x - 2)^2 + 4x - 8$
c. $x^2 - a^2 + 3x - 3a$
d. $4x^3 - 32$

35. a. $(z + 1)^2 + 3(z + 1) + 2$
b. $2(s + 1)^2 + 8(s + 1) + 6$
c. $s^2t^2 - 2st^3 - 63t^4$
d. $w^4 - 10w^2 + 9$

By completing the square, write each of the following in either of the forms $m(x + p)^2 - k$, $k - m(x + p)^2$, or $m(x + p)^2 + k$.

36. **a.** $x^2 + 6x + 4$
b. $x^2 - 8x + 3$
c. $x^2 - 2x - 5$
d. $x^2 + 10x + 5$

37. **a.** $1 + 4x - x^2$
b. $6 - 2x - x^2$
c. $6x - x^2$
d. $3 + 8x - x^2$

38. **a.** $x^2 - 3x + 1$
b. $x^2 + 5x - 4$
c. $x^2 + x - 2$
d. $x^2 - 5x + 1$

39. **a.** $2x^2 + 5x - 1$
b. $2x^2 + 4x - 3$
c. $3x^2 - x - 4$
d. $2x^2 + x - 6$

40. **a.** $x^2 + x + 2$
b. $2x^2 - 3x + 4$
c. $x^2 + 4x + 3$
d. $3x^2 - 3x + 1$

Section 4—Quotients One topic in elementary algebra with which most students have difficulty is that of finding and simplifying the sum and product of quotients. It is on these ideas, then, that we now focus our attention.

The basic concepts involved are expressed in the remark which follows theorem M2 and in the following statements:

THEOREM Q1. *If x is a number, then* $\frac{x}{1} = x$.

THEOREM Q2. *If each of a and b is a number and if each of x and y is a number distinct from* 0, *then* $\frac{a}{x} \cdot \frac{b}{y} = \frac{ab}{xy}$.

THEOREM Q3. *If each of a and b is a number and if x is a number different from* 0, *then* $\frac{a}{x} + \frac{b}{x} = \frac{a+b}{x}$.

THEOREM Q4. *If x is a number and if y is a number different from* 0, *then* $-\frac{x}{y} = \frac{-x}{y} = \frac{x}{-y}$.

THEOREM Q5. *If each of x and y is a number distinct from* 0, *then* $\dfrac{\frac{1}{x}}{y} = \frac{y}{x}$.

Since quotients are troublesome enough in any case, it is wise to know how to simplify any quotient that can be simplified before using it in any computation. In fact, as a mathematician you are expected to be able to simplify quotients and find their sum and product whenever the glorious opportunity arises. In part, at least, the ability to do these things is based upon the above theorems and is best reviewed by considering examples.

Example. Find and simplify

$$\frac{3}{y+2} - \frac{2}{y-2} + \frac{y}{(y-2)(y+2)},$$

$$\frac{3}{y+2} = \frac{3(y-2)}{(y+2)(y-2)},$$

and

$$\frac{2}{y-2} = \frac{2(y+2)}{(y-2)(y+2)}.$$

Thus the given sum equals the sum

$$\frac{3y-6}{y^2-4} + \frac{-2y-4}{y^2-4} + \frac{y}{y^2-4} = \frac{2y-10}{y^2-4}.$$

Example. Find and simplify

$$\frac{x+4}{(x-2)(x+1)} + \frac{3}{x+1}.$$

Note that

$$\frac{3}{x+1} = \frac{3(x-2)}{(x+1)(x-2)}$$

hence the given sum is equivalent to the sum

$$\frac{x+4}{(x-2)(x+1)} + \frac{3x-6}{(x-2)(x+1)} = \frac{4x-2}{(x-2)(x+1)}.$$

Example. $\dfrac{x-4}{x^2-5x+6} + \dfrac{2x}{x^2-2x-3}$

$$= \frac{x-4}{(x-6)(x+1)} + \frac{2x}{(x-3)(x+1)}$$

$$= \frac{(x-4)(x-3) + 2x(x-6)}{(x-6)(x-3)(x+1)}$$

$$= \frac{x^2-7x+12+2x^2-12x}{(x-6)(x-3)(x+1)}$$

$$= \frac{3x^2-19x+12}{(x-6)(x-3)(x+1)}$$

Example.
$$\frac{5}{xy + y^2} - \frac{3}{x^2 + xy} = \frac{5}{y(x + y)} - \frac{3}{x(x + y)}$$
$$= \frac{5x}{xy(x + y)} + \frac{-3y}{xy(x + y)}$$
$$= \frac{5x - 3y}{xy(x + y)}$$

Example. Simplify the quotient

$$\frac{x + \dfrac{x}{x - 1}}{1 + \dfrac{1}{x^2 - 1}}$$

Now $x + \dfrac{x}{x - 1} = \dfrac{x^2}{x - 1}$ and $1 + \dfrac{1}{x^2 - 1} = \dfrac{x^2}{x^2 - 1}$. Hence the given quotient can be simplified to the product

$$\frac{x^2}{x - 1} \cdot \frac{1}{\dfrac{x^2}{x^2 - 1}}$$

which, by theorem Q5, is the product $\dfrac{x^2}{x - 1} \cdot \dfrac{x^2 - 1}{x^2}$. At this in the example, most students would "cancel" x^2 and $x - 1$ and finish rather rapidly by stating that the given quotient is the number $x + 1$. Let us briefly examine what is meant by this universally used terminology "cancel." Begin by noting that

$$\frac{x^2}{x - 1} \cdot \frac{(x - 1)(x + 1)}{x^2} = x^2 \cdot \frac{1}{x - 1} \cdot (x - 1)(x + 1) \cdot \frac{1}{x^2}$$
$$= [x^2 \cdot 1/x^2] \cdot [(x - 1) \cdot 1/(x - 1)] \cdot (x + 1)$$

and that, in each case, the number in the square bracket is 1 by axiom M5. Therefore, let us cancel for we understand what this means.

Example. $(2 + 1/x) \div (2x^2 + x) = \dfrac{2x + 1}{x} \cdot \dfrac{1}{x(2x + 1)} = 1/x^2$

Example. $\dfrac{x^2 - x - 6}{x^2 - 4} = \dfrac{(x - 3)(x + 2)}{(x - 2)(x + 2)} = \dfrac{x - 3}{x - 2}$

Example. $$\frac{x^3 - y^3}{x + y} \div \frac{x^2 + xy + y^2}{ax + ay}$$

$$= \frac{(x - y)(x^2 + xy + y^2)}{x + y} \cdot \frac{a(x + y)}{x^2 + xy + y^2}$$

$$= a(x - y)$$

PROBLEMS FOR REVIEW

Group III: Quotients

Simplify each of the following:

1. $1/x + 1/y$

2. $\dfrac{3t - 1}{10} + \dfrac{5 - 2t}{15}$

3. $5 - \dfrac{5}{x + 3}$

4. $\dfrac{3x - 1}{5} + \dfrac{4 - 5x}{6}$

5. $\dfrac{3}{y - 2} - \dfrac{2}{y + 2} - \dfrac{y}{y^2 - 4}$

6. $\dfrac{3}{a - 3} + \dfrac{a^2 + 2}{a^3 - 27}$

7. $\dfrac{2xy}{x^3 + y^3} - \dfrac{x}{x^2 - xy + y^2}$

8. $\dfrac{4}{x^2 - 4x - 5} + \dfrac{2}{x^2 - 1}$

9. $\dfrac{x^2 - 2xy + y^2}{x^3 - y^3} \cdot \dfrac{x^2 + xy + y^2}{x - y}$

10. $\dfrac{x - x/t}{z - z/t}$

11. $\dfrac{1}{1 - \dfrac{1}{1 + \dfrac{1}{x}}}$

12. $\dfrac{x}{1 - \dfrac{1}{1 + \dfrac{x}{y}}}$

13. $\dfrac{x + y}{\dfrac{1}{x} + \dfrac{1}{y}}$

14. $\left(y + \dfrac{2y}{y - 2}\right) \div \left(1 + \dfrac{4}{y^2 - 4}\right)$

15. $\dfrac{y^2 - 2y - 15}{y^2 - 9} \div \dfrac{12 - 4y}{y^2 - 6y + 9}$

16. $\dfrac{\dfrac{s^2 - 3s + 2}{s^2 + 4s - 21}}{\dfrac{4 - 4s + s^2}{9 - s^2}}$

17. $\dfrac{\dfrac{1}{x} + \dfrac{1}{y}}{\dfrac{1}{x^2} - \dfrac{1}{y^2}}$

18. $\dfrac{t^2 - \dfrac{1}{t}}{\dfrac{1}{t} + t + 1}$

19. $\dfrac{x/w + w^2/x^2}{w/x^2 - 1/x + 1/w}$

20. $(x - 1/x) \div (x + 2 + 1/x)$

21. $\dfrac{x^2 - 16}{x - 4}$

22. $\dfrac{x^3 - 8}{x - 2}$

23. $\dfrac{2x^3 + 54}{x + 3}$

24. $\dfrac{\dfrac{x}{2x + 1} - \dfrac{3}{5}}{x + 3}$

25. $\dfrac{1/x - 1/a}{x - a}$

26. $\dfrac{1/x^2 - 1/a^2}{x - a}$

27. $\left[\dfrac{1}{\sqrt{x}} - \dfrac{1}{2}\right] \div (x - 4)$

In each of the following, rationalize the numerator and simplify

28. $\dfrac{\sqrt{x} - \sqrt{a}}{x - a}$

29. $\dfrac{\sqrt{(x + 1)} - \sqrt{2}}{x - 1}$

30. $\dfrac{\sqrt{(x^2 - 1)} - \sqrt{3}}{x - 2}$

Section 5—Equations Study the examples given in this section; their purpose is to help you recall the details in finding the solution set of certain statements about numbers called equations. Now, the solution set of a statement about numbers (equation or not) is the number set to which a number belongs if and only if the statement is true for that number.

I Quadratic equations: If $a \neq 0$, then $ax^2 + bx + c = 0$ is called a quadratic equation. Some quadratic equations can be quickly and easily solved by factoring. This is illustrated in the examples which immediately follow.

Example. Solve $x^2 + 3x = 0$.

Solution. Since $x^2 + 3x$ is $x(x + 3)$, it quickly follows that $x^2 + 3x$ is 0 if and only if $x = 0$ or $x = -3$. Hence the solution set is $\{0, -3\}$.

Example. Solve $2x^2 - 11x + 5 = 0$.

Solution. $2x^2 - 11x + 5$ is $(2x - 1)(x - 5)$ and so $2x^2 - 11x + 5 = 0$ if and only if $2x - 1 = 0$ or $x - 5 = 0$. The solution set therefore is $\{1/2, 5\}$.

But not all quadratic equations can be so easily solved. For those quadratic equations which we cannot easily factor, the technique of completing the square can be used to find the solution set. Consider the following examples.

Example. Solve $x^2 + 8x - 3 = 0$.

Solution. $x^2 + 8x = 3$. Since $2p$ is 8, p is 4, and p^2 is 16; hence, $x^2 + 8x + 16 = 19$ or $(x + 4)^2 = 19$. Thinking about a difference of two squares, which can be easily factored, we go on to write $(x + 4)^2 - (\sqrt{19})^2 = 0$ so that

$$[(x + 4) - \sqrt{19}][(x + 4) + \sqrt{19}] = 0$$

therefore, the solution set is $\{-4 + \sqrt{19}, -4 - \sqrt{19}\}$. It is permissible to write the solution set as

$$\{-4 \pm \sqrt{19}\}.$$

Example. Solve $3x^2 - 5x - 2 = 0$. While this may be factored, let us suppose that the factoring escapes us and resort to the completion of the square. Since the coefficients of x^2 is not one, we begin with $x^2 - 5/3 \cdot x = 2/3$. We now think $2p$ is $-5/3$ so that p is $-5/6$; hence p^2 is $25/36$. Adding p^2 to both sides, we get

$$x^2 - 5/3 \cdot x + 25/36 = 49/36$$
$$(x - 5/6)^2 - 49/36 = 0$$
$$(x - 5/6)^2 - (7/6)^2 = 0$$
$$[(x - 5/6) + 7/6][(x - 5/6) - 7/6] = 0$$
$$(x + 1/3)(x - 2) = 0$$

therefore, the solution set is $\{-1/3, 2\}$.

Example. Solve $2x^2 - 4x + 1 = 0$.

Solution. $x^2 - 2x = -1/2$. Since $2p = -2$, $p^2 = 1$, thus

$$x^2 - 2x + 1 = 1/2$$
$$(x - 1)^2 - (\sqrt{2}/2)^2 = 0 \quad \text{Note: } 1/2 \text{ is } 2/4.$$
$$[x - 1] + \sqrt{2}/2][x - 1 - \sqrt{2}/2] = 0$$

and the solution set is

$$\left\{\frac{2 \pm \sqrt{2}}{2}\right\}.$$

Since we confine ourselves to the set of real numbers, not every quadratic equation will have a solution set. This is illustrated in the next example.

Example. Solve $x^2 + 6x + 10 = 0$.

Solution.

$x^2 + 6x = -10.$	(here $2p = 6$ so $p^2 = 9$)
$x^2 + 6x + 9 = -1$	
$(x + 3)^2 = -1$	and we should stop at this point and realize why there is no number x for which this is true.

II Radical equations: If each of r and s is a positive number, then every solution of the radical equation $\sqrt{r} = s$ is also a solution of the equation $r = s^2$. The converse of this statement, however, is not true as the first example illustrates.

Example. Solve $\sqrt{(2x^2 - 2x + 1)} = 2x - 3$.

Solution. Square both sides and we get

$$2x^2 - 2x + 1 = 4x^2 - 12x + 9, \text{ hence}$$
$$2x^2 - 10x + 8 = 0 \text{ or } x^2 - 5x + 4 = 0.$$

Thus $(x - 4)(x - 1) = 0$ and so $x = 4$ or $x = 1$. Now, if $x = 1$, then $\sqrt{(2x^2 - 2x + 1)} = 1$ but $2x - 3 = -1$, therefore 1 is not a solution of the given equation. Replacing x with 4 in the given equation shows that $x = 4$ is a solution. Thus the solution set is $\{4\}$.

Example. Solve $\sqrt{(4x + 1)} = 3 - \sqrt{(x - 2)}$.

Solution. Square both sides and get

$$4x + 1 = 9 - 6\sqrt{(x - 2)} + x - 2$$

which simplifies to

$$x - 2 = -3\sqrt{(x - 2)}$$

squaring again, we get

$$x^2 - 4x + 4 = 9(x - 2)$$

and simplify this to get

$$x^2 - 13x + 22 = 0$$

so $(x - 11)(x - 2) = 0$, hence $x = 11$ or $x = 2$.

Replacing x with 11 in the given equation, we have $\sqrt{(4x + 1)} = \sqrt{45}$ and $3 - \sqrt{(x - 2)} = 3 - \sqrt{9} = 0$ but $\sqrt{45} \neq 0$, hence 11 is not a solution. If $x = 2$, then $\sqrt{(4x + 1)} = 3$ and $3 - \sqrt{(x - 2)} = 3$ so that the solution set is $\{2\}$.

Example. Solve $3\sqrt{(x + 5)} = 5 + \sqrt{(3x + 4)}$.

Solution. If we square both sides we get

$$9(x + 5) = 25 + 10\sqrt{(3x + 4)} + (3x + 4).$$

This simplifies to $6x + 16 = 10\sqrt{(3x + 4)}$ or, dividing by 2, to

$$3x + 8 = 5\sqrt{(3x + 4)}.$$

Squaring both sides of this equation and simplifying, we obtain the equation

$$x^2 - 3x - 4 = 0 \text{ so that } (x - 4)(x + 1) = 0.$$

Replacing x with 4 in the given equation shows that 4 is a solution. The given equation is also true if x is -1, thus the solution set is $\{-1,4\}$.

PROBLEMS FOR REVIEW

Group IV: Equations

If it exists, give the solution set for each of the following statements (equations).

1. **a.** $\dfrac{2x + 1}{x} + \dfrac{x - 4}{x + 1} = 3$

b. $\dfrac{2x - 1}{x + 3} - \dfrac{x + 2}{x} = 1$

c. $\dfrac{1 + 3x}{x - 1} - \dfrac{8}{x} = 3$

2. **a.** $\dfrac{x}{x - 2} = a$

b. $\dfrac{x + 1}{x - 4} = a$

c. $\dfrac{x - 2}{x + 3} = a$

3. **a.** $x - \dfrac{2x}{x + 1} = \dfrac{5}{x + 1} - 1$

b. $\dfrac{2x + 1}{x + 2} + \dfrac{x + 2}{2x - 1} = \dfrac{10}{3}$

c. $\dfrac{x - 5}{x + 3} - \dfrac{3x + 9}{x - 5} = 2$

4. **a.** $\sqrt{(x + 4)} = \sqrt{(2x - 1)}$

b. $\sqrt{(2x - 9)} = \sqrt{(x + 8)}$

c. $\sqrt{(x + 1)} = \sqrt{(2x - 7)}$

5. **a.** $x^2 - 5x - 14 = 0$
b. $x^2 - x - 20 = 0$
c. $x^2 + 8x + 15 = 0$

6. **a.** $2x^2 + 7x + 3 = 0$
b. $2x^2 - 9x + 4 = 0$
c. $2x^2 - 11x + 12 = 0$

In each of problems 7–10, solve by completing the square.

7. **a.** $x^2 - 6x + 1 = 0$
b. $x^2 - 4x - 3 = 0$
c. $x^2 + 2x - 2 = 0$

8. **a.** $4x^2 - 4x - 5 = 0$
b. $x^2 - 2x - 12 = 0$
c. $x^2 - 6x - 17 = 0$

9. **a.** $8x^2 - 10x + 3 = 0$
b. $6x^2 + 7x + 2 = 0$
c. $8x^2 - 14x + 3 = 0$

10. **a.** $6x^2 + 3x + 1 = 0$
b. $2x^2 - 2x + 3 = 0$
c. $x^2 + 2x + 5 = 0$

11. **a.** $\sqrt{(3x - 5)} = x - 1$
b. $\sqrt{(3x + 7)} = x + 1$
c. $\sqrt{(9x - 5)} = x + 1$

12. **a.** $\sqrt{(2x + 7)} = 2 + \sqrt{x}$
b. $\sqrt{(2x + 10)} + \sqrt{(x + 9)} = 2$
c. $\sqrt{(2x - 7)} = 1 + \sqrt{(x - 4)}$

13. **a.** $\sqrt{(9 - 2x)} = x - 3$
b. $\sqrt{(2x + 5)} = 2x - 1$
c. $\sqrt{(2x + 5)} = x + 1$

14. **a.** $\dfrac{x}{2a} = \dfrac{4a}{x + 2a}$
b. $2x^2 - rx + 2sx = rs$
c. $2x^3 - 5x^2 + 2x = 0$

In problem 15, you are warned that it is best to check your work before giving the solution set.

15. **a.** $\dfrac{x}{x + 1} + \dfrac{9}{(x + 1)(x - 2)} = \dfrac{3}{x - 2}$

b. $\dfrac{9}{x(x - 3)} - \dfrac{x}{x - 3} = 2 - \dfrac{5}{x}$

c. $\dfrac{3(2 + x)}{1 - x} - \dfrac{2x}{2 - x} = \dfrac{4}{(1 - x)(2 - x)}$

CHAPTER 1
Order of the Real Numbers

Section 1—Order Axioms We begin our study of the order of the real numbers by stating several assumptions or axioms. These axioms involve the symbol $<$ (read *is less than*). Thus, if x is a number and y is a number, then $x < y$ is read "x is less than y." The denial may be expressed by "$x \nless y$, x is not less than y."

AXIOM 1. *If each of x and y is a number, then one and only one of the following statements is true: $x = y$, $x < y$, $y < x$.*

AXIOM 2. *If each of x, y, and z is a number, $x < y$, and $y < z$, then $x < z$.*

AXIOM 3. *If each of x, y, and z is a number and $x < y$, then $x + z < y + z$.*

AXIOM 4. *If each of x, y, and z is a number, $x < y$, and $0 < z$, then $z \cdot x < z \cdot y$.*

AXIOM 5. 0 *is not* 1.

AXIOM 6. 1 *is a counting number and, if x is a number such that $x < 1$, then x is not a counting number; if x is a counting number, then $x + 1$ is a counting number and, if z is a number such that $x < z$ and $z < x + 1$, then z is not a counting number.*

In order to improve our vocabulary, we give the following definitions:

DEFINITIONS. *If each of* x *and* t *is a number, then* $x \nless t$ *means that "*x *is not less than* t*";* $t > x$ *(read as "*t *is more than* x*" or "*t *is greater than* x*") means that* $x < t$*;* $t \ngtr x$ *means that "*t *is not more than* x*."*

Notation. In light of the above definitions, let us agree that we can use the following notations:

$x \nless t$ may be denoted by $t \leq x$ or $x \geq t$
$x \ngtr t$ may be deoted by $x \leq t$.

DEFINITIONS. *The statement that "*x *is a positive number" means that* $0 < x$*; "*x *is a negative number" means that* $x < 0$*; "*x *is an integer" means that* x *is a counting number, the negative of a counting number, or* 0*; "*x *is a rational number" means that there exists an integer* m *and an integer* n*,* $(n \neq 0)$*, such that* x *is the quotient* m/n*.*

Axiom 2 is sometimes called the "transitive axiom" and it may help to think of it as saying—"if the first number is smaller than the second and the second is smaller than a third, then the first is smaller than the third." Statements about the order of numbers are usually called inequalities, and in solving inequalities one should treat multiplication with care. Notice that axiom 4 specifies multiplication by a positive number only. In contrast to this, note that axiom 3 permits the addition of any number to both sides of an inequality but, we repeat, treat multiplication of an inequality with caution.

We now try to show how additional order properties follow as consequences of these axioms.

Suppose that $t < x$, then it follows that

$$t + [(-x) + (-t)] < x + [(-x) + (-t)] \quad \text{by axiom 3}$$

so that

$$[t + (-t)] + (-x) < [x + (-x)] + (-t) \quad \text{by axioms A2 and A3}$$

Therefore $(-x) < (-t)$ by axioms A4 and A5.

The above argument establishes the first of the three theorems listed below.

THEOREM 1. *If* t *and* x *are numbers and* $t < x$*, then* $-x < -t$*.*

THEOREM 2. $0 < 1$.

THEOREM 3. *If x is a number, then $x < x + 1$.*

Now, theorem 2 is so difficult to prove that we shall, for the present at least, assume that it is true. The third theorem follows so easily from the second that its proof is left to the student in the exercises. The importance of these three theorems perhaps can be explained by the fact that they are enough to establish the order of the set of integers. Just how this is done will be shown or at least indicated by the following example:

Example. $4 < 4 + 1$, that is, $4 < 5$ by theorem 3, thus $-5 < -4$ by theorem 1. Since $5 < 5 + 1$ by theorem 3, it follows that $4 < 6$ by axiom 2.

The next example serves not only as another sample proof, but it also establishes results which we shall use later with a great deal of zest and effectiveness.

Example. Show that each of the following statements is a theorem.

T 1. If each of p and q is a positive number, then $0 < pq$.

H 2. If each of p and q is a negative number, then $0 < pq$.

M 3. If $p < 0$ and $0 < q$, then $pq < 0$.

4

Proofs.

1. Since $0 < p$, $0 \cdot p < q \cdot p$ by axiom 4, thus $0 < pq$.
2. Since $p < 0, 0 < -p$ by theorem 1, hence $(-p) \cdot q < 0 \cdot (-p)$ by axiom 4. (Note the care taken in multiplying by positive numbers only.) Hence $-(pq) < 0$ and therefore,by theorem 1, $0 < pq$.
3. Since $p < 0$, $p \cdot q < 0 \cdot q$ by axiom 4 so that $pq < 0$.

These results will be of such use to us later, that we shall refer to them collectively as theorem 4.

Any admiration that we have felt for what we have done so far is rather rudely dealt with when we attempt to solve even a simple inequality such as $px + 3 < 7$. We can begin easily enough, for $px < 4$ by axiom 3, but now we have reached an impasse. It may seem innocent and reasonable to use axiom 4 to assert that $x < 4/p$, but it may also be false. The point to be made here is that we do not know enough about the number $1/p$ to use axiom 4. Even if we knew that $0 < p$, there is no assurance that $1/p$ is also a positive number. This deplorable situation should certainly indicate a need for the following:

THEOREM 5. *If* $0 < p$, *then* $0 < 1/p$.

The proof of this theorem provides an opportunity to come to grips with denials of statements and "indirect proof," and the pleasure of proving it will not be denied you by giving a proof in these pages. You have, of course, the right to demand that your instructor give a proof. While we are waiting for you either to give a proof or to demand your rights, let us accept the risk of believing that theorem 5 is true. If we accept this risk, then we can deduce other useful notions about reciprocals. For example, if t and x are positive numbers, $t < x$, then theorem 5 assures us that $1/t$ and $1/x$ are positive numbers. So, by multiplying both sides of the given inequality $t < x$ by the positive numbers $1/t$ and $1/x$, we can show that the following statement is true:

THEOREM 6. *If t and x are positive numbers and $t < x$, then* $1/x < 1/t$.

In the next two examples, we intend to indicate how some of the statements in this section can be used to "solve" some simple inequalities. Thus, in the examples that follow, our intentions are honorable.

Example. The following two statements shall mean the same thing:

1. Solve the inequality $3x + 2 < 6$.
2. Find $\{x \mid 3x + 2 < 6, x \in R\}$

Solution. If $3x + 2 < 6$, then, by axiom 3, $3x < 4$.

Now $0 < 3$ so that, by theorem 5, $0 < 1/3$, hence, using axiom 4, it follows that $1/3 \cdot 3x < 4 \cdot 1/3$ which simplifies to $x < 4/3$.

Now, if $x < 4/3$, then multiplying by 3 and adding 2 (using axioms 4 and 3) shows that $3x + 2 < 6$. Therefore

$$\{x \mid 3x + 2 < 6, x \in R\} = \{x \mid x < 4/3\}.$$

Example. Solve $\dfrac{1}{x+5} > 0$ or (the same thing) find

$$\left\{x \mid \frac{1}{x+5} > 0, x \in R\right\}.$$

Solution. $\dfrac{1}{x+5} > 0$ implies that $x + 5 > 0$ by virtue of theorem 5 and the "flopover" theorem, better known as theorem Q5 (see Section 4 in the previous chapter if you must). Hence $x > -5$ by axiom 3. Thus the solution set of the statement $\frac{1}{x+5} > 0$ is

$$\{x \mid x > -5\}.$$

We now move on to consider the idea that between any pair of numbers there is another. Suppose that t and x are numbers and that $t < x$. Axiom 3 assures us that

$$t + t < x + t \text{ and that } t + x < x + x$$

that is, $2t < t + x$ and $t + x < 2x$. But $0 < 2$, hence $0 < 1/2$ by theorem 5. Using axiom 4 is all that remains for us to do in order to prove the following theorem:

THEOREM 7. *If t and x are numbers and $t < x$, then*

$$t < \frac{t+x}{2} \quad \text{and} \quad \frac{t+x}{2} < x.$$

It can't hurt anyone to know that $\frac{t+x}{2}$ is called the arithmetic mean or average of t and x. It should help everyone to know that if t, x, and z are numbers and $t < x$, then the statement that "z is between t and x" means that $t < z$ and that $z < x$. We shall denote this statement by $t < z < x$.

PROBLEM SET 1.1

1. The following two order properties are easy to prove if you first read axiom 2. Prove the following:

a. If $0 < x$ and $x < 1$, then $x^2 < 1$.
b. If t and x are positive numbers and $t < x$, then $t^2 < x^2$.

2. a. Prove theorem 3. *Hint:* assume theorem 2.
b. Prove theorem 6. *Hint:* see discussion in text.

3. Show that

a. $1 < 2$ **b.** $2 < 3$ **c.** $0 < 3$ **d.** $-3 < -2$.

4. Use theorem 4 to help you explain why the following statement is true:

If $x \neq 0$, then $x^2 > 0$.

5. **a.** Solve $2x - 3 < 4 - x$.
b. Find $\{x \mid 5x + 1 < 2, x \in R\}$.

6. **a.** Solve $3x - 2 < 1 + x$.
b. Find $\{x \mid 4x - 1 < x + 5, x \in R\}$.

7. Solve the following inequalities:

a. $\dfrac{1}{x-4} > 0$ **b.** $\dfrac{1}{2x-3} > 0$ **c.** $\dfrac{1}{2-x} > 0$

8. Use theorem 7 to show that

a. $0 < 1/2 < 1$ **b.** $1/2 < 3/4 < 1$ **c.** $-2 < -3/2 < -1$

9. If $1 < z < 4$, take a look at theorem 6 and find a number m and a number M so that $m < 1/z < M$.

10. Suppose that $\dfrac{b-a}{b} < \dfrac{b-a}{a}$, give a number z that is between this pair of numbers.

11. Read this. *FOR STUDENT EYES ONLY:* A straight line drawn on a writing surface and imagined to extend indefinitely can serve as a model for the set of all numbers. Suppose this line is drawn horizontally and extends indefinitely both to the left and to the right. Each dot on this line is associated with a number and each number with a dot on the line in such a manner that if t and x are numbers and $t < x$, then the dot associated with t is to the left of the dot associated with x. Further, if $b - a = x - t$, then that piece of the line whose ends are identified with t and x shall have the same measure (length) as that piece of the line whose ends are identified with the numbers a and b. Refer to figure 1.1 as the *straight line model of the numbers.*

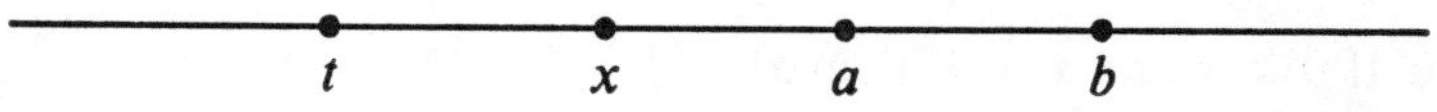

Fig. 1.1

We shall borrow from this model the notions *closed interval* and *open interval.* The statement that $[a,b]$ is a closed interval means that a and b are numbers $a < b$, and that $[a,b]$ is the number set to which x belongs if and only if x is a number between a and b, x is a, or x is b. The statement that $\langle a,b\rangle$ is an open interval means that a and b are numbers $a < b$, and that $\langle a,b\rangle$ is the number set to which x belongs if and only if x is a number between a and b. The numbers a and b are often called the *end points* of the interval and the positive number $b - a$ is the *length* of the interval.

[12.] If we pick a dot on a straight line model and associate it with 0 and pick another dot to the right of the first and associate it with 1, we can then obtain a model for the integers $-2, -1, 0, 1, 2$. *Do so.*

13. Suppose $A = \langle 1,3 \rangle$, $B = [3,5]$, and $C = \langle 5,6 \rangle$.

a. Give $A \cup B$. What symbol might be used for this set?
b. Give $B \cup C$. Suggest a symbol.
c. Explain why there is not much interest in $A \cap B$.

14. Recalling that if $x \neq 0$, then $x^2 > 0$, show that each of the following statements is true:

a. If $t > 0$ and $(t - 1) \neq 0$, then $2 < t + 1/t$.
b. If r and s are positive numbers, then $2 < r/s + s/r$.
c. If r and s are positive numbers, then $\dfrac{4rs}{r + s} < r + s$.

15. If p and q are positive numbers, find:

a. a number which is between p and q.
b. a number which is greater than poth p and q.
c. a positive number which is less than both p and q, that is, is between 0 and p *and* between 0 and q.

16. a. Suppose x, y, and z are numbers, $x < y$, and that, for some number t between 0 and 1, $z = tx + (1 - t)y$. Show that z is between x and y.
b. Illustrate the above using $x = 3$, $y = 7$, and $t = 1/2$.

Section 2—Rational Inequalities In this section, we will try to show how the statements of the previous sections can be utilized to solve certain inequalities. In particular, considerable use will be made of the ideas that (1) the product of two positive (negative) numbers is a positive number, (2) the product of a positive number and a negative number is a negative number, and (3) theorem 5. Each of the examples will be accompanied by a straight-line model of the numbers solely for the purpose of helping the student visualize the solution set and to give him some simple means of recording the results of pertinent investigations.

Example. Solve $x^2 - 2x - 3 < 0$.

Analysis. Note that the given problem is equivalent to $(x - 3)(x + 1) < 0$. Clearly $x \neq 3$ and $x \neq -1$. Why? Note how this information is conveyed in the drawing below. Now, if $x < 3$, then $x - 3 < 0$ but if $x > 3$, then $x - 3 > 0$. If $x < -1$, then $x + 1 < 0$ while $x + 1$

>0 if $x > -1$. Note how this is recorded in figure 1.2. Thus the solution set is $\{x \mid -1 < x < 3\}$ or, recalling earlier remarks about intervals, we can write $\langle -1,3 \rangle$.

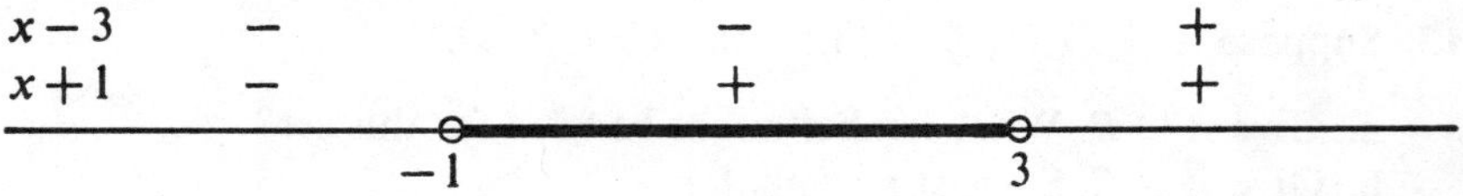

Fig. 1.2

Example. Solve $x(x - 2) \geq 0$.

Analysis. Surely 0 and 2 are elements of the solution set. Contrast with previous example. If $x < 2$, then $x - 2 < 0$. If $x > 2$, then $x - 2 > 0$. Recalling idea (1), the solution set is $\{x \mid x \leq 0 \text{ or } x \geq 2\}$. Note how our analysis is recorded in figure 1.3.

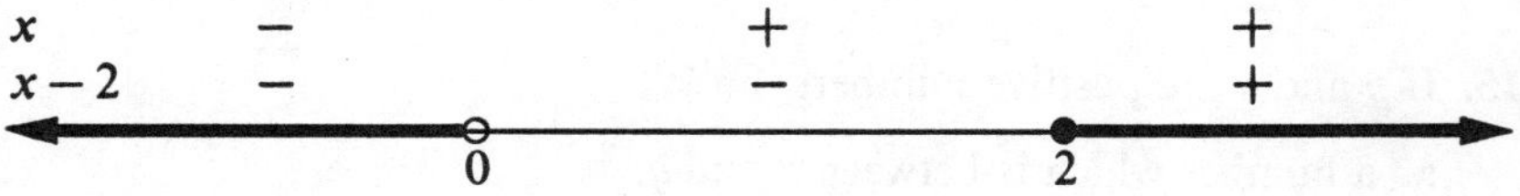

Fig. 1.3

Example. Solve $\dfrac{2}{x - 4} \geq 1$.

Analysis. Begin by writing $\dfrac{2}{x - 4} - 1 \geq 0$. This will simplify to $(6 - x) . \dfrac{1}{x - 4} \geq 0$. Clearly $x \neq 4$ but $x = 6$ is a solution. Further, if $x > 6$, $6 - x < 0$ and if $x < 6$, then $6 - x > 0$. If $x < 4$, then $x - 4 < 0$, hence $\dfrac{1}{x - 4} < 0$. If $x > 4$, $x - 4 > 0$ so that $\dfrac{1}{x - 4} > 0$. The point to be noted here is that one needs only to consider the number $x - 4$ in order to gain information about its reciprocal. Therefore the solution set is given as $\{x \mid 4 < x \leq 6\}$. (See figure 1.4.) Alternate notation $\langle 4,6]$.

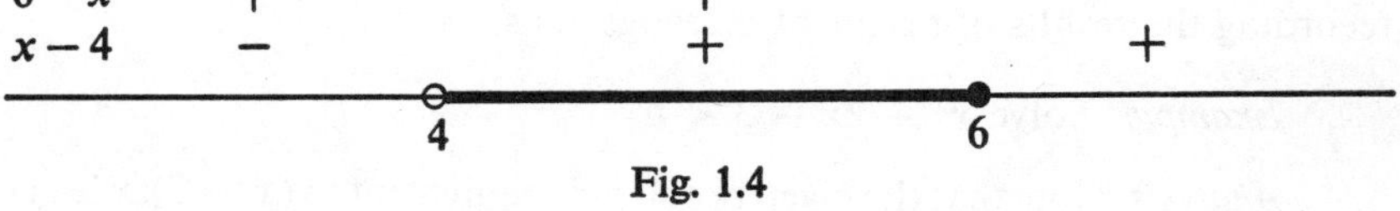

Fig. 1.4

Example. Solve $\dfrac{3}{x - 2} + \dfrac{1}{x + 1} > 0$.

Analysis. One can show that $\dfrac{3}{x-2} + \dfrac{1}{x+1} = \dfrac{4x-1}{(x-2)(x+1)}$ so the given problem is equivalent to $(4x+1)\cdot\dfrac{1}{x-2}\cdot\dfrac{1}{x+1} > 0.$ This example is more of the same. (See figure 1.5.) The solution set is therefore given by

$$\{x \mid -1 < x < -1/4 \text{ or } x > 2\}.$$

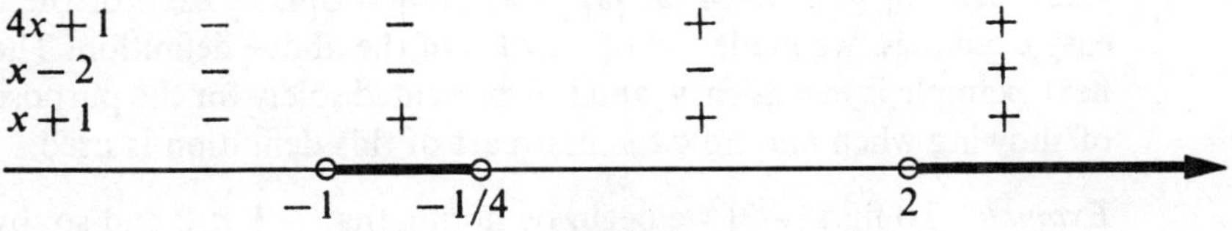

Fig. 1.5

PROBLEM SET 1.2

Solve the following inequalities.

1. $(x+2)(x-1) < 0$

2. $(x-3)(x-1) < 0$

3. $x^2 + x - 2 > 0$

4. $x^2 - 4x + 3 > 0$

5. $x^2 + 4x \le 0$

6. $2x^2 - x \le 0$

7. $4x^2 + 3x - 1 < 0$

8. $3x^2 - 4x + 1 < 0$

9. $\dfrac{1}{x-3} < 0$

10. $\dfrac{1}{x+2} > 0$

11. $\dfrac{x-3}{x+5} \ge 0$

12. $\dfrac{x+2}{x-5} \le 0$

13. $\dfrac{x+1}{x-4} < 2$

14. $\dfrac{2x-1}{x-3} < 1$

15. $\dfrac{2x-3}{x+2} \le -1$

16. $\dfrac{x-3}{2x-1} \ge 1$

17. $\dfrac{4}{x+2} - \dfrac{1}{x-1} \ge 0$

18. $\dfrac{6}{x+3} + \dfrac{1}{x-2} \ge 0$

19. $\dfrac{3}{x-4} < \dfrac{1}{x+2}$

20. $\dfrac{2}{5-x} < \dfrac{1}{x+4}$

21. $\dfrac{x}{2x-5} \le 0$

22. $\dfrac{2x-3}{x} \ge 0$

23. $\dfrac{x}{2x-3} < 0$

24. $\dfrac{x}{1-2x} \le 0$

Section 3—Absolute Value Our attention and efforts are now directed toward the following definition and to some of its consequences.

DEFINITION. *If z is a number, then $|z|$ is a number such that* (1) $|z| = -z$ *if* $z < 0$ *and* (2) $|z| = z$ *if* $z \nless 0$. *Read* $|z|$ *as the absolute value of* z.

Examples. $|3| = 3$; $|5| = 5$; $|\pi| = \pi$; $|3/4| = 3/4$. In each of these easy examples, we made use of part (2) of the above definition. The next example is just as easy, and it is presented solely for the purpose of showing when and how the first part of this definition is used.

Example. To find $|-3|$ we begin by noting that $-3 < 0$ and so, by the definition, $|-3| = -(-3)$. This is exactly what the definition says and nothing more. However, we are aware that by a previous result (see review-theorem A3) $-(-3) = 3$. Thus we shall write that $|-3| = 3$ and when asked "why?" we shall simply reply "by definition" even though we realize that this answer is incomplete. Most mathematicians and engineers will tolerate this incomplete reply, and we are not particularly concerned about anyone else.

Examples. $|7| = 7$ and $|-7| = 7$.

Question. If $x - c$ is a number, when does $|x - c| = c - x$? When does $|x - c| = x - c$?

Another question. Can you see why we are not especially interested in the task of finding a number z such that $|z| < 0$?

One simple consequence of this definition is seen by noting that if p is a positive number, then $|z| = p$ only when $z = p$ or $z = -p$. The solution of some equations depends upon this simple consequence as the following examples will show.

Example. Solve the equation $|2x + 3| = 5$. From the above we know that $2x + 3 = 5$ or $2x + 3 = -5$, hence $x = 1$ or $x = -4$.

Example. If $|1 - 3x| = 2$, then $1 - 3x = 2$ or $1 - 3x = -2$. Thus $x = -1/3$ or $x = 1$.

Another consequence can be surmised by considering questions such as "for what z is $|z| < 3$?" The answer to this question and others of a similar nature should result in our being able to conjecture the following statement:

(A) Suppose $p > 0$. If $|z| < p$, then $-p < z < p$.

We shall temporarily assume this statement and, for the present at least, not give a proof.

In the next two examples, we examine how this statement can be useful in solving inequalities involving absolute value. In these same examples we shall recall what we have read about intervals and even draw a number line model of the solution set.

Example. If $|3x - 5| < 1$, then $-1 < 3x - 5 < 1$. From this it follows that $4 < 3x < 6$, hence, $4/3 < x < 2$. (See figure 1.6.)

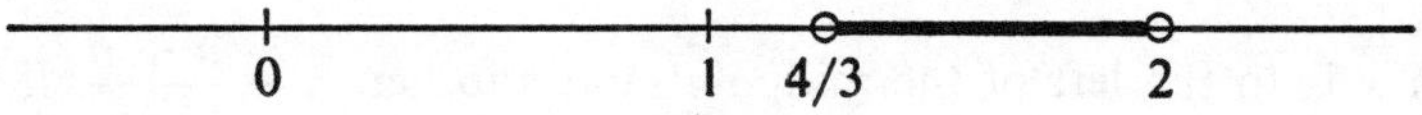

Fig. 1.6

Example. If $|1 - 2x| < 5$, then $-5 < 1 - 2x < 5$. It follows that $-6 < -2x < 4$ and so, by theorem 1, $-4 < 2x < 6$. (See figure 1.7.) Therefore $-2 < x < 3$.

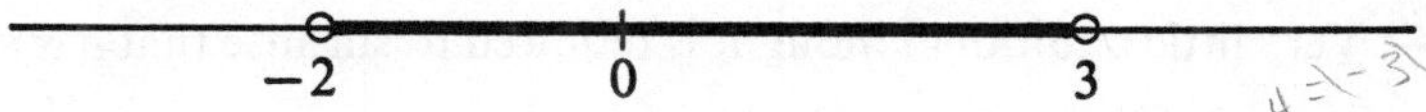

Fig. 1.7

If you reexamine these two examples, you should be simultaneously struck by two ideas:

1. If $|z| \leq p$, then $-p \leq z \leq p$, and
2. there is a connection between absolute value inequalities and intervals open and closed.

In the next paragraph, we shall consider further this second idea.

Consider the straight line models shown below. One figure is the model of the open interval $\langle a,b \rangle$ while the other models the closed interval $[a,b]$. (See figure 1.8.) Either has length $b - a$ and, since it is easy to show that

$$b - \frac{a+b}{2} = \frac{a+b}{2} - a$$

either interval has $\frac{a+b}{2}$ as its midpoint. We now consider a remark directed primarily at $\langle a,b \rangle$. If x is between a and b, that is, if $x \in \langle a,b \rangle$, it

Fig. 1.8

should cause no brain strain to observe that, for any such number x, *the length of the segment from x to the midpoint is less than one-half the positive number $b - a$.* (In the event $x = \frac{a+b}{2}$, we say that this length is 0 which is not incompatible with our underlined remark.) A difficulty arises, however, whenever we try to put this remark down on paper in the concise form so loved by the mathematician. The difficulty arises from the generally held view that length ought to be a positive number. This bothers us, you see, for we would have to write one thing, $\frac{a+b}{2} - x$, when x is to the left of the midpoint and another, $x - \frac{a+b}{2}$, if to the right. The difficulty vanishes as soon as we realize that the answer to the first question posed in this section is relevant to this problem. For, with this realization comes the resolution of our difficulty and we can write the underlined statement as follows: if $x \in \langle a,b \rangle$ then $x - \left|\frac{a+b}{2}\right| < \frac{b-a}{2}$. Very little additional thought is required to surmise that if $x \in [a,b]$, then $\left|x - \frac{a+b}{2}\right| \leq \frac{b+a}{2}$. Using this as a guide, the interval $\langle -2,4 \rangle$ has length 6 and midpoint 1, thus if x belongs to this interval, then $|x - 1| < 3$. The interval $-p,p$ has length $2p$ and midpoint 0, thus (as a companion to statement A) we have the statement

(B) If $-p < z < p$, then $|z| < p$.

Hoping that this paragraph has helped, we now return our attention to the last of the examples which followed statement A.

> *Puzzler.* Student Terrence Xelent O'Leary (being sound of mind but clumsy afoot) miscopied the example as $|2x - 1| < 5$. He then used theorem 8 to deduce that $-5 < 2x - 1 < 5$ so that $-4 < 2x < 6$; hence $-2 < x < 3$, that is, $x \in \langle -2,3 \rangle$. Tis not often that we copy problems wrong but get right answers. Is it merely the luck of the Irish?

In the remaining examples we emphasize the idea that

If $p > 0$ and $|z| > p$, then $z > p$ or $z < -p$.

This idea, like the others, is a consequence of the definition of absolute value. This definition tells us that if $z < 0$, then $|z| > p$ implies that $-z < p$, that is, $z < -p$. It also tells us that when $z \not< 0$ that $|z| > p$ implies that $z > p$.

Example. If $|2x + 3| > 5$, then $2x + 3 > 5$ or $2x + 3 < -5$. From this it follows that $2x > 2$ or $2x < -8$, hence

$$x > 1 \text{ or } x < -4.$$

The straight line model of this number set is not an interval but looks like figure 1.9.

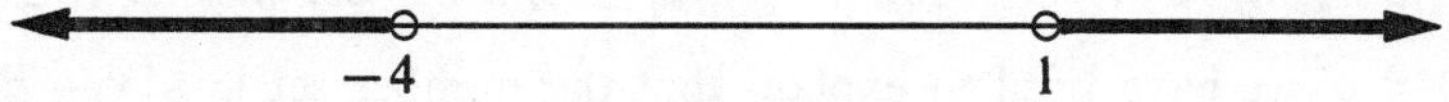

Fig. 1.9

Example. Solve $|x - 5| \geq 2$.

Solution. $x \geq 7$ or $x \leq 3$. (See figure 1.10.)

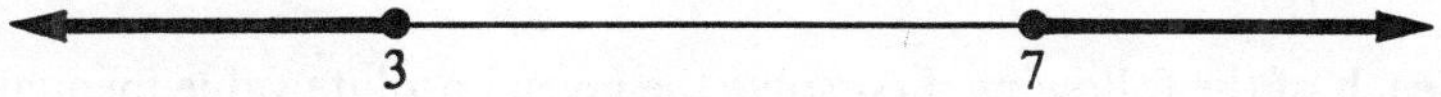

Fig. 1.10

Solely for reference purposes, we shall label the ideas given in statements A and B as theorem 8 and we write these two statements in the compact form shown.

THEOREM 8. *Suppose* $p > 0$. $|z| < p$ iff $-p < z < p$.

PROBLEM SET 1.3

1. Since it is easy, let us show that $b - \dfrac{a+b}{2} = \dfrac{a+b}{2} - a$ for then we will all feel better about calling $\dfrac{a+b}{2}$ the midpoint of either $\langle a,b \rangle$ or $[a,b]$.

2. Briefly explain each of the following:

a. There is not much interest in solving equations like $|x - 1| = -2$ or inequalities such as $|x + 2| < -3$.

[b.] If x is a number, the statement $\sqrt{x^2} = x$ is false but the statement $\sqrt{x^2} = |x|$ is true.

Solve the following equations:

3. $|4x - 6| = 2$ **4.** $|3x + 1| = 7$ **5.** $|3 - 2x| = 2$

6. $|1 - 2x| = 3$ **7.** $\dfrac{1}{|2x + 5|} = 1$ **8.** $\dfrac{1}{|2x - 5|} = 1$

Solve the following inequalities:

9. $|3x - 4| < 7$ **10.** $|2x + 5| < 3$ **11.** $|x - 3| > 2$

12. $|x + 2| > 2$ **13.** $|2 - 5x| \le 3$ **14.** $|3 - 7x| \le 10$

15. $|3x - 1| < 4$ **16.** $|2x - 3| < 2$ **17.** $|3 - 2x| < 5$

18. $|2 - 3x| < 6$ **19.** $|x + 1/2| \le 1/4$ **20.** $|x - 3/4| \le 3/2$

If $0 < p$, we have tried to explain that the number set $\{x \mid |x - a| < p\}$ can be described as an interval. Describe each of the following number sets as intervals.

21. $\{x \mid |x/4 - 1| < 1/2\}$ **22.** $\{x \mid |x + 4| < 2\}$

23. $\left\{x \,\middle|\, \left|x - \frac{a+b}{2}\right| \le \frac{b-a}{2}\right\}$ **24.** $\{x \mid |3 - 2x| \le 4\}$

In each of the following, first solve the given absolute value inequality and then find m and M. *Hint.* See theorem 6.

25. If $|x - 1| < 2$, then $m < \dfrac{1}{x + 4} < M$.

26. If $|2 - x| < 1/2$, then $m < 1/x < M$.

The following two problems do not have anything to do with absolute value but they may be pleasant and informative.

27. If $0 < t < 4$, show that $\sqrt{(4 + t)} - 2 < 2 - \sqrt{(4 - t)}$. *Hint.* Rationalize each side and compare numerators and denominators.

28. If $0 < t < 1/3$, show that $3 - \dfrac{3}{1 + 3t} < \dfrac{3}{1 - 3t} - 3$. *Hint:* Simplify the RHS and the LHS.

Section 4—Triangle Law Up to this point in our study of absolute value, we have made no mention or use of two important consequences of the definition. If we suppose that $t < 0$ and that $0 < x$, then $tx < 0$, hence $|tx| = -tx$. Also $|t| = -t$, $|x| = x$, so that $|t|\cdot|x| = -tx$. Considering other possibilities should lead us to the idea embodied in the following statement:

THEOREM 9. *If each of t and x is a number, then*

$$|t|\cdot|x| = |t\cdot x|.$$

Armed with this statement, one may be mildly surprised to learn that $|t| + |x|$ may not be equal to $|t + x|$. For example, if $t = -3$ and

$x = 7$, then $t + x = 4$ so that $|t + x| = 4$. Surely $|t + x| \neq |t| + |x|$. The idea hinted at in this discussion is of prime importance and is often called the *triangle law*. In this text, however, it will be called theorem 10.

THEOREM 10. *If each of t and x is a number, then*

$$|t + x| \leq |t| + |x|.$$

For the present, let us assume these two statements, realizing that a proof may be required later. Let us also note, however, that whereas the party of the first part (the student) may not be required to prove these two theorems, and whereas many of the problems in problem set 1.4 are intended solely for the edification and amusement of the party of the first part, let us resolve to acquaint ourselves with some ideas associated with these two theorems by doing the rather brief exercise below.

Brief exercise. Using $t = -5$ and $x = 8$, evaluate each of the following: $|t| + |x|$, $|t + x|$, $|tx|$, $|t|\cdot|x|$, $|t| - |x|$, $|t - x|$, and $|x - t|$. Answer the following question: Under what conditions does $|t + x| = |t| + |x|$?

Let us now study some examples of a problem that is rather common in the calculus.

Example. By theorem 9 we know that $|x^2 + 3x - 10| = |x + 5|\cdot|x - 2|$, so if $|x - 2| < 1$, that is, $1 < x < 3$, then $6 < x + 5 < 8$, and so $|x + 5| < 8$. This permits us to make the following statement:

$$\text{if } |x - 2| < 1, \text{ then } |x^2 + 3x - 10| \leq 8\cdot|x - 2|.$$

Example. Find a positive number M so that whenever $|x - 4| < 1$, then $|x^2 - 16| \leq M\cdot|x - 4|$. To do this, note that if $|x - 4| < 1$, then $3 < x < 5$ so that $7 < x + 4 < 9$, hence $|x + 4| < 9$. Since $|x^2 - 16| = |x - 4|\cdot|x + 4|$, whenever $|x - 4| < 1$, $|x^2 - 16| \leq 9\cdot|x - 4|$, that is, $M = 9$. You should try to see that any number greater than 9 would have been an acceptable choice for M; but no apologies are required for choosing the smallest possible value for M.

Example. Find a positive number M such that whenever $|x + 5| < 1$, then $|x^2 + 4x - 5| \leq M\cdot|x + 5|$.

Solution. If $|x + 5| < 1$, then $-6 < x < -4$, hence $-7 < x - 1 < -5$ and so $|x - 1| < 7$. Since $|x^2 + 4x - 5| = |x - 1|\cdot|x + 5|$, we know that if $|x + 5| < 1$, then $|x^2 + 4x - 5| \leq 7\cdot|x + 5|$, so $M = 7$ suffices.

Example. Suppose that x, t, and p are numbers and that $|x-2| + |2-t| < p$. One could use theorem 10 to observe that $|x-2+2-t| \leq |x-2| + |2-t|$ and conclude (after simplifying the left hand side) that p had better be a positive number greater than $|x-t|$.

PROBLEM SET 1.4

1. Prove theorem 8. You are informed that *iff* is an authorized abbreviation for *if and only if* and that to prove theorem 8 you must show that each of the following is true:

a. If $|x| < p$, then $-p < x < p$.
b. If $-p < x < p$, then $|x| < p$.

2. Prove theorem 9.

3. If r and s are numbers, show that $|r-s| = |s-r|$.

4. Using theorem 10, show that $|x-2| + |x-3 \not< 1$. *Hint:* See problem 3, it should help.

5. Use theorem 10 to show that $|x-a| + |x-b| \not< |b-a|$. *Hint:* See problem 3.

6. If x is a number, show that $-|x| \leq x \leq |x|$. *Hint:* Consider the cases $x < 0$ and $x \not< 0$.

7. Suppose that each of x, y, and t is a number and that $0 < p$. If $|x-t| < p/2$ and $|y-t| < p/2$, show that $|x-y| < p$.

8. Realizing that $|t| = |x + (t-x)|$, use theorem 10 to show that $|t| - |x| \leq |t-x|$.

9. If each of t and x is a number, then, by problem 6, we have that $-|x| \leq x \leq |x|$ and that $-|t| \leq t \leq |t|$. Use this information together with theorems 4 and 8 to prove the "triangle law."

10. *Bonus Problem.* If $0 < p$ and if

$$\left|k - \sum_{i=1}^{n} F(z_i)\Delta_i x\right| + \left|J - \sum_{i=1}^{n} F(z_i)\Delta_i x\right| < p,$$

show that $p \not< |J-K|$. The wonderful thing about this is that you do not even need to know what $\sum_{i=1}^{n} F(z_i)\Delta_i x$ means except that you hope it is a number.

In each of problems 11–15, find a positive number M such that:

11. If $|x - 1| < 1$, then $|x^2 + 3x - 4| \leq M \cdot |x - 1|$.

12. If $|x - 3| < 1$, then $|x^2 - 2x - 3| \leq M \cdot |x - 3|$.

13. If $|x + 2| < 1$, then $|x^2 - x - 6| \leq M \cdot |x + 2|$.

14. If $|x + 2| < 1/2$, then $|x^2 - x - 6| \leq M \cdot |x + 2|$.

15. If $|x + 3| < 1$, then $|x^2 - 9| \leq M \cdot |x + 3|$.

Section 5—Bounded Number Sets If S is a finite set of numbers, then S contains a smallest and a largest number. However, if S is an infinite set of numbers, it may or may not contain a smallest or a largest element. For example, the interval $[a,b]$ has a smallest element a and a largest element b but the interval $\langle a,b \rangle$ has neither a smallest or a largest element. For example, suppose c is the largest number in the interval $\langle a,b \rangle$, then surely $a < c < b$. However, theorem 7 assures us that $c < \frac{b + c}{2} < b$ so that the number $\frac{b + c}{2}$ belongs to the interval and is larger than c. This shows that the supposition is ridiculous, that is, false. One should also note that the set of counting numbers N contains a smallest element, namely 1, but has no largest element. All of this is designed to direct your attention to the idea that some number sets are bounded above—that there is a number exceeded by no number in the set.

> DEFINITION. *If S is a number set the statement that "b is an upper bound of S" or "S is bounded above by b" means that b is a number and that if $x \in S$, then $x \leq b$.*

The assumption is made that the student can define what is meant by the statement that "S is bounded below by b" and rush on to point out that if b is an upper bound to S, then so is $b + 1, b + 2$, or any number larger than b. Thus upper bounds, when they exist, are not unique. Let us further note that if S is bounded above, then the number set to which $-x$ belongs, only if $x \in S$, will be bounded below; and if S is not bounded above, then this second set is not bounded below. From this, one can see that the set of integers is not bounded above nor below.

Since the existence of an upper bound ensures the existence of many upper bounds, we may ponder the fact that, for the interval $\langle a,b \rangle$, b is the smallest of the upper bounds and is therefore called the *least upper bound* of $\langle a,b \rangle$. This is to serve notice, therefore, that not only will we concern ourselves with the idea of determining the existence of an upper bound but, when it exists, we shall also be concerned with finding the least upper bound.

DEFINITION. *If S is a number set, the statement that "b is the least upper bound of S" (denote this by* LUB *of* $S = b$*) means that b is an upper bound of S and that if* $c < b$ *then c is not an upper bound of S.*

Again we assume that the student can define the *greatest lower bound* and give him the notation GLB of S and point out that GLB of $\langle a,b \rangle =$ GLB of $[a,b] = a$.

Example. If $A = \{x \mid |2x - 1| < 3\}$, we can show that $A = \langle -1,2 \rangle$, hence LUB of $A = 2$ and GLB of $A = -1$. Note $2 \notin A$ and $-1 \notin A$.

Example. If $B = \{x \mid 3 \leq x < 7\}$, then LUB of $B = 7$ and GLB of $B = 3$. Note $7 \notin B$ but $3 \in B$.

One important question, ignored so far, is: If a number set is bounded above, is there a least upper bound? We answer this question with the next (and last) axiom.

AXIOM OF COMPLETENESS. *If S is a number set and S is bounded above, then there is a least upper bound of S.*

It is this axiom which distinguishes the set of real numbers from the rest of rational numbers for, without this axiom, one could not show the existence of irrational numbers such as $\sqrt{2}$. Up to this point however, we have assumed that such numbers exist and, since the proofs are difficult, we shall continue to so assume.

Example. Let $C = \left\{x \,\middle|\, x = \dfrac{n-1}{2n}, n \in N\right\}$. We indicate C by the roster method $C = \{0, 1/4, 2/6, \ldots\}$ for this should help you to see that C is bounded below and that the GLB of C is 0. It also shows that $0 \in C$. If we also note, by dividing $n - 1$ by $2n$, that $\dfrac{n-1}{2n} = \dfrac{1}{2} - \dfrac{1}{2n}$, then we see that each number in C is less than 1/2. This in turn, helps us to see that the LUB of $C = 1/2$ and that $1/2 \notin C$.

Remark. If we were compelled to call this number set C either a *strictly increasing* or a *strictly decreasing* set, we should select the former. Nevertheless, it is interesting to note that C is bounded above.

Example. Suppose that $D = \left\{x \,\middle|\, x = \frac{4n-3}{2n-1}, n \in N\right\}$. If you indicate D by the roster method, then you will see that GLB of $D = 1$ and that $1 \in D$. If we divide $4n - 3$ by $2n - 1$, we can see that $\frac{4n-3}{2n-1} = 2 - \frac{1}{2n-1}$. Since $\frac{1}{2n-1} > 0$ for every counting number n, we can conclude that LUB of $D = 2$ and that $2 \notin D$.

Example. Suppose that $F = \{x \mid x(x-2) > 0\}$. While we will not even try to indicate this number set by the roster method, we will make use of our previous experience in solving inequalities to proceed as follows: Clearly $0 \notin F$ and $2 \notin F$. A further analysis, outlined in figure 1.11, assures us that $F = \{x < 0 \text{ or } x > 2\}$, thus F is not bounded either above or below, hence we lose no time in searching for LUBs or GLBs.

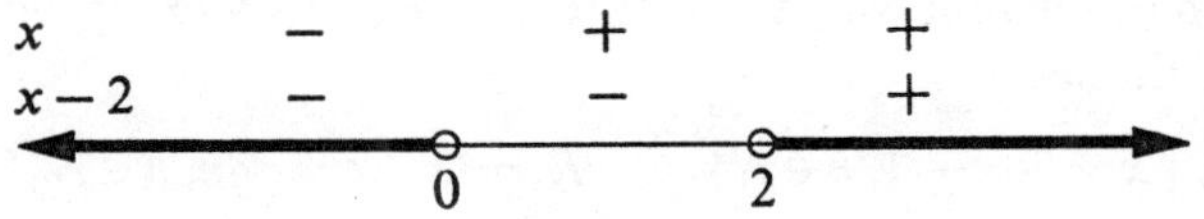

Fig. 1.11

Example. Suppose that $G = \{x \mid x^2(x-3) \geq 0\}$. Note that $0 \in G$ and that $3 \in G$. Our analysis (see figure 1.12) shows us that

$$G = \{x \mid x \geq 3\} \cup \{0\}$$

hence, GLB of $G = 0$ but G is not bounded above.

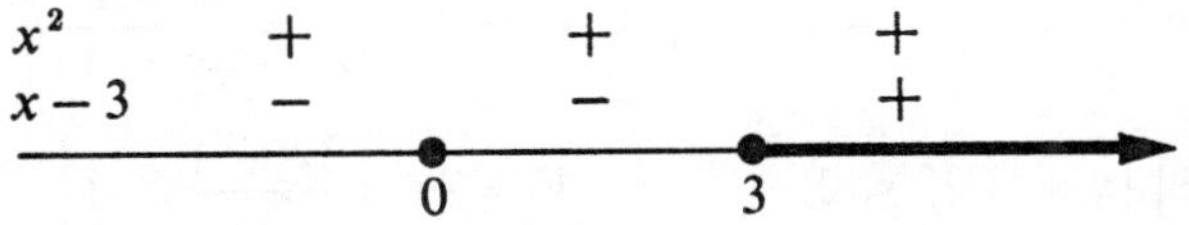

Fig. 1.12

Example. Suppose that $W = \left\{x \,\middle|\, x = (-1)^n\left\{1 + \frac{1}{n}\right\}, n \in N\right\}$. By roster, $W = \{-2, 3/2, -4/3, \ldots\}$. If you can determine that W is bounded above or below, do so and give (if they exist) the LUB or GLB of W.

If you can't determine the boundedness of W, let us observe that

$$(-1)^n\left(1 + \frac{1}{n}\right) = \begin{cases} 1 + \frac{1}{n} & \text{when } n \text{ is even} \\ -\left(1 + \frac{1}{n}\right) & \text{when } n \text{ is odd} \end{cases}$$

and recall that $2m$ is even for every counting number m while $2m - 1$ is odd. We can then break W up and write it as $W = S_1 \cup S_2$ where

$$S_1 = \left\{x \,\middle|\, x = 1 + \frac{1}{2m}, m \in N\right\}$$

and

$$S_2 = \left\{x \,\middle|\, x = -\left[1 + \frac{1}{2m - 1}\right], m \in N\right\}$$

for this may help determine the boundedness of W. Tiring of the example, I leave it in your capable hands.

PROBLEM SET 1.5

1. For each of the following number sets, give (if they exist) the LUB and GLB of the set.

$A = \{x \mid x = 2n - 1, n \in N\}$ $\qquad$ $K = \{x \mid x = 1/n, n \in N\}$

$B = \{x \mid 3 - x \nless 0\}$ $\qquad$ $M = \{x \mid x = 1 - 1/n^2, n \in N\}$

$C = \{x \mid |3x - 1| \ngtr 2\}$ $\qquad$ $S = \left\{x \,\middle|\, x = \dfrac{n + 1}{2n}, n \in N\right\}$

$D = \{x \mid |6 - x| < 2\}$ $\qquad$ $T = \left\{x \,\middle|\, x = \dfrac{3n - 2}{n}, n \in N\right\}$

$F = \left\{x \,\middle|\, x = \dfrac{3n}{2n - 1}, n \in N\right\}$ $\qquad$ $U = \left\{x \,\middle|\, x = \dfrac{n - 1}{3n - 1}, n \in N\right\}$

$G = \{x \mid |x + 3/2| \le 2\}$ $\qquad$ $V = \left\{x \,\middle|\, \dfrac{2x + 1}{x} > 1\right\}$

$W = \left\{x \,\middle|\, \dfrac{x}{x - 2} \le 0\right\}$ $\qquad$ $Z = \{1/x \mid |2x - 1| < 1/2\}$

2. List, by the capital letter designating the set, those that a) contain their LUB, and b) contain their GLB.

3. Suppose that Y is a number set and that LUB of $Y = b$. If $c < b$, show that there is at least one number in Y which exceeds c.

4. Same directions as were given in problem 1 for the following number sets.

$\mathscr{A} = \left\{x \,\middle|\, x = \dfrac{2n^2 - 1}{n}, n \in N\right\}$ $\qquad$ $\mathscr{F} = \{x \mid |2x + 3| > 1\}$

$\mathscr{B} = \{x \mid x^2 + 3x \leq 0\}$ $\mathscr{G} = \{x \mid x^3 + 2x > 0\}$

$\mathscr{C} = \left\{x \,\middle|\, \left|\frac{3}{4}x - \frac{2}{3}\right| < \frac{1}{2}\right\}$ $\mathscr{H} = \{x \mid x^4 - 16 \leq 0\}$

$\mathscr{D} = \left\{x \,\middle|\, x = (-1)^n \cdot \left(\frac{1}{n}\right), n \in N\right\}$ $\mathscr{Z} = \{x \mid x = (-1)^n, n \in N\}$

$\mathscr{E} = \{x \mid x = (-1)^{n+1}(2n - 1), n \in N\}$

5. *Bonus Problem.* Look back at the last example in this section and give (if they exist) the LUB and GLB of W.

Section 6—Ordering Rationals In this section, we shall raise a question and scrutinize some concepts whose proofs range from mildly to unbelievably difficult. We begin with the question, and we shall try to answer it in greater detail than we have so far.

Question. How can the rational numbers be ordered?

Answer. If r and s are rational numbers and $r < 0$ while $0 < s$, then the task of ordering r and s is easy for axiom 2 does this. If r and s are negative rational numbers, we postpone the task of ordering them for we realize that if we can order the positive rationals, then theorem 1 will do the job for us. Thus, you see, the only question we need to consider is the following: If each of j, k, m, and n is a counting number, then how can we order j/k and m/n?

Observe: if $j/k < m/n$, then $n \cdot (j/k) < m$ so that $j \cdot n < m \cdot k$. Note the use made of theorem 5 and of axiom 4 in this argument. On the other hand, if $jn < mk$, then $(jn) \cdot 1/k < m$ so that $j \cdot 1/k < m \cdot 1/n$ or $j/k < m/n$.

To summarize and generalize this discussion, we state the following theorem.

THEOREM 11. *If each r, s, t, and z is a positive number, then* $\frac{r}{s} < \frac{t}{z}$ iff $rz < ts$.

Example. Since $4 \cdot 5 < 3 \cdot 7$, it follows that $\frac{5}{7} < \frac{3}{4}$. Furthermore, $\frac{2}{3} < \frac{7}{8}$ since $2 \cdot 8 < 3 \cdot 7$.

In the study of any topic in mathematics, it is usually helpful to be able to draw diagrams that have some bearing on the topic being considered. These diagrams may not be an essential part of the mathematical theory but are rather "visual aids" and, as such, can be useful. One such visual aid is the representation of the real numbers as points on a line. We look now at a way of associating rational numbers with points on a line.

With a fixed unit of length we locate a point P a unit distance to the right of O, a point Q two units to the right, and so forth. Then, with the unit length divided into n equal parts, we can conceivably associate, by measurement, a point with each rational number m/n. In geometry we studied a method of construction for dividing a line segment in any ratio. This construction is illustrated in figure 1.13.

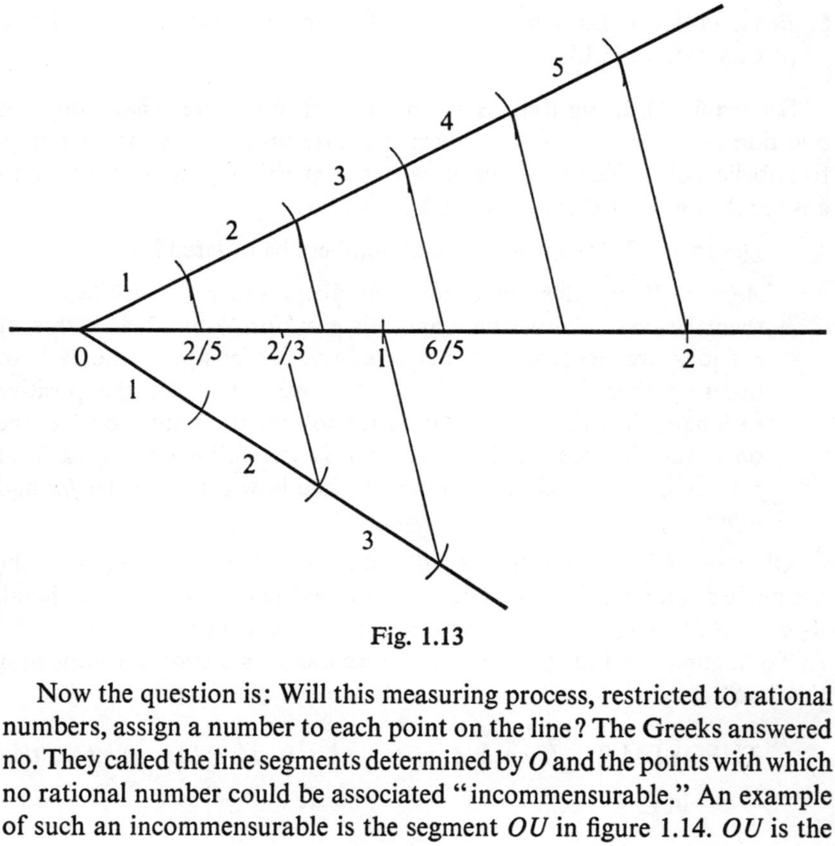

Fig. 1.13

Now the question is: Will this measuring process, restricted to rational numbers, assign a number to each point on the line? The Greeks answered no. They called the line segments determined by O and the points with which no rational number could be associated "incommensurable." An example of such an incommensurable is the segment OU in figure 1.14. OU is the

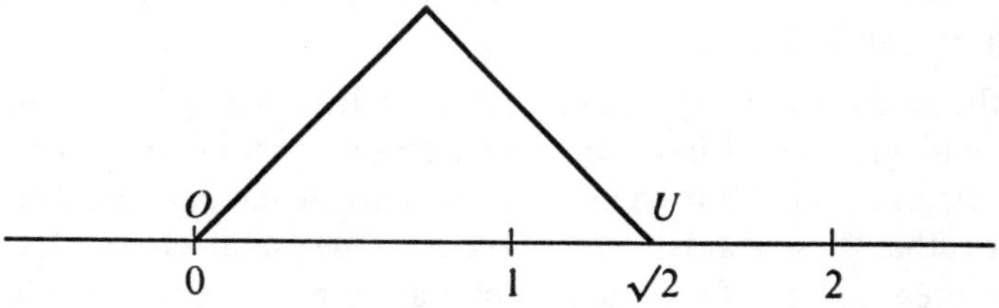

Fig. 1.14

hypotenuse of an isosceles right triangle with sides of unit length. The number which should be associated with U is the number whose square is 2.

The Pythagoreans—who believed numbers to be the basic elements in all of nature—discovered that the hypotenuse of an isosceles right triangle with sides of unit length could not be measured by a rational number. They believed this to be a flaw in God's handiwork or, just maybe, a flaw in their theory. Not being sure whose fault this was, they decided to remain silent about this discovery.

In order to show that there is no rational number whose square is 2, we shall suppose that m/n is a rational number, where m and n are not both even, such that $(m/n)^2$ is 2. But this assumption implies that $m^2 = 2n^2$ so that m^2 is even, hence (see problem 11 of problem set 1.6) m is even. Thus there is some integer k such that $m = 2k$ and $m^2 = 2n^2$ becomes $4k^2 = 2n^2$ or $2k^2 = n^2$. But this implies that n^2 is even and so n is even. This contradicts our assumption that m and n are not both even. Thus we conclude that there is no rational number whose square is 2.

It is important to understand just what this argument shows. It shows that there is no rational number r such that $r^2 = 2$. This assertion is usually expressed by saying that $\sqrt{2}$ is an *irrational number*. However, the use of the symbol $\sqrt{2}$ implies the existence of some real number whose square is 2 and we have not proven that such a number exists. We will confidently (truthfully) assert that, at present, a proof is difficult for us, yea verily, impossible without the axiom of completeness. Yet the existence of the square root of any positive number is suggested by the fact that there is a ruler and compass construction (see figure 1.15) in which, given a unit

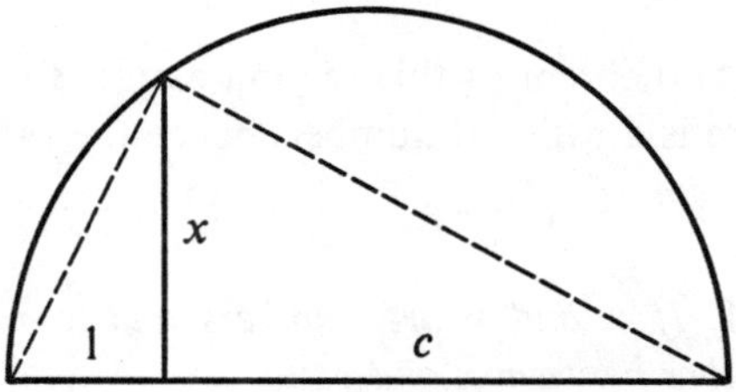

Fig. 1.15

segment and a segment of length c, we can produce a segment of length x where $x^2 = c$. (In the figure, the indicated right triangles are similar, hence $\frac{1}{x} = \frac{x}{c}$ so that $x^2 = c$.)

In the appendix, we give a proof of the existence of $\sqrt{c}$, $c > 0$, along with several other items for those who might be interested. However, the remainder of this section will be devoted to stating some interesting and useful theorems that are consequences of the axiom of completeness. The first of these theorems is usually called

THE ARCHIMEDIAN PROPERTY. *If t and x are positive numbers, then there is a counting number n such that $nt > x$.*

If $t > x$, there is no sweat, for $n = 1$ will do. Geometrically this theorem asserts that a given line segment, no matter how long, may be covered by a finite number of line segments of a given length, no matter how small (see figure 1.16). Thus a small ruler used often enough can measure extremely

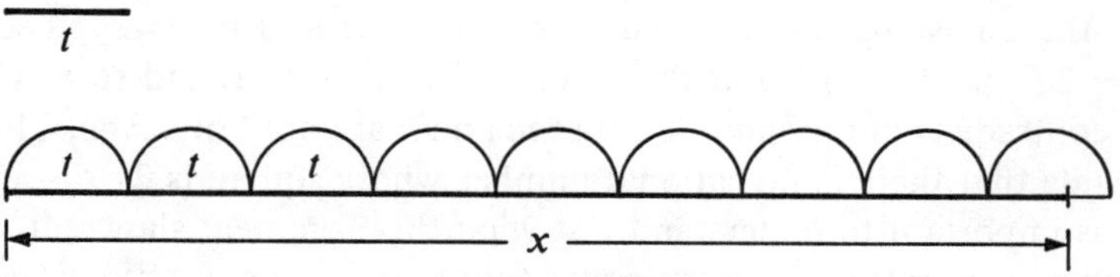

Fig. 1.16

large distances. This property, in turn, provides a means of showing another relation between real numbers and the counting numbers. This relation is the topic of the next theorem and it probably seems obvious.

THEOREM. *If x is a number, then there is an integer m such that $m \leq x < m + 1$.*

The next (and last) theorem of this section asserts that between any two real numbers there is a rational number, hence the rationals are dense in the reals.

THEOREM. *If x and y are numbers and $x < y$, then there is a rational number between x and y.*

The proof of the above theorem relies upon the previous theorem and on problem 13 of the next problem set.

Remark about this last theorem. We can be proud of the fact that if x and y are rational numbers, then theorem 7 assures us that the number $\frac{x + y}{2}$ is between them and $\frac{x + y}{2}$ is certainly rational.

PROBLEM SET 1.6

In problems 1–8 order the given rational numbers.

1. $\frac{5}{4}$ and $\frac{6}{5}$ **2.** $\frac{5}{9}$ and $\frac{3}{5}$

3. $\frac{27}{100}$ and $\frac{3}{11}$ **4.** $\frac{2251}{500}$ and $\frac{446}{99}$

5. $\frac{13}{6}, \frac{13}{7}$, and $\frac{7}{3}$ **6.** $-\frac{21}{13}$ and $-\frac{7}{5}$

7. $\frac{14}{33}, \frac{21}{50}$, and $\frac{19}{45}$ **8.** $\frac{3}{4}$ and $\frac{677}{900}$

9. If p and q are counting numbers, show that

a. $\frac{pq}{p+q} < p$ **b.** $\frac{pq}{p+q} < q$.

10. Verify that $1.41 < \sqrt{2} < 1.42$ and that $2.23 < \sqrt{5} < 2.24$.

11. If p is an integer and p^2 is even, show that p is even. *Hint:* If p is even, then $p = 2k$ for some integer k, and if p is odd, then $p = 2k + 1$ for some integer k.

12. Prove theorem 11.

13. Prove the following corollary of the Archimedian property.

COROLLARY. *If* $0 < p$, *then there is a counting number* n *such that* $1/n < p$.

Hint: Let $t = p$ and $x = 1$.

14. Give a rational number between 1.41 and $\sqrt{2}$.

15. Give a rational number between $\sqrt{2}$ and $\sqrt{3}$.

16. For each given number x, find an integer m such that

$m < x < m + 1$.

a. $x = \frac{20}{3}$ **b.** $x = -\frac{7}{5}$ **c.** $x = 1.8$ **d.** $x = \frac{3}{2}$ **e.** $x = \sqrt{7}$

SUMMARY

In this chapter we finished the construction of the foundation started in the review sections. A foundation upon which our work in mathematics will rest—the axioms for the real number system. In the course of this construction, many of the properties of the real numbers were stated and proved. Not because it was thought that the student was unfamiliar with these properties but rather to indicate how they may be derived from the given set of axioms.

It was (and is) hoped that the student gained some appreciation of the sheer beauty of mathematics as a deductive system. It is true that the proofs for some of these properties were perhaps a "bit much" at this time and they were relegated to the appendix.

Naturally, this chapter also devoted some time to the very practical task of using these properties for the solution of some problems. In order to check your grip on many of these properties, the following chapter review should be considered.

Review Problems

1. Give the first four order axioms.
2. What is meant by the statements (a) "k is an integer"? (b) "r is a rational number"?
3. When does $|x - 3| = 3 - x$?
4. State the relationship between each of the given pair of numbers.
   a. $|x + t|; |x| + |t|$ **b.** $|x|\cdot|t|; |xt|$ **c.** $|x - t|; |t - x|$ **d.** $|x - t|; |x| - |t|$ **e.** $0; |x|$.
5. Solve the following *and* draw number line models of their solution sets.
   a. $\dfrac{x+2}{x} \leq 0$ **b.** $\dfrac{x-4}{x+1} > 0$ **c.** $\dfrac{2x-5}{x} < 1$ **d.** $|x + 5| < 7$
   e. $|x - 3| > 2$.
6. Express each of the given intervals in terms of an absolute value inequality.
   a. $\langle -4,4\rangle$ **b.** $[-2,2]$ **c.** $[-1,3]$ **d.** $\langle 2,10\rangle$ **e.** $\langle -6,-1\rangle$
7. If S is a number set, what is meant by each of the following statements?
   a. "S is bounded below."
   b. "M is the least upper bound of S."
8. If they exist, give the LUB and the GLB for each of the number sets given below.
   $A = \left\{x \,\middle|\, x = \dfrac{3n-1}{2n-1}, n \in N\right\}$ $B = \{x \mid x^2(x + 4) > 0\}$
   $C = \{x \mid x = (-1)^{n+1}\cdot n^2, n \in N\}$ $D = \{x \mid |2x + 3| \leq 5\}$

9. Explain how the positive rational numbers may be ordered.

10. **a.** If $t = 0.2$ and $x = 4$, find the smallest counting number n that will satisfy the Archimedian property.

b. Find an integer m such that $m \leq \sqrt{7} < m + 1$.

CHAPTER 2
Ordered Pairs of Numbers

In the previous chapter, we considered several important notions pertaining to number sets. In so doing, we saw some truth in that ancient Gaelic proverb about the relative merits of many words and a single diagram. Surely our study of these notions was at least facilitated by the assumed one-to-one correspondence of the real numbers and the points on a line. Hoping to profit from it, we shall now consider some means of establishing a correspondence between pairs of real numbers and points in a plane. Any means of establishing such a one-to-one correspondence is called a *coordinate system.* The means that we shall use is usually credited to the mathematician René Descartes (1596–1650) and the resulting coordinate system is called a *rectangular* or *Cartesian* system.

Although our immediate interest shall be focused upon "ordered pairs" of real numbers, one may later wish to speak of "ordered pairs" of objects other than real numbers. At such a time, we may need the following:

DEFINITION. *If each of A and B is a set, then the Cartesian product of A and B, denoted by $A \times B$, is the set of all pairs (a,b) where $a \in A$, $b \in B$, and where equality of pairs (a,b) and (x,y) in $A \times B$ is defined by $(a,b) = (x,y)$* iff *a is x and b is y. The elements of $A \times B$ are called* ordered pairs.

Examples.

1. If $A = \{\varepsilon,\delta,\theta\}$ and $B = \{r,s\}$, then $A \times B = \{(\varepsilon,r), (\varepsilon,s), (\delta,r), (\delta,s), (\theta,r), (\theta, s)\}$.

2. If $C = \{i,\pi\}$ and $D = \{i,0,1\}$, then $C \times D = \{(i,i), (i,0), (i,1), (\pi,i), (\pi,0), (\pi,1)\}$.

3. If $E = \{\text{fair}\}$ and $F = \{\text{ball,dinkum}\}$, then $E \times F = \{(\text{fair,ball}), (\text{fair,dinkum})\}$.

For the present however, we are to be concerned only with the Cartesian product of the set of real numbers R with itself. Thus the elements of $R \times R$ are to be called "ordered pairs of real numbers." We certainly hope to consider some subsets of $R \times R$, and we conclude this introduction with the warning that these subsets are quite different from the number sets we so valiantly dealt with in the previous chapter.

Section 1—Basic Ideas of Cartesian Geometry The basic idea in Cartesian geometry is the establishment of a one-to-one correspondence between the points of an Euclidean plane and number pairs. While this can probably be done in other ways, we shall elect to make this correspondence in a most informal manner.

A plane writing surface, for example, a sheet of paper, imagined to be indefinite in extent may serve as a model for the set of all points. Select one of the horizontal lines on this surface and denote it by X. Select a vertical line and denote it by Y. Regard each of these lines as a model for the set of all numbers as described in the early part of chapter 1, the number 0 being at the intersection of X and Y, the positive numbers to the right on the horizontal line and the positive numbers upward on the vertical line. If x is a number on X, denote by $x|$ the vertical line containing x and if y is a number on Y, denote by y the horizontal line containing y, then we shall identify the number pair (x,y) with the point of intersection of $x|$ and y (see figure 2.1). Further, see figure 2.2, if P is a point, then the vertical line through P will intersect X at some number h and the horizontal line through P will intersect Y at some number k. Hence, the number pair (h,k) is associated with the point P and we shall write $P = (h,k)$.

In this manner, we assign a pair of real numbers (x,y) to each point in the plane and each pair of real numbers has one and only one point assigned to it. We will identify this association of a number pair (x,y) with the point P by writing $P = (x,y)$ and we will call the number pair the *coordinates of P*.

One useful consequence of this association is the notion that the distance from a point P to a point Q, denoted by $d(P,Q)$, can be given in terms of the coordinates of the two points. To see this, let $P = (h,k)$ and

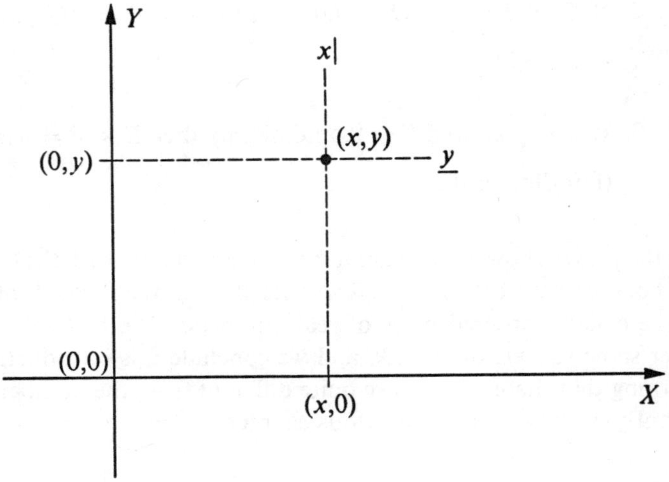

Fig. 2.1

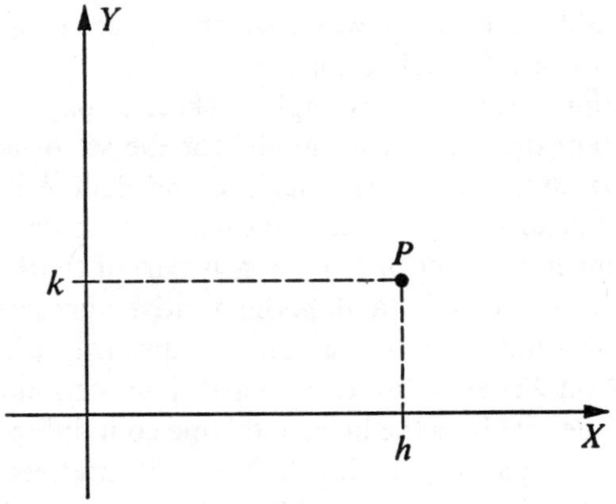

Fig. 2.2

$Q = (r,s)$. The vertical line through P and the horizontal line through Q (see figure 2.3) will intersect at the point U. Clearly $U = (h,s)$. Now

$$d(P,U) = |s - k| = |k - s| \quad \text{and}$$
$$d(U,Q) = |r - h| = |h - r|.$$

The Pythagorean theorem* of geometry tells us that

$$\begin{aligned}[d(P,Q)]^2 &= [d(P,U)]^2 + [d(U,Q)]^2 \\ &= |h - r|^2 + |k - s|^2\end{aligned}$$

* Named after the Irish mathematician Patty G. O'Ryan who resided for a while in Greece.

hence,

$$d(P,Q) = \sqrt{[(h - r)^2 + (k - s)^2]}.$$

The above statement is called the *distance formula.*

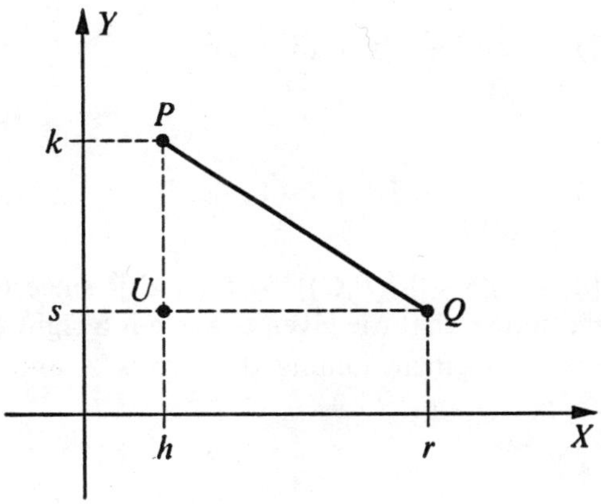

Fig. 2.3

The examples that follow illustrate how some geometric problems can be solved by the association of points in the Euclidean plane with pairs of numbers and the use of the distance formula.

Example 1. If the vertices of a triangle are the points A, B, and C where $A = (4,1)$, $B = (3,5)$, and $C = (-1,4)$, show that the triangle is isosceles but that it is not an equilateral triangle.

Solution.

$$d(A,B) = \sqrt{[(4 - 3)^2 + (1 - 5)^2]} = \sqrt{17}$$
$$d(B,C) = \sqrt{[(3 + 1)^2 + (5 - 4)^2]} = \sqrt{17}.$$

This pair of computations show that the triangle is an isosceles triangle. To continue, since

$$d(A,C) = \sqrt{[(4 + 1)^2 + (1 - 4)^2]} = \sqrt{34}$$

we know that the triangle is not an equilateral triangle.

Example 2. Use the converse of the Pythagorean theorem to show that the triangle whose vertices are the points $(-1,5)$, $(6,-1)$, and $(7,3)$ is a right triangle.

Solution. Label the vertices A, B, and C respectively. Applying the distance formula, we get:

$$\begin{aligned} d(A,B) &= \sqrt{[(-1-6)^2 + (5+1)^2]} \\ &= \sqrt{85}. \end{aligned}$$

Similarly,

$$\begin{aligned} d(A,C) &= \sqrt{[(7+1)^2 + (3-5)^2]} \\ &= \sqrt{68}, \end{aligned}$$

while

$$\begin{aligned} d(B,C) &= \sqrt{[(6-7)^2 + (-1-3)^2]} \\ &= \sqrt{17}. \end{aligned}$$

Note that $[d(A,C)]^2 + [d(B,C)]^2 = [d(A,B)]^2$ since $68 + 17$ is 58. Thus we have shown that the given triangle is a right triangle whose hypotenuse is the segment joining the points A and B. (See figure 2.4.)

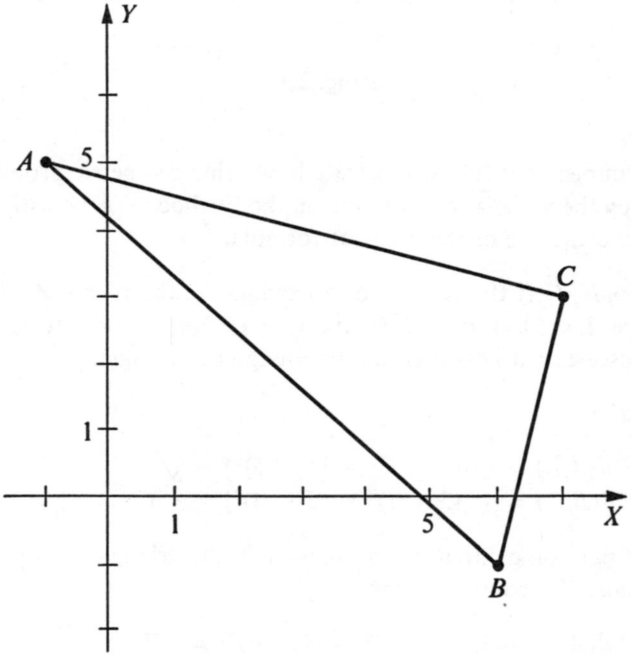

Fig. 2.4

Example 3—the midpoint formula. Suppose that $P = (h,k)$ and $Q = (r,s)$ are two points in the plane and that $h \neq r$. Let x be the midpoint of the interval $[h,r]$, $x = \frac{h+r}{2}$, then the vertical line through $(x,0)$ will intersect the segment joining P and Q at the point $M = (x,y)$. Now, if M is the midpoint of the segment joining P and Q, then it should follow that $d(P,M) = d(M, Q)$ and so $[d(P,M)]^2 = [d(M,Q)]^2$. (See figure 2.5.) Using the distance formula, we have the equation

1. $(x - h)^2 + (y - k)^2 = (x - r)^2 + (y - s)^2.$

Let us pause and consider the following computations. Since $x = \frac{h+r}{2}$, $(x - h)^2 = \frac{(r-h)^2}{4}$ and $(x - r)^2 = \frac{(r-h)^2}{4}$, thus $(x - h)^2$ is $(x - r)^2$. The beauty of these computations is that they allow us to simplify equation (1) above to

2. $(y - k)^2 = (y - s)^2$ or $y^2 - 2ky + k^2 = y^2 - 2sy + s^2$

hence, continuing to simplify,

3. $2sy - 2ky = s^2 - k^2$ or $y(s - k) = \frac{s^2 - k^2}{2}.$

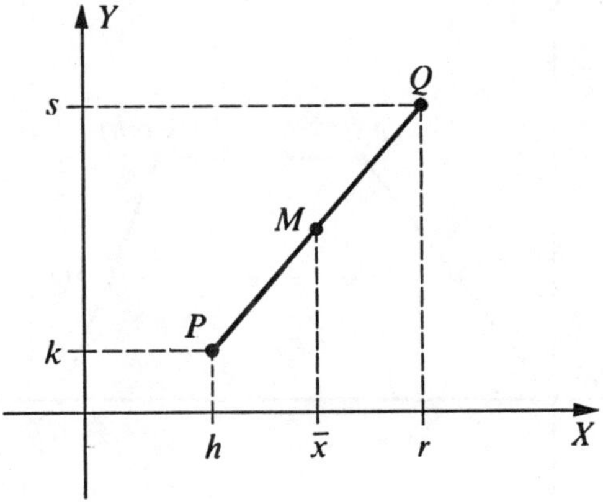

Fig. 2.5

Therefore, if we further suppose that $s \neq k$, then we can solve this equation for y and get $y = \frac{s + k}{2}$. However, in order to feel better about calling M the midpoint of the segment joining P and Q, it is necessary that we go on to show, with $M = \left[\frac{h + r}{2}, \frac{s + k}{2}\right]$ that $d(P,M) = d(M,Q)$, but we shall assume that the student can and will do this in his spare time.

To recap the preceding, if $P = (h,k)$ and $Q = (r,s)$, then the midpoint M of the segment joining P and Q is given by

$$M = \left(\frac{h + r}{2}, \frac{k + s}{2}\right).$$

In the next example, we try to indicate how the use of coordinates and algebra can help us with proofs of certain geometric theorems. Such a use of coordinates and algebra is called *analytical geometry*. Notice the convenient placement of the coordinate axis in this example, for quite often this leads to a simple proof of the desired property.

Example 4. Show that the line segment joining the midpoints of the nonparallel sides of any trapezoid has a length equal to half the sum of the lengths of the parallel sides. (The parallel sides of a trapezoid are called the bases of the trapezoid. See figure 2.6.)

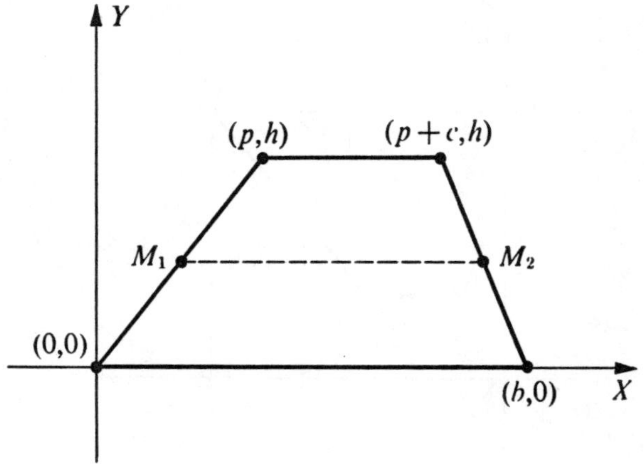

Fig. 2.6

Proof. Denote the lengths of the bases by b and c. Let the altitude of the trapezoid have length h. The midpoint formula states that

$$M_1 \text{ is } \left(\frac{p}{2}, \frac{h}{2}\right) \text{ and that } M_2 \text{ is } \left(\frac{p + c + b}{2}, \frac{h}{2}\right),$$

hence, by the distance formula,

$$d(M_1,M_2) = \sqrt{\left[\frac{c + b}{2}\right]^2} = \frac{b + c}{2} \qquad \text{QED}$$

Remark to a point. In the interest of establishing communications with each other about problems that are to come in this and later chapters of the textbook, let us agree to the following terminology when referring to the elements of $R \times R$.

> The statement that "P is a point" means that P is an ordered number pair (x,y) whose first element x is called the *abscissa* of P and whose second element y is called the *ordinate* of P. The two numbers are collectively called the *coordinates* of P, and we shall write $P = (x,y)$. Further, $(x,y) = (r,s)$ if and only if x is r and y is s.

PROBLEM SET 2.1

1. Plot the points (1,0), (0,1), (3,2), (2,3), (5,−2), (3,−5), and (−5,−3).

2. Verify graphically, that the points $(1 - 6t, 2 + 3t)$ corresponding to $t = 0, \pm 1/3, \pm 1/2, \pm 2/3, \pm 1$ lie on a line.

3. Verify graphically, that the points $(-3 + 2t, 1 + 2t)$ corresponding to $t = 0, \pm 1/2, \pm 1, \pm 2$ lie on a line.

4. Give the distance between the following pairs of points:
a. (0,0) and (3,4) **b.** (4,−5) and (−2,3)
c. (4,−3) and (−1,−2) **d.** (1,1) and (1,−4)

5. Same as problem 4 for the following pairs of points:

a. (4,−1) and (2,6) **b.** (5,0) and (−2,−1)
c. (1,1.5) and (−2,0.5) **d.** (1/2,−2) and (−3/2,4)

6. Same as problem 4 for the following pairs of points:

a. $(\sqrt{k},0)$ and $(0,\sqrt{m})$ **b.** $(\sqrt{k},0)$ and $(0,k)$
c. $(-k,-k)$ and (k,k) **d.** $(\sqrt{k},0)$ and $(\sqrt{m},0)$

In each of problems 7–12, use the distance formula to determine if the given points are the vertices of an equilateral triangle, a right triangle, *or* neither of these. State which and if it is an equilateral triangle, give the length of a side; if it is a right triangle, give the length of the hypotenuse.

7. a. $(0,5), (-3,4), (-1,2)$
b. $(3,-6), (8,-2), (-1,-1)$

8. a. $(0,2), (4,-2), (1,3)$
b. $(1,-1), (4,2), (6,-3)$

9. a. $(1,-2), (-4,2), (1,6)$
b. $(-3,0), (1,0), (-1,2\sqrt{3})$

10. a. $(-3,2), (1,-1), (-2,-5)$
b. $(1,2), (3,7), (-2,1)$

11. a. $(5,6), (13,6), (5,12)$
b. $(1,\sqrt{2}), (-5,3\sqrt{2}), (-\sqrt{2},-3)$

12. a. $(3,2), (1,1), (-1,5)$
b. $(0,0), (1,0), (1/2,\sqrt{3}/2)$

13. Give the midpoint of the segment joining the given points:
a. $(4,8)$ and $(6,3)$ **b.** $(-5,4)$ and $(7,-2)$

14. Give the midpoint of the segment joining the given points:
a. $(-3,-1)$ and $(-6,4)$ **b.** $(3,2)$ and $(7,8)$

15. The midpoint of the segment joining the points $(-4,-3)$ and (r,s) is $(8,-5)$. Find r and s.

16. A median of a triangle is a segment joining a vertex of the triangle and the midpoint of the side opposite this vertex. Find the lengths of all three of the medians of the triangle whose vertices are at $A = (4,3)$, $B = (-1,-4)$, and $C = (-3,2)$.

17. If the vertices of a triangle are the points $(0,6)$, $(0,-4)$, and $(5\sqrt{3},1)$, show that the triangle is equilateral *and* give its area. *Hint:* A sketch of the triangle might help you to see what to do if you don't know.

18. Find a point $P = (x,y)$ that is equidistant from the points $A = (-3,1)$, $B = (2,-4)$, and $C = (1,1)$. *Hint:* Start by setting $d(P,A) = d(P,C)$ and solve for x. Then consider how you can use the point B to help find y.

19. Using the methods of analytic geometry to show that the distance between the midpoints of two sides of a triangle is one-half the length of the third side. *Hint:* Let the vertices be at $(0,0)$, $(a,0)$, and (b,h).

20. Use the methods of analytic geometry to show that the midpoint of the hypotenuse of a right triangle is equidistant from the three vertices. *Hint:* Let the vertices be the points $(0,0)$, $(a,0)$, and $(0,b)$.

Section 2—Subsets of the Plane Each collection or set of ordered number pairs (points) is a subset of $R \times R$, and in this section we shall devote our attention to two such subsets, one called a *circle* and the other we shall call a *disk*. The reasons for turning our attention to these particular point sets is that we do not wish to tackle anything very difficult at this time, and the circle and disk are easy consequences of the distance formula just mastered. Thus, light of heart and in good cheer, we advance to the task.

If r is a positive number and C is a given point, then a circle is the set of all points whose distance from the given point C is the number r. The given point C is called the *center of the circle* and the positive number r is called the *radius of the circle*. Now, the first task that we assign ourselves is that of giving some analytic method of describing a circle. (See figure 2.7.) So let us begin. Suppose $C = (h,k)$. If $P = (x,y)$ is a point on the circle whose center is C and whose radius is r, $r > 0$, then

$$d(P,C) \text{ is } r.$$

By the distance formula, this means that

$$\sqrt{[(x - h)^2 + (y - k)^2]} = r$$

and, squaring both sides gives

$$(x - h)^2 + (y - k)^2 = r^2.$$

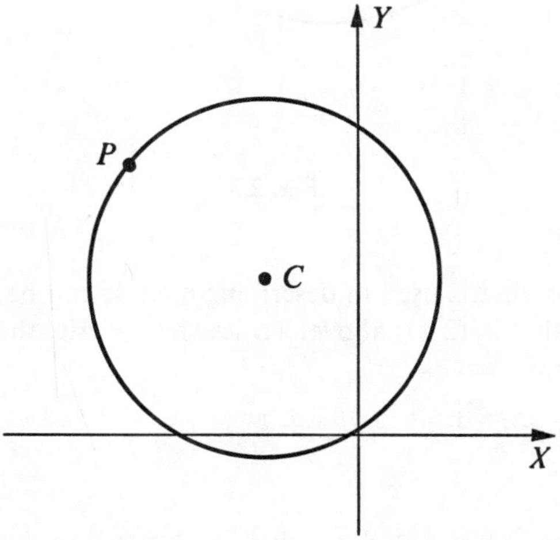

Fig. 2.7

Therefore the circle with center (h,k) and radius r is the point set to which a point (x,y) belongs only if $(x - h)^2 + (y - k)^2$ is r^2. This is precisely what we shall mean by writing

(2.21) $$\{(x,y) \mid (x - h)^2 + (y - k)^2 = r\}.$$

Example 1. The circle with center at $(2, -1)$ and radius 3 is the set of points described by

$$\{(x,y) \mid (x-2)^2 + (y+1)^2 = 9\}.$$

The graph of this point set is indicated by figure 2.8.

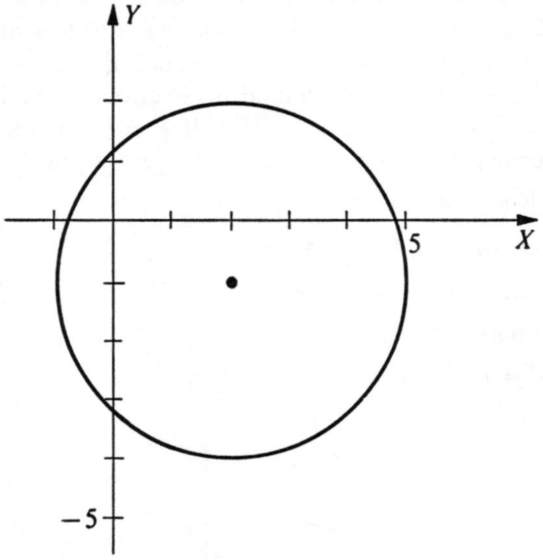

Fig. 2.8

The equation that is used in describing a circle can be given in a form different from that in (2.21) above. To see this, rewrite the equation given there in the form

$$x^2 - 2hx + h^2 + y^2 - 2ky + k^2 = r^2$$

and then as

$$x^2 + y^2 - 2hx - 2ky + (h^2 + k^2 - r^2) = 0,$$

the latter being an equation of the form

$$x^2 + y^2 + Dx + Ey + F = 0$$

where $D = -2h$, $E = -2k$, and $F = h^2 + k^2 - r^2$.

So, if a point set (*containing at least two points*) has an equation of this form, then it is a circle. The reason for the underlined condition, parenthetically inserted, will not be explained here but will be left to the conclusion of this section. The student is assured that, in the problem sets, those point sets described by the notation in (2.22) are circles.

(2.22) $$\{(x,y) \mid x^2 + y^2 + Dx + Ey + F = 0\}$$

Example 2. Describe the point set denoted by

$$\{(x,y) \mid x^2 + y^2 - 6x + 4y - 3 = 0\}$$

Solution. Rewrite the equation in the form

$$x^2 - 6x + (\ ) + y^2 + 4y + (\ ) = 3$$

and use the technique of completing the square to write this in the form

$$x^2 - 6x + 9 + y^2 + 4y + 4 = 3 + 9 + 4$$

or

$$(x - 3)^2 + (y + 2)^2 = 16$$

for then we see that the given set is a circle with center at $(3, -2)$ and radius 4.

Example 3. Discuss and sketch the graph of the point set

$$\{(x,y) \mid (x - 2)^2 + y^2 < 8\}.$$

Solution. Now $(x - 2)^2 + y^2$ is $(x - 2)^2 + (y - 0)^2$, so the graph of the equation $(x - 2)^2 + y^2 = 8$ would be a circle with center $(2,0)$ and radius $\sqrt{8}$. (Approximately 2.8.) Furthermore, $(x - 2)^2 + y^2$ is the square of the distance from (x,y) to $(2,0)$, thus (x,y) must be interior to the circle. The given set, then, could be described as the interior of a circle but we shall call it a *disk*. In sketching the graph of this point set, we adopt the convention of drawing the circle as a broken arc to denote the fact that points on the circle *do not* belong to the set described. (See figure 2.9.)

Example 4. The set of points denoted by

$$\{(x,y) \mid (x - h)^2 + (y - k)^2 \leq r^2\}$$

is the union of two point sets. One is the circle with center (h,k) and radius r, and the other is the interior of this circle, that is, a *disk*. Figure 2.10 indicates the conventions we shall use in drawing the graph of such a point set.

Remarks. To show the need of the underlined assumption in the discussion pertaining to the notation in (2.22), we begin with the equation

$$x^2 + y^2 + Dx + Ey + F = 0.$$

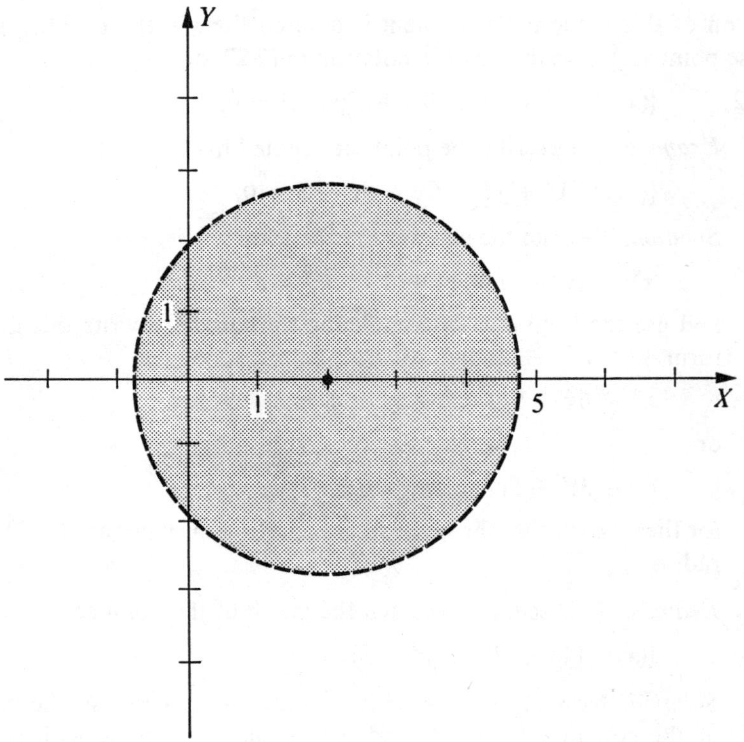

Fig. 2.9. Graph of $\{(x,y) \mid (x - 2)^2 + y^2 < 8\}$

By completing the square, we can write this equation as

$$\left(x^2 + Dx + \frac{D^2}{4}\right) + \left(y^2 + Ey + \frac{E^2}{4}\right) = \frac{D^2}{4} + \frac{E^2}{4} - F$$

so that

$$\left(x + \frac{D}{2}\right)^2 + \left(y + \frac{E}{2}\right)^2 = \frac{D^2 + E^2 - 4F}{4}.$$

If $D^2 + E^2 - 4F$ is positive, then the equation is an equation of a circle with center at $\left(-\frac{D}{2}, -\frac{E}{2}\right)$ and radius $1/2 \cdot \sqrt{(D^2 + E^2 - 4F)}$. If $D^2 + E^2 - 4F$ is 0, then both terms on the left are also zero so that the only point in the set is the point $\left(-\frac{D}{2}, -\frac{E}{2}\right)$. Finally, if $D^2 + E^2 - 4F < 0$, then the set is empty since neither term on the left can be negative.

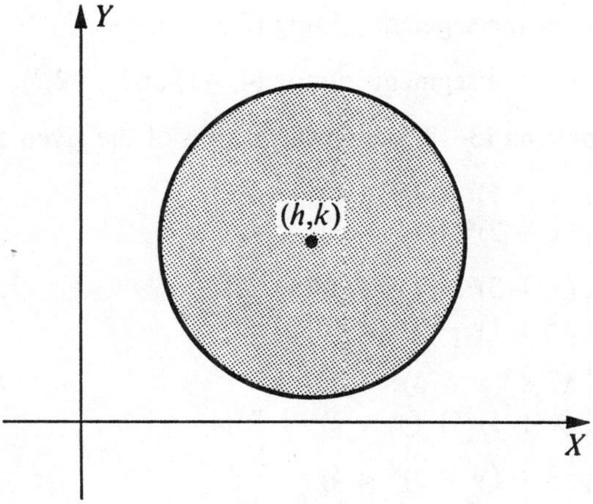

Fig. 2.10

PROBLEM SET 2.2

In each of problems 1–12, use the notation of (2.21) to denote the circle that satisfies the given conditions *and* sketch its graph.

1. **a.** Center (3,2) and radius 3.
   b. Center (−3,−1) and radius 6.

2. **a.** Center (1,4) and radius 4.
   b. Center (−2,−5) and radius 4.

3. **a.** Center (−1,1) and radius 5/2.
   b. Center (−3,0) and radius 5.

4. **a.** Center (−1,3) and radius 3/2.
   b. Center (0,4) and radius $\sqrt{10}$.

5. **a.** Center (2,−2) and radius $\sqrt{2}$.
   b. Center (−3,−4) and radius 2/3.

6. **a.** Center (−3,3) and radius $\sqrt{3}$.
   b. Center (4,−2) and radius 3/4.

7. Center (3,−1) and passing through (1,4).

8. Center (1,−1) and passing through (0,2)

9. Center (0,0) and passing through $(1/2,\sqrt{3}/2)$.

10. Center (−2,3) and passing through (4,4).

11. A diameter is the segment joining (2,3) and (−2,3).

12. A diameter is the segment joining (4, −1) and (−2,3).

In each of problems 13–18, sketch the graphs of the given circles or disks.

13. **a.** $\{(x,y) \mid (x - 5)^2 + (y + 1)^2 = 49\}$
b. $\{(x,y) \mid (x + 2)^2 + (y - 3)^2 < 4\}$

14. **a.** $\{(x,y) \mid (x + 3)^2 + (y + 2)^2 = 25\}$
b. $\{(x,y) \mid x^2 + (y - 3)^2 < 16\}$

15. **a.** $\{(x,y) \mid x^2 + (y + 4)^2 = 36\}$
b. $\{(x,y) \mid (x + 1)^2 + (y - 2)^2 < 25/4\}$

16. **a.** $\{(x,y) \mid x^2 + (y - 5)^2 < 4\}$
b. $\{(x,y) \mid (x + 1)^2 + (y + 2)^2 \leq 4\}$

17. **a.** $\{(x,y) \mid x^2 + y^2 \leq 1\}$
b. $\{(x,y) \mid (x - 1)^2 + y^2 < 81/4\}$

18. **a.** $\{(x,y) \mid x^2 + y^2 < 5\}$
b. $\{(x,y) \mid (x - 2)^2 + y^2 \leq 16\}$

In each of problems 19–27, give the center and the radius of the given circle. No graph required.

19. $\{(x,y) \mid x^2 + y^2 + 2x - 4y - 11 = 0\}$

20. $\{(x,y) \mid x^2 + y^2 - 14x - 6y + 9 = 0\}$

21. $\{(x,y) \mid x^2 + y^2 - 8x + 7 = 0\}$

22. $\{(x,y) \mid x^2 + y^2 + 2y - 3 = 0\}$

23. $\{(x,y) \mid x^2 + y^2 - 10x - 24y = 0\}$

24. $\{(x,y) \mid x^2 + y^2 - 8x + 6y = 0\}$

25. $\{(x,y) \mid x^2 + y^2 + 10y - 75 = 0\}$

26. $\{(x,y) \mid x^2 + y^2 + x + 2y - 1 = 0\}$

27. $\{(x,y) \mid x^2 + y^2 + 2x - 3y - 3 = 0\}$

28. *Small bonus problem.* Sketch the graph of the point set denoted by

$$\{(x,y) \mid 4 < x^2 + (y - 2)^2 < 9\}.$$

Section 3—Straight Line In this section, we shall consider another subset of the plane which we shall call a line. The word *line* shall not mean anything other than a straight line.

If each of A, B, and C is a number and at least one of A or B is distinct from zero, then an equation of the form $Ax + By = C$ shall be called the equation of a line. The solution set $S = \{(x,y) \mid Ax + By = C\}$ is an infinite subset of $R \times R$ and one of the simpler ways of describing this set is to plot its graph. The customary procedure for plotting the graph of a line is to find two pairs (x,y) which satisfy the equation, to plot them, and then to draw the line through them. To simplify our language, we often speak of "the line $Ax + By = C$," when we mean the line $\mathscr{L}$ which is the graph of this equation. It is hoped that this shortcut will not cause any confusion. It may be in order, though, to point out that we are dealing with three separate but related concepts:

1. The equation $Ax + By = C$.
2. The solution set S of this equation, which is a set of ordered pairs (points) of real numbers.
3. The line $\mathscr{L}$, where $\mathscr{L}$ is that subset of the plane which contains only those points (x,y) that are elements of S. This line $\mathscr{L}$ is called the graph of the equation $Ax + By = C$.

Thus it should cause no confusion if we write

$$\mathscr{L} = \{(x,y) \mid Ax + By = C\}$$

and speak of the equation as the *general equation of* $\mathscr{L}$.

Example. Sketch the graph of $\{(x,y) \mid 3x - 2y = 6\}$.

Solution. If $x = 0$, then $y = -3$ and if $x = 2$, $y = 0$ so this line contains the points (0,3) and (2,0). The graph can now be easily drawn. Note that if $x = 4$, $y = 3$ so that the point (4,3) also belongs to this line. (See figure 2.11.)

As an interesting by-product of this discussion, consider the following example:

Example. Sketch the graph of the point set S where

$$S = \{(x,y) \mid x + 2y < 3\}.$$

Solution. First consider the equation $x + 2y = 3$ and its graph $\mathscr{L}$. Of course, no point of $\mathscr{L}$ belongs to S but this line does divide the plane into two mutually exclusive regions, one of which is S. But which? The question is easily answered. Select any point not on $\mathscr{L}$, (0,0) will do, and evaluate $x + 2y$ at this point. Since $0 + 2 \cdot 0 < 3$, S is that region of the plane which contains the point (0,0) and is bounded by $\mathscr{L}$. (See figure 2.12.)

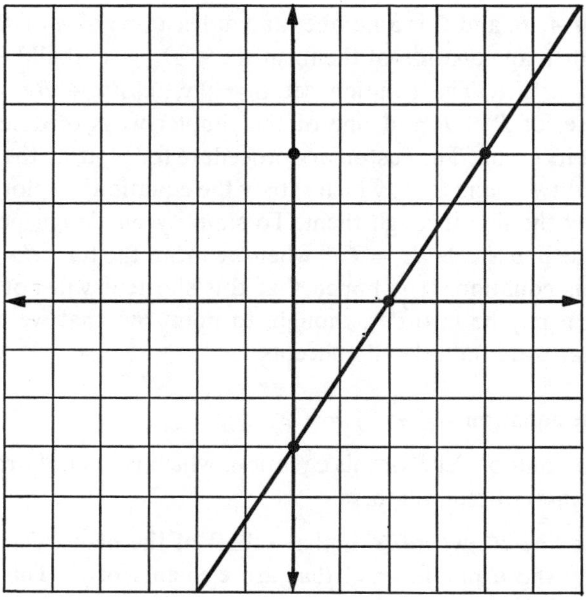

Fig. 2.11

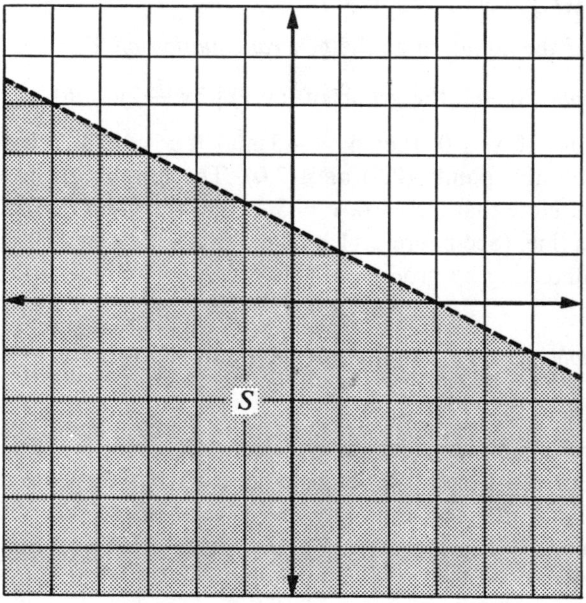

Fig. 2.12

Two special cases of the general equation of a line need to be called to your attention.

Case I (the vertical line). If $B = 0$, the general equation of a line could be written as $Ax + 0y = C$ so that, for every number y, $Ax = C$. In the present context, therefore, the equation $x =$ constant will be considered as describing a vertical line.

Case II (the horizontal line). If $A = 0$, the equation $0x + By = C$ or simply $By = C$ describes a horizontal line since, for every number x, $y = C/B$. Hence, the equation of a horizontal line may be given by an equation of the form $y =$ constant.

The Slope of a Line Suppose that $\mathscr{L}$ is a nonvertical line (note that this is equivalent to saying that $B \neq 0$). The general equation of such a line can be written in the form $y = -\frac{A}{B}\cdot x + \frac{C}{B}$ so that for each nonvertical line we can associate the number $-\frac{A}{B}$. This number is called "the slope" of the line. If (h,k) is a point on this line, then $Ah + Bk = C$ so that its general equation can be written as

$$Ax + By = Ah + Bk \text{ and hence}$$
$$By - Bk = Ah - Ax$$
$$B(y - k) = -A(x - h)$$
$$y - k = -\frac{A}{B}(x - h).$$

Since it is customary to denote the slope of a line by m, the equation of a nonvertical line containing the point (h,k) with slope m, is written in the form

$$y - k = m(x - h).$$

This equation is called the "point-slope equation" of the line.

Example. Give an equation of the line containing $(3,-4)$ whose slope is $-2/3$ and sketch its graph.

Solution. The point-slope equation is given by

$$y - (-4) = -2/3\cdot(x - 3)$$

or

$$y + 4 = -2/3\cdot(x - 3).$$

To sketch the graph, we get a second point by selecting x arbitrarily, say $x = 0$, substitute into the above equation and find $y = -2$, thus $(0,-2)$ is another point on the line. By simplifying the above equation, we obtain the general equation of this line $2x + 3y = -6$. (See figure 2.13.)

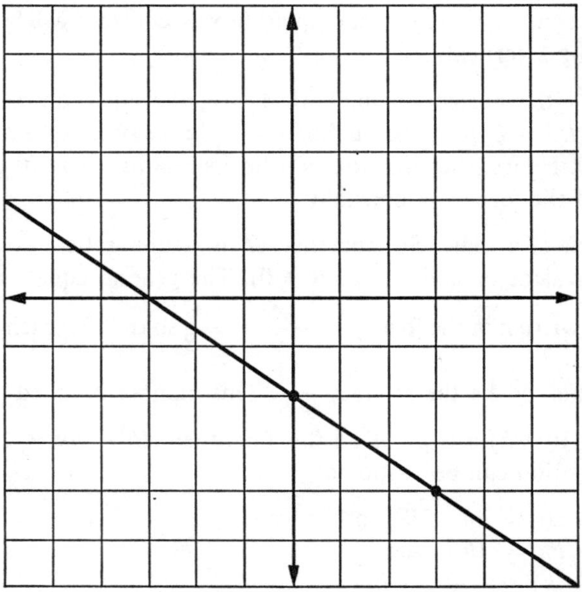

Fig. 2.13

It is also of interest to note that the slope of a nonvertical line can be determined from the coordinates of any two points on the line. To see this, let (h,k) and (p,q) be two points on the line. Using the first point, we write $y - k = m\cdot(x - h)$ and, since (p,q) must satisfy this equation, $q - k = m\cdot(p - h)$ thus, solving for m,

$$m = \frac{q - k}{p - h}.$$

Parallel lines Suppose that $\mathscr{L}_1$ and $\mathscr{L}_2$ are two parallel lines. Since it is obvious that any pair of vertical or any pair of horizontal lines are parallel, let us further suppose that $\mathscr{L}_1$ and $\mathscr{L}_2$ are neither horizontal nor vertical. Then, from plane geometry,

$$\text{angle } \theta_1 = \text{angle } \theta_2$$

so that the two right triangles indicated in figure 2.14 are similar and, hence, the ratios of their corresponding sides are equal. That is,

$$\frac{q}{p-a} = \frac{n}{m-c}.$$

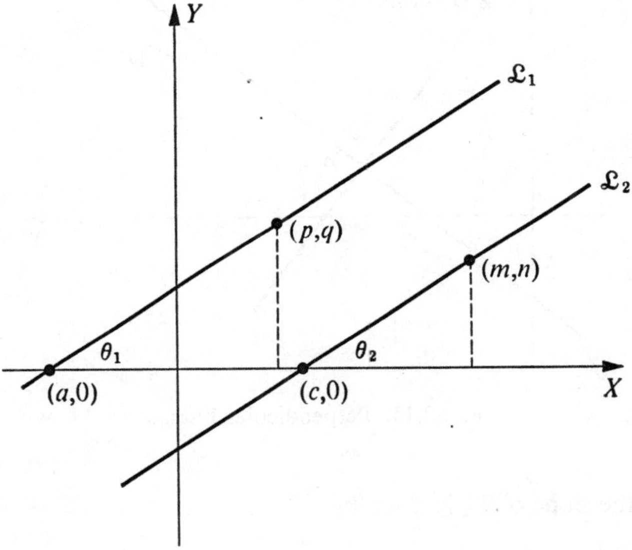

Fig. 2.14. Parallel lines.

The above leads us to make the following definition.

DEFINITION. *If $\mathscr{L}_1$ and $\mathscr{L}_2$ are two lines with slopes m_1 and m_2 respectively, then the statement that "$\mathscr{L}_1$ is parallel to $\mathscr{L}_2$" means that* $m_1 = m_2$.

Perpendicular lines Now, let us suppose that $\mathscr{L}_2$ is perpendicular to $\mathscr{L}_1$ and that $P = (a,b)$ is their point of intersection. Let $T = (p,q)$ be a point on $\mathscr{L}_1$ distinct from P and let $Q = (h,k)$ be a point on $\mathscr{L}_2$ distinct from P. It follows, from the theorem of Pythagoras, that (see figure 2.15)

$$[d(P,Q)]^2 + [d(P,T)]^2 = [d(Q,T)]^2.$$

Before this goes any further, the student should observe that the slope of $\mathscr{L}_1$ is given by

$$m_1 = \frac{q-b}{p-a}$$

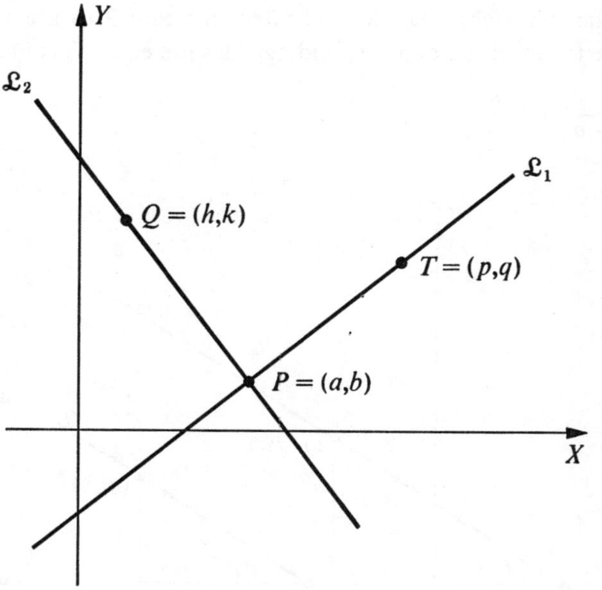

Fig. 2.15. Perpendicular lines.

and that the slope of $\mathscr{L}_2$ is given by

$$m_2 = \frac{k - b}{h - a}.$$

Now, let us return to the equation that resulted from the use of the Pythagorean theorem. By the distance formula, this means that

$$(h - a)^2 + (k - b)^2 + (p - a)^2 + (q - b)^2 = (k - q)^2 + (h - p)^2.$$

Expanding each of the squared terms, the above can be simplified to (check this)

$$-2ah + 2a^2 + 2hp - 2ap + 2kq - 2bk - 2bq + 2b^2 = 0$$

dividing both sides by 2 and factoring, we have

$$-a(h - a) + p(h - a) + k(q - b) - b(q - b) = 0$$

which can be written as

$$(h - a)(p - a) + (q - b)(k - b) = 0 \quad \text{or}$$

$$(h - a)(p - a) = -(q - b)(k - b)$$

thus

$$\frac{h-a}{k-b} = -\frac{q-b}{p-a} \quad \text{or} \quad \frac{1}{m_2} = -m_1.$$

DEFINITION. *If $\mathscr{L}_1$ and $\mathscr{L}_2$ are two lines with slopes m_1 and m_2 respectively, then the statement that "$\mathscr{L}_1$ is perpendicular to $\mathscr{L}_2$" means that $\frac{1}{m_2} = -m_1$, that is, that $m_1 \cdot m_2 = -1$.*

PROBLEM SET 2.3

1. Suppose that $\mathscr{L}$ is a nonvertical line containing the points (h,k) and (p,q) and that $h < p$.

a. If $m > 0$, show that $k < q$.
b. If $m < 0$, show that $k > q$.

2. a. Explain why a vertical line has no slope.
b. Explain why a horizontal line has slope 0.

The graph of each of the following is a line. Give three of its points, its slope, and sketch the line.

3. $\{(x,y) \mid 3x + y = 4\}$ **4.** $\{(x,y) \mid 2x + y = 0\}$

5. $\{(x,y) \mid y = 5\}$ **6.** $\{(x,y) \mid x - 3 = 0\}$

7. $\{(x,y) \mid x = -2\}$ **8.** $\{(x,y) \mid y + 4 = 0\}$

9. $\{(x,y) \mid 4x - 3y = 6\}$ **10.** $\{(x,y) \mid 3x - y = 7\}$

11. $\{(x,y) \mid x + 5y = 8\}$ **12.** $\{(x,y) \mid 3x - 2y = 0\}$

In each of the following, give an equation of a line satisfying the given conditions *and* sketch the line.

13. Contains (1,1) and has slope 2.

14. Contains (2,−4) and has slope −1/2.

15. Contains (−3,1) and (−3,−4).

16. Contains (0,2) and (3,2).

17. Contains (−1,2) and has slope −2/3.

18. Contains (−1,2) and has slope 2/3.

19. Contains (4,3) and has slope 0.

20. Contains $(-3,3)$ and has no slope.

Sketch the graph of each of the following:

21. $\{(x,y) \mid x - y < 1\}$

22. $\{(x,y) \mid 3x - y < 6\}$

23. $\{(x,y) \mid 2x + 3y \geq 6\}$

24. $\{(x,y) \mid 2x + y \geq 0\}$

In each of the following, give an equation of the line containing the given point that is (a) parallel to the given line *and* (b) is perpendicular to the given line.

25. $(3,1)$; $2x + 3y = 1$

26. $(2,-3)$; $3x + y = 5$

27. $(-5,3)$; $x - 2y = 4$

28. $(0,0)$; $2x - y = 3$

29. *Small bonus problem.* Suppose that

$D = \{(x,y) \mid x^2 + y^2 < 9\}$ and $U = \{(x,y) \mid 2x - y < 2\}$, show graphically $D \cap U$.

Section 4—Two-dimensional Vectors In section 1 we gave a description of a rectangular coordinate system in a plane. The coordinates (x,y) are an ordered pair of real numbers and locate a point in the plane. It is our intention now to define and study algebraic operations on ordered pairs of real numbers and to introduce some of the basic vocabulary generally associated with such a study. Our initial effort consists of the following definitions.

DEFINITION. (*Addition of ordered pairs of real numbers*)

$$(x,y) + (a,b) = (x + a, y + b).$$

DEFINITION. (*Multiplication of an ordered pair of real numbers by a real number*)

$$r \cdot (x,y) = (rx,ry).$$

Examples.

$$(3,5) + (1,-2) = (4,3)$$
$$3 \cdot (4,-2) = (12,-6)$$

The set $R \times R$ of ordered pairs of real numbers together with the operations of addition and multiplication as defined above is an example of a *two-dimensional vector* space and is denoted by V_2. The elements of this vector space—the ordered pairs of real numbers—are called vectors or points.

The numbers a and b are called the 1st and 2nd components, respectively, of the vector $\mathbf{v} = (a,b)$. Our definitions concerning vectors (ordered pairs of real numbers) can then be expressed:

1. "Two vectors are equal if their corresponding components are equal."
   The vector $\mathbf{u} = (3,1)$ is not equal to the vector $\mathbf{v} = (1,3)$.
2. "The sum of two vectors is the vector obtained by adding the corresponding components."
   If $\mathbf{u} = (3,5)$ and $\mathbf{v} = (4,-2)$, then $\mathbf{u} + \mathbf{v} = [3 + 4, 5 + (-2)] = (7,3)$.
3. "A real number r times a vector $\mathbf{v}$ is the vector obtained by multiplying each component of $\mathbf{v}$ by the number r."
   If $\mathbf{v} = (3,-2)$, then $4\cdot\mathbf{v} = (12,-8)$ and $-1\cdot\mathbf{v} = (-3,2)$.

Though certain additional results can be rather easily developed from the definitions already given, we shall simply assume the following:

There is a unique element $\mathbf{0}$ in V_2—called the "zero vector"—with the property that $\mathbf{v} + \mathbf{0} = \mathbf{v}$ for all $\mathbf{v}$ in V_2. $[\mathbf{0} = (0,0)]$.

For every $\mathbf{v} = (a,b)$ in V_2 there is a unique element $-\mathbf{v} = (-a,-b)$ in V_2—called "the negative of $\mathbf{v}$"—with the property that $\mathbf{v} + (-\mathbf{v}) = \mathbf{0}$.

In a manner similar to that used in subtraction of real numbers, subtraction of vectors is defined by the addition of the negative.

DEFINITION: (subtraction of vectors). *For each* $\mathbf{u}$ *and* $\mathbf{v}$ *in* V_2, $\mathbf{u} - \mathbf{v} = \mathbf{u} + (-\mathbf{v})$.

Example.
$$\begin{aligned}(3,5) - (2,-2) &= (3,5) + (-2,2)\\ &= (3 - 2, 5 + 2)\\ &= (1,7).\end{aligned}$$

Thus "two vectors are subtracted by subtracting corresponding components."

The next example illustrates how one can establish certain properties of vectors.

Example. If each of r and s is a real number, show that

$$(rs)\cdot\mathbf{v} = r\cdot(s\cdot\mathbf{v}).$$

Solution. Let $\mathbf{v} = (a,b)$. It follows that

$(rs)\cdot\mathbf{v} = (rsa,rsb)$. Also $s\cdot\mathbf{v} = (sa,sb)$.

Thus $r\cdot(s\cdot\mathbf{v}) = (rsa,rsb)$. This shows that

$(rs)\cdot\mathbf{v} = r\cdot(s\cdot\mathbf{v})$ as required.

PROBLEM SET 2.4

In each of problems 1–5, give the vectors $\mathbf{u} + \mathbf{v}$ and $\mathbf{u} - \mathbf{v}$.

1. $\mathbf{u} = (3,5)$, $\mathbf{v} = (1,-2)$

2. $\mathbf{u} = (-6,1)$, $\mathbf{v} = (0,4)$

3. $\mathbf{u} = (-2,-7)$, $\mathbf{v} = (-8,3)$

4. $\mathbf{u} = (-3,3/2)$, $\mathbf{v} = (12,-1/2)$

5. $\mathbf{u} = (0.32,4)$, $\mathbf{v} = (0.6,1.3)$

6. If $\mathbf{v} = (6,-3)$, give the following vectors:

a. $-\mathbf{v}$ **b.** $-1/2\cdot\mathbf{v}$ **c.** $2\cdot\mathbf{v}$ **d.** $0.23\cdot\mathbf{v}$

7. If $\mathbf{z} = (\sqrt{6},\sqrt{2})$, give the following vectors:

a. $\sqrt{2}\cdot\mathbf{z}$ **b.** $\dfrac{1}{\sqrt{2}}\cdot\mathbf{z}$ **c.** $-\sqrt{3}\cdot\mathbf{z}$ **d.** $(1 + \sqrt{3})\cdot\mathbf{z}$

8. Suppose that $\mathbf{i} = (1,0)$ and that $\mathbf{j} = (0,1)$. For each given vector $\mathbf{v}$, find a number x and a number y such that $\mathbf{v} = x\cdot\mathbf{i} + y\cdot\mathbf{j}$.

a. $\mathbf{v} = (-5,3)$ **b.** $\mathbf{v} = (2,-7)$
c. $\mathbf{v} = (1/2,3/4)$ **d.** $\mathbf{v} = (0,6)$

9. If each of r and s is a number and if $\mathbf{u} = (p,q)$ and $\mathbf{v} = (a,b)$, use the definitions to show that:

a. $\mathbf{u} + \mathbf{v}$ is $\mathbf{v} + \mathbf{u}$
b. $r\cdot(\mathbf{u} + \mathbf{v})$ is $r\cdot\mathbf{u} + r\cdot\mathbf{v}$
c. $(r + s)\cdot\mathbf{v}$ is $r\cdot\mathbf{v} + s\cdot\mathbf{v}$

10. If $\mathbf{z} = (m,n)$ and $\mathbf{v} = (a,b)$, you are first to recall that $\mathbf{z} = \mathbf{v}$ iff m is a and n is b, and then to use this recall to solve the following "vector equations" for $\mathbf{z}$.

a. $\mathbf{z} + (3,-1) = (5,4)$ **b.** $\mathbf{z} + (1,6) = (1,6)$
c. $\mathbf{z} - (0,5) = 2\cdot(-1,1)$ **d.** $\mathbf{z} + (3,2) = (0,0)$
e. $3\cdot\mathbf{z} + (4,-1) = (-5,1)$

11. Suppose that $\mathbf{u} = (-2.5,4)$ and that $\mathbf{v} = (1,2)$.

a. If t is a number and $\mathbf{z} = \mathbf{u} + t\cdot\mathbf{v}$, express $\mathbf{z}$ as a two-dimensional vector whose components are given in terms of t.
b. Use $t = 0, \pm 1/2, \pm 1$, and ± 3 and, for each value of t, plot $\mathbf{z}$ as a point in the plane.

12. Suppose that $\mathbf{u} = (3,-2)$ and that $\mathbf{v} = (-1,2)$.

a. If t is a number and $\mathbf{z} = t \cdot \mathbf{u} + (1 - t) \cdot \mathbf{v}$, express $\mathbf{z}$ as a two dimensional vector whose components are given in terms of t.

b. Use $t = 0, \pm 1/2, \pm 1$, and $\pm 3/2$ and, for each value of t, plot $\mathbf{z}$ as a point in the plane.

13. If $\mathbf{v} = (a,b)$, show that $-1 \cdot \mathbf{v} = -\mathbf{v}$.

Section 5—Geometric Representation of Vectors Although we may have accepted, in the previous sections, that there are reasons for studying ordered pairs of real numbers, it may not be clear why we should be interested in the algebra of ordered pairs (vectors). It seems appropriate at this time, then, to present the intuitive geometric reasons and ideas that will guide us and to describe how we shall represent vectors by "arrows."

In a rectangular coordinate system, as described in section 1 of this chapter, an "arrow" representing the vector $\mathbf{v} = (a,b)$ is constructed in the following manner (see figure 2.16):

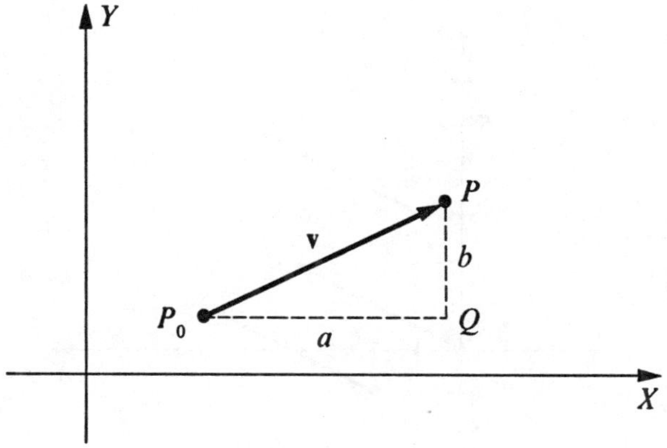

Fig. 2.16

1. Start with any point P_0.

2. Move horizontally from P_0 a distance a (to the right if $a > 0$ to the left if $a < 0$) and thus locate a point Q.

3. Move vertically from Q a distance b (up if $b > 0$, down if $b < 0$) and thus locate a point P.

4. Draw a line segment joining P_0 with P placing the arrow head at P; this is the arrow representing the vector $\mathbf{v} = (a,b)$.

The point P_0 is called the *initial* point of **v** and the point P is labeled as the *terminal* point of **v**.

The above figure and a knowledge of Pythagoras' theorem suggests that we can associate with each vector **v** a nonnegative number which is called the *length of* **v** and which we shall denote by $|\mathbf{v}|$. Hence,

$$\text{if } \mathbf{v} = (a,b), \text{ then } |\mathbf{v}| = \sqrt{(a^2 + b^2)}.$$

Since the point P_0 is selected arbitrarily, several arrows can be constructed in the manner just described, each representing a given vector v. For example, the vector $\mathbf{v} = (4,2)$ can be represented by any of the arrows drawn in figure 2.17. We can be consoled, however, by the fact that all of the arrows which can be drawn representing **v** have the same length, are parallel to each other, point in the same direction, and they all correspond to the same ordered pair of real numbers. For reasons that may strike you as being obvious, the arrow drawn from the origin can serve as a useful representation of the vector **v**. If this is not obvious, be patient for we shall comment further on such vectors.

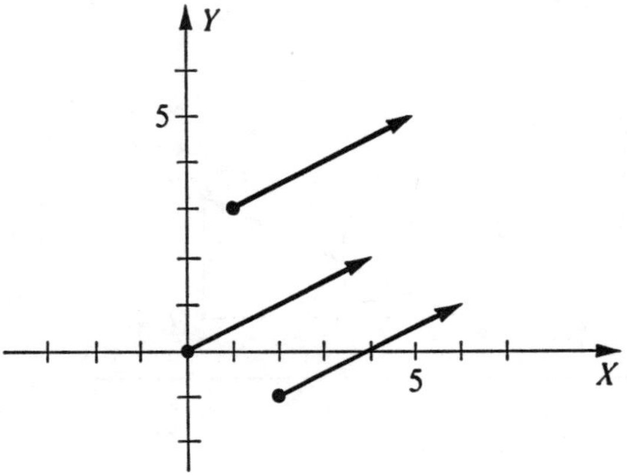

Fig. 2.17

A geometric description of the addition of the vectors $\mathbf{v} = (a,b)$ and $\mathbf{u} = (s,t)$ can be seen in figure 2.18 and can be constructed as follows:

1. Select any point P_0 and construct the arrow **v** from P_0, thus locate the point P_1.
2. Construct the arrow **u** from P_1 and thus locate the point P_2; the arrow drawn from P_0 to P_2 corresponds to the vector $\mathbf{v} + \mathbf{u}$.

If we redraw figure 2.18 omitting the dashed lines and construct the parallelogram defined by the arrows **v** and **u**, we can see that **v** + **u** is a diagonal of this parallelogram (see figure 2.19).

Figure 2.19 illustrates a principle which is called the *parallelogram law.* It also illustrates that vector addition is commutative—that is, **v** + **u** is **u** + **v**.

In figure 2.20, we illustrate the subtraction of vectors. For simplicity, the vectors **u** and **v** are drawn with the same initial point. Clearly, the vector **z** is that vector such that **z** + **v** = **u** thus **z** = **u** − **v**. Note that z is drawn from the terminal point of v to the terminal point of u.

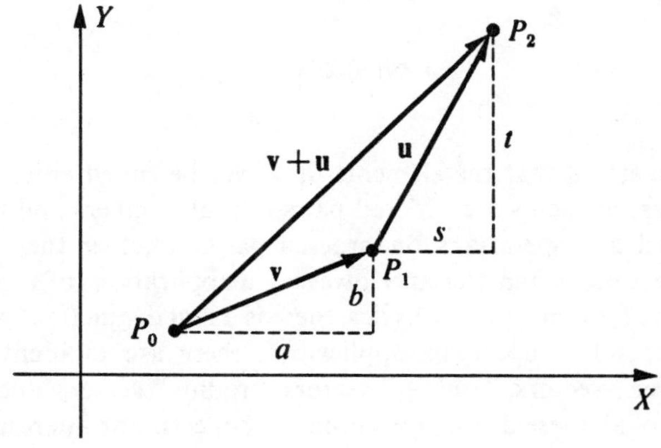

Fig. 2.18

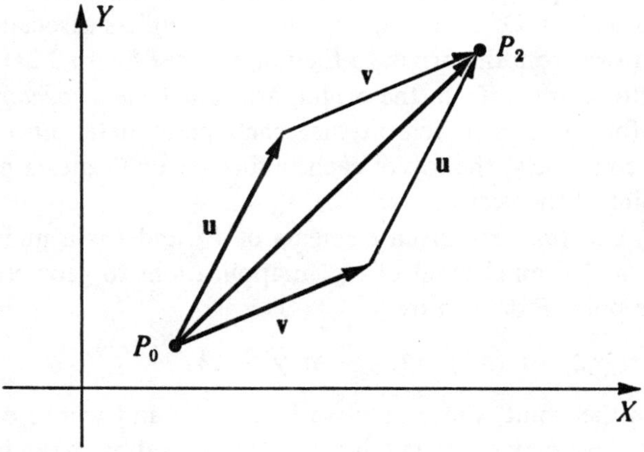

Fig. 2.19

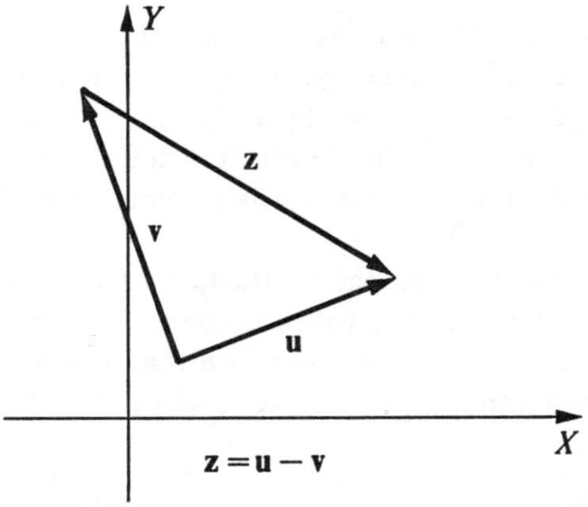

Fig. 2.20

We have stated that the elements of V can be called either *points* or *vectors*. These elements are ordered pairs of real numbers and the operations defined are operations on ordered pairs; whether they are called points or vectors, is immaterial. However, in applications of vector spaces to problems of geometry and physics, there is a real distinction to be made. Further, depending upon the application, there are different types of vectors: "free" vectors, "sliding" vectors, "radius" vectors, and so forth. The nature of all these distinctions need not concern nor interest us at the present. However, we should be interested in knowing that when we call the ordered pair (a,b) of real numbers a *point*, we will write $P = (a,b)$ and have in mind the construction given in section 1 of this chapter; when we call (a,b) a *vector*, we will write $\mathbf{v} = (a,b)$ and have in mind the construction given in this section. Of course, the point $P = (a,b)$ has associated with it the arrow drawn from the origin to the point P (see figure 2.21).

Such vectors, drawn from the origin, are called *radius vectors* and are of interest for the simple reason that each point determines a radius vector and, conversely, the top of each radius vector locates a point—the terminal point of the vector.

If (x_0,y_0) and (a,b) are given elements of V_2 and t is a number, then $(x_0,y_0) + t\cdot(a,b)$ is an element of V_2. In applications to geometry we will speak of the point P defined by

$$P = (x_0,y_0) + t\cdot(a,b) = (x_0 + ta, y_0 + tb).$$

So P is then the point whose abscissa is $x_0 + ta$ and whose ordinate is $(y_0 + tb)$, and we may say that P is the point located by the radius vector $(x_0 + ta, y_0 + tb)$.

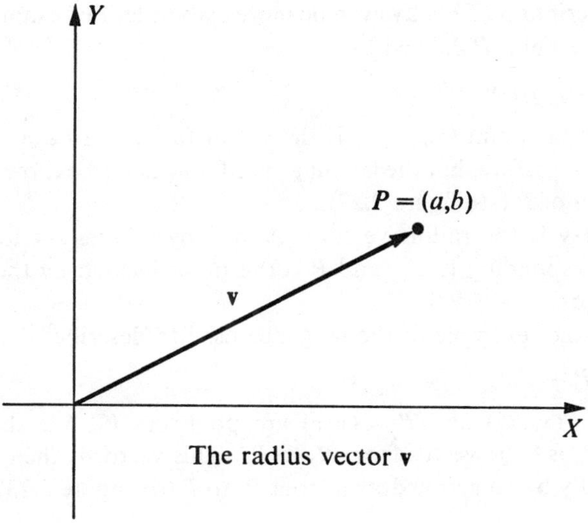

Fig. 2.21

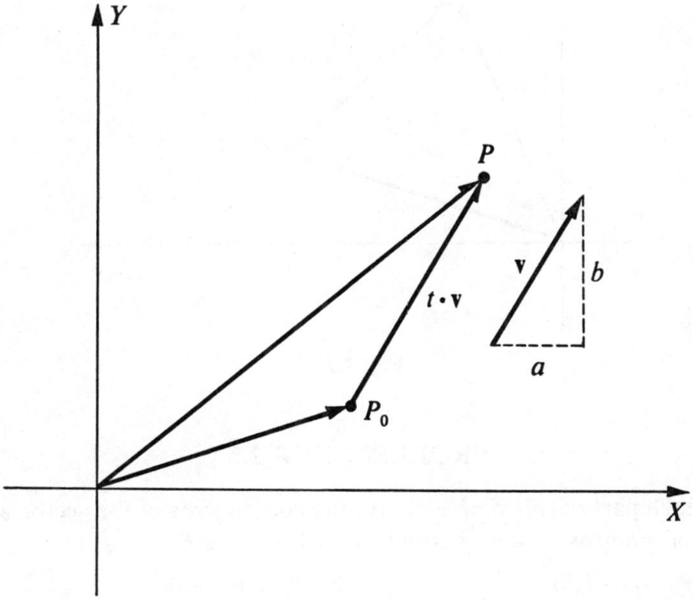

Fig. 2.22

The description of P may even be more elaborate. For example, we may speak of the point P defined by

$$P = P_0 + t \cdot \mathbf{v}$$

where P_0 is the point (x_0,y_0), $\mathbf{v}$ is the vector (a,b), and $t \in R$. The point P is the same as before, but the language used now describes how we visualize the location of P (see figure 2.22).

"$P_0 + t \cdot \mathbf{v}$ is the radius vector obtained by adding $t \cdot \mathbf{v}$ to the radius vector corresponding to P_0, and P is the point located by the tip of the radius vector $P_0 + t \cdot \mathbf{v}$."

For another example of the language used to describe

$$\mathbf{u} = P - P_0$$

where $P_0 = (x_0,y_0)$ and $P = (x,y)$ are points in V_2. We shall say that $\mathbf{u} = P - P_0$ is "the vector from P_0 to P." This vector is then represented geometrically by an arrow drawn from P_0 to P (see figure 2.23).

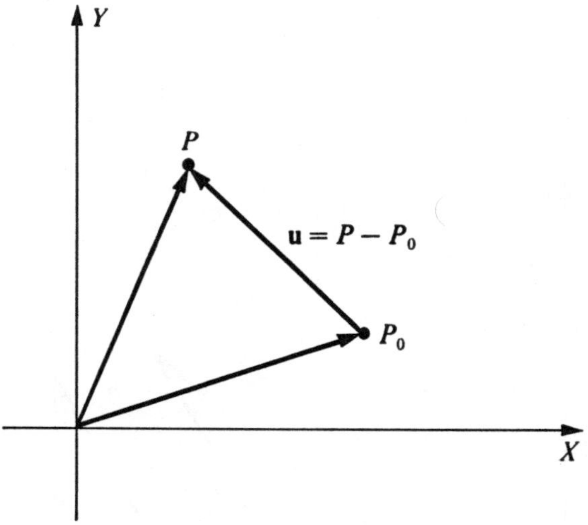

Fig. 2.23

PROBLEM SET 2.5

1. In each part of this problem, give the components of the vector whose initial point is P_0 and whose terminal point is P.

   a. $P_0 = (-1,5)$
   $P = (3,2)$
   call this vector $\mathbf{u}$

   b. $P_0 = (4,-6)$
   $P = (1,2)$
   call this vector $\mathbf{v}$

c. $P_0 = (3,3)$
$P = (-2,0)$
call this vector **w**

d. $P_0 = (0,2)$
$P = (5,7)$
call this vector **x**

e. $P_0 = (1,-3)$
$P = (4,-1)$
call this vector **z**

2. Represent each of the vectors given in problem 1 by an arrow drawn from P_0 to P *and* by a radius vector.

3. Find the length of each of the vectors given in problem 1.

4. Represent each of the following vectors by an arrow whose initial point is $(-1,-3)$. Use a separate figure for each.

$\mathbf{u} = (3,2)$ $\mathbf{v} = (4,0)$ $\mathbf{w} = (-4,4)$ $\mathbf{z} = (0,-2)$

5. Represent each of the following vectors by an arrow whose initial point is $(2,-3)$. Use a separate figure for each.

$\mathbf{u} = (2,5)$ $\mathbf{v} = (-2,-1)$ $\mathbf{w} = (-5/2,5)$ $\mathbf{z} = (0,5)$

6. Find $\mathbf{u} + \mathbf{v}$ and $\mathbf{u} - \mathbf{v}$ algebraically and then, on the same coordinate axis, draw the vectors $\mathbf{u}$, $\mathbf{v}$, $\mathbf{u} + \mathbf{v}$, and $\mathbf{u} - \mathbf{v}$ as arrows whose initial point is $(1,1)$.

a. $\mathbf{u} = (-3,2)$
$\mathbf{v} = (5,6)$

b. $\mathbf{u} = (4,7/4)$
$\mathbf{v} = (-2,1/2)$

7. Find $\mathbf{u} + \mathbf{v}$ and $\mathbf{u} - \mathbf{v}$ algebraically and then, on the same coordinate axis, draw the vectors $\mathbf{u} + \mathbf{v}$, and $\mathbf{u} - \mathbf{v}$ as radius vectors.

a. $\mathbf{u} = (-3,6)$
$\mathbf{v} = (-1,-4)$

b. $\mathbf{u} = (1,5)$
$\mathbf{v} = (3,2)$

In each of problems 8–12, let $\mathbf{w} = (2,4)$ and $z = (1,-3)$.

8. Represent $\mathbf{w} - \mathbf{z}$ and $\mathbf{z} - \mathbf{w}$ as radius vectors.

9. Represent $\mathbf{w}$, $2\mathbf{z}$, and $\mathbf{w} + 2\mathbf{z}$ as radius vectors.

10. Represent $\mathbf{w}$, $2\mathbf{z}$, and $2\mathbf{z} - \mathbf{w}$ as radius vectors.

11. Represent $2\mathbf{w}$, $3\mathbf{z}$, and $2\mathbf{w} + 3\mathbf{z}$ as radius vectors.

12. Represent $\frac{1}{2}\cdot\mathbf{w}$, $\mathbf{z}$, and $\frac{1}{2}\cdot\mathbf{w} + \mathbf{z}$ as radius vectors.

In problems 13–16, no diagrams are required but you must recall that if $\mathbf{z} = (x,y)$, then $|\mathbf{z}| = \sqrt{(x^2 + y^2)}$.

13. If $\mathbf{u} = (0,3)$ and $\mathbf{v} = (-5,-3)$, give $|\mathbf{u} - \mathbf{v}|$ and $|\mathbf{u} + \mathbf{v}|$.

14. If $\mathbf{u} = (\sqrt{3},0)$ and $\mathbf{v} = (2,3\sqrt{3})$, give $|\mathbf{u} - \mathbf{v}|$ and $|\sqrt{3}\mathbf{u} + \mathbf{v}|$

15. If $\mathbf{u} = (1/2,1/3)$ and $\mathbf{v} = (-1/3,-1/2)$, give $|\mathbf{u} - \mathbf{v}|$ and $|\mathbf{u} + \mathbf{v}|$.

16. If $\mathbf{u} = (-4,-5/2)$ and $\mathbf{v} = (1,-2)$, give $|2\mathbf{u} + \mathbf{v}|$ and $|\mathbf{u} - 3\cdot\mathbf{v}|$.

17. Suppose that $\mathbf{i} = (1,0)$ and that $\mathbf{j} = (0,1)$. Using a different figure for each, represent the following vectors as radius vectors:

a. $\mathbf{i} + \mathbf{j}$ **b.** $3\cdot\mathbf{i} + \mathbf{j}$ **c.** $2\mathbf{i} + 3\mathbf{j}$
d. $2\sqrt{2}\cdot\mathbf{i} + 2\cdot\mathbf{j}$ **e.** $\mathbf{i} - \mathbf{j}$ **f.** $\mathbf{j} - \mathbf{i}$

Bonus problem. An airplane is headed NE with an air speed of 400 km/hr. The wind is blowing due south at 90 km/hr. The velocity $\mathbf{v}_g$ of the plane with respect to the ground is the sum of the above two vectors. Determine $\mathbf{v}_g$ graphically.

Section 6—Parallel and Orthogonal Vectors So far, we have considered some of the basic ideas in the algebra of two-dimensional vectors, and we have seen how we can represent vectors by arrows. This representation of vectors by arrows is certainly not in conflict with the intuitive notion of using arrows to indicate direction, nor does the direction indicated by an arrow depend upon the length of that arrow. These same intuitive notions, aided and abetted by diagrams such as the figure below, provide us with all the evidence we need in order to establish what we shall mean by the notions of the direction of a vector and of parallel vectors. In figure 2.24, $\mathbf{v} = (3,4)$ so that $\mathbf{a} = 2\cdot\mathbf{v} = (6,8)$, $\mathbf{b} = -1/2\cdot\mathbf{v} = (-1.5,-2)$, and $\mathbf{c} = 3/2\cdot\mathbf{v} = (4.5,6)$.

A glance at this figure should suggest that all of these vectors are parallel and that $\mathbf{a}$ and $\mathbf{c}$ are in the same direction as $\mathbf{v}$, while $\mathbf{b}$ is in the opposite direction. We now feel safe in making the following definitions:

DEFINITION. *Two nonzero vectors are said to be parallel if one of them is a real number times the other.*

DEFINITION. *Two nonzero vectors are said to be in the same direction if one of them is a positive number times the other.*

What do you think should be the meaning of the statement that two nonzero vectors are in opposite directions?

The word *orthogonal* means "right angles" and is synonymous with "perpendicular." Once again we shall make use of some figures to suggest and to make reasonable a definition we wish to make. Let $\mathbf{u}$ and $\mathbf{v}$ be the

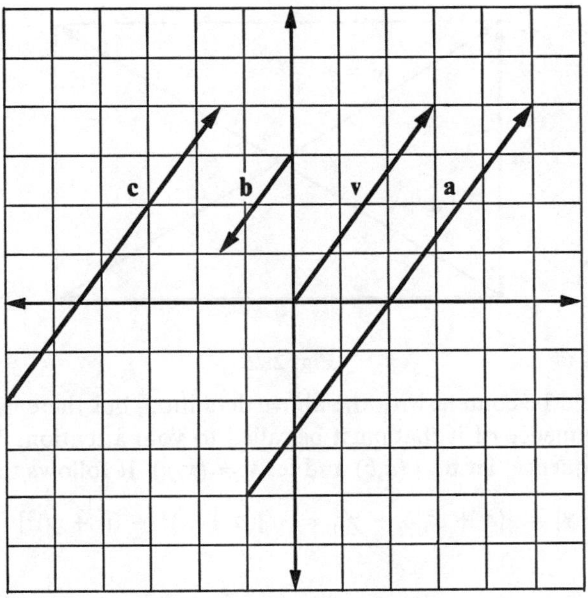

Fig. 2.24

sides of a parallelogram (figure 2.25). The vectors **u** + **v** and **u** − **v** are the diagonals of this parallelogram. So, if the vectors **u** and **v** are to be orthogonal (figure 2.26), then the diagonals must have the same length, that is, the parallelogram is a rectangle. So it should at least be reasonable to assert that

> if **u** and **v** are vectors, then the statement that "**u** is orthogonal to **v**" means that $|\mathbf{u} + \mathbf{v}| = |\mathbf{u} - \mathbf{v}|$.

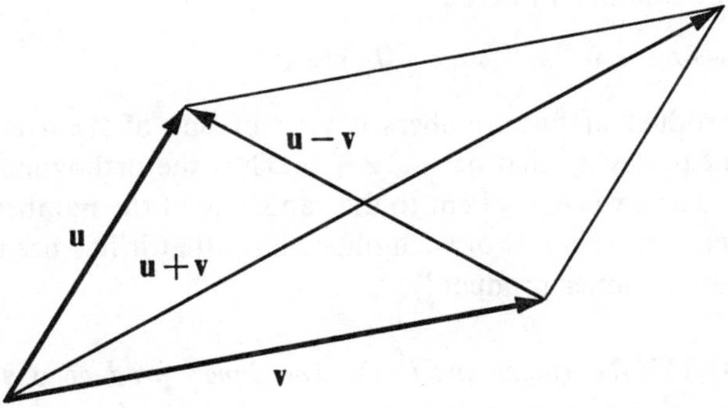

Fig. 2.25

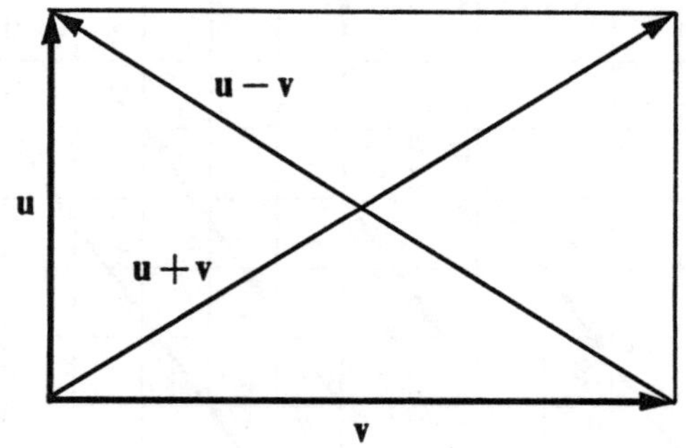

Fig. 2.26

We could be content with the above definition, but there is one important consequence of it that must be called to your attention. To establish this consequence, let $\mathbf{u} = (a,b)$ and let $\mathbf{v} = (x,y)$. It follows that

$$|\mathbf{u} + \mathbf{v}| = |(a + x, b + y)| = \sqrt{[(a + x)^2 + (b + y)^2]}$$

and that

$$|\mathbf{u} - \mathbf{v}| = \sqrt{[(a - x)^2 + (b - y)^2]}.$$

So, if we set $|\mathbf{u} + \mathbf{v}| = |\mathbf{u} - \mathbf{v}|$ and square both sides, we obtain the following equations:

$$|\mathbf{u} + \mathbf{v}|^2 = |\mathbf{u} - \mathbf{v}|^2$$

hence

$$(a + x)^2 + (b + y)^2 = (a - x)^2 + (b - y)^2$$

or

$$a^2 + 2ax + x^2 + b^2 + 2by + y^2 = a^2 - 2ax + x^2 + b^2 - 2by + y^2$$

which can be simplified to read

$$4ax + 4by = 0 \quad or \quad 4(ax + by) = 0.$$

Since the product of two numbers is zero iff one of them is zero, this is equivalent to saying that $ax + by = 0$. Thus the orthogonality of the two vectors **u** and **v** is equivalent to the vanishing of the number $ax + by$. This number, $(ax + by)$, is of such importance that it has been given the special name of "inner product."

DEFINITION. (*inner product*). *The inner product* $\mathbf{u}\cdot\mathbf{v}$*—read as "u dot v"—of two vectors* $\mathbf{u} = (a,b)$ *and* $\mathbf{v} = (x,y)$ *is defined by* $\mathbf{u}\cdot\mathbf{v} = ax + by$.

Remarks. Note that the inner product of two vectors is not a vector; it is a real number. Other names for this same concept are "scalar product" and "dot product."

As a result of this discussion, we state the following:

DEFINITION. *Two vectors,* **u** *and* **v**, *are orthogonal* iff $\mathbf{u}\cdot\mathbf{v}$ *is* 0.

This last statement can be used to find a vector which is orthogonal to a given vector as the following examples will show.

Example. Given $\mathbf{u} = (-2,5)$, find a vector **v** orthogonal to **u**.

Solution. Let $\mathbf{v} = (x,y)$. In order that **v** be orthogonal to **u**, it is necessary that $\mathbf{u}\cdot\mathbf{v} = 0$ hence

$$-2x + 5y = 0$$

and there are many values of x and y for which this is true, for example, $(-5,-2)$ or $(0,0)$, or even $\mathbf{v} = (5,2)$ will suffice. Note that this example also indicates that the zero vector $(0,0)$ is orthogonal to any given vector, a result that we will seldom, if ever, use.

Example. Find a vector **v** which is orthogonal to $\mathbf{u} = (1,3)$ *and* has length 1.

Solution. Let $\mathbf{v} = (x,y)$. The two requirements—(1) that $\mathbf{u}\cdot\mathbf{v} = 0$, and (2) that $|\mathbf{v}| = 1$—are equivalent to the two equations $x + 3y = 0$ and $x^2 + y^2 = 1$. From the first equation we can see that $x = -3y$. Substituting this into the second equation gives us $9y^2 + y^2 = 1$, hence $y^2 = \frac{1}{10}$ or $\frac{10}{100}$ so that $y = \frac{\sqrt{10}}{10}$ or $y = -\frac{\sqrt{10}}{10}$. Using these values of y, the first equation tells us that $x = -\frac{3\sqrt{10}}{10}$ or $x = \frac{3\sqrt{10}}{10}$. So there are two vectors which satisfy the requirements of the problem, one of them being

$$\mathbf{v} = \left[\frac{3\sqrt{10}}{10}, -\frac{\sqrt{10}}{10}\right].$$

Remark. When a unit vector (vector with length 1) is obtained from a given vector by multiplying the given vector by an appropriate real number, the given vector is said to be "normalized" (see problem 12 in problem set 2.6).

PROBLEM SET 2.6

1. Which of the following pairs of vectors are parallel?
a. (2,4) and $(-1/2,-1)$ **b.** $(-3,1)$ and $(6,-2)$
c. (1,1) and (3,2) **d.** (2,5) and (1,2.5)

2. Are the following vectors orthogonal?
a. (2,4) and $(2,-1)$ **b.** $(-8,-4)$ and $(2/3,-4/3)$
c. $(1/3,\sqrt{2})$ and $(-6,\sqrt{2})$ **d.** $(3,-5)$ and (5,3)

3. If $\mathbf{u} = (1,4)$ and $\mathbf{v} = (-2,3)$, find the length of each of the following vectors:

a. $\mathbf{u} + \mathbf{v}$ **b.** $\mathbf{u} - \mathbf{v}$ **c.** $\mathbf{v} - \mathbf{u}$ **d.** $\frac{1}{|\mathbf{u}|}\cdot\mathbf{u}$ **e.** $\frac{1}{|\mathbf{v}|}\cdot\mathbf{v}$

In each of problems 4–11, give a vector $\mathbf{v}$ that is orthogonal to *and* has the same length as the given vector $\mathbf{u}$.

4. $\mathbf{u} = (1,2)$ **5.** $\mathbf{u} = (-1,2)$

6. $\mathbf{u} = (1,0)$ **7.** $\mathbf{u} = (0,1)$

8. $\mathbf{u} = (2,-3)$ **9.** $\mathbf{u} = (5,-4)$

10. $\mathbf{u} = (3,3)$ **11.** $\mathbf{u} = (m,1)$

12. If $\mathbf{u} = (a,b)$, find the vector $\frac{1}{|\mathbf{u}|}\cdot\mathbf{u}$ *and* give its length.

In each of problems 13–15, give a vector $\mathbf{v}$ that is orthogonal to the given vector $\mathbf{u}$ *and* has length 1.

13. $\mathbf{u} = (4,3)$ **14.** $\mathbf{u} = (2,2)$ **15.** $\mathbf{u} = (1,\sqrt{3})$

16. If $\mathbf{u} = (a,b)$ and $\mathbf{v} = (x,y)$, give some evidence why the following statement is true:

$$(\mathbf{u}\cdot\mathbf{v})^2 \leq (\mathbf{u}\cdot\mathbf{u})(\mathbf{v}\cdot\mathbf{v}).$$

17. Let $\mathbf{v} = (1,2)$ and $P_0 = (0,2)$. If, for every number t, $P = P_0 + t\cdot\mathbf{v}$, find and plot the points P which correspond to the values of $t = 0, 1/2, 1, -1, -2$. On the same set of coordinate axis, also plot $\mathbf{v}$ as a radius vector.

Section 7—Lines In plane geometry, the notion of a line was usually left undefined. Here, we shall give a definition of a straight line (line) in terms of another abstraction called a real number. In making our definition of a line, we will employ vector addition and multiplication of a vector by a number. Our definition of a line stems from an intuitive realization that a line is determined by a point P_0 and a direction $\mathbf{v}$ ($\mathbf{v}$ a nonzero vector).

Since, for every number t, $t\cdot\mathbf{v}$ is parallel to $\mathbf{v}$ and so every vector $t\cdot\mathbf{v}$ with P_0 as initial point is parallel to $\mathbf{v}$. Thus the points P on the line $\mathscr{L}$ through P_0 in the direction parallel to $\mathbf{v}$ (see figure 2.27) are all points of the form

$$P = P_0 + t\cdot\mathbf{v}$$

for every number t. A more formal statement of these ideas is given in the following definition.

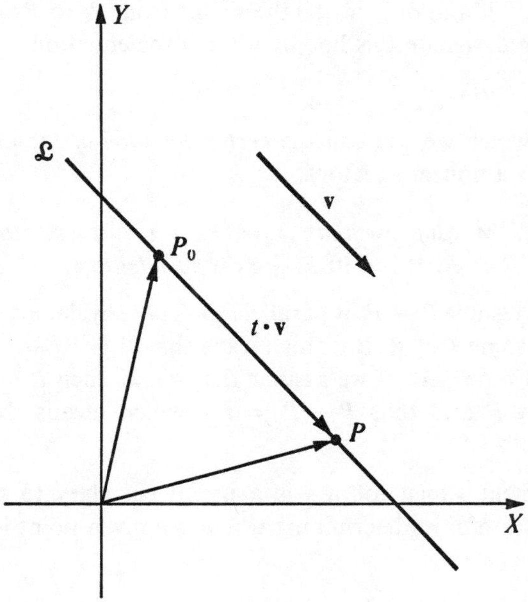

Fig. 2.27

DEFINITION. *A set* $\mathscr{L}$ *of points in* $R \times R$ *is called a line if there is a point* P_0 *in* $R \times R$ *and a nonzero vector* $\mathbf{v}$ *in* V_2 *such that*

$$\mathscr{L} = \{P_0 + t\cdot\mathbf{v} \mid t \in R\}.$$

The line $\mathscr{L}$ just defined is called *the line through* P_0 *parallel to the nonzero vector* $\mathbf{v}$.

Example 1. Give an equation for the line $\mathscr{L}$ through $P_0 = (2,-1)$ parallel to $\mathbf{v} = (2,3)$.

Solution. By definition,

$$\mathscr{L} = \{(2,-1) + t\cdot(2,3) \mid t \in R\}$$

which can be simplified to

$$\mathscr{L} = \{(2 + 2t, 3t - 1) \mid t \in R\}.$$

Either of the above equations serve to identify $\mathscr{L}$. By letting $t = -1$, we see that $(0,-4) \in \mathscr{L}$; P_0 corresponds to the choice of $t = 0$; for $t = 1$, we see that $(4,2) \in \mathscr{L}$.

The definition of a line states—as the following theorem shows—that a point P is in $\mathscr{L}$ if and only if (iff) the vector from P_0 to P is parallel to $\mathbf{v}$. Hence we can designate this line by writing the equation

$$P = P_0 + t\cdot\mathbf{v},$$

an equation which we will call the *vector equation of the line through* P_0 *parallel to* $\mathbf{v}$ ($\mathbf{v}$ a nonzero vector).

THEOREM. *Suppose that* $\mathscr{L} = \{P_0 + t\cdot\mathbf{v} \mid t \in R\}$ *and that* P *is a point in* $R \times R$. $P \in \mathscr{L}$ iff $P - P_0$ *is parallel to* $\mathbf{v}$.

Proof. Assume $P - P_0$ is parallel to $\mathbf{v}$. Then, by definition, $P - P_0 = t\cdot\mathbf{v}$ for some t in R. But this means that $P = P_0 + t\cdot\mathbf{v}$ and hence, $P \in \mathscr{L}$. Conversely, if we assume that $P \in \mathscr{L}$, then $P = P_0 + t\cdot\mathbf{v}$ for some $t \in R$ and thus $P - P_0 = t\cdot\mathbf{v}$ which means that $P - P_0$ is parallel to $\mathbf{v}$.

The statement which follows is a useful corollary to the preceding theorem. It is useful in determining whether a given point is on a line or not.

COROLLARY. *Suppose that* $\mathbf{u}$ *is orthogonal to* $\mathbf{v}$ *and that* $\mathscr{L} = \{P_0 + t\cdot\mathbf{v} \mid t \in R\}$. $P \in \mathscr{L}$ *if and only if*

$$\mathbf{u}\cdot(P - P_0) \text{ is } 0.$$

Example. If $\mathscr{L} = \{(5,2) + t\cdot(2,1) \mid t \in R\}$, are the points $(3,1)$ and $(0,-3)$ on $\mathscr{L}$?

Solution. $\mathbf{u} = (-1,2)$ is orthogonal to $\mathbf{v} = (2,1)$. Since

$$(-1,2)[(5,2) - (3,1)] = (-1,2)\cdot(2,1) = 0$$

and

$$(-1,2)[(5,2) - (0,-3)] = (-1,2)\cdot(5,-1) \neq 0$$

we conclude that $(3,1) \in \mathscr{L}$ but $(0,-3) \notin \mathscr{L}$.

We also expect lines to have the property that two points should determine a line. If a line passes through the points P_0 and P_1, we use $P_1 - P_0$ for the direction vector **v** and write the vector equation of this line as

$$P = P_0 + t(P_1 - P_0)$$

or

$$P = t \cdot P_1 + (1 - t) \cdot P_0$$

Example. Find the line determined by the points $P_0 = (-2,4)$ and $P_1 = (3,0)$.

Solution. $P_1 - P_0 = (3 + 2, 0 - 4) = (5,-4)$, hence

$$\mathscr{L} = \{(-2,4) + t(5,-4) \mid t \in R\} = \{(5t - 2, 4 - 4t)\}.$$

Note that this line surely contains P_0 and P_1. To see this, let $t = 0$ and then let $t = 1$.

PROBLEM SET 2.7

In problems 1–8, give an equation for the given line.

1. The line $\mathscr{L}$ through (5,0) and parallel to $\mathbf{v} = (1,1)$.

2. The line $\mathscr{L}$ through (−1,1) and parallel to $\mathbf{v} = (0,1)$.

3. The line $\mathscr{L}$ through (1,3) and (4,2).

4. The line $\mathscr{L}$ through (0,2) and (1,3).

5. The line $\mathscr{L}$ through (0,1) and parallel to $\mathbf{v} = (-3,2)$.

6. The line $\mathscr{L}$ through (−2,0) and parallel to $\mathbf{v} = (-1,-1)$.

7. The line $\mathscr{L}$ through (3,4) and (5,−2).

8. The line $\mathscr{L}$ through (2,3) and (7,8).

9. The line $\mathscr{L}$ contains the points $P_0 = (-3,1)$ and $P_1 = (1,1)$. Determine which of the following points are on $\mathscr{L}$.

a. (0,0) **b.** (0,1) **c.** (1,2)

d. (2,1) **e.** (−2,1)

In problems 10–14, give an equation for the given line.

10. The line $\mathscr{L}$ through $(a,0)$ and $(0,b)$.

11. The line $\mathscr{L}$ through (h,k) and parallel to $\mathbf{v} = (1,m)$.

12. The line $\mathscr{L}$ through $(2,-1)$ and orthogonal to $\mathbf{v} = (1,4)$.

13. The line $\mathscr{L}$ through $(2,-1)$ and parallel to $\mathbf{v} = (1,4)$.

14. The line $\mathscr{L}$ through $(-2,0)$ and orthogonal to the line given in problem 13.

15. Determine the line through $(1,-6)$ that is parallel to the line given in
a. problem 1 **b.** problem 2
c. problem 3 **d.** problem 4

16. Let $\mathscr{L}_1$ be the line through $(1,4)$ and $(3,0)$, $\mathscr{L}_2$ be the line through $(0,0)$ and parallel to $\mathbf{v} = (3,2)$, and $\mathscr{L}_3 = \{(2t - 3, t + 2) \mid t \in R\}$. Are the following pairs of lines orthogonal? (Yes or no will do.)

a. $\mathscr{L}_1$ and $\mathscr{L}_2$ **b.** $\mathscr{L}_1$ and $\mathscr{L}_3$ **c.** $\mathscr{L}_2$ and $\mathscr{L}_3$

Section 8—Slope of a Line The main purpose of this section is to specify the direction of a line by a number m which we shall call *the slope of the line*. This shall be done by appealing to the notion that the direction of a line is determined by a nonzero vector.

Suppose that $\mathscr{L}$ is the line through $P_0 = (x_0,y_0)$ parallel to $\mathbf{v} = (p,q)$. According to the theorem in the previous section, $P = (x,y)$ is an element of $\mathscr{L}$ iff, for some number t, $P - P_0 = t\cdot\mathbf{v}$. If we substitute for P, P_0, and $\mathbf{v}$, this implies that

$$(x,y) - (x_0,y_0) = (tp,tq)$$

or that

$$(x,y) = (x_0 + tp, y_0 + tq).$$

But this implies that $x = x_0 + tp$ and that $y = y_0 + tq$ and solving the first of these two equations for t we have that $t = \dfrac{x - x_0}{p}$ if $p \neq 0$. Using this result in the second equation gives us $y = y_0 + \dfrac{x - x_0}{p}\cdot q$ which we rewrite as

$$y - y_0 = \frac{q}{p}(x - x_0).$$

Remarks. The useful corollary in the previous section can be used to provide another path to the same results. Let $\mathbf{u} = (-q,p)$, (it is easy see that $\mathbf{u}\cdot\mathbf{v}$ is 0), noting that $P - P_0 = (x - x_0, y - y_0)$, we think about the corollary and deduce that $(-q,p)\cdot(x - x_0, y - y_0) = 0$ so that, by definition of "inner product," $-q(x - x_0) + p(y - y_0) = 0$, hence $p(y - y_0) = q(x - x_0)$. From this last result, we can get the same equation given above. Let us now return to this equation for some further observations.

Your attention is now directed to the fact that p and q are the components of the vector **v** which specifies the direction of $\mathscr{L}$. Also note that if $p = 0$, that is, $\mathbf{v} = (0,q)$, then $\mathscr{L}$ is parallel to the Y axis. So, the assumption that p is not 0 is equivalent to saying that $\mathscr{L}$ is a nonvertical line. But, if $\mathscr{L}$ is a nonvertical line, then $\frac{1}{p}\cdot\mathbf{v} = (1,q/p)$ is the unique vector of the form $(1,m)$ parallel to $\mathscr{L}$.

Since the direction of $\mathscr{L}$ is determined by this unique vector, we feel free to make the following definition.

DEFINITION. *The slope m of a nonvertical line $\mathscr{L}$ is the number $m = \frac{q}{p}$ where $\mathbf{v} = (p,q)$ is a nonzero vector parallel to $\mathscr{L}$.*

Point-slope Equation If $\mathscr{L}$ is a nonvertical line through (x_0,y_0) with slope m, then

(8.1) $$\mathscr{L} = \{(x,y) \mid y - y_0 = m(x - x_0)\}.$$

The equation $y - y_0 = m(x - x_0)$ is called the *point-slope equation of $\mathscr{L}$.*

Thus, if m is the slope of a line $\mathscr{L}$, then $\mathscr{L}$ is parallel to the vector $(1,m)$, and we can see (figure 2.28) that m is a measure of the steepness of $\mathscr{L}$.

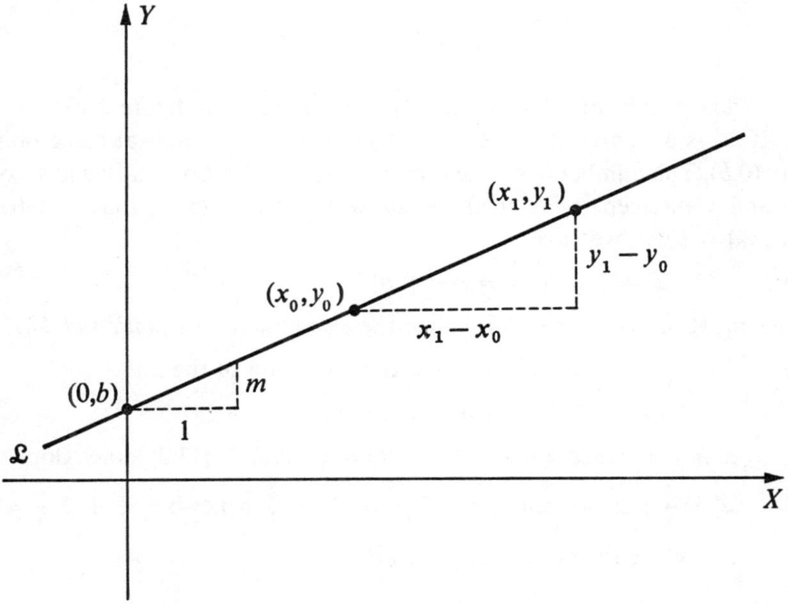

Fig. 2.28

The slope m is the number of units that $\mathscr{L}$ rises or falls per horizontal unit. If m is a positive (negative) number, then $\mathscr{L}$ rises (falls) m units per horizontal unit in going from left to right.

If $m = 0$, then $\mathscr{L}$ is said to be a horizontal line. If two points (x_1,y_1) and (x_0,y_0) on the line are known, then $\mathscr{L}$ is parallel to $(x_1 - x_0, y_1 - y_0)$; the line is not vertical if $x \neq x_0$, and

(8.2) $$m = \frac{y_1 - y_0}{x_1 - x_0}.$$

Of course, we shall not make any comments about the slope of a vertical line since it does not have a slope.

Example. Give the point-slope equation of the line

$$\mathscr{L}_1 = \{(1,-3) + t\cdot(2,4) \mid t \in R\}.$$

Solution. $m = 4/2 = 2$ and $P_0 = (1,-3)$, hence, the desired equation is $y + 3 = 2(x - 1)$ and we can denote $\mathscr{L}_1$ by

$$\mathscr{L}_1 = \{(x,y) \mid y + 3 = 2(x - 1)\}.$$

Example. Give the point-slope equation of the line $\mathscr{L}_2$ through (1,1) and (3,−2).

Solution. Using (8.2) we find $m = \dfrac{-2 - 1}{3 - 1} = -\dfrac{3}{2}$ so the desired equation is given by

$$y - 1 = -\frac{3}{2}(x - 1).$$

The graphs of the two lines $\mathscr{L}_1$ and $\mathscr{L}_2$ appear in figure 2.29.

If $\mathscr{L}$ is a nonvertical line, then it intersects the Y axis at some point, say $(0,b)$. The number b is called the *y intercept of b*. So, if a line has slope m and y intercept of b, then, by substituting into (8.1), that is, letting $(x_0,y_0) = (0,b)$, we have

(8.2) $$\mathscr{L} = \{(x,y) \mid y = mx + b\}.$$

The equation $y = mx + b$ is called the *slope-intercept equation of* $\mathscr{L}$.

Example. Give the slope-intercept equation of the line

$$\mathscr{L}_3 = \{(3t - 2, 1 + 2t) \mid t \in R\}.$$

Solution. Since $(3t - 2, 1 + 2t) = (-2,1) + t(3,2)$, the slope of $\mathscr{L}_3$ is $\frac{2}{3}$. Furthermore, $3t - 2 = 0$ iff $t = \frac{2}{3}$, hence $b = 1 + 2\cdot\frac{2}{3} = \frac{7}{3}$. Therefore the desired equation is

$$y = \frac{2}{3}x + \frac{7}{3}.$$

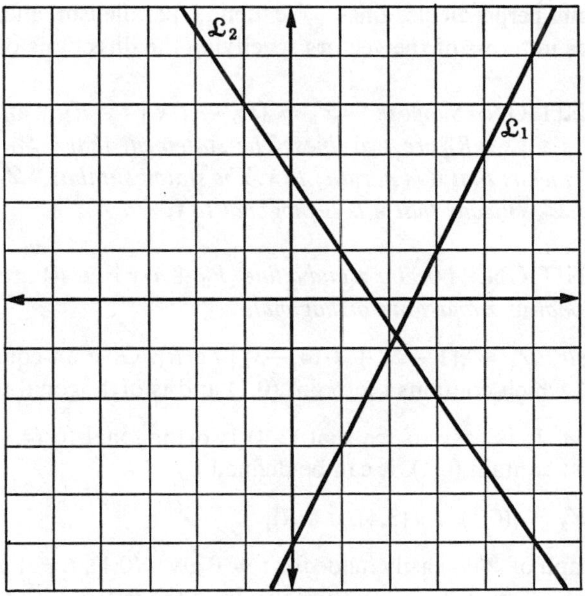

Fig. 2.29

The student should look back at the last three examples and make the calculations which support the contention that the equations of these three lines can be written in the form $Ax + By = C$. He should, in other words, verify that

$$\mathscr{L}_1 = \{(x,y) \mid 2x - y = 5\}$$
$$\mathscr{L}_2 = \{(x,y) \mid 3x + 2y = 5\}$$
$$\mathscr{L}_3 = \{(x,y) \mid 2x - 3y = -7\}$$

Remarks. The student should also note that the slope of each of these lines equals $-\dfrac{A}{B}$. We shall call the equation $Ax + By = C$ the *general equation of the line.*

Thus a line is determined if one point on the line and its slope are known (or we know that it has no slope). For, if a line $\mathscr{L}$ is vertical, then it has no slope and

$$\mathscr{L} = \{(x_0,y_0) + t\cdot(0,1) \mid t \in R\}$$

or

$$\mathscr{L} = \{(x_0,y) \mid y \in R\}.$$

Parallel and perpendicular lines We define parallelism and orthogonality of lines in terms of the vectors specifying the directions of the lines.

DEFINITION. *Suppose* $\mathscr{L}_1 = \{P_0 + t\cdot\mathbf{v} \mid t \in R\}$ *and* $\mathscr{L}_2 = \{Q_0 + s\cdot\mathbf{u} \mid s \in R\}$ *are two lines. The statement that "*$\mathscr{L}_1$ *is parallel to* $\mathscr{L}_2$*" means that* **u** *is parallel to* **v**. *The statement that "*$\mathscr{L}_1$ *is orthogonal to* $\mathscr{L}_2$*" means that* **u** *is orthogonal to* **v**.

DEFINITION. *A vector* **z** *and a line* $\{P_0 + t\cdot\mathbf{v} \mid t \in R\}$ *are said to be orthogonal* iff **z** *and* **v** *are orthogonal.*

Example. $\mathscr{L}_1 = \{(1,-2) + t\cdot(4,-3) \mid t \in R\}$. Give an equation of a line $\mathscr{L}_2$ which contains the point (0,1) and is orthogonal to $\mathscr{L}_1$.

Solution. It is easy to see that (3,4) is orthogonal to (4, −3). Since $\mathscr{L}_2$ must contain (0,1), it can be defined by

$$\mathscr{L}_2 = \{(0,1) + t(3,4) \mid t \in R\}.$$

The graph of $\mathscr{L}_2$ is easily made for $t = 0$ gives (0,1), $t = 1$ gives (3,5), and $t = -1$ gives (−3, −3) and we draw the graph with these three points. (See figure 2.30.) Similar computations allow us also to

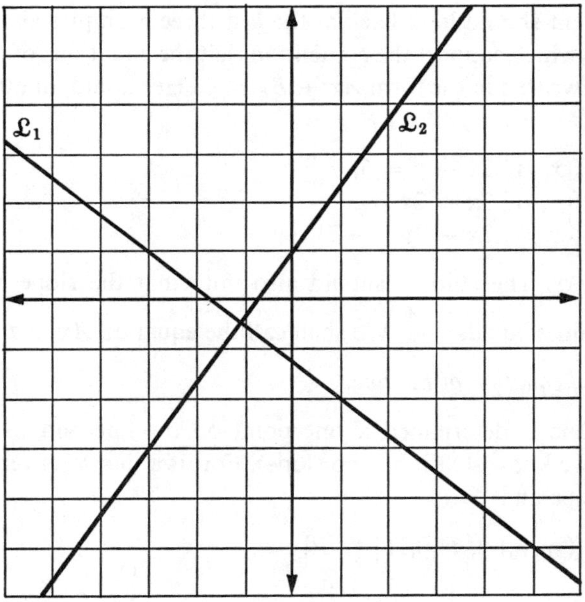

Fig. 2.30

quickly and easily draw the graph of $\mathscr{L}_1$. The equation that we have given for $\mathscr{L}_2$ is its vector equation. We could have given other forms of this equation by noting that the slope of $\mathscr{L}_2 = \frac{4}{3}$ and that it contains (0,1). Hence, the slope-intercept equation is

$$\mathscr{L}_2 = \left\{(x,y) \middle| \ y = \frac{4}{3}x + 1\right\}$$

or, by simplifying this equation, we can give its general equation

$$\mathscr{L}_2 = \{(x,y) \mid 4x - 3y = -3\}.$$

PROBLEM SET 2.8

1. Suppose $\mathscr{L}$ and $\mathscr{L}_1$ are lines with slopes m and m_1 respectively;
  a. if $\mathscr{L}$ is parallel to $\mathscr{L}_1$, show that $m = m_1$.
  b. if $\mathscr{L}$ is orthogonal to $\mathscr{L}_1$, show that $m = -1/m_1$.

For each of the lines given in problems 2–5, give (a) the point-slope equation, (b) slope-intercept equation, and (c) the general equation of the line.

2. $\mathscr{L} = \{(-2,2) + t(-4,3) \mid t \in R\}$.

3. $\mathscr{L} = \{(1 - t, 3t + 2) \mid t \in R\}$.

4. $\mathscr{L} = \{(3,5) + t(1,1) \mid t \in R\}$.

5. $\mathscr{L} = \{(2t, t - 3) \mid t \in R\}$.

6. Determine the slope and y intercept of each of the following lines.
  a. The line through (0,2) and (2,5)
  b. The line through (−1,4) and (1,3).
  c. The line whose general equation is $3x - y = 6$.

7. Give the general equation of each of the following lines.
  a. The line through (5,−2) whose slope is −1/2.
  b. The line whose slope is 3 and y intercept is −4.
  c. The line through the points (−2,−4) and (1,0).

8. Give the general equation of the line through (−1,5) which is parallel to the line $\{(1,3) + t\cdot(1,-2) \mid t \in R\}$.

9. Give the slope-intercept equation of the line through (−2,1) which is parallel to the line $\{(3,0) + t(1,1) \mid t \in R\}$.

10. Give the slope-intercept equation of the line through $(-2,1)$ which is orthogonal to $\{(3,0) + t(1,1) \mid t \in R\}$.

11. Give an equation for the line through (2,6) and parallel to the line whose equation is $5x + 2y = 12$.

12. Give an equation for the line through $(3,-2)$ and parallel to the line whose equation is $3x - 7y = -4$.

13. Give an equation for the line through (0,0) and orthogonal to the line whose equation is $3x + 4y = -6$.

14. Give an equation for the line through $(-1,0)$ and orthogonal to the line whose equation is $3x - y = 4$.

Section 9—Conic Sections The point sets known as the *conic sections*, or simply *conics*, were originally defined and studied by the ancient Greeks in terms of the literal meaning of the words *conic* and *sections*, that is, as plane sections of right circular cones. They did not, of course, use the algebraic descriptions that we will shortly develop. Instead, the Greeks developed their theory of the conics by purely geometrical methods. They started with a circular cone (a figure they considered as being obtained by rotating a line about another intersecting line) and considered what curves would be obtained by intersecting this cone by a plane. As illustrated in figures 2.31, 2.32, 2.33, and 2.34, they determined that, if the plane does not pass through the vertex (tip) of the cone, then the intersection will be either a circle, an ellipse, a parabola, or a hyperbola.

Their results are summarized in the books written by Appollonius (260–170 B.C.) and represent one of the truly remarkable achievements of the Greek mathematicians.

We shall not adopt this historic view in our discussion of the conic sections. Instead, we will give geometric definitions of the ellipse, hyperbola, and parabola and from them deduce or derive equations for these point sets by choosing suitable coordinate axes.

Fig. 2.31. Plane parallel to the base of the cone cuts it in a *circle*.

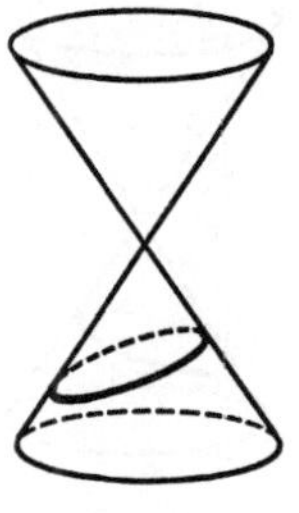

Fig. 2.32. Plane which intersects only one nappe of the cone and is not parallel to a generator cuts the cone in an *ellipse.*

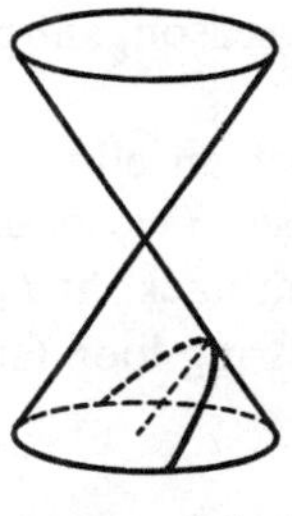

Fig. 2.33. Plane parallel to a generator of the cone cuts the cone in a *parabola.*

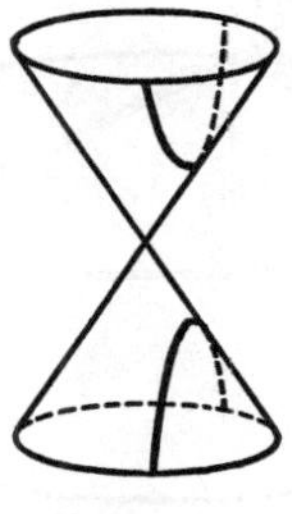

Fig. 2.34. Plane which intersects both nappes of the cone but does not pass through the vertex cuts the cone in a *hyperbola.*

Remark. By analyzing the large amount of data on planetary motion known by 1600, the German astronomer Johannes Kepler realized that the six planets known at the time moved in paths that were ellipses with the sun at one focus.

Ellipse (See figure 2.35.) In making our definition we shall let F_1 and F_2 be two given points in the plane, and we shall let the length of the segment F_1F_2 be denoted by the positive number $2c$. Letting a denote a positive number greater than c, the set of points to which the point P belongs only if

$$d(PF_1) + d(PF_2) = 2a$$

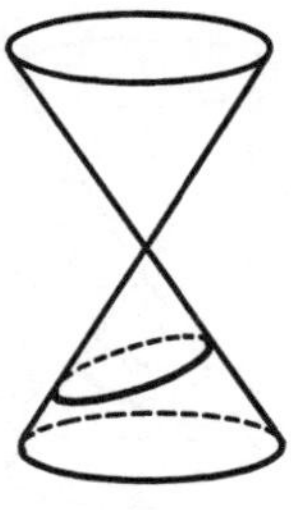

Fig. 2.35

is called an ellipse. (The main reason, simplicity, for using $2c$ and $2a$ will become clear.)

This informal description of an ellipse suggests a simple method by which we may draw an ellipse. Two ends of a string whose length is $2(a + c)$ are looped about thumbtacks at F_1 and F_2 and a pencil traces the curve as it is held taut in the string loop (see figure 2.36).

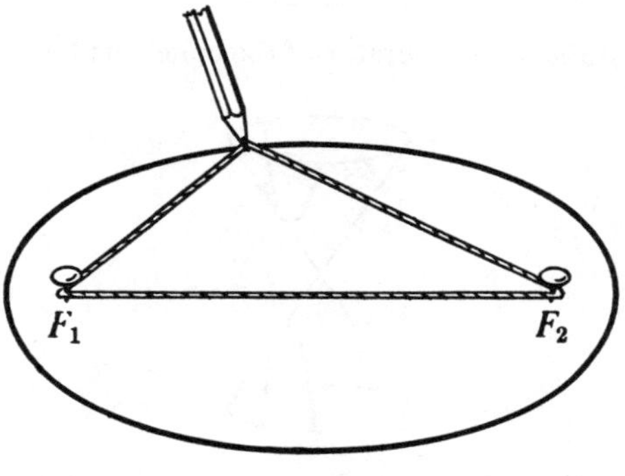

Fig. 2.36

DEFINITION. *The statement that "$\mathscr{E}$ is an ellipse" means that $\mathscr{E}$ is the set of all points with the property that the sum of the distances of each point in $\mathscr{E}$ from two given points F_1 and F_2 is a constant.*
The given points F_1 and F_2 are called foci, *and the midpoint of the segment from F_1 to F_2 is called the* center *of the ellipse.*

To derive an equation for an ellipse, we choose the line through the foci as the X axis and the perpendicular bisector of the segment F_1 to F_2 as the Y axis (see figure 2.37). Let $F_1 = (-c,0)$ and $F_2 = (c,0)$. Hence, the length of the segment from F_1 to F_2 is $2c$.

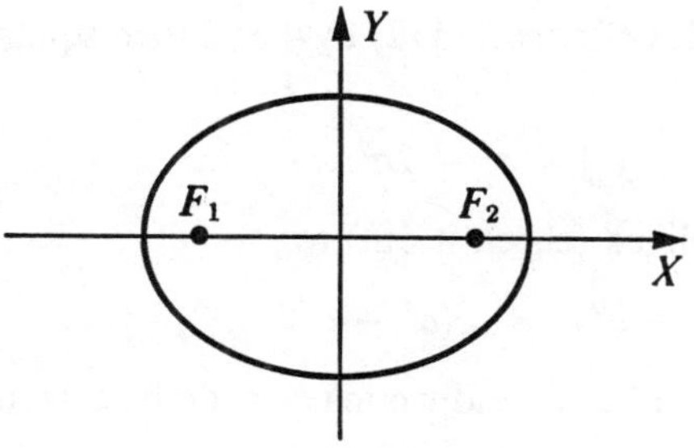

Fig. 2.37

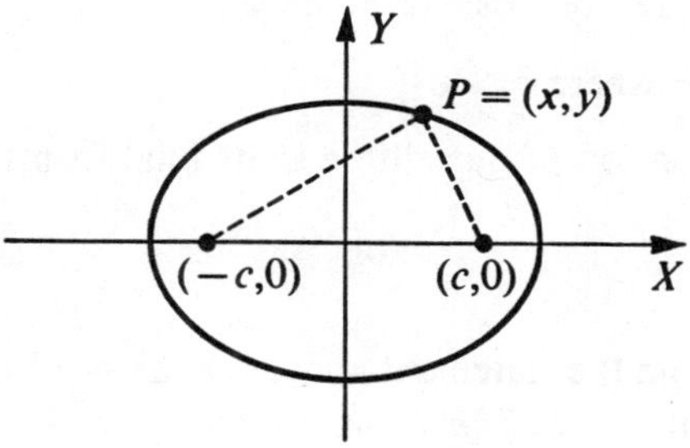

Fig. 2.38

If $P = (x,y)$ is any point in (see figure 2.38), then

$$d(PF_1) + d(PF_2) = 2a,\ a > 0$$

which can be written in the form, using the distance formula,

$$(1)\quad \sqrt{[(x + c)^2 + y^2]} + \sqrt{[(x - c)^2 + y^2]} = 2a.$$

Since, in any triangle, the sum of the lengths of any two sides must exceed the length of the third side, it is clear that $d(PF_1) + d(PF_2) > d(F_1F_2)$. Thus $2a > 2c$ and so $a > c$.

Equation (1) may be written

$$\sqrt{[(x + c)^2 + y^2]} = 2a - \sqrt{[(x - c)^2 + y^2]}.$$

Squaring both sides, we get

$$(x + c)^2 + y^2 = 4a^2 - 4a\sqrt{[(x - c)^2 + y^2]} + (x - c)^2 + y^2$$

which we can simplify and write as

$$(2)\quad 4a\sqrt{[(x - c)^2 + y^2]} = 4a^2 - 4cx.$$

Dividing both sides of equation (2) by 4 and then squaring each side gives us

$$a^2[(x - c)^2 + y^2] = a^4 - 2a^2cx + c^2x^2.$$

This can be simplified (check it out) to

(3) $(a^2 - c^2)x^2 + a^2y^2 = a^2(a^2 - c^2).$

Since $a > c$, $a^2 - c^2 > 0$, and we may divide by this number to get

$$\frac{x^2}{a^2} + \frac{y^2}{a^2 - c^2} = 1.$$

For simplicity, we set the positive number

$$a^2 - c^2 = b^2 \text{ where } b > 0$$

and obtain the equation of our ellipse in its final form:

$$\frac{x^2}{a^2} + \frac{y^2}{b^2} = 1.$$

Remark. From the statement above that $a^2 - c^2 = b^2$, we can easily conclude that $a > b$. *This is important.*

Armed with this algebraic knowledge, we can describe the *ellipse whose center is at the origin and whose foci are on the X axis* by the eye pleasing notation

$$\mathscr{E} = \left\{(x,y) \left| \frac{x^2}{a^2} + \frac{y^2}{b^2} = 1, a > b > 0 \right.\right\}.$$

Since each point of this ellipse must satisfy the equation and conversely, it follows that the points $(a,0)$, $(-a,0)$, $(0,b)$, and $(0,-b)$ must belong to $\mathscr{E}$ (see figure 2.39). Each of these points is called a *vertex*. The lengths of the

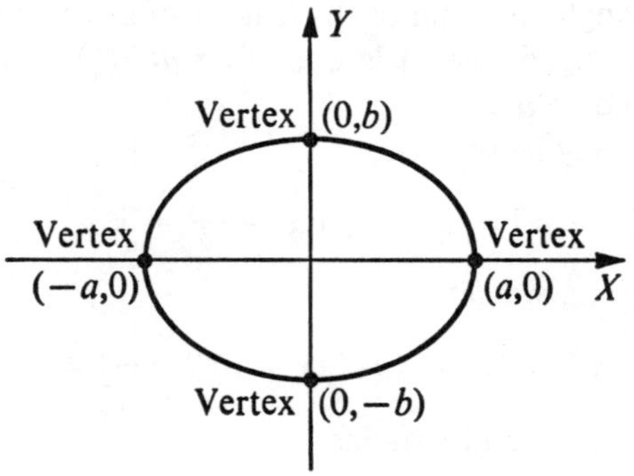

Fig. 2.39

two line segments opposite vertices are $2a$ and $2b$. The longer of these two line segments is called the *major axis* and the shorter is called the *minor axis* (see figure 2.40).

The statements that we have just finished reading regarding the vertices and the major and minor axes of the ellipse are extremely useful in the task of sketching its graph. To illustrate, we give an example.

The unfinished example. (See figure 2.41.) Graph the ellipse

$$\mathscr{E} = \left\{(x,y) \,\middle|\, \frac{x^2}{16} + \frac{y^2}{9} = 1\right\}.$$

We know the following: Center (0,0), $a = 4$, $b = 3$, vertices (4,0), (−4,0), (0,3), and (0,−3). We plot the vertices. We are thus asssured that the graph will be bounded by the indicated rectangle, that is,

$x \in [-4,4]$, $y \in [-3,3]$.

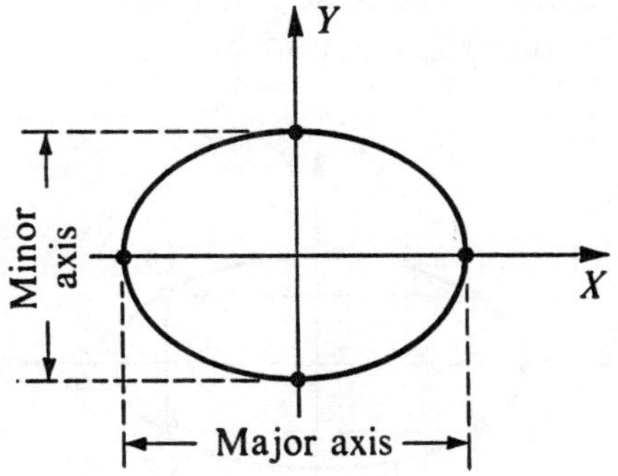

Fig. 2.40

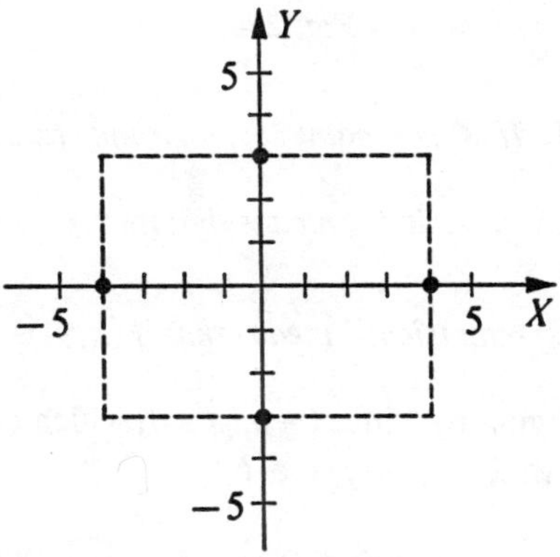

Fig. 2.41

We could continue by calculating other values of x and y that satisfy the equation for $\mathscr{L}$, but let us stop here and note something else about our ellipse that is not only eye pleasing but has the practical benefit of reducing the required calculations by at least 50%.

The ellipse is characterized by the equation

$$\frac{x^2}{a^2} + \frac{y^2}{b^2} = 1$$

so, if $(p,q) \in \mathscr{E}$, it is easy to see that

$$\frac{(-p)^2}{a^2} + \frac{q^2}{b^2} = \frac{p^2}{a^2} + \frac{q^2}{b^2} = 1$$

hence $(-p,q) \in \mathscr{E}$. Likewise, we can show that $(p,-q) \in \mathscr{E}$ (see figure 2.42). These two observations—rather easily made, you will agree—are said to mean that the graph of an ellipse has the property of being symmetrical about the Y and X axes respectively. Just for the sake of the very sound of the phrases, let us agree to call these nice properties *Y symmetrical* and *X symmetrical*. More formally if you wish:

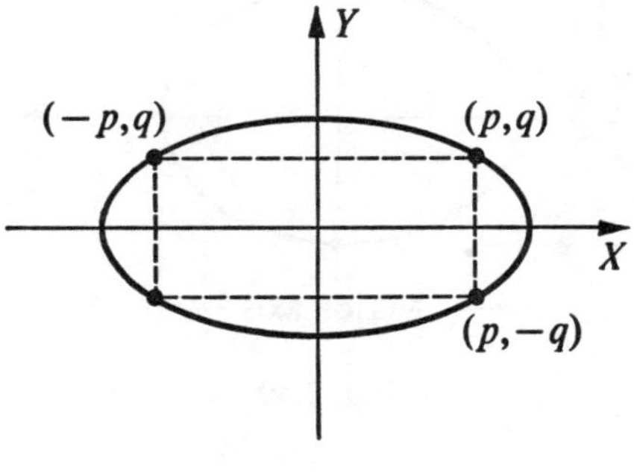

Fig. 2.42

DEFINITION. *If $\mathscr{S}$ is a point set, then the statements that:*

1. *"$\mathscr{S}$ is Y symmetrical" means that if $(x,y) \in \mathscr{S}$, then $(-x,y) \in \mathscr{S}$; and*
2. *"$\mathscr{S}$ is X symmetrical" means that if $(x,y) \in \mathscr{S}$, then $(x,-y) \in \mathscr{S}$.*

Question (see figure 6). Since $(-p,q) \in \mathscr{E}$, which of the above properties assures us that $(-p,-q) \in \mathscr{E}$?

With less formality, we could explain that if the graph of a point set is of such a nature that if we fold its graph along the Y axis and that part of the graph to the right of the Y axis coincides with that part on the left, we would say that the point set is Y symmetrical. If folding along the X axis has that part of the graph above the X axis coinciding with that part below the X axis, then the point set is X symmetrical.

Thus we are pleased with our ellipse for it has both kinds of symmetry; but so does a circle whose center is at the origin.

Now, let us return to the unfinished example and make use of these properties.

Example. Sketch the graph of the ellipse $\frac{x^2}{16} + \frac{y^2}{9} = 1$.

Recall that we had already located the vertices. Now, let $x = 3$, then $\frac{y^2}{9} = 1 - \frac{9}{16} = \frac{7}{16}$, hence $y = \frac{3\sqrt{7}}{4}$. So we instantly or sooner plot the four points shown in figure 2.43a. For $x = 2$, $y = \frac{3\sqrt{3}}{2}$ and four more points are plotted (see figure 2.43b). Sooner or later, we will conclude that the graph of this ellipse resembles figure 2.43c and we are finished with the example.

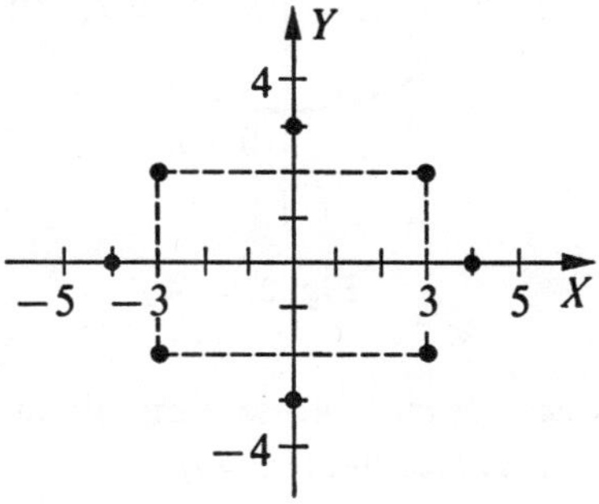

Fig. 2.43a

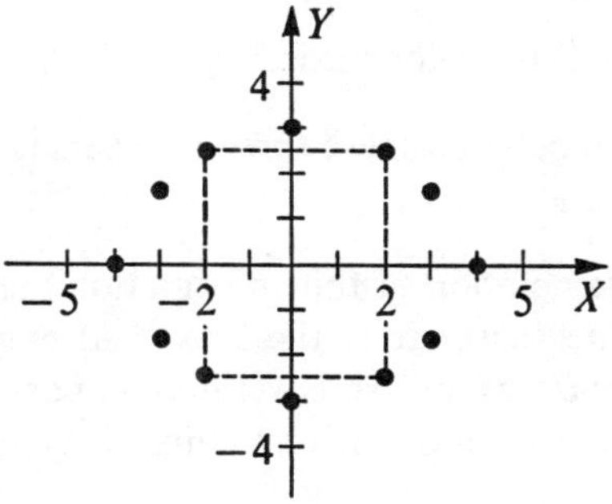

Fig. 2.43b

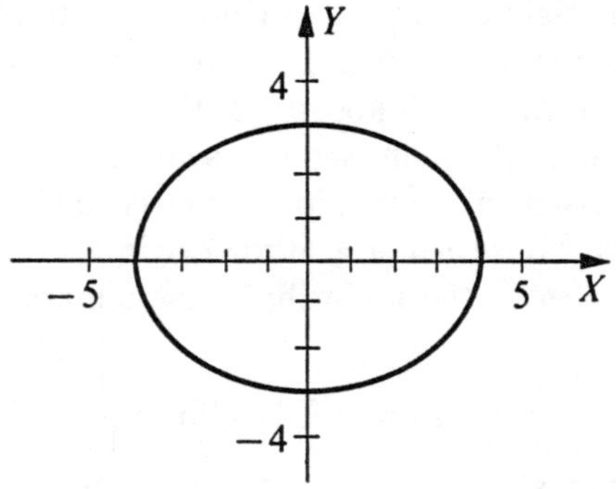

Fig. 2.43c

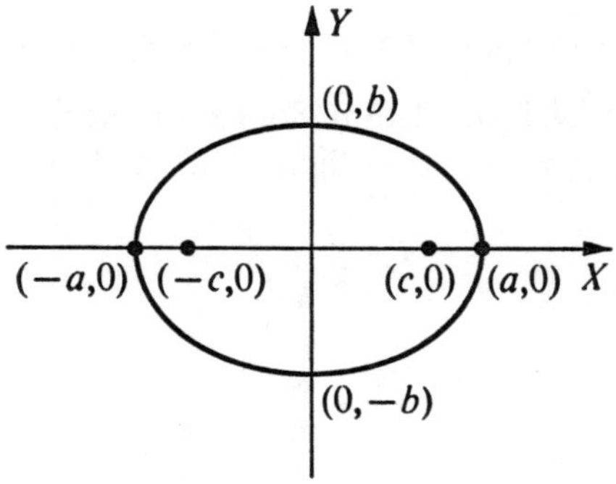

Fig. 2.44

We pause to collect some information about important distances associated with the ellipse (see figure 2.44).

1. c = distance from center to a focus.
2. a = one-half the major axis.
3. b = one-half the minor axis.

and the relationship between these three numbers is given by

4. $b^2 = a^2 - c^2$.

Our serene contemplation of this information may be shattered by the thought—suppose the foci were on the Y axis! After glancing at figure 2.45, we shall leave it to the student to recover his composure and to show (in the exercises) that the equation of such an ellipse is given by

$$\frac{x^2}{b^2} + \frac{y^2}{a^2} = 1.$$

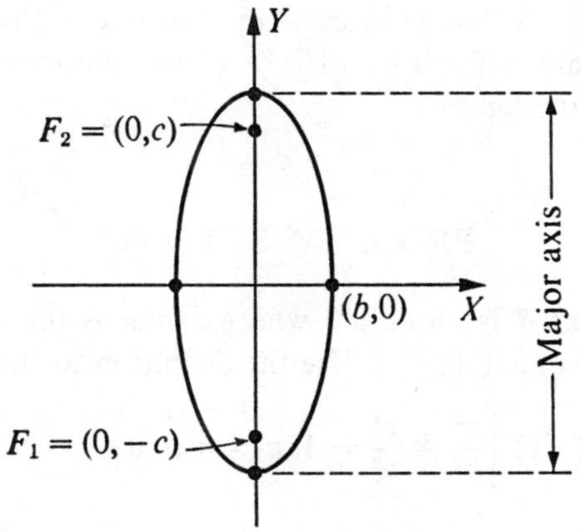

Fig. 2.45

Example. The ellipse described by the equation

$$\frac{x^2}{2} + \frac{y^2}{16} = 1$$

has its foci on the Y axis (the major axis). Its vertices are $(\sqrt{2},0)$, $(-\sqrt{2},0)$, $(0,4)$, and $(0,-4)$. Since it is both X and Y symmetrical, it also contains the points $(1,2\sqrt{2})$, $(-1,2\sqrt{2})$, $(-1,2\sqrt{2})$, and $(-1,-2\sqrt{2})$, Its graph is shown in figure 2.46.

Question. What are the foci?

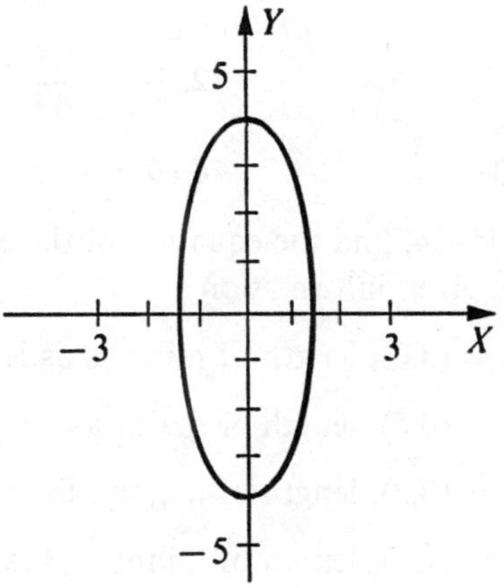

Fig. 2.46

Answer. Recall that $b^2 = a^2 - c^2$, hence $2 = 16 - c^2$ so $c^2 = 14$. So the foci are at $(0,\sqrt{14})$ and $(0, -\sqrt{14})$. Approximate $\sqrt{14}$ and plot the foci in the figure.

PROBLEM SET 2.9$\mathscr{E}$

[1]. Suppose that $\mathscr{E}$ is an ellipse whose center is the origin and whose foci are at $(0,c)$ and $0,-c)$. Use the definition to show that

$$\mathscr{E} = \left\{(x,y) \,\middle|\, \frac{x^2}{b^2} + \frac{y^2}{a^2} = 1, a > b > 0\right\}.$$

2. Show that each of the equations given below describe an ellipse and give its foci.

a. $25x^2 + 9y^2 = 225$ **b.** $3x^2 + 5y^2 - 15 = 0$

In each of problems 3–14, sketch the graph of the given ellipse and plot its foci.

3. $\frac{x^2}{25} + \frac{y^2}{9} = 1$ **4.** $\frac{x^2}{25} + \frac{y^2}{16} = 1$

5. $\frac{x^2}{64} + \frac{y^2}{36} = 1$ **6.** $\frac{x^2}{9} + \frac{y^2}{5} = 1$

7. $\frac{x^2}{4} + \frac{y^2}{25} = 1$ **8.** $\frac{x^2}{9} + \frac{y^2}{25} = 1$

9. $\frac{x^2}{7} + \frac{y^2}{9} = 1$ **10.** $\frac{x^2}{16} + \frac{y^2}{36} = 1$

11. $\frac{x^2}{100} + \frac{y^2}{81} = 1$ **12.** $\frac{x^2}{64} + \frac{y^2}{49} = 1$

13. $3x^2 + 10y^2 = 30$ **14.** $9x^2 + 20y^2 = 180$

In each of problems 15–24, find the equation of the ellipse whose center is (0,0) that satisfies the given information.

15. $F_1 = (-3,0)$, $F_2 = (3,0)$, length of minor axis is 8.

16. $F_1 = (0,-6)$, $F_2 = (0,6)$, length of major axis is 20.

17. $F_1 = (0,-2)$, $F_2 = (0,2)$, length of major axis is 8.

18. $F_1 = (-1,0)$, $F_2 = (1,0)$, length of minor axis is 6.

19. One focus at (4,0), one vertex at (0,2).

20. One focus at $(0,\sqrt{2})$, one vertex at $(1,0)$.

21. Length of major axis is 10, length of minor axis is 6, major axis on the Y axis.

22. Length of major axis is 10, length of minor axis is 8, major axis on the X axis.

23. One vertex at $(5/2,0)$, one focus at $(0,\sqrt{5})$.

24. One focus at $(\sqrt{6},0)$, length of minor axis is 4.

25. Show that the equation $10x^2 + 28y^2 = 35$ describes an ellipse whose center is $(0,0)$ by giving the numbers a, b, and c.

Bonus problem. Suppose that $a > 0$, that Q is the point $(ae,0)$ where $0 < e < 1$, and that $\mathscr{L}$ is the vertical line described by the equation $x = a/e$. Show that if $\mathscr{S}$ is a point set such that $P = (x,y)$ belongs to $\mathscr{S}$ if and only if the distance from P to Q equals e times the length of the horizontal line segment from P to $\mathscr{L}$, then

$$\mathscr{S} = \{(x,y) \mid (1 - e^2)x^2 + y^2 = a^2(1 - e^2)\}.$$

Additional bonus. By substituting $c = ae$ and $b^2 = a^2(1 - e^2) = a^2 - c^2$, go on and show that $\mathscr{S}$ is an ellipse with center $(0,0)$.

Remark. The number e is called the *eccentricity* of $\mathscr{E}$ and the vertical line $\mathscr{L}$ is called its *directrix*.

Hyperbola (See figure 2.47.) Because the definition and many of the properties of the hyperbola very closely resemble those of the ellipse, many of the details will be left for the student to work out.

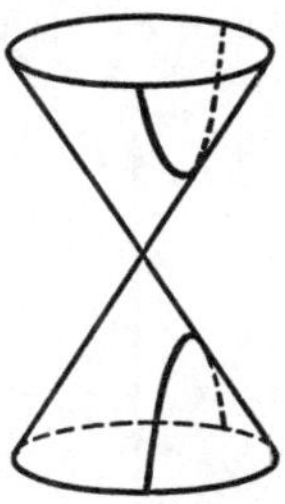

Fig. 2.47

DEFINITION. *The statement that "$\mathscr{H}$ is a hyperbola" means that $\mathscr{H}$ is the set of all points with the property that the absolute value of the difference of the distances of each point in $\mathscr{H}$ from two given points F_1 and F_2 is a constant.*

Again, the points F_1 and F_2 are called the *foci* and the midpoint of the segment from F_1 and F_2 is the *center*. The points V_1 and V_2 (see figure 2.48) are called the *vertices* and the line segment V_1V_2 is called the *transverse axis.*

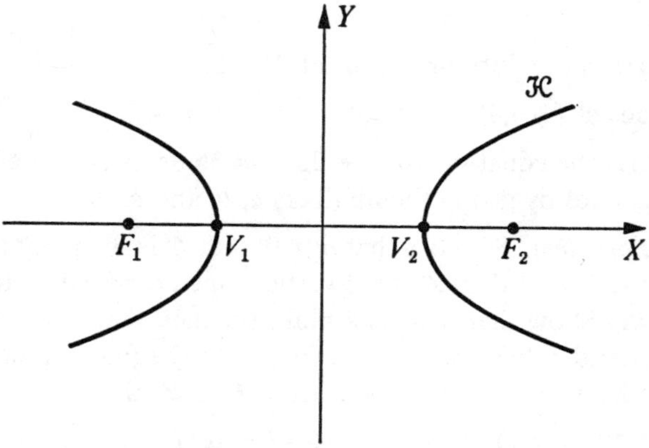

Fig. 2.48

Choosing the coordinate axes in the same manner as used for the ellipse, we again let $F_1 = (-c,0)$ and $F_2 = (c,0)$ (see figure 2.49). So, if $P = (x,y)$ is any point in $\mathscr{H}$, then $|d(PF_1) - d(PF_2)| = 2a$. From the distance formula we get

$$|\sqrt{[(x + c)^2 + y^2]} - \sqrt{[(x - c)^2 + y^2]}| = 2a \text{ where } a > 0$$

or

$$\sqrt{[(x + c)^2 + y^2]} - \sqrt{[(x - c)^2 + y^2]} = \pm 2a.$$

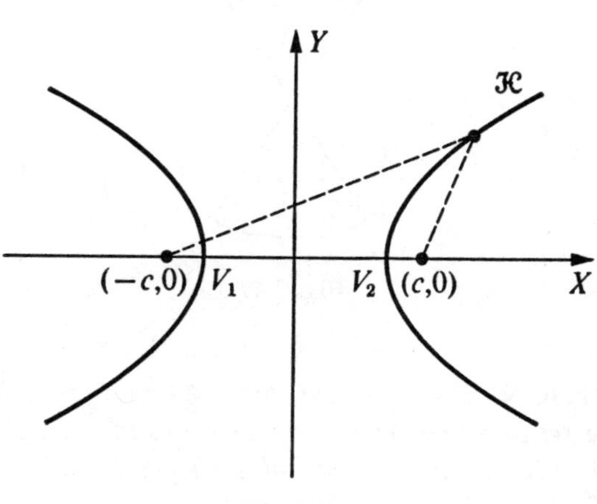

Fig. 2.49

By operations similar to those used in obtaining the equation for the ellipse, we can obtain the equation

$$\frac{x^2}{a^2} + \frac{y^2}{a^2 - c^2} = 1.$$

This is the same equation we obtained for the ellipse! In this case, however, by going back to figure 2.49 and noting that $V_1 = (-a,0)$ and $V_2 = (a,0)$, we can see that $2a < 2c$ so $a < c$ and therefore $a^2 - c^2 < 0$. *Take the time to compare this information with the similar information for the ellipse.* To continue, set $a^2 - c^2 = -b^2$, $b > 0$ and rewrite the equation in its final form

$$\frac{x^2}{a^2} - \frac{y^2}{b^2} = 1.$$

The fact that $a < c$ should be taken as a warning that a hyperbola is not an ellipse. Not only does $(0,b)$ not belong to $\mathscr{H}$, but its graph does not intersect the Y axis. In fact, since

$$\frac{x^2}{a^2} = 1 + \frac{y^2}{b^2} > 1,$$

it follows that $x^2 > a^2$, and there are no points on the graph in the region $-a < x < a$. Furthermore. there is no reason to assume that $a > b$.

Thus the hyperbola whose *foci are on the X axis and whose center is the origin* is described by

$$\mathscr{H} = \left\{(x,y) \,\middle|\, \frac{x^2}{a^2} - \frac{y^2}{b^2} = 1\right\}.$$

We now undergo a ruthless interrogation.

Question. Is $\mathscr{H}$ Y symmetrical? X symmetrical?

Answer (both times). *Ja.*

Question. What significance can we attach to the number b in light of the previous discussion?

Answer. While the point $(0,b)$ does not belong to the hyperbola, the segment from $(0,b)$ to $(0,-b)$ is called the *conjugate axis* and its length is $2b$.

Additional comments.

$$\frac{y^2}{b^2} = \frac{x^2}{a^2} - 1 < \frac{x^2}{a^2},$$

hence

$$y^2 < \frac{b^2x^2}{a^2}.$$

So, for every point $(x,y) \in \mathscr{H}$, we have

$$-\frac{b}{a}x < y < \frac{b}{a}x \text{ for } x > 0$$

and

$$\frac{b}{a}x < y < -\frac{b}{a}x \text{ for } x < 0.$$

Thus every point in $\mathscr{H}$ lies between the two lines $y = \frac{b}{a}x$ and $y = -\frac{b}{a}x$. Examine figure 2.50 to see that these two lines contain the diagonals of the

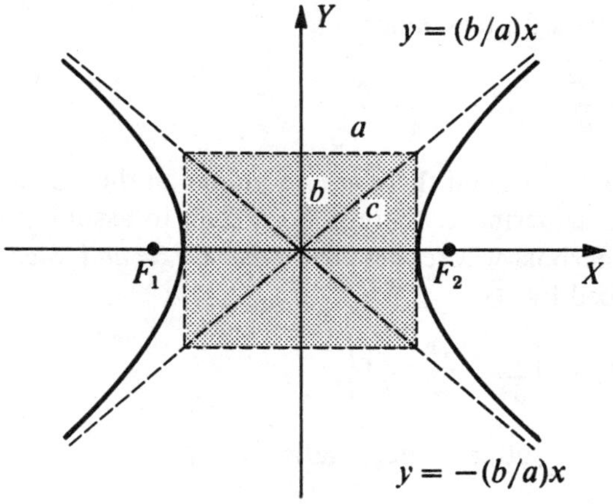

Fig. 2.50

rectangle shown in figure 2.50. This figure should also serve to remind us of the Pythagorean relationship between the numbers a, b, and c.

$$c^2 = a^2 + b^2.$$

These lines *are not* part of the graph of the hyperbola but they are called the *asymptotes of the hyperbola.*

Since the equation of a hyperbola can also be written as

$$y = \pm\frac{bx}{a}\sqrt{(1 - a^2/x^2)}$$

we can observe that the larger $|x|$ becomes, the closer the point (x,y) on the graph of the hyperbola is to one of these asymptotes.

Though no details are given, it is hoped that our vision is clear enough to see that *if the foci are on the Y axis,* then

$$\mathcal{H} = \left\{(x,y)\ \middle|\ \frac{y^2}{a^2} - \frac{x^2}{b^2} = 1\right\}.$$

It is easy to see that this hyperbola is both X and Y symmetrical. Its graph is shown in figure 2.51, and you are encouraged to note that the

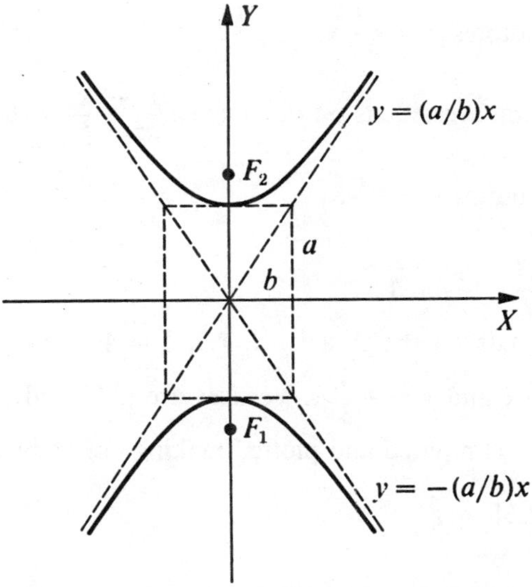

Fig. 2.51. $\dfrac{y^2}{a^2} - \dfrac{x^2}{b^2} = 1$

asymptotes contain the diagonals of a rectangle and that their equations are $y = \dfrac{a}{b}x$ and $y = -\dfrac{a}{b}x$.

A Recap on the Hyperbola

The hyperbola, like the ellipse, is both Y symmetrical and X symmetrical. Unlike the ellipse, there are two lines, called *the asymptotes,* associated (not part of) the hyperbola that sort of "fence it in." The major distances associated with the hyperbola are:

1. c = distance from center to a focus
2. b = one-half the length of the conjugate axis
3. a = one-half the length of the transverse axis

where

4. $a < c$ and $b^2 = c^2 - a^2$.

I Foci and transverse axis on the X axis: $\frac{x^2}{a^2} - \frac{y^2}{b^2} = 1$

Asymptotes $y = \pm\frac{b}{a}x$.

II Foci and transverse axis on the Y axis: $\frac{y^2}{a^2} - \frac{x^2}{b^2} = 1$

Asymptotes $y = \pm\frac{a}{b}x$.

Example. $\frac{x^2}{9} - \frac{y^2}{16} = 1$.
Transverse axis on the X axis. $a = 3$, $b = 4$. Asymptotes are the lines $y = \frac{4}{3}x$ and $y = -\frac{4}{3}x$. Vertices are (3,0) and (−3,0). Additional points computed and plotted making use of symmetry.

$$\left[4,\frac{4\sqrt{7}}{3}\right], (6,4\sqrt{3}).$$

Naturally, use was made of the table of square roots given in the appendix to approximate the ordinates of these points before they were plotted on the graph. (See figure 2.52.)

Example. $\frac{y^2}{9} - \frac{x^2}{4} = 1$.
Transverse axis on the Y axis. $a = 3$, $b = 2$. Asymptotes are the lines $y = \frac{3}{2}x$ and $y = -\frac{3}{2}x$. Vertices are (0,3) and (0,−3). Additional points plotted $(2,3\sqrt{2})$ and $(4,3\sqrt{5})$. Symmetry was used to obtain other points shown on the graph. (See figure 2.53.)

Question. What are the foci?

Answer. Well, $b^2 = c^2 - a^2$ so $4 = c^2 - 9$ which means that $c = \pm\sqrt{13} \approx 3.8$. So you can plot the foci.

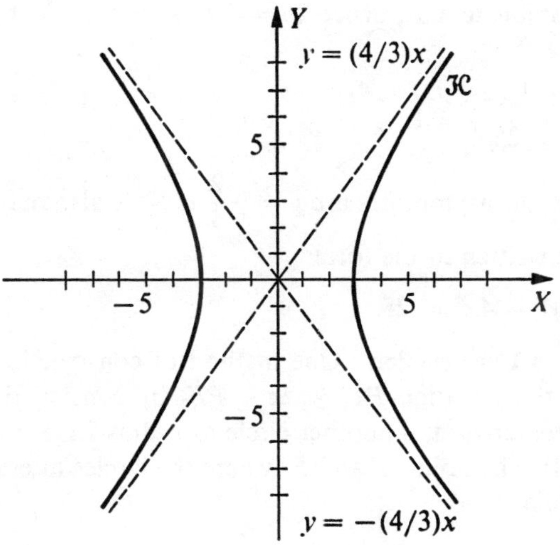

Fig. 2.52

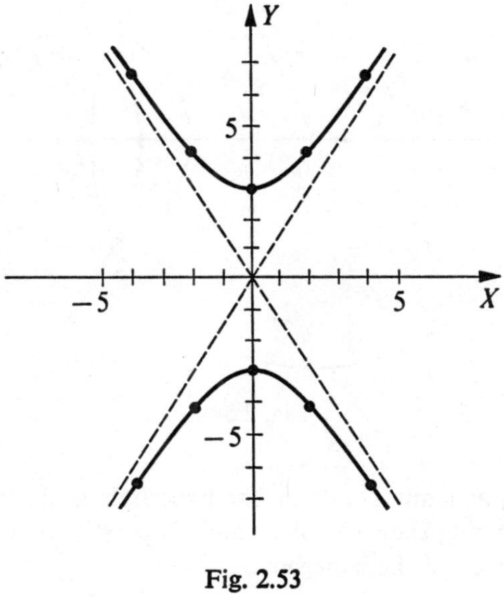

Fig. 2.53

Example. Find an equation for the hyperbola whose vertices are (0,1) and (0, −1) and whose conjugate axis has length 5.

Solution. Since the vertices are given, we know that the transverse axis is on the Y axis and that $a = 1$. We also have given the length of the conjugate axis, hence $2b = 5$ or $b = 5/2$. That is all we need to write

$$\frac{y^2}{1} - \frac{x^2}{25/4} = 1.$$

Note that the asymptotes are $y = \pm\frac{2}{5}x$. Note also that this equation could be written in the form

$$25y^2 - 4x^2 = 25.$$

Hyperbola—a Construction One method of constructing a hyperbola makes use of the equation $PF_1 = 2a + PF_2$ by constructing a circle of radius r with center F_2 and another circle of radius $2a + r$ with center F_1 (see figure 2.54). The points P and P' where the circles intersect are points on the hyperbola.

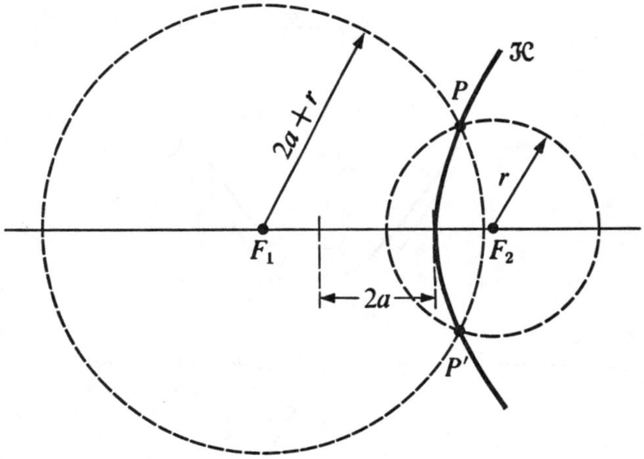

Fig. 2.54

By varying r, as many points on the hyperbola as desired may be obtained. Interchanging the roles of F_1 and F_2 permits us to obtain points on the other branch of the hyperbola.

PROBLEM SET 2.9$\mathcal{H}$

1. Start with the equation

$$\sqrt{[(x + c)^2 + y^2]} - \sqrt{[(x - c)^2 + y^2]} = \pm 2a$$

and develop the equation

$$\frac{x^2}{a^2} + \frac{y^2}{a^2 - c^2} = 1.$$

2. Show that each of the following equations defines a hyperbola whose center is (0,0) *and* give its vertices.

a. $5x^2 - 20y^2 = 80$ **b.** $25x^2 - 3y^2 = 75$

In each of problems 3–12, sketch the graph of the given hyperbola *and* sketch (with dotted lines) its asymptotes.

For Student Eyes Only Asymptotes are a useful aid in drawing hyperbolas. After the numbers a and b are found, the most convenient steps in drawing an *approximate graph* are: (*a*) draw the rectangle indicated in figures 2.51 and 2.52, (b) draw the asymptotes, (c) plot the vertices, and (d) sketch the hyperbola.

3. $\frac{x^2}{9} - \frac{y^2}{4} = 1$ **4.** $\frac{x^2}{16} - \frac{y^2}{9} = 1$

5. $\frac{y^2}{25} - \frac{x^2}{4} = 1$ **6.** $\frac{y^2}{36} - \frac{x^2}{36} = 1$

7. $\frac{x^2}{16} - \frac{y^2}{25} = 1$ **8.** $\frac{x^2}{9} - \frac{y^2}{16} = 1$

9. $\frac{x^2}{16} - \frac{y^2}{16} = 1$ **10.** $\frac{y^2}{4} - \frac{x^2}{9} = 1$

11. $\frac{y^2}{9} - \frac{x^2}{9} = 1$ **12.** $\frac{x^2}{9} - \frac{y^2}{36} = 1$

13. For each of problems 3–12 that you were assigned, give the lengths of the transverse and conjugate axes.

In each of problems 14–21, give the equation of a hyperbola with center (0,0) that satisfies the given conditions.

14. $F_1 = (-5,0)$, $F_2 = (5,0)$, length of transverse axis is 6.

15. $F_1 = (-4,0)$, $F_2 = (4,0)$, length of conjugate axis is 2.

16. $F_1 = (0,-4)$, $F_2 = (0,4)$, length of conjugate axis is 6.

17. Vertices at $(-1,0)$ and $(1,0)$, one focus at $(2,0)$.

18. Transverse axis on X axis, length of transverse axis is 10, length of conjugate axis is 6.

19. $F_1 = (0, -\sqrt{41})$, $F_2 = (0, \sqrt{41})$, length of conjugate axis is 10.

20. $F_1 = (-2\sqrt{13}, 0)$, $F_2 = (2\sqrt{13}, 0)$, length of transverse axis is 12.

21. Transverse axis on the X axis, equations of asymptotes are $y = 3x$ and $y = -3x$, length of conjugate axis is 12.

22. Give a *brief* argument to support the following statement: "If the asymptotes of a hyperbola are perpendicular, then $a = b$." Now, look back at that subset of problems 6, 9, or 11 that you were assigned.

23. Suppose that $\mathscr{S}$ is the point set described by the notation

$$\mathscr{S} = \{(x,y) \mid (1 - e^2)x^2 + y^2 = a^2(1 - e^2)\}.$$

If $e > 1$, let $c = ae$ and let $b^2 = -a^2(1 - e^2)$ and show that $\mathscr{S}$ is a hyperbola whose center is (0,0). When finished, read the remarks that follow the bonus problems in problem set $9\mathscr{E}$.

24. *Bonus problem.* Show that a line segment drawn from a focus of a hyperbola perpendicular to an asymptote has length b. *Hint:* Write the equations of the asymptote and the line containing the focus which is perpendicular to the asymptote and then find their intersection.

25. *Bonus problem.* Suppose we have a hyperbola whose center is at (2,3) and whose transverse axis is parallel to the X axis (a sketch might help). Construct a new set of coordinate axes, U and V, with its origin at (2,3) and with the V axis as the vertical axis. *Now,*

a. write the equation of the given hyperbola in this *new* coordinate system $(u,v) \in U \times V$.

b. write the equation of the given hyperbola in the original coordinate system $X \times Y$.

Parabola (See figure 2.55.) The remaining conic, the parabola, differs in several aspects from either the ellipse or hyperbola. It has only one focus, one vertex, and has no center. There are other differences but perhaps we had best just get on with its definition.

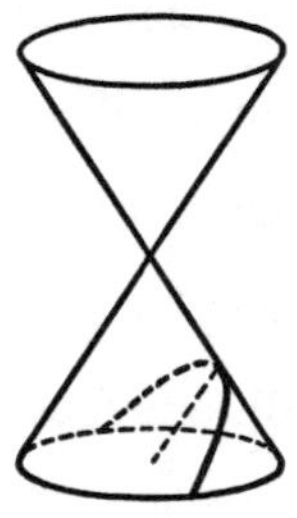

Fig. 2.55

DEFINITION. *Given a point F and a line $\mathscr{L}$, $F \notin \mathscr{L}$, the statement that "$\mathscr{P}$ is a parabola" means that $\mathscr{P}$ is the set of all points with the property that the distance of each point in $\mathscr{P}$ from the given point F is equal to the length of a line segment drawn from the point in $\mathscr{P}$ perpendicular to the given line $\mathscr{L}$.*
The given point F is called the focus. *The given line $\mathscr{L}$ is called the* directrix.

To obtain an equation for a parabola we let the X axis be the line containing the focus and perpendicular to the directrix. For simplicity, we then construct the Y axis at the midpoint of the line segment from F to $\mathscr{L}$ (see figure 2.56). Let $F = (c,0)$ for then the directrix is the vertical line $x = -c$. One immediate gain of this simplicity is that the origin, being equidistant from F and $\mathscr{L}$, belongs to the parabola. Call (0,0) the *vertex* of $\mathscr{L}$.

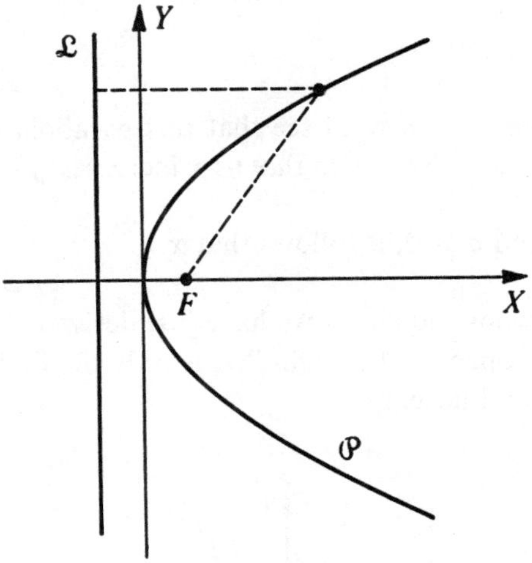

Fig. 2.56

So, if $P = (x,y)$ is a point on the parabola (see figure 2.57), then $d(PQ) = d(PF)$, that is

$$\sqrt{[(x + c)^2]} = \sqrt{[(x - c)^2 + y^2]}.$$

Squaring both sides we have

$$x^2 + 2cx + c^2 = x^2 - 2cx + c^2 + y^2$$

which is simplified to

$$y^2 = 4cx \text{ where } c > 0.$$

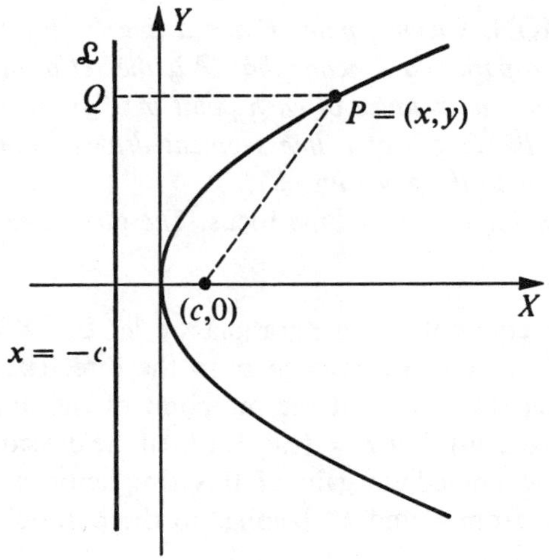

Fig. 2.57

From this equation it is easy to see that this parabola is X symmetrical but not Y symmetrical. Note also that as x increases, y^2 will increase.

Since $y^2 \geq 0$ and $c > 0$, it follows that $x \geq 0$.

The parabola whose equation we have just derived might be described by saying that it "opens to the right." It is only the first of four possible cases which we could develop.

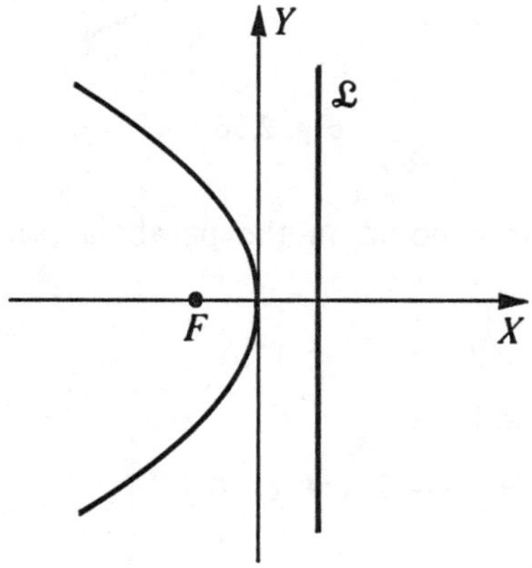

Fig. 2.58

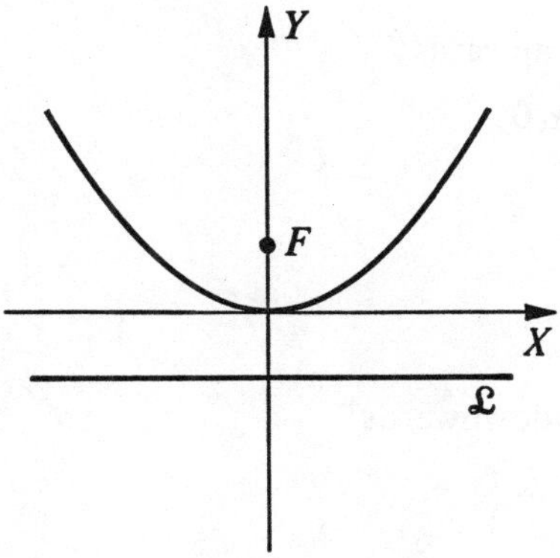

Fig. 2.59

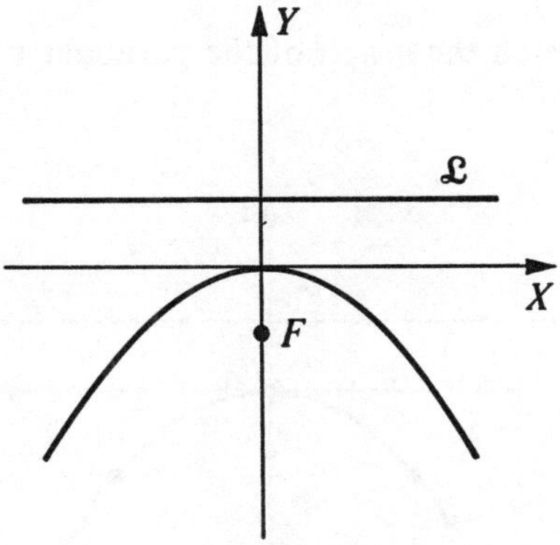

Fig. 2.60

Remark. Since it is very similar to that already shown, the algebraic development of equations for the other three cases will be omitted.

The remaining three cases (figures 2.58, 2.59, and 2.60) are as follows (assume $c > 0$ in all cases):

CASE II: "opens to the left"

$$y^2 = -4cx, \; x \leq 0$$

Focus: $(-c,0)$
Directrix: $x = c$
Vertex: (0,0)
X symmetrical

CASE III: "opens upwards"

$$x^2 = 4cy, y \geq 0$$

Focus: (0,c)
Directrix: $y = -c$
Vertex: (0,0)
Y symmetrical

CASE IV: "opens downwards"

$$x^2 = -4cy, y \leq 0$$

Focus: (0,c)
Directrix: $y = c$
Vertex: (0,0)
Y symmetrical

Example. Sketch the graph of the parabola $x^2 = -6y$. (See figure 2.61.)

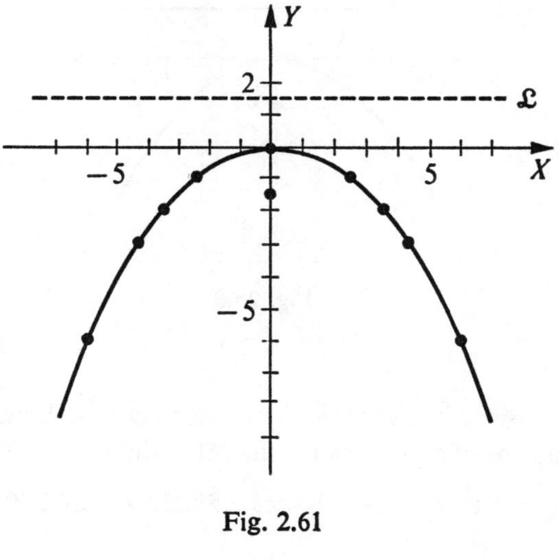

Fig. 2.61

Discussion. This parabola "opens downwards." Since $x^2 \geq 0$, $y \leq 0$. Since $-4c = -6$, $c = 3/2$. Thus focus is $(0, -3/2)$, vertex is (0,0), directrix is the horizontal line $y = 3/2$, graph is Y symmetrical.

If $y = -1$, $x = \pm\sqrt{6}$
If $y = -2$, $x = \pm 2\sqrt{3}$
If $y = -3$, $x = \pm 3\sqrt{2}$
If $y = -6$, $x = \pm 6$

PROBLEM SET 2.9𝒫

[1] Take a look at the figure for Case III and use the definition to develop the equation of this parabola.

In each of problems 2–9, find the coordinates of the focus, the equation of the directrix, and sketch the graph.

2. $y^2 = 12x$	**3.** $y^2 = -2x$
4. $x^2 = 4y$	**5.** $x^2 = -y$
6. $y^2 = -x$	**7.** $y^2 = 4x$
8. $y^2 = -8x$	**9.** $x^2 = 16y$

In each of problems 10–14, give an equation for a parabola that satisfies the given conditions and sketch its graph.

10. Focus at (1,0), directrix $x = -1$.

11. Focus at $(0,-2)$, directrix $y = 2$.

12. Focus at $(0,-3)$, directrix $y = 3$.

13. Focus at $(-4,0)$, directrix $x = 4$.

14. Focus at (0,3/4), directrix $y = -3/4$.

15. The line segment with endpoints on the parabola and perpendicular to the axis of symmetry at the focus is called the *focal chord.* Suppose $y^2 = -4cx$ is the parabola, find the length of its focal chord.

16. Use the definition to develop an equation for a parabola whose focus is $(2,-2)$ and whose directrix is $y = -4$. *Give its vertex.* Now, construct a *new* set of coordinate axes U and V, the U axis parallel to the X axis, whose origin is the vertex of this given parabola. What are the *new* coordinates of the focus of the given parabola?

Section 10—Translation of Axes By choosing a Cartesian coordinate system, we establish a one-to-one correspondence between ordered pairs of real numbers and points in the plane. But the choice of the coordinate system is abitrary. The same point will, of course, have different coordinates in different coordinate systems. We will consider only the simplest case of this situation.

First we choose one coordinate system and then we select another or "new" one, with a different origin, but with the "new" axes (X' and Y') parallel and in the same direction as the "old" axes (X and Y). Thus (see figure 2.62), if $P = (x,y)$ is a point in the old system whose coordinates in the new system are denoted by (x',y'), and if (h,k) are the old coordinates of the new origin $(0',0')$, then we see that

$$x' = x - h \quad \text{and}$$
$$y' = y - k.$$

Figure 2.62 shows the case (the easiest to understand) where the new origin falls in the first quadrant and P is in the new first quadrant. The student ought to look at figure 2.63 to convince himself that the equations given above also hold in other cases.

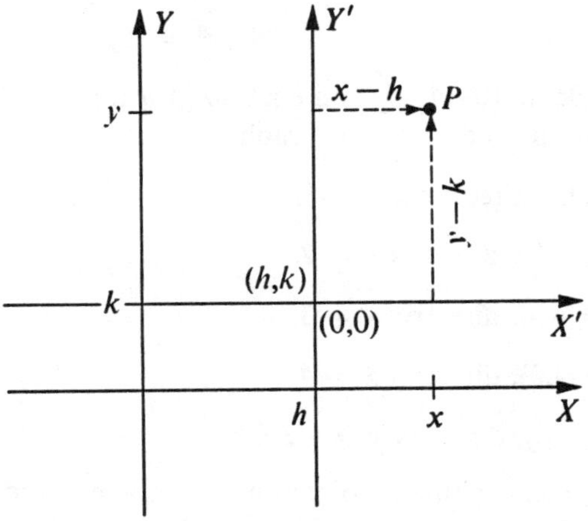

Fig. 2.62

In order to start learning about advantages to be gained by introducing "new" coordinates, we can look at an example so simple that the new coordinate system is not needed. If we wish to learn about the point set defined by the equation

$$x^2 + y^2 - 4x - 10y + 13 = 0$$

we can rewrite it as

$$x^2 - 4x + y^2 - 10y = -13$$

and, by completing the square in x and in y, write the given equation in the form

$$(x - 2)^2 + (y - 5)^2 = 16.$$

It is now easy to see that the point set is a circle with center at (2,5) and

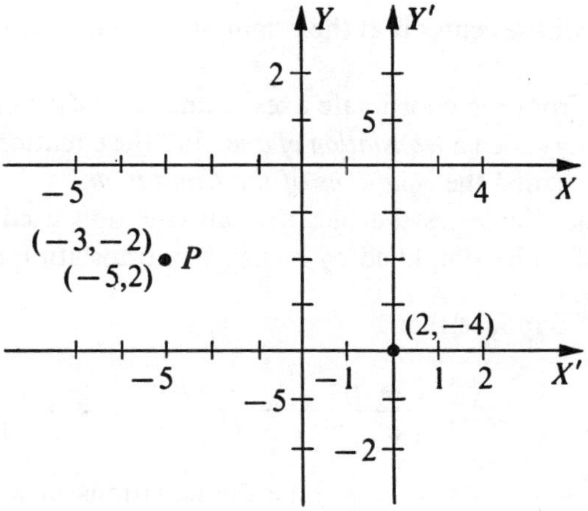

$$x' = x - h \qquad y' = y - k$$
$$= -3 - 2 \qquad = -2 - (-4)$$
$$= -5 \qquad = 2$$

Fig. 2.63

radius 4. Its graph is shown in figure 2.64. If, in this last equation, we set $x' = x - 2$ and $y' = y - 5$, then the equation has the simpler form

$$x'^2 + y'^2 = 16.$$

If we think of x' and y' as new coordinates, then this equation looks like

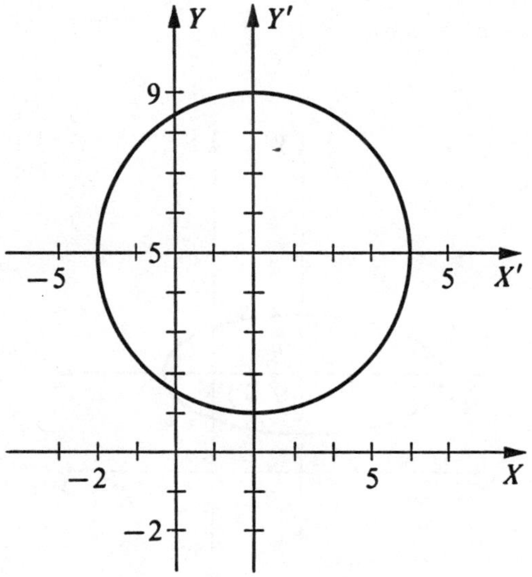

Fig. 2.64

that of a circle whose center is at the origin of the new "prime" coordinate system.

The change from the coordinate axes X and Y to the new axes X' and Y' is customarily called a *translation of axes* and the equations $x' = x - 2$, $y' = y - 5$ are called the *equations of the translation.*

This example illustrates the fact that an equation used to describe a point set can often be simplified by means of a translation of axes.

Example. Suppose that

$$\mathscr{S} = \left\{(x,y) \,\middle|\, \frac{(x-h)^3}{a^2} + \frac{(y-k)^2}{b^2} = 1,\ a > b\right\}.$$

Let $x' = x - h$ and $y' = y - k$ be the equations of a translation of the axis to the new origin $(0',0') = (h,k)$. The equation which defines $\mathscr{S}$ can then be simplified to the form

$$\frac{x'^2}{a^2} + \frac{y'^2}{b^2} = 1$$

which we instantly recognize as the equation of an ellipse whose center is at the origin of the new coordinate system and whose major axis is on the X' axis. (See figure 2.65.) Recalling what we have already mastered, we see that in the new system

Center: $(0,0)$
Vertices: $(-a,0)$, $(a,0)$.

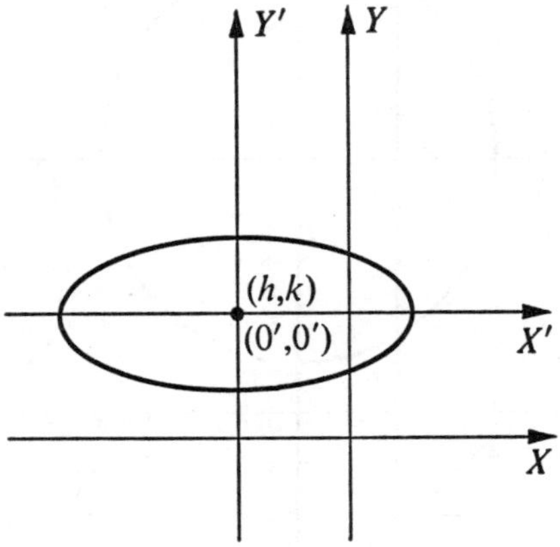

Fig. 2.65

Using the equations of the translation, we see that in the old system

Center: (h,k)
Vertices: $(h - a,k)$, $(h + a,k)$.

Example. By a suitable translation of the axes, simplify the equation of $\mathscr{S}$ where $\mathscr{S} = \{(x,y) \mid 2x^2 + y + 8x + 2 = 0\}$.

Solution. We first write

$$y + 2 = -2x^2 - 8x$$

and then

$$\frac{y + 2}{-2} = x^2 + 4x.$$

By completing the square on x we obtain

$$\frac{y + 2}{-2} = (x + 2)^2 - 4$$

and, multiplying by -2 and transposing the constant, get

$$y - 6 = -2(x + 2)^2.$$

Let $x' = x + 2 = x - (-2)$ and $y' = y - 6$ and we have the equation

$$y' = -2x'^2$$

which is an equation of a parabola that opens downwards with its vertex at the origin. (See figure 2.66.)

Example. Suppose $\mathscr{S} = \{(x,y) \mid 4x^2 + y^2 + 40x + 84 = 0\}$. Identify this point set and sketch its graph.

Solution. We plan on using a translation of axes to simplify the equation with the hope that we can then easily recognize $\mathscr{S}$. By completing the square in x and y we will also give the equations of the translation. Let us begin by writing $4x^2 + 40x + y^2 = -84$ and then rewrite this as $4(x^2 + 10x) + y^2 = -84$ to make the completion of the square on x and y easier to see. Thus we can write

$$4(x^2 + 10x + 25) + (y + 0)^2 = -84 + 100$$

or

$$4(x + 5)^2 + (y + 0)^2 = 16.$$

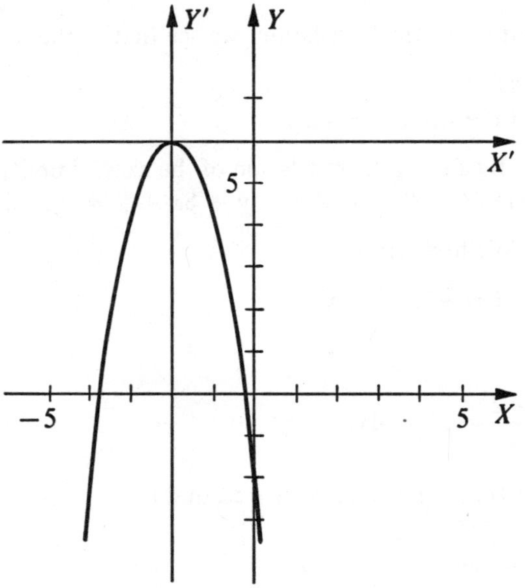

Fig. 2.66

Dividing by 16 we have

$$\frac{(x+5)^2}{4} + \frac{(y+0)^2}{16} = 1.$$

Let the translation have the equations

$$x' = x + 5 \quad \text{and} \quad y' = y + 0$$

which simplifies the last equation to the form

$$\frac{x'^2}{4} + \frac{y'^2}{16} = 1.$$

Therefore $\mathscr{S}$ is an ellipse whose major axis is parallel to the Y axis, $a = 4$ and $b = 2$. We sketch the graph as shown in figure 2.67.

Example. Find an equation for a hyperbola whose center is at $(-1,3)$, transverse axis horizontal and of length 8, and distance between foci equal to $2\sqrt{41}$.

Solution. We know how to handle hyperbola whose transverse axis is on the horizontal axis and whose center is at the origin. Since the transverse axis has length 8 we know that $2a = 8$ so $a = 4$. Since the distance between the foci is $2\sqrt{41}$, we know that $c = \sqrt{41}$. With this information and the knowledge that $b^2 = c^2 - a^2$ we determine that $b = 5$. We will let the origin of a new coordinate system

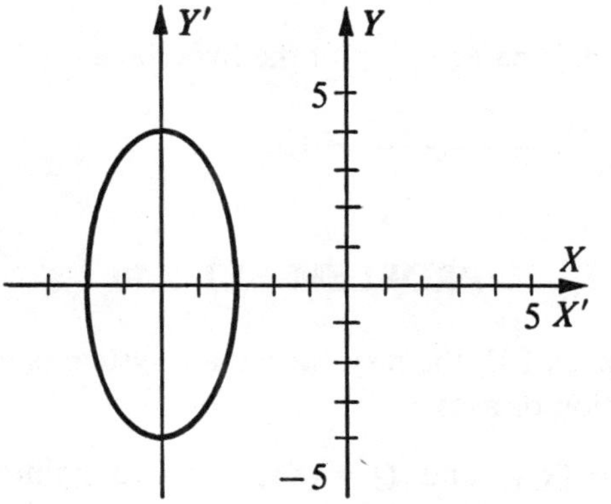

Fig. 2.67

be at $(-1,3)$ which suggests the translation defined by the equations

$$x' = x + 1 \quad \text{and} \quad y' = y - 3.$$

So, we draw our new coordinate axes X' and Y' as shown and (recalling some eyes only remarks) we sketch a certain rectangle, draw the asymptotes, plot the vertices, and then, with a flourish and a drum roll, sketch the graph of this hyperbola *in the* $X'Y'$ coordinate system. (See figure 2.68.) Making the substitutions suggested by the equations of the translation into

$$\frac{x'^2}{16} - \frac{y'^2}{25} = 1$$

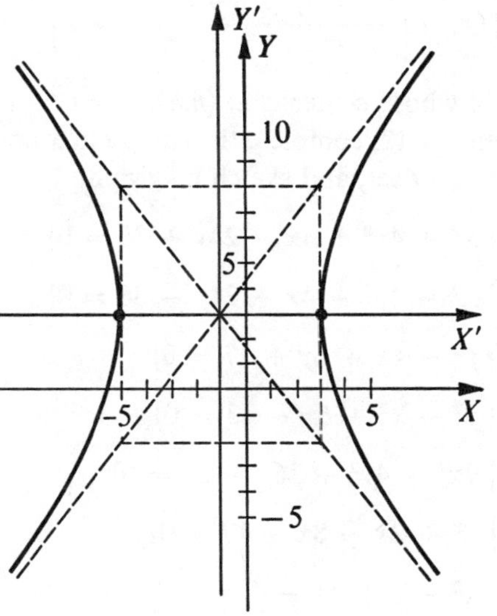

Fig. 2.68

we get the desired equation of the hyperbola

$$\frac{(x+1)^2}{16} - \frac{(y-3)^2}{25} = 1.$$

PROBLEM SET 2.10

In each of problems 1–3, the new coordinate system is obtained from the old by a translation of axes.

1. Suppose $P = (x,y)$ and $Q = (r,s)$ are two points in the old XY coordinate system. Find $d(PQ)$. Now, translate the axes so that the new coordinate system has its origin at P. Find the new coordinates of Q and then show that, in this new coordinate system, $d(PQ)$ is the same as in the old.

2. If the coordinates of the old origin are $(-3,2)$ in the new system, what are the coordinates of the new origin in the old system? Illustrate by a sketch.

3. If the coordinates of the new origin are $(0,3)$ in the old system and the coordinates of P in the old system are $(-1,4)$, what are the coordinates of P in the new system? Illustrate by a sketch.

4. Select a translation of the axes that will show that

$$\mathscr{S} = \left\{(x,y) \,\middle|\, \frac{(x-h)^2}{a^2} - \frac{(y-k)^2}{b^2} = 1\right\}$$

is a hyperbola whose center is at (h,k).

In each of problems 5–12, complete the square in x and/or y, translate the axes, identify the point set, and sketch its graph.

5. $\mathscr{S} = \{(x,y) \mid x^2 + 4y^2 + 4x - 24y + 36 = 0\}$

6. $\mathscr{S} = \{(x,y) \mid x^2 - 4y^2 + 4x + 24y - 36 = 0\}$

7. $\mathscr{S} = \{(x,y) \mid y^2 - 4x + 6y + 17 = 0\}$

8. $\mathscr{S} = \{(x,y) \mid y^2 - x^2 + 6x - 13 = 0\}$

9. $\mathscr{S} = \{(x,y) \mid 9x^2 + 4y^2 + 16y - 20 = 0\}$

10. $\mathscr{S} = \{(x,y) \mid x^2 + 2x - 8x + 17 = 0\}$

11. $\mathscr{S} = \{(x,y) \mid x^2 - y + 3x - 4 = 0\}$

12. $\mathscr{S} = \{(x,y) \mid y^2 - x^2 + 6x + 4y - 14 = 0\}$

In problems 13–15 use a translation of axes to simplify the problem but give the desired equation in the XY coordinate system.

13. Give the equation of a hyperbola whose foci are at (6,3) and (−2,3) and whose transverse axis has length 4.

14. Give the equation of a parabola whose focus is at (−4,0) and whose directrix is the line $x = 6$.

15. Give the equation of an ellipse whose foci are at (−3,8) and (−3,2) and whose major axis has length 12.

SUMMARY

In this chapter we have studied a model of the Euclidean plane $R \times R$ which we have called a *Cartesian coordinate system.* We have seen by way of examples and problems that this is a powerful tool for the study of geometry. Also, we have studied the basic elements of the algebra of two-dimensional vectors. Our vectors are set in boldface type. In physics and engineering you will meet with what are called *vector quantities*, quantities that have both magnitude and direction. Such quantities are represented geometrically by arrows and analytically by ordered pairs of real numbers. When the time comes, you will find that the vector algebra we studied will be applicable to the study of such quantitities.

The analytic geometry that we have learned here serves another purpose. In the next chapter we begin a study of some important concepts in analysis, and it will surely prove helpful to be able to interpret them geometrically.

We have also in this chapter spent a considerable amount of time in the study of certain point sets—the line and the conic sections. For we have derived their equations and sketched their graphs until we are "blue in the face," and we have studied some properties (slope, intercepts, and symmetry) of their graphs which could be determined from their equations.

For convenience, we tabulate below some of the results which we have noted in our study of the conic sections.

Circle

$(x - h)^2 + (y - k)^2 = r^2$
Center: (h,k) Radius: r, $r > 0$

Ellipse

Major axis:

parallel to X axis	parallel to Y axis
$\frac{(x-h)^2}{a^2} + \frac{(y-k)^2}{b^2} = 1$	$\frac{(y-k)^2}{a^2} + \frac{(x-h)^2}{b^2} = 1$

$$b^2 = a^2 - c^2, a > b, a > c.$$

Foci:

F_1	$(h - c, k)$	$(h, j - c)$
F_2	$(h + c, k)$	$(h, k + c)$

Length of major axis = $2a$
Length of minor axis = $2b$
Center: (h,k)

Hyperbola

Transverse axis:

parallel to X axis	parallel to Y axis
$\frac{(x-h)^2}{a^2} - \frac{(y-k)^2}{b^2} = 1$	$\frac{(y-k)^2}{a^2} - \frac{(x-h)^2}{b^2}$

$$b^2 = c^2 - a^2, c > a.$$

Foci:

F_1	$(h - c, k)$	$(h, k - c)$
F_2	$(h + c, k)$	$(h, k + c)$

Length of transverse axis = $2a$
Length of conjugate axis = $2b$
Center: (h,k)

Parabola

Directrix:

parallel to X axis	parallel to Y axis
$(x - h)^2 = 4c(y - k)$	$(y - k)^2 = 4c(x - h)$
opens up	opens right
$(x - h)^2 = -4c(y - k)$	$(y - k)^2 = -4c(x - h)$
opens down	opens left

$$c > 0$$

Focus:

$(h, k \pm c)$ $(h \pm c, k)$

Vertex: (h,k)

Review Problems

1. **a.** A vertical line has no slope but a horizontal line has slope ______.
   b. If a line contains the points (a,b) and (m,n), $m \neq a$, then the slope of this line is ______.
2. If a line contains the points $(-2,3)$ and $(1,-2)$, give its slope-intercept equation.
3. Suppose that $\mathscr{L} = \{(x,y) \mid x - 2y + 3 = 0\}$, give an equation for a line that contains the point $(2,2)$ and is

   a. parallel to $\mathscr{L}$. **b.** perpendicular to $\mathscr{L}$.
4. Suppose that $\mathscr{S}$ is the set of all points that are twice as far from $(2,-1)$ as from $(6,6)$. Write an equation whose graph is $\mathscr{S}$, hence showing that $\mathscr{S}$ is a circle. Find its center and radius.
5. Find equation of a parabola with vertex at $(1,1)$ and focus at $(2,1)$.
6. Find equation of ellipse with center $(6,1)$, a vertex at $(1,1)$, and a focus at $(2,1)$.
7. Find equation of hyperbola with foci at $(2,0)$ and $(2,-6)$ and a vertex at $(2,-1)$.
8. Identify the point set described by each of the given equations. If it is a line, give slope and y intercept; for an ellipse or a hyperbola, give center, and direction of major or transverse axis; if the point set is a parabola, give vertex and direction in which it opens.

   a. $\dfrac{(x+3)^2}{4} - \dfrac{(y-1)^2}{4} = 1$ **b.** $\dfrac{(x-2)^2}{25} + \dfrac{(y+1)^2}{9} = 1$

   c. $x^2 + 2y + 3 = 0$ **d.** $x + 2y + 3 = 0$

   e. $3x^2 - 5y^2 + 12x + 10y + 5 = 0$ **f.** $3x - 2y - 6 = 0$

   g. $2x^2 + y^2 + 8x + 4y = 0$

CHAPTER 3
Functions

The idea of a function is one of the basic concepts in mathematics and the basic idea of a function is that of a correspondence from a set A of objects to a set B of objects. Now a correspondence from A to B is equivalent to a set S of ordered pairs (a,b) of the Cartesian product $A \times B$: if the element b in B corresponds to the element a in A, then the ordered pair (a,b) will be in S, and if (a,b) is in S, then b corresponds to a. If we limit this correspondence by the requirement that one and only one element of B should correspond to each element of A, then the set S would be called a function. Thus the meaning of function is essentially this: Given two sets A and B, a *function* is a correspondence which associates with each element of A one and only one element of B. The set A is called the domain of the function. Those elements of B associated with the elements in A are a set called the range of the function (this may be all of B, but it need not be).

Although the function idea places no restrictions on the nature of the objects in the sets A and B, we shall confine ourselves, at least until further notice, to functions whose domain and range are sets of real numbers. Such functions are called *real-valued functions of one real variable*. Sometimes, this title is shortened to just *real function*. Until further notice, we may be even more informal (but not disrespectful) and simply call such a function simply a *function*.

Section 1—Real Functions From the set $R \times R$ of all ordered pairs of real numbers we select a special kind of subset called a function and formally defined by

DEFINITION. *The statement that "F is a real function" means that F is a set of ordered pairs of real numbers such that no two pairs in F have the same first number.*

There is no apparent harm in our using some of the terminology of the previous chapter and describing a function F as a point set in which no two points have the same abscissa. The student should note, however, that not all point sets are functions. For example, the vertical line and the circle are not functions. Why not? For additional clarification, suppose that U, V, and W are the point sets (sets of ordered pairs of real numbers) given below:

$$U = \{(0,2), (0,3), (2,-2)\}$$
$$V = \{(0,-1), (2,-1), (3,2)\}$$
$$W = \{(0,0), (1,3), (2,-1)\}.$$

U is not a function but V and W are functions. Special attention is invited to V for the definition says nothing about the second number (ordinate) of the pairs (points).

If F is a function, then we associate with F two number sets called the *domain of F* and the *range of F* which are defined as follows:

DEFINITION. *The* domain of *F is the number set to which x belongs* iff *x is a number and x is the abscissa of a point in F.*

DEFINITION. *The* range of *F is the number set to which y belongs* iff *y is a number and y is the ordinate of a point in F.*

Remark. If F is a function, then $F(x)$ (read as "F at x") denotes the ordinate of the point in F whose abscissa is x. So, if (x,y) is in F, then y is $F(x)$.

Example. Suppose $G = \{(-1,3), (0,1), (2,5), (3,8)\}$ then G is a function. The domain of G is the number set $\{-1, 0, 2, 3)$. The range of G is the number set $\{1, 3, 5, 8\}$. Further, $G(-1)$ is 3, $G(0)$ is 1, $G(2)$ is 5, and $G(3)$ is 8.

The preceding definitions and other statements exhaust what we shall say at this time about real functions in general. The rest of our efforts should now be directed toward particular functions and to specify (define) any particular function there are two things that must be given:

1. the domain of the function, and
2. a rule of correspondence, that is, a statement or rule by which, for each number in the domain, we can find the corresponding second element of the pair.

Usually this rule of correspondence will be an equation into which we substitute any number in the domain and compute or evaluate the corresponding ordinate.

Example. Suppose f is the function whose rule of correspondence is the equation $f(x) = x^2 - |x|$. By substituting $x = 3$ into this rule, we find that $f(3) = 3^2 - |3| = 6$. For $x = -5$ we see that $f(-5) = (-5)^2 - |-5| = 20$. In rapid succession then, we can see that $f(0) = 0$, $f(1) = 0$, and that $f(4) = 12$.

It is important to note that these computations mean that each of the following ordered pairs belong to f

$$(3,6), (-5,20), (0,0), (1,0), \text{ and } (4,12).$$

It is a common practice to describe only partly a function by giving just its rule of correspondence. To illustrate this malpractice, consider the following.

Example. Suppose that F, G, and Q are the functions for which $F(x) = \sqrt{x}$, $G(x) = x^2 - x$, and $Q(x) = \dfrac{x}{x-2}$. Each of these rules tell how to determine the second number of the pair in the function whose abscissa is x. Thus $F(4) = 2$, $G(3) = 6$, while $Q(1) = -1$. Such a rule is sometimes said to define the function but it is not the function, for the function itself is the set of ordered number pairs. But we have just pointed out the two things that must be given in order to define a function and we have given only the rule of correspondence. What are the domains of these functions? To answer the question:

1. The domain of F is the set of nonnegative numbers.
2. The domain of G is the set of all numbers.
3. The domain of Q is the set of all numbers distinct from 2.

The suggested notation for the above is given below.

1. dom $F = \{x \mid x \geq 0\}$.
2. dom $G = R$.
3. dom $Q = \{x \mid x \neq 2\}$.

The preceding example illustrates the assumption that we shall make and employ to combat this practice of defining a function by giving only its rule of correspondence. This assumption is spelled out in the following "gentlemen's agreement."

AGREEMENT ON DOMAIN. *If F is a function, then the domain of F (if not otherwise specified) will be assumed to be the number set to which x belongs only if $F(x)$ is a number.*

The following examples might help you to understand what we are agreeing to.

Example. $F(x) = x^3 + 2x$.

Now x^3 is certainly a number for every number x. So is $2x$, thus $F(x)$ is a number for every $x \in R$. Therefore we agree that dom $F = R$.

Example. $g(x) = \sqrt{(4 - x^2)}$.

Since $\sqrt{(4 - x^2)}$ is a number only when $4 - x^2 \geq 0$, we agree that the domain of g will contain each number x for which $(2 - x)(2 + x) \geq 0$. Solving this inequality we find that

$$\text{dom } g = [-2,2].$$

Example. $U(x) = \dfrac{x + 2}{x}$.

Since $\dfrac{x + 2}{x}$ is a number for every x except $x = 0$, we agree that dom $U = \{x \mid x \neq 0\}$.

Question. Is $U(-2)$ a number?

Answer. Most assuredly, for $U(-2) = 0$.

Example. Suppose that P is the function for which

$$P(x) = x^2 + 3x, \text{ dom } P = x \mid 0 \leq x < 3.$$

It is irrelevant that the sum $x^2 + 3x$ can be evaluated for $x = 4$ for the domain of P *is given and is not left to our agreement.* Note that dom $P = [0,3\rangle$.

Though the main aim of the examples given in this section is to clarify an agreement on the domain of a function, maybe this is just as good a time as any to make the following remarks.

Remarks on the range of a function. Given $f(x)$, if we can see (because of either good vision or previous knowledge) what the range of the function is, there is no good reason why we should not do so. For example, for $F(x) = \sqrt{x}$ and $A(x) = |x|$, we should see that

$$\text{rng } F = \text{rng } A = \{y \mid y \geq 0\}.$$

If our vision of the range of a function is blurred, the following routine might improve it. Given $f(x)$, set $y = f(x)$ and try to solve for x in terms of y. For example, take the function U where $U(x) = \frac{x+2}{x}$. Set $y = \frac{x+2}{x}$. It then follows that $xy = x + 2$, hence $xy - x = 2$. So $x(y - 1) = 2$ and we solve for x and get $x = \frac{2}{y-1}$, which makes it clear that $y \neq 1$. Therefore rng $U = \{y \mid y \neq 1\}$. For the example in which $g(x) = \sqrt{(4 - x^2)}$, we know enough about square roots to realize that $g(x) = y \geq 0$. Thus setting $y = \sqrt{(4 - x^2)}$ and squaring we get $y^2 = 4 - x^2$. So $x^2 = 4 - y^2$ or $x = \pm\sqrt{(4 - y^2)}$. Hence $y^2 \leq 4$. Remembering that $0 \leq y$, we can now see that

$$\text{rng}\, g = \{y \mid 0 \leq y \leq 2\}, \text{ that is, } [0,2].$$

End of remarks on the range of a function.

Example. The primary purpose of this example is to dismiss the notion that the rule of correspondence must always be a single equation.

$$\text{Let } Z(x) = \begin{cases} 2 & \text{if } 0 \leq x < 2. \\ x - 1 & \text{if } 2 \leq x \leq 4. \end{cases}$$

Note that the domain is given as the interval [0,4] so don't worry about it. Note also that $Z(0) = Z(1) = 2$ while $Z(2) = 1$, $Z(3) = 2$, and $Z(4) = 3$. These computations assure us that the points shown in figure 3.1 certainly belong to the graph of the point set Z.

Question. What is the rng Z?

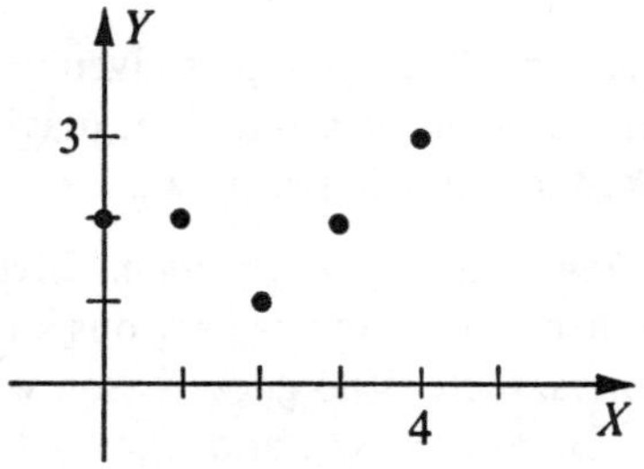

Fig. 3.1

PROBLEM SET 3.1

1. Which of the following point sets are *not* functions?

$F = \{(-1/2,0), (0,1), (\sqrt{2},2), (10,\pi)\}$
$f = \{(1,2), (3,0), (4,-2), (1,-3)\}$
$G = \{(0,1), (1,1), (\pi,\pi)\}$
$g = \{(2,2)\}$
$Z = \{(x,y) \mid x + y = 5\}$
$U = \{(x,y) \mid y = 2\}$
$V = \{(x,y) \mid 2x - y < 0\}$
$C = \{(x,y) \mid x^2 + y^2 = 1\}$

In each of problems 2–11, the rule of correspondence only is given. You are to give the domain of the function defined by this rule and thus practice using the agreement we have made.

2. $f(x) = x^3$ **3.** $g(x) = x^2 - 2x$

4. $F(x) = \sqrt{(x - 1)}$ **5.** $G(x) = \sqrt{(2 - x)}$

6. $Q(x) = \dfrac{x - 1}{x}$ **7.** $U(x) = \dfrac{x}{x + 2}$

8. $Z(x) = \dfrac{x}{1 - x^2}$ **9.** $W(x) = \dfrac{x + 1}{x^2 - 2x}$

10. $V(x) = \sqrt{(x^2 - 1)}$ **11.** $F(x) = \sqrt{(9 - x^2)}$

12. If $g(x) = 2x + 3$, give $g(-1)$, $g(3)$, $g(2) - g(0)$, $g(t - 3)$, and $g(b) - g(a)$.

13. If $g(x) = x + 1$, give $g(2)$, $g(-2)$, $g(1/2)$, $\dfrac{1}{g(2)}$, $g(a + b)$, and $g(a) + g(b)$.

14. If $f(z) = z^2 - 2z$, give $f(2)$, $f(1/2)$, $f(0)$, $f(1)\cdot f(3)$, and $f(t + 1) - f(1)$.

15. If $f(x) = 2x - x^2$, give $f(2)$, $f(1/2)$, $f(3/2)$, and $\dfrac{f(t) - f(2)}{t - 2}$.

16. If $F(s) = \dfrac{s}{s - 1}$, give $F(0)$, $F(1/2)$, $F(3) + F(2)$, and $F(1/a)$.

17. If $h(t) = 1/t$, give $h(1)$, $h(2/3)$, $a \cdot h(a)$, and $h(x) - h(a)$.

18. If $P(m) = m^3$, give $P(-1)$, $P(\sqrt{2})$, $P(2/3)$, $P(t^2)$, and $P(2c)$.

19. If $F(x) = \sqrt{(x + 2)}$, give $F(-2,)$, $F(-1/2)$, $F(1/4)$, and $F(6)$.

20. If $f(x) = x - |x|$, give $f(-2), f(-1/2), f(\pi)$, and $f(5)$.

21. Let $G(x) = \begin{cases} 1 - x & \text{if } 0 \le x < 2 \\ 2x - 5 & \text{if } 2 \le x \le 4 \end{cases}$

Give $G(1)$, $G(1.9)$, $G(2)$, $G(2.1)$, and $G(4)$.

22. Let $T(x) = \begin{cases} |x - 1| & \text{if } -2 \le x < 1 \\ \sqrt{(x - 1)} & \text{if } 1 \le x \le 2 \end{cases}$

Give $T(-2)$, $T(0.9)$, $T(1)$, $T(1.01)$, and $T(2)$.

23. Suppose $F(x) = x^2$ for all real x. Verify each of the following statements. In each case, if the statement is not true for all real numbers, give the set of real t for which the statement is true.

a. $F(t) = F(-t)$ **b.** $F(2t) = 4 \cdot F(t)$
c. $F(t^2) = [F(t)]^2$ **d.** $\sqrt{[F(t)]} = t$
e. $\sqrt{[F(t)]} = |t|$

[24] While we have said several things about the domain of a function, very little has been said about its range. One way of finding a subset of the range (maybe even the range) can be described as follows: replace $F(x)$ by y and (if you can) solve for x. Use this idea to find the ranges of the functions given in

a. Prob. 3 **b.** Prob. 5 **c.** Prob. 6 **d.** Prob. 7.

Section 2—Graphs and Special Functions Many of the properties of functions have a geometric interpretation, so one important task in the study of functions is the problem of sketching its graph. It may even be true that the graph of a function may enable us to understand better the additional properties that we shall later encounter in the study of calculus.

We have assumed that an ordered number pair corresponds to a point and, since a function is a particular type of subset of $R \times R$, the graph of a function will consist of all of the points in that function. Thus the graph of a function may be some line or curve drawn on our writing surface. Further, no vertical line will intersect the graph of a function in more than one point.

Suppose that figure 3.2 is the graph of some function F whose domain is $[a,b]$. Let P be that point in F whose abscissa is a, then $P = (a,F(a))$.

Question. $Q = ?$

Now let us suppose that G is a function whose domain is the set of all real numbers. As mortals, we can manage to plot only a finite number of points. Since this is the best we can do, then all we can do is to plot the number of points sufficient to indicate, with a reasonable degree of accuracy, the nature of the graph.

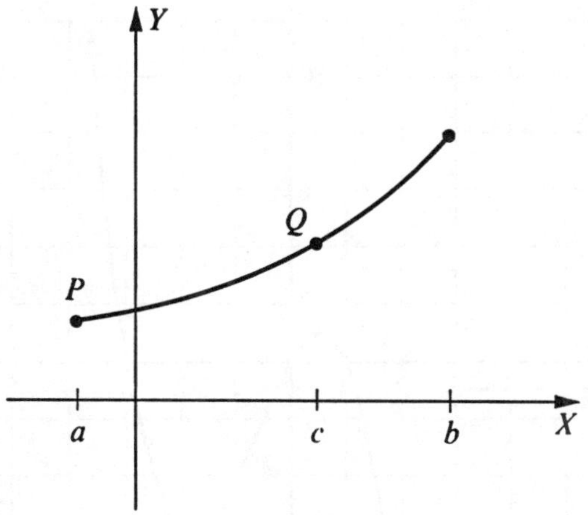

Fig. 3.2

In finding and plotting the points from which we guess at the nature of the graph of some function, we should include those whose ordinate is zero, if there are any. These points, of course, are the points where the graph intersects the X axis and even if there are none, this itself is certainly revealing insofar as the graph is concerned. Finding these points, when they exist, is thought to be so important that a particular bit of terminology is used.

THE ZERO(S) OF A FUNCTION. *If F is a function and x is a number in the domain of F such that $F(x) = 0$, then we will call x* a zero of F.

Example. $F(x) = x^3 - 2x^2 - 3x$, dom $F = R$.

Factoring, we have $F(x) = x(x - 3)(x + 1)$ so the zeros of F are 0, 3, and -1. Further, $F(-2) = -10$, $F(0) = 0$, $F(1) = -4$, and $F(2) = 6$. The graph of this function is shown in figure 3.3 and from this figure one can conjecture that the range of F is the set of all real numbers.

Example. $Q(x) = \dfrac{x}{x - 2}$, dom $Q = \{x \mid x \neq 2\}$.

It is easy to see that $Q(x)$ is 0 iff $x = 0$, thus Q has only one zero. For other points in Q, $Q(-2) = 1/2$, $Q(-1) = 1/3$, $Q(1) = -1$, $Q(3/2) = -3$. The vertical line $x = 2$ is drawn (dashed) only to indicate that there is no point in Q with abscissa 2. (See figure 3.4.)

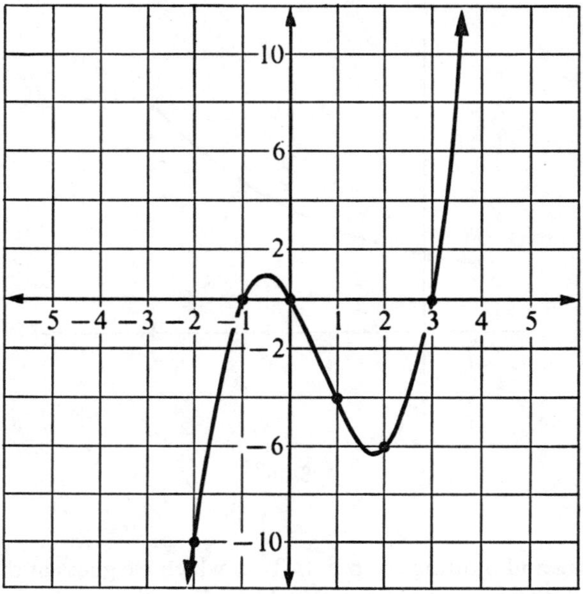

Fig. 3.3

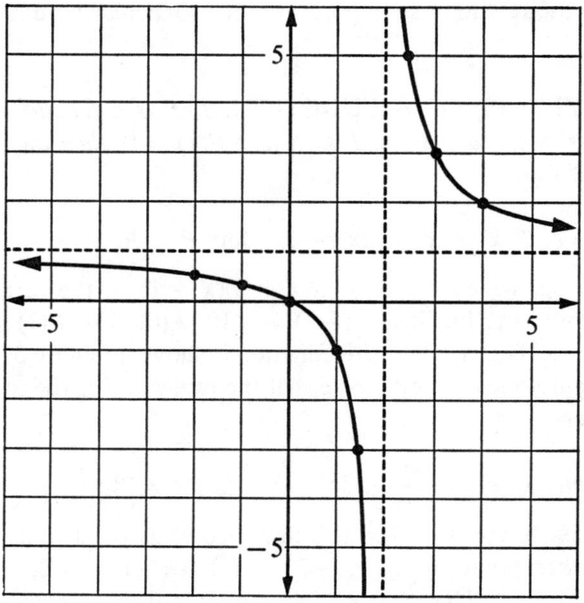

Fig. 3.4

To continue our computations, $Q(3) = 3$, $Q(4) = 2$, and $Q(6) = 3/2$. If we let $y = Q(x)$ and solve for x we get $xy - 2y = x$, so that $xy - x = 2y$ and so $x = \frac{2y}{y-1}$. Hence rng $Q = \{y \mid y \neq 1\}$. The horizontal line $y = 1$, serves only to indicate this fact. That is, there is no point in Q with ordinate 1.

The next example presents the idea that a function need not be described by only one rule of correspondence.

Example. (See figure 3.5.) Suppose that J is the funct ion for which

$$J(x) = \begin{cases} 2 & \text{if } x < 1. \\ x - 1 & \text{if } x \nless 1. \end{cases}$$

Note that the domain of J is given as the set of all real numbers and is not, therefore, left to our agreement. Note also the manner by which we indicate that $(1,2) \notin J$ for $J(1) = 0$. Study also the reason why rng $J = \{y \mid y \geq 0\}$.

Many of the functions presented thus far in the examples or problems were given primarily to help develop an understanding of the concept of a function and provide practice in the use of the functional notation. Most

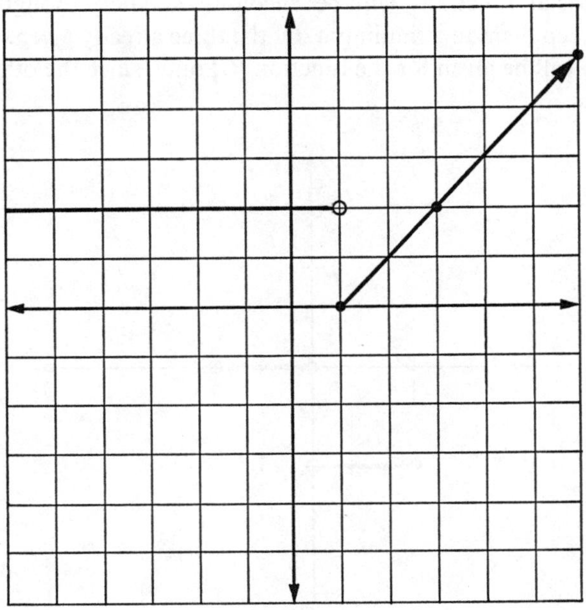

Fig. 3.5

of these functions in themselves are not particularly significant otherwise. At this time, however, attention is to be called to a few special functions in which we shall have more than just a passing interest. Some of these are so useful that special symbols will be introduced for them, and these symbols will not be used in this text for any other function. We now introduce some of these special functions.

DEFINITION. *I is the function, called the* identity function, *whose domain is the set of all numbers and for which*

$$I(x) = x.$$

DEFINITION. *$[G]$ is the function, called the* greatest integer function, *whose domain is the set of all numbers and for which, if k is an integer and $k \leq x < k + 1$, then*

$$[G](x) = k.$$

DEFINITION. *H is the function, call it what you will, whose domain is the set of positive numbers and for which*

$$H(x) = 1/x.$$

Other special functions will be encountered later, though we shall attempt to keep them to a minimum. Of the three already given, additional information will be given for the function $[G]$ only, since the other two are very simple.

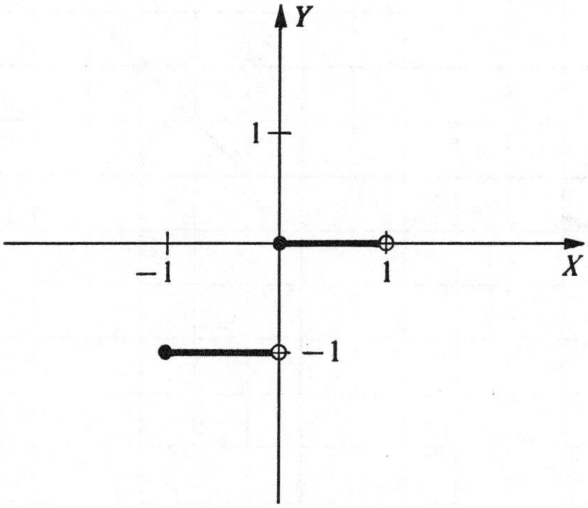

Fig. 3.6

Let us begin by noting that $-1 \leq -1/2 < 0$, hence $[G](-1/2) = -1$. As a matter of fact, for every x between -1 and 0, $[G](x) = -1$. We can also see that $[G](0) = 0$. Furthermore, if $0 < x < 1$, $[G](x) = 0$, hence the function $[G]$ has infinitely many zeros. Armed only with the computations mentioned here, if we were to start to draw the graph of this special function, our efforts would result in something like figure 3.6.

PROBLEM SET 3.2

1. Sketch the graphs of the special functions I and H.

2. a. Give $[G](-3)$, $[G](-1.5)$, $[G](2.5)$, $[G](2.9)$, and $[G](3)$.
b. Sketch the graph of this special function.

In the next two problems, we consider a simple type of function known as a linear function. If $F(x) = mx + b$ for all real x, then F is called a linear function and the number m is called *the slope of F*.

3. a. Sketch the graph of $F(x) = -1/2 \cdot x + 3$.
b. Choose any two pairs in F, (x_1, y_1) and (x_2, y_2), and show that

$$\frac{y_2 - y_1}{x_2 - x_1} = m.$$

4. Same as problem 3 for $F(x) = 2x - 1$.

In each of problems 5–16, give the domain *and* range of the given function *and* sketch its graph.

5. $f(x) = |x - 1|$ **6.** $f(x) = |x + 2|$ **7.** $F(x) = x + |x|$

8. $F(x) = x - |x|$ **9.** $p(x) = x^2 - x$ **10.** $P(x) = x^2 + 2x$

11. $Z(x) = \sqrt{x}$ **12.** $z(x) = \sqrt{(x + 1)}$ **13.** $G(x) = \sqrt{(2 - x)}$

14. $g(x) = 4 - x^2$ **15.** $Q(x) = \dfrac{x}{x - 1}$ **16.** $Q(x) = \dfrac{x}{x + 1}$

In each of problems 17–21, sketch the graphs of the given functions.

17. $F(x) = \begin{cases} 3 & \text{if } x < 0 \\ x - 2 & \text{if } x \nless 0 \end{cases}$ **18.** $f(x) = \begin{cases} -1 & \text{if } -3 \leq x < 1 \\ 1 - 2x & \text{if } 1 \leq x \leq 3 \end{cases}$

19. $U(x) = \begin{cases} 1/2 \cdot x^2 & \text{if } -2 \leq x \leq 0 \\ 2 - x & \text{if } 0 < x \leq 4 \end{cases}$ **20.** $f(x) = \begin{cases} -x & \text{if } x < 0 \\ 2 & \text{if } x = 0 \\ x + 3 & \text{if } x > 0 \end{cases}$

21. $W(x) = \dfrac{|x|}{x}$ if $x \neq 0$ and $W(0) = 0$.

In problems 22–25, give the zeros of the function *and* sketch its graph.

22. $F(x) = x^3 + x^2 - 2x$ **23.** $F(x) = x^3 - 2x^2$

24. $G(x) = x^3 - 3x^2 - 4x$ **25.** $G(x) = x^3 - 4x^2 + 4x$

Section 3—Algebra of Functions In this section we consider a task that could be described as follows: given a pair of functions, construct some new functions.

We enter this construction business with the following blueprints and specifications:

DEFINITION. *If each of F and G is a function, then each of* $[F + G]$, $[F \cdot G]$, and $[F/G]$ is a function and

1. $[F + G](x) = F(x) + G(x)$; dom $[F + G]$ = dom $F \cap$ dom G
2. $[F \cdot G](x) = F(x) \cdot G(x)$; dom $[F \cdot G]$ = dom $F \cap$ dom G
3. $[F/G](x) = \dfrac{F(x)}{G(x)}$; dom $[F/G]$ = dom $F \cap$ dom G and $G(x) \neq 0$.

Note that the above specify the domain of the functions $[F + G]$, $[F \cdot G]$, and $[F/G]$, hence they are not left to the whims of our agreement on domain.

The next two examples will illustrate the nature of these constructions and some of the associated problems.

Example. Suppose that F and G are the functions given below. Note that each is a collection of ordered pairs, no two of which have the same first element.

$$F = \{(-2,-3), (-1,-2), (0,2), (1,3), (3,5)\}$$
$$G = \{(-2,-2), (0,-1), (1,0), (2,3)\}$$

In constructing the function $[F + G]$, please note that

$[F + G](x) = F(x) + G(x)$ for then it would be required that $[F + G](2) = F(2) + G(2)$. Now $G(2)$ is defined ($G(2) = 3$) but $F(2)$ is not. Thus $[F + G](2)$ is not defined. For similar reasons $[F + G](3)$ does not exist. So it is crucial that we first determine the domain of $[F + G]$. A look at the definition reveals that

$$\text{dom } [F + G] = \text{dom } F \cap \text{dom } G = \{-2, 0, 1\}.$$

Since this is also the domain of $[F \cdot G]$, the construction of these two functions can proceed on schedule.

$$[F + G] = \{(-2,-5), (0,-3), (1,3)\}$$
$$[F \cdot G] = \{(-2,6), (0,2), (1,0)\}$$

A small delay is encountered in the construction of $[F/G]$, for there is some difficulty in trying to find $[F/G](1)$.

Question. What is the trouble?

Answer. $G(1) = 0$.

So we must delete 1 from $\{-2,0,1\}$ and thus decide that

$$\text{dom } [F/G] = \{-2,0\}.$$

Now everything is in order and we finish with our construction with

$$[F/G] = \{(-2,3/2), (0,2)\}.$$

The next example is more typical of the constructions you will be asked to do than the previous example.

Example. Suppose that $F(x) = x^2 - x$ and that $G(x) = x - 1$. Our agreement on domains tells us that dom F = dom G = R, hence

$$[F + G](x) = (x^2 - x) + (x - 1) = x^2 - 1; \text{ dom } [F + G] = R$$
$$[F \cdot G](x) = (x^2 - x)(x - 1) = x^3 - 2x^2 + x; \text{ dom } [F \cdot G] = R.$$

(A mall difficulty must be overcome before we can construct $[F/G]$ for $G(x) = 0$ when $x = 1$.)

$$[F/G](x) = \frac{x^2 - x}{x - 1} = x; \text{ dom } [F/G] = \{x \mid x \neq 1\}.$$

Since the graph of at least one of these newly constructed functions may be interesting, figure 3.7 shows the graphs of the functions $[F + G]$ and $[F/G]$. Note the "hole" in the graph of $[F/G]$. Can you account for this "hole"?

Example. If $F(x) = \sqrt{(x + 2)}$ and $G(x) = \sqrt{(3 - x)}$, then dom F $= \{x \mid x \geq -2\}$ and dom $G = \{x \mid x \leq 3\}$ so that dom $F \cap$ dom G $= [-2,3]$. It follows that

$$[F + G](x) = \sqrt{(x + 2)} + \sqrt{(3 - x)}, \text{ dom } [F + G] = [-2,3]$$
$$[F \cdot G](x) = \sqrt{[(x + 2)(3 - x)]}, \text{ dom } [F \cdot G] = [-2,3]$$

$$[F/G](x) = \frac{\sqrt{(x + 2)}}{\sqrt{(3 - x)}}$$

and, since $G(x) = 0$ iff $x = 3$, we find that dom $[F/G] = [-2,3\rangle$.

Assuming that these examples are enough to illustrate how we construct the sum, product, and quotient functions, we now consider a somewhat more subtle technique for constructing a function from a given pair of functions. This technique is known as *composition* and will require a

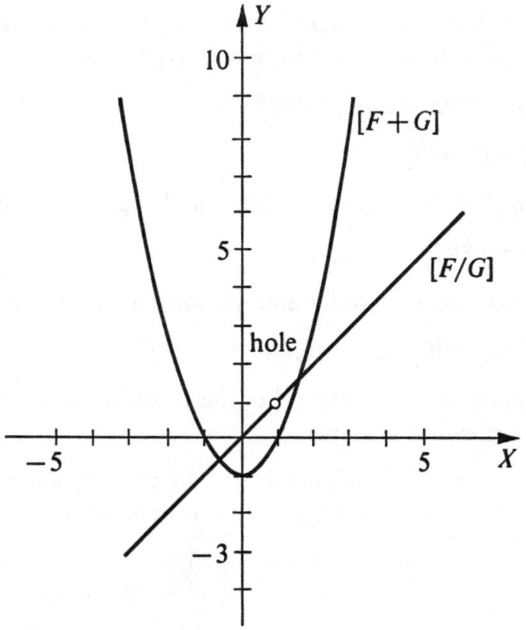

Fig. 3.7

very careful study of its definition at a later time (the next section as a matter of fact).

Prior to this, however, let us try to see what the construction known as composition might entail, at least in a general way.

First, a notation that we shall use: given a pair of functions F and G, the "composition or composite of F with G" is to be denoted by $[F \circ G]$. Read this as "F circle G" or, as some wag has suggested, as "fog." "F circle G" is the preferred pronunciation.

Secondly, the general nature of composition might be clarified by a diagram and the following remarks: suppose that A and B are number sets (see figure 3.8a) and that the function G has its domain in A and its range

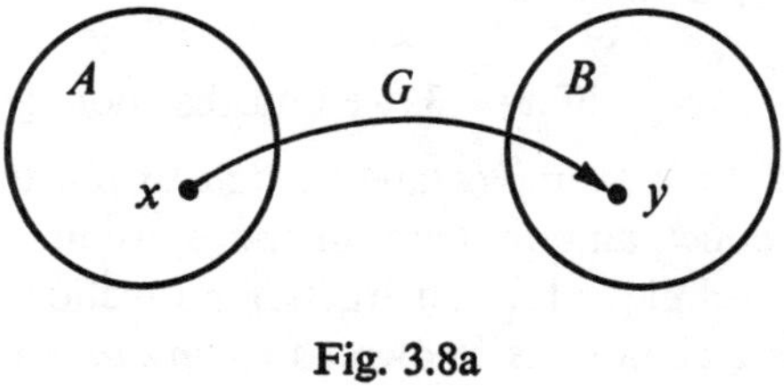

Fig. 3.8a

in B. Thus we might envisage the function G as a mapping of some number $x \in A$ into some number $y \in B$. Obviously $x \in \text{dom } G$ and $y = G(x)$. If we further suppose that C is another number set and that the function F has its domain in B and its range in C, then if $s \in \text{dom } F$, a subset of B, we can think of F as mapping the number s into some number $t \in C$ (see figure 3.8b). So, if the function F also maps y into $z \in C$, that is, $F(y) = z$, then the composite function "F circle G" is that function (see figure 3.8c) which will map $x \in A$ directly into z.

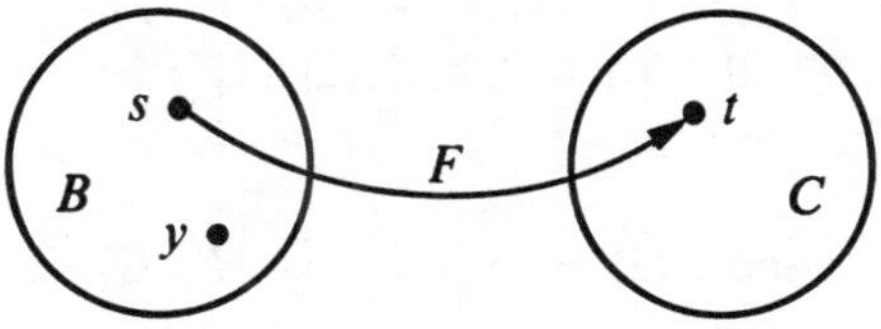

Fig. 3.8b

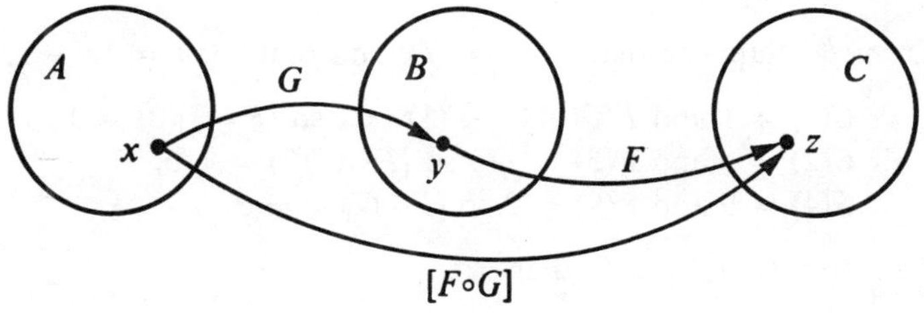

Fig. 3.8c

Remark. Go back over these figures and note that $G(x) = y$ and that $F(y) = z$, for then we conjecture that

$$[F \circ G](x) = F(G(x)).$$

Now that the general nature of compo——. *Wait just a bloody minute!!* Suppose that $y \notin \text{dom } F$, then F won't map it anywhere so that $F(y)$ does not exist, so $F[G(x)]$ does not exist, and we have big trouble. Let us get rid of this messy affair as gracefully as we can by simply admitting that $x \notin \text{dom } [F \circ G]$ and hope that this doesn't happen too often.

In the following example, we are going to use the same functions F and G that were used in the very first example of this section. To save you time, we will begin the example by repeating these functions.

Example.

$$F = \{(-2,-3), (-1,-2), (0,2), (1,3), (3,5)\}$$
$$G = \{(-2,-2), (0,-1), (1,0), (2,3)\}$$

Problem. To construct $[F \circ G]$.

Solution.

$G(-2) = -2$ and $F[G(-2)] = F(-2) = -3$, so $(-2,-3) \in [F \circ G]$

$G(0) = -1$ and $F(-1) = -2$, so $(0,-2) \in [F \circ G]$

$G(1) = 0$ and $F(0) = 2$, so $(1,2) \in [F \circ G]$

and, finally,

$G(2) = 3$ and $F(3) = 5$, so $(2,5) \in [F \circ G]$.

Therefore $[F \circ G] = \{(-2,-3), (0,-2), (1,2), (2,5)\}$.

Question. Since $F(-1)$ exists, why is there no point in $[F \circ G]$ whose abscissa is -1?

Example. Suppose that $F(x) = \sqrt{x}$ and that $G(x) = 2x + 1$, then

$G(0) = 1$ and $F[G(0)] = F(1) = 1$ so $[F \circ G](0) = 1$.

$G(2) = 5$ and $F(5) = \sqrt{5}$ so $[F \circ G](2) = \sqrt{5}$.

$G(4) = 9$ and $F(9) = 3$ so $[F \circ G](4) = 3$.

If $t \in \text{dom } G$ and $t > 0$, then

$G(t) = 2t + 1$ and $F[G(t)] = F(2t + 1) = \sqrt{(2t + 1)}$ so that $[F \circ G](t) = \sqrt{(2t + 1)}$.

$G(-2) = -3$ but $F(-3)$ does not exist, hence there is no point in the graph of "F circle G" whose abscissa is -2.

PROBLEM SET 3.3

1. $F = \{(-2,3), (-1,1), (0,4), (1,2), (2,5)\}$
   $G = \{(-1,2), (0,-1), (1,0), (2,-2)\}$
   Give $[F + G]$, $[FG]$, and $[F/G]$.

In each of problems 2–5, give and simplify (a) $[F + G](x)$, (b) $[F \cdot G](x)$, (c) $[F/G](x)$, and (d) the domains of these new functions.

2. $F(x) = x - 1/x$
   $G(x) = \dfrac{1}{x + 1}$

3. $F(x) = 2x - 1$
   $G(x) = 2 - x$

4. $F(x) = x + |x|$
$G(x) = x - |x|$

5. $F(x) = x^2 + 3x - 10$
$G(x) = x - 2$

6. Give $[H \cdot I](x)$ and the domain and range of $[H \cdot I]$.

7. Give $[H + I](x)$ and the domain and range of $[H + I]$.

8. Give $[I/H](x)$ and the domain and range of $[I/H]$.

9. If $F(x) = \dfrac{x^2 - x}{x + 2}$ and $G(x) = \dfrac{1}{x + 2}$, give and simplify

a. $[F + G](x)$ **b.** $[F/G](x)$.

10. If $F(x) = \dfrac{2}{x + 1}$ and $G(x) = \dfrac{x}{x - 1}$, give and simplify

a. $[F + G](x)$ **b.** $[F \cdot G](x)$.

11. If $f(x) = 1 - \sqrt{x}$ and $g(x) = 1 + \sqrt{x}$, give and simplify

a. $[f \cdot g](x)$ **b.** $[f + g](x)$.

12. If $f(x) = 1 - |x|$ and $g(x) = 1 + |x|$, give and simplify

a. $[f \cdot g](x)$ **b.** $[f + g](x)$.

13. Sketch the graph of the function U whose domain is R and for which $U(x) = x \cdot |x|$.

14. Sketch the graph of the function W whose domain is R and for which $W(x) = x \cdot |x - 1|$.

15. a. What does the statement "G is a function" mean?
b. If G is a function, what does $G(a)$ denote?

16. If F and G are the same functions given in problem 1, give $[F \circ G]$.

In each of problems 17–19, evaluate $[F \circ G](x)$ for each of the following values of $x \mid x, -1, 0, 1/2, 1, 2$. If, for any of these values of x, $[F \circ G](x)$ does not exist, say so.

17. $F(x) = \sqrt{x}$
$G(x) = x - 1$

18. $F(x) = x^2 + x$
$G(x) = 2 - x$

19. $F(x) = \dfrac{x}{1 - x}$
$G(x) = \dfrac{x}{x - 1}$

20. Using the functions F and G given in problem 18 above and assuming that δ is a number so that $[F \circ G](\delta)$ exists, give $[F \circ G](\delta)$.

21. Using the functions F and G given in problem 19 above, and assuming that δ is a number so that $[F \circ G](\delta)$ exists, give $[F \circ G](\delta)$.

22. If $F(x) = |x|$ and $G(x) = 2x$, give $[F \circ G](\delta)$ assuming that it exists.

23. *Bonus problem.* In figure 3.9 we have sketched the graphs of the functions F, G, and I. Give the coordinates of all of the indicated points A, B, C, and D. *Hint:* $A = [c,G(c)]$.

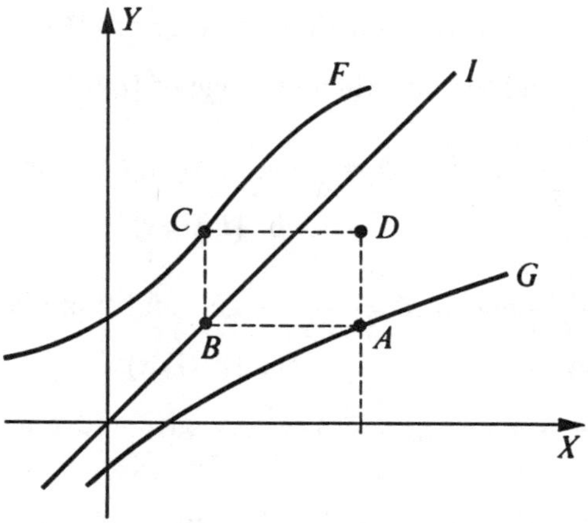

Fig. 3.9

Section 4—Composite Function Our task in this section is to define formally the composite function $[F \circ G]$. The hope is entertained that our work with it in the last section will make our task less formidable. It may be cheerful to learn that, because of some difficulties hinted at in the last section, we shall leave to another day the associated task of finding the domain and range of the composite function.

You are urged to study very carefully the following definition and to try to relate it to what was done in the last section.

> DEFINITION. *If each of F and G is a function, the statement that "$[F \circ G]$ is the composite function of F with G" means that there is a number x in the domain of G such that $G(x)$ belongs to the domain of F and that, for each such x, $[F \circ G](x) = F[G(x)]$, that is, the ordinate of that point in $[F \circ G]$ whose abscissa is x is the ordinate of a point in F whose abscissa is $G(x)$.*

> *Remark.* By simply interchanging F and G in the above definition, we define the function $[G \circ F]$ and we shall allow the examples to clarify the notion that $[F \circ G]$ and $[G \circ F]$ may be different functions.

Before we consider any examples, let us remind ourselves that we are going to assume that $[F \circ G](x)$ exists and, for the present at least, "sweep under the rug" any x that gives us trouble.

Example. Given $F(x) = \sqrt{(x - 1)}$ and $G(x) = 2x + 1$, find $[F \circ G](x)$.

Let us start by recalling that $[F \circ G](x) = F(G(x))$ so if we were asked (we were not) to find $[F \circ G](2)$, we would realize that $G(2) = 5$ and that $F(5) = 2$, so $[F \circ G](2) = 2$. If were asked to find $[F \circ G](\delta)$, we could write that $G(\delta) = 2\delta + 1$, plug $2\delta + 1$ into the rule of correspondence for F and find $F(2\delta + 1) = \sqrt{[(2\delta + 1) - 1]} = \sqrt{2\delta}$, hence $[F \circ G](\delta) = \sqrt{2\delta}$.

This should allow you to follow the solution to the task that we were asked to do, namely find $[F \circ G](x)$.

$$[F \circ G](x) = F[G(x)] = \sqrt{[G(x) - 1]} = \sqrt{2x}$$

Now, let us find $[G \circ F](x)$.

$$[G \circ F](x) = G[F(x)] = 2 \cdot F(x) + 1 = 2\sqrt{(x - 1)} + 1$$

Any questions?

Remark. Judging from the above example, it should be clear that $[F \circ G]$ and $[G \circ F]$ are different functions.

Example.

$$F(x) = |x| \text{ and } G(x) = \frac{x}{x - 1}$$

$$[F \circ G](x) = F[G(x)] = |G(x)| = \left|\frac{x}{x - 1}\right|.$$

$$[G \circ F](x) = G[F(x)] = \frac{F(x)}{F(x) - 1} = \frac{|x|}{|x| + 1}.$$

Example.

$$F(x) = x^2 - 2x \text{ and } G(x) = \frac{x + 1}{x}.$$

$$[F \circ G](x) = F[G(x)] = \left(\frac{x + 1}{x}\right)^2 - 2 \cdot \frac{x + 1}{x}$$

$$= \frac{x^2 + 2x + 1}{x^2} - \frac{2x + 2}{x}$$

$$= \frac{1 - x^2}{x^2}.$$

$$[G \circ F](x) = \frac{F(x) + 1}{F(x)} = \frac{x^2 - 2x + 1}{x^2 - 2x} = \frac{(x - 1)^2}{x(x - 2)}.$$

PROBLEM SET 3.4

1. Suppose that F and G are the functions given by

$F = \{(2,3), (3,4), (4,6), (5,2)\}$
$G = \{(2,5), (3,6), (4,5), (5,3)\}$.

Give the function $[F \circ G]$ *and* sketch its graph.

2. Using the same functions F and G as given in problem 1, give the function $[G \circ F]$ *and* sketch its graph.

For Student Eyes Only: In problems 1 and 2, if you have drawn some "smooth" curve through the points you obtained, you have made a mistake in your graph.

In each of problems 3–8, give and simplify $[F \circ G](x)$ *and* $[G \circ F](x)$.

3. $F(x) = \sqrt{x}$
$G(x) = x - 1$

4. $F(x) = x^2 + x$
$G(x) = 2 - x$

5. $F(x) = \dfrac{x}{1 - x}$

$G(x) = \dfrac{x}{x - 1}$

6. $F(x) = |x|$

$G(x) = 2x + 1$

7. $F(x) = x^2 - 2x$

$G(x) = x + 1$

8. $F(x) = \dfrac{x + 2}{x}$

$G(x) = \dfrac{x - 1}{x + 2}$

9. If $F(x) = \sqrt{x}$, give $[H \circ F](x)$.

10. If $F(x) = 3x$, give $[H \circ F](x)$.

11. If $F(x) = x^3 + x$, give $[I \circ F](x)$ *and* $[F \circ I](x)$.

12. Suppose that F and F^* are functions, that $(3,2) \in F$, and that $(2,3) \in F^*$, find $[F \circ F^*](2)$ and $[F^* \circ F](3)$.

13. *Bonus problem.* If $F(x) = x + 1$, sketch the graph of the function $[[G] \circ F]$ over the interval $[0,3]$.

Section 5—Composites and Some Geometry In the last section we defined the function $[F \circ G]$. In this section your attention is directed to

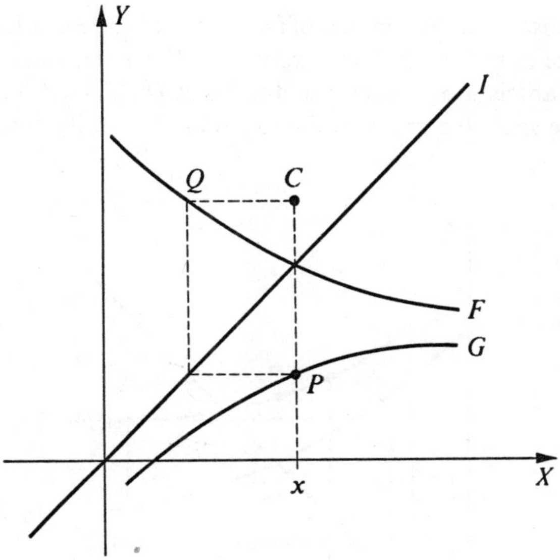

Fig. 3.10

figure 3.10 and to some simple geometry. In this figure, F and G are functions and I is the identity function, while P is the point $[x, G(x)]$. Through P, a horizontal line is drawn intersecting the graph of I at the point $[G(x),G(x)]$. Through this point, a vertical line is drawn intersecting the graph of F at the point Q. Finally, through Q a horizontal line is drawn and its intersection with a vertical line through P is labeled as the point C.

Question. What number pair is associated with the point C?

Answer. $C = (x, [F \circ G](x))$ and so C is a point in the composite function $[F \circ G]$.

By choosing other points in G and repeating this simple geometry at each such point, we can construct the graph of $[F \circ G]$ as indicated in figure 3.11.

The purpose of the sequence of sketches below, and the examples that follow, is to shed some light on the task of determining the domain of the composite function. This is a task that we have studiously avoided so far. In figure 3.12 the dom G is shown as the interval $[a,b]$. Note also, that the dom F has been projected onto 0| as the interval $[c,d]$. This projection was made in order to see that there is a number $x_1 \in \text{dom } G$ such that $G(x_1) \in \text{dom } F$. Thus P_1 is the point in $[F \circ G]$ with x_1 as abscissa. However, in this same figure, note that $G(a) \notin \text{dom } F$ so there is no point in

$[F \circ G]$ with abscissa a. As a matter of fact (see figure 3.13), x_2 is the smallest number in the dom G such that $G(x_2) \in \text{dom } F$. Hence, P_2 is the point in $[F \circ G]$ with abscissa x_2. Note also the point $(b, [F \circ G](b))$. Finally, in figure 3.14 we show the graph of the function $[F \circ G]$, its domain, and its range.

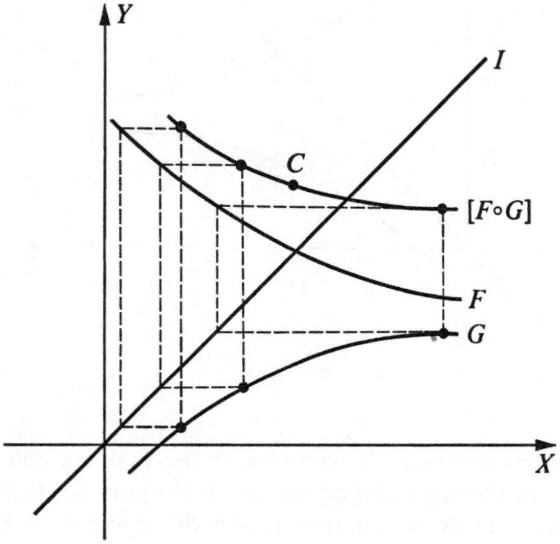

Fig. 3.11

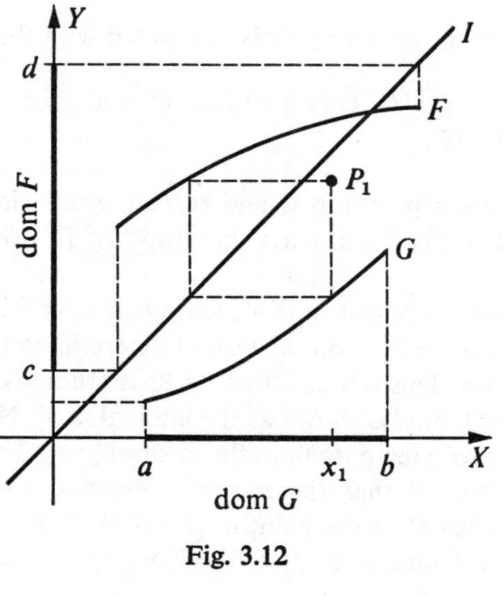

Fig. 3.12

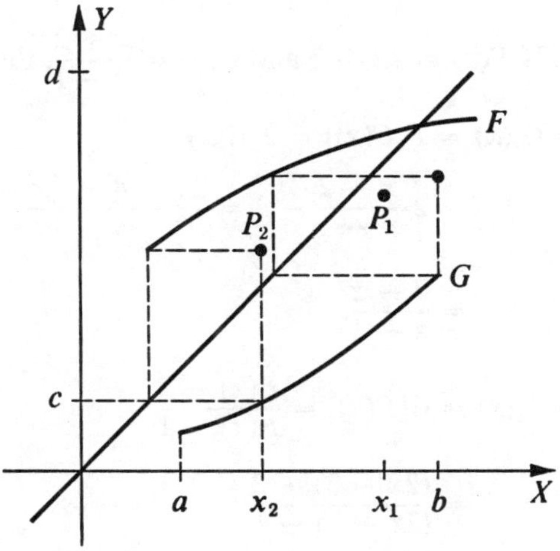

Fig. 3.13

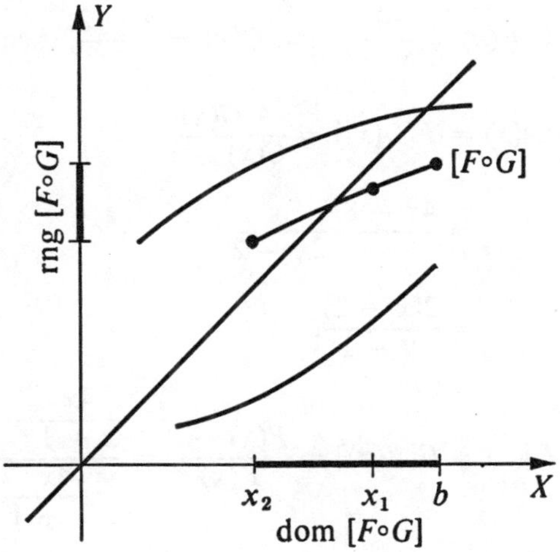

Fig. 3.14

Given the graphs of the functions F and G, these sketches show how to construct, point by point, the graph of the composite function by "running around some rectangles." It is also possible to see in these sketches the prestigious use we have made of the definition for the purpose of finding the domain of $[F \circ G]$.

Now that we have the geometry of composition under our belt, so to speak, some time can be profitably spent in reviewing the problem of finding $[F \circ G](x)$ and $[G \circ F](x)$.

Example. If $F(x) = 2x - 3$ and $G(x) = \dfrac{x+2}{x-1}$, then

$$[F \circ G](x) = F[G(x)] = 2 \cdot G(x) - 3$$

$$= 2 \cdot \frac{x+2}{x-1} - 3 = \frac{2x + 4 - 3x + 3}{x - 1}$$

$$= \frac{7 - x}{x - 1}.$$

$$[G \circ F](x) = G[F(x)] = \frac{F(x) + 2}{F(x) - 1}$$

$$= \frac{(2x - 3) + 2}{(2x - 3) - 1}$$

$$= \frac{2x - 1}{2(x - 1)}.$$

Example. If $F(x) = \dfrac{4x}{x+1}$ and $G(x) = \dfrac{x-2}{x}$, then

$$[F \circ G](x) = F[G(x)] = \frac{4 \cdot G(x)}{G(x) - 1}$$

$$= \frac{4x - 8}{x} \cdot \frac{x}{2x - 2}$$

$$= \frac{2(x - 2)}{x - 1}.$$

$$[G \circ F](x) = G[F(x)] = \frac{F(x) - 2}{F(x)} = \frac{\dfrac{4x}{x+1} - 2}{\dfrac{4x}{x+1}}$$

$$= \frac{2(x - 1)}{x + 1} \cdot \frac{x + 1}{4x}$$

$$= \frac{x - 1}{2x}$$

PROBLEM SET 3.5

In each of problems 1–5, (see figures 3.15 through 3.19) sketch the graph of the function $[F \circ G]$ and, as shown in Section 5, figure 3.14, indicate the dom $[F \circ G]$ and the rng $[F \circ G]$.

1.

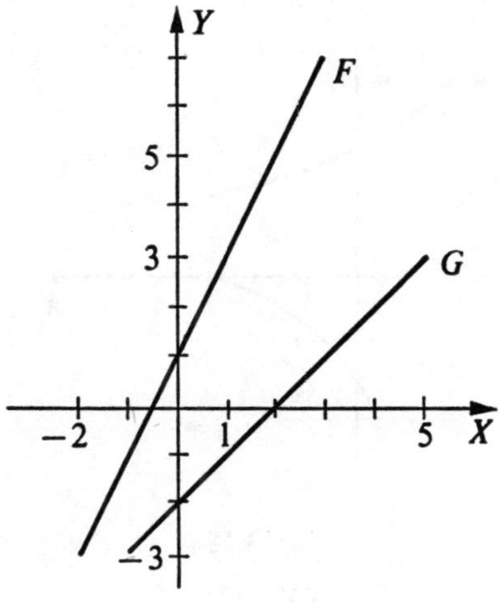

Fig. 3.15

2.

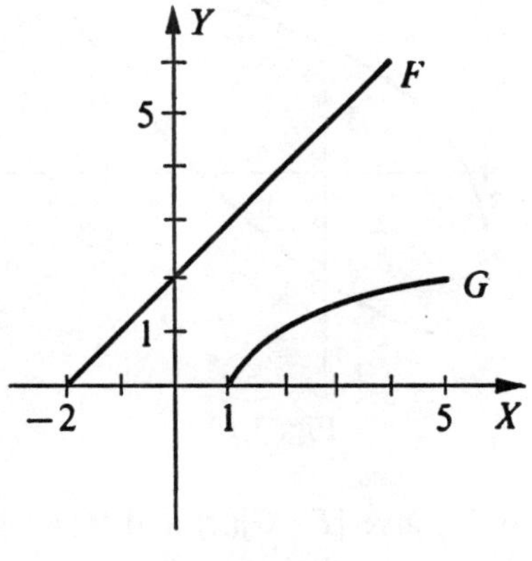

Fig. 3.16

3.

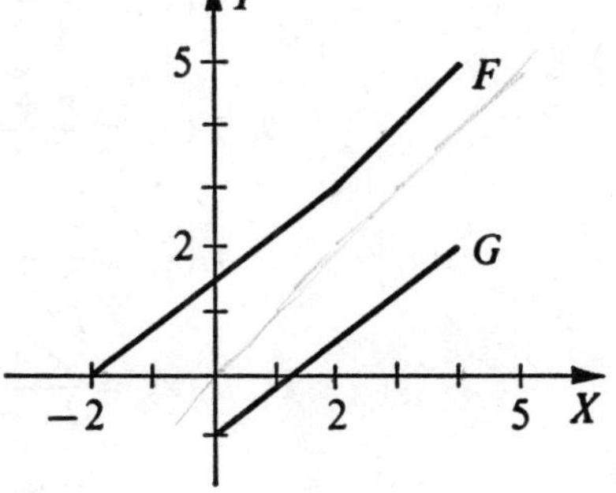

Fig. 3.17

4.

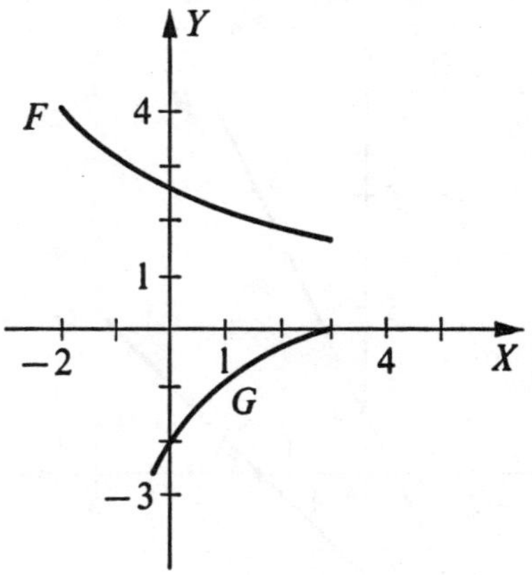

Fig. 3.18

5.

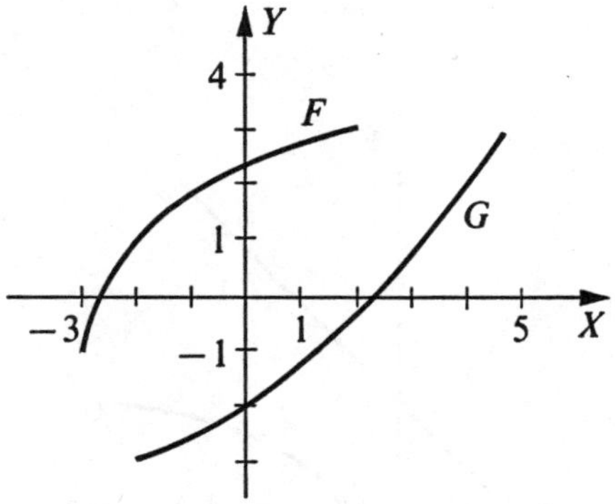

Fig. 3.19

In each of problems 6–13, give $[F \circ G](x)$ and $[G \circ F](x)$.

6. $F(x) = 2x + 3$

$G(x) = \frac{3}{2}x - 1$

7. $F(x) = x^2 + 3x - 4$

$G(x) = x + 4$

8. $F(x) = \frac{x - 3}{2x}$

$G(x) = 2x + 1$

9. $F(x) = \frac{x + 1}{2x}$

$G(x) = \frac{x}{2} - 2$

10. $F(x) = \frac{2x - 1}{x + 2}$

$G(x) = \frac{x}{x - 2}$

11. $F(x) = \frac{x + 3}{2x - 1}$

$G(x) = \frac{x + 1}{x}$

12. $F(x) = x^2 - 2$
$G(x) = \sqrt{(x + 2)}$

13. $F(x) = x^2 + 2$
$G(x) = \sqrt{(x - 2)}$

In problems 14–19, give the point-slope equation of the line containing the given points.

14. $(2, -1)$ and $(3,2)$

15. $(0,4)$ and $(4,2)$

16. $(-4,2)$ and $(0, -1)$

17. $(-2, -3)$ and $(2/3, -1/2)$

18. $(1, -3/4)$ and $(-2,1/3)$

19. $(a,k - 1)$ and $(-a,k)$

20. Using the functions F and G given in problem 7 above, describe the functions $[F + G]$ and $[F/G]$. Recall that this requires you to give the domains.

Section 6—More on $[F \circ G]$ As a sequel to the geometry of the previous section, we now consider the analytic process for determining the domain of the composite function; for the domain of $[F \circ G]$ is not left to the "whims" of our earlier agreement on domains but rather is specified by that portion of the definition of the function $[F \circ G]$ which reads as follows: "there is a number x in the domain of G such that $G(x) \in$ dom F. . . ." The examples which follow are intended to help us see and understand how the definition is employed for the purpose of finding the domain of the composite function.

Example. Suppose $F(x) = \sqrt{x}$ and $G(x) = x - 1$. It follows that dom $F = \{x \mid x \geq 0\}$ and that dom $G = R$. Therefore the statements

$x \in$ dom G implies that $x \in R$, and
$G(x) \in$ dom F implies that $x - 1 \geq 0$, that is, $x \geq 1$.

From this we conclude that

$$\text{dom } [F \circ G] = \{x \mid x \geq 1\}.$$

While the primary purpose of this and the other examples is finding this domain, let us go on by noting that

$$[F \circ G](x) = \sqrt{[G(x)]} = \sqrt{(x - 1)}$$

for then we can observe that

$$\text{rng } [F \circ G] = \{y \mid y \geq 0\}.$$

Example. Suppose that F and G are the same functions used in the preceding example, then $[G \circ F](x) = \sqrt{(x)} - 1$ and

$x \in$ dom F implies that $x \geq 0$, while
$F(x) \in$ dom G implies that $\sqrt{(x)} \in R$, that is, $x \geq 0$.

Thus, from the definition, we have argued that

$$\text{dom } [G \circ F] = \{x \mid x \geq 0\}.$$

We can also see, from the rule of pairing above, that

$$\text{rng } [G \circ F] = \{y \mid y \geq -1\}.$$

Observe that, in each example, we have described the composite function, for we have given its domain as well as its rule of pairing.

Example. Suppose that $F(x) = \dfrac{x+1}{x}$ and that $G(x) = \dfrac{x-1}{x}$, by agreement, dom F = dom $G = \{x \mid x \neq 0\}$. Hence,

$x \in$ dom G implies that $x \neq 0$, and

$G(x) \in$ dom F implies that $\dfrac{x-1}{x} \neq 0$.

Since $\dfrac{x-1}{x}$ is 0 iff x is 1, we can conclude that

$$\text{dom } [F \circ G] = \{x \mid x \neq 0 \text{ and } x \neq 1\}.$$

Now $[F \circ G](x) = \dfrac{2x-1}{x-1}$ and this should make it extremely clear why $1 \notin$ dom $[F \circ G]$. To find the range of this composite function, we begin by setting $y = [F \circ G](x) = \dfrac{2x-1}{x-1}$ and solve for x. Thus, $y(x-1) = 2x-1$ and so $xy - y = 2x - 1$. From this, we have $xy - 2x = y - 1$ and finally $x = \dfrac{y-1}{y-2}$.

Armed with this solution, it is clear that $y \neq 2$. While it is not quite so obvious, if we recall that $x \neq 0$, we can see that $y \neq 1$. Therefore we conclude,

$$\text{rng } [F \circ G] = \{y \mid y \neq 1 \text{ and } y \neq 2\}.$$

Example. Suppose that F and G are the same functions used in the preceding example.

$x \in$ dom F implies that $x \neq 0$, and

$F(x) \in$ dom G implies that $\dfrac{x+1}{x} \neq 0$.

Since $\dfrac{x+1}{x}$ is 0 iff $x = -1$, we have shown that

$$\text{dom } [G \circ F] = \{x \mid x \neq 0 \text{ and } x \neq -1\}.$$

Continuing,

$$[G \circ F](x) = \frac{F(x) - 1}{F(x)} = \frac{1}{x + 1}.$$

So, by setting $y = \dfrac{1}{x + 1}$ and solving this equation for x, we can show that $x = \dfrac{1 - y}{y}$. This makes it clear that $y \neq 0$. If we recall that $x \neq 0$, then we can also see that $y \neq 1$. Therefore

$$\text{rng}\,[G \circ F] = \{y \mid y \neq 0 \text{ and } y \neq 1\}.$$

The graphs of the functions $[F \circ G]$ and $[G \circ F]$ of these last two examples, are shown in figures 3.20 and 3.21.

Example. If $Q(x) = x^2 + 2x$ and $W(x) = x + 3$, then dom Q = dom $W = R$ so that $x \in \text{dom } W$ and $W(x) \in \text{dom } Q$ implies that dom $[Q \circ W] = R$. Also,

$$[Q \circ W](x) = [W(x)]^2 + 2 \cdot W(x) = (x + 3)^2 + 2(x + 3)$$
$$= x^2 + 8x + 15.$$

In finding the range of this composite function, we shall make use of an old technique called *completing the square*. We begin with $y = [Q \circ W](x) = x^2 + 8x + 15$ so that

$$y - 15 = x^2 + 8x.$$

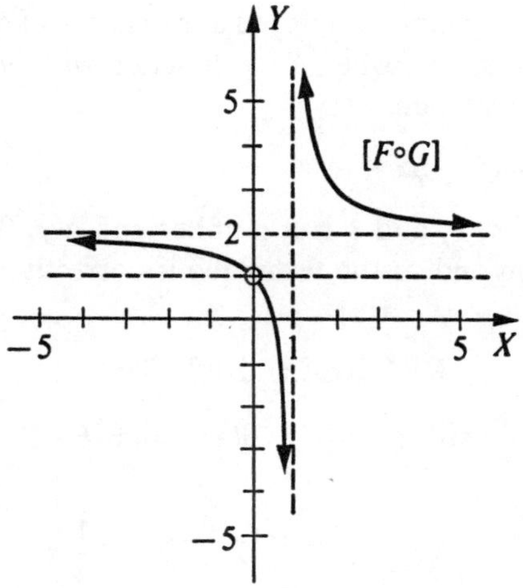

Fig. 3.20

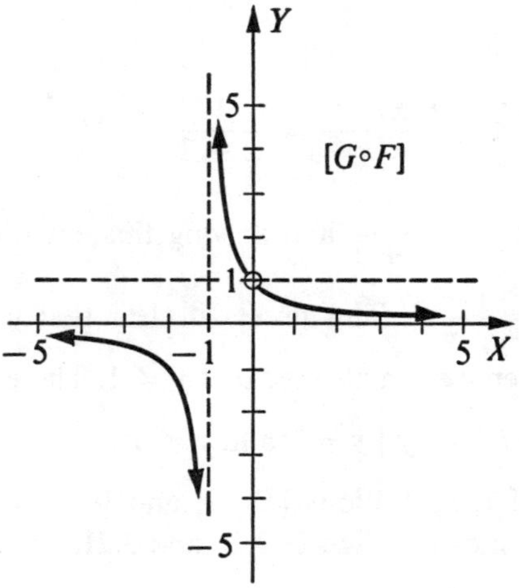

Fig. 3.21

We now think $2p$ is 8, hence p is 4 and so p^2 is 16. We then go on to write

$$y - 15 + 16 = x^2 + 8x + 16, \text{ that is, } y + 1 \text{ is } (x + 4)^2.$$

$$\text{Hence, } y = (x + 4)^2 - 1.$$

Now, if x is -4, y is -1. But, if x is not -4, $y \geq -1$. Thus, for every x, $y \geq -1$ and this tells us that

$$\text{rng } [Q \circ W] = \{y \mid y \geq -1\}.$$

Of course, the technique employed above is useful only for quadratic functions, that is, only when F is a function with domain R and rule of pairing the equation

$$F(x) = ax^2 + bx + c$$

where each of a, b, and c is a number and $a \neq 0$. But $[Q \circ W]$ is such a function and so the technique is correctly used.

PROBLEM SET 3.6

In each of problems 1–10, give $[F \circ G](x)$, dom $[F \circ G]$, rng $[F \circ G]$, and sketch its graph.

1. $F(x) = 2x - 3$

$G(x) = 4 - x$

2. $F(x) = \frac{1}{2}x + 1$

$G(x) = 3 - x$

3. $F(x) = x^2 + x$
$G(x) = x - 2$

4. $F(x) = x^2 - 2x$
$G(x) = x + 3$

5. $F(x) = x^2 - 1$
$G(x) = 1 - \sqrt{x}$

6. $F(x) = x + 2$
$G(x) = \sqrt{(x - 2)}$

7. $F(x) = x^2 - 1$
$G(x) = \sqrt{(x + 1)}$

8. $F(x) = x^2$
$G(x) = \sqrt{x}$

9. $F(x) = \dfrac{x}{2 - x}$

$G(x) = \dfrac{1}{x + 1}$

10. $F(x) = \dfrac{x}{x - 3}$

$G(x) = \dfrac{1}{x - 1}$

11. If $P(x) = \sqrt{(2x + 1)}$ find $[P \circ I](x)$ and give the domain and range of $[P \circ I]$.

12. If $G(x) = \dfrac{x - 1}{x}$ find $[G \circ I](x)$ and give the domain and range of $[G \circ I]$.

In each of problems 13–16, give $[G \circ F](x)$, dom $[G \circ F]$, rng $[G \circ F]$, and sketch its graph.

13. $F(x) = x^2 - x$
$G(x) = |x|$

14. $F(x) = 1 - x^2$
$G(x) = |x|$

15. F and G are the same functions as in problem 3.

16. F and G are the same functions as in problem 4.

17. If $F(x) = x - 1$, then $[G_i \circ F]$ is the function so that if k is an integer and $k \leq x < k + 1$, then $[G_i \circ F](x) = k - 1$. Sketch the graph of this composite function.

18. Give the domain, range, rule of pairing, and sketch the graph of the function $[H \circ I]$.

[19.] *Bonus problem.* Begin this problem by drawing a line segment through the points (1,1) and $(a,1/a)$ and another line segment through the points $(c,1/c)$ and $(ca,1/ca)$. Each of these points belongs to the graph of H. Now, gather some evidence which indicates that the area of region S equals the area of region S_1. (See figure 3.22.)

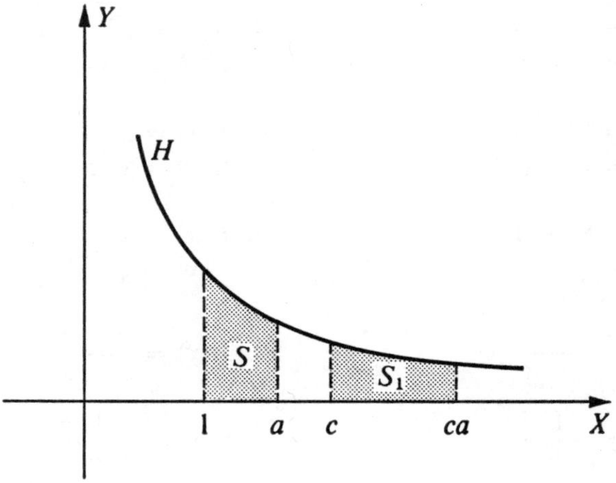

Fig. 3.22

SUMMARY

In this chapter, the important concept of a function was introduced. It will be remembered that we considered only *real functions*, that is a set of ordered pairs of real numbers. This is equivalent to saying that a function is a correspondence or mapping from some subset of the reals, called the *domain*, to some subset of the reals called the *range*. The representation of such functions by a graph as a set of points was a useful picture of a function. The next chapter will stress just how useful the graph can be.

We also defined certain algebraic operations with functions which included *composition*. This is an important notion and we hope that you have learned it well.

Review Problems

1. a. If f is a function, what does $f(x)$ denote?
b. What is meant by the domain of a function? Range?

2. Define the functions I and H. Recall that you must give not only the rule of pairing but also give the domain of each.

3. If $F(x) = x - 3$ and $G(x) = x^2 + 4x$, determine each of the functions $[F + G]$, $[F \cdot G]$, $[F \circ G]$, and $[G \circ F]$.

4. If $f(x) = \begin{cases} 3x + 4 & \text{if } x \in [0,2] \\ 1 - x & \text{if } x \in \langle 2,4] \end{cases}$ and $g(x) = \begin{cases} x^2 & \text{if } x \in [0,3\rangle \\ 2 & \text{if } x \in [3,4], \end{cases}$

evaluate each of the functions $[f+g]$, $[f \cdot g]$, $[f/g]$, and $[f \circ g]$ for each of the following values of x.

$$x \in \{1, 5/2, 3\}$$

5. Sketch the graph of the function g given in problem 4.

6. If z is a number and if the notation $[z]$ denotes the number $[G](z)$, as it does in many textbooks, show that if x and y are numbers, then $[x] + [y] \leq [x + y]$.

CHAPTER 4
General Properties of Functions

At this time and place, we are reminded that functions whose domain and range are subsets of the real numbers are properly called *real functions.* However, we also remember that we have no immediate intentions of studying any other kind of function. For this reason, then, we shall continue to call such a function simply a *function.*

Now, functions have an overwhelming number of properties, most of which will come under your scrutiny as you continue the study of mathematics. For the present, we shall look only at a few which have a fairly simple geometric interpretation and are easy to comprehend.

Section 1—Increasing, Decreasing, and 1-to-1 Functions We shall first examine what is meant by saying that a function is increasing (decreasing) over or on some subset of its domain. Because it is the simplest way,* we shall allow the graph of a function to indicate where the function is increasing (decreasing) by simply observing where the graph of the function rises (falls). Figure 4.1 may not exhaust the possibilities, but it does show enough of them for us to examine what we shall mean by saying that a function is increasing or decreasing on a subset of its domain. Look at this figure and then consider the following remarks:

* When you study the calculus, you will learn another method for determining where a function is increasing or decreasing over some subset of its domain. It cannot be stated here that this other method is easier or harder than the one we shall use. It can be stated, however, that it is not presently available.

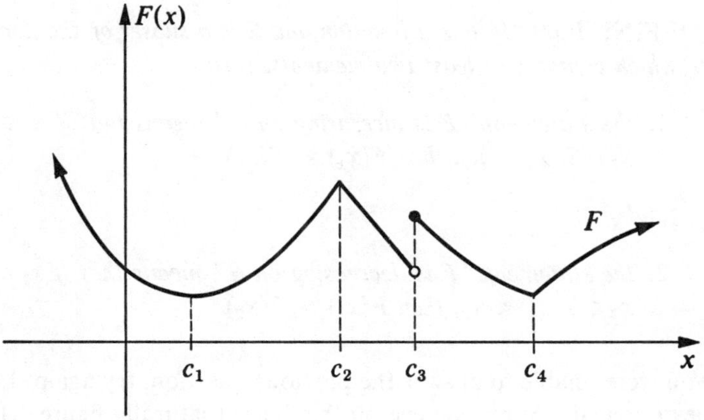

Fig. 4.1

1. F is decreasing on $\{x \mid x < c_1\}$
2. F is increasing on $\{x \mid c_1 < x < c_2\}$
3. F is decreasing on $\{x \mid c_2 < x < c_3\}$
4. F is decreasing on $\{x \mid c_3 < x < c_4\}$
5. F is increasing on $\{x \mid x > c_4\}$

Being interested in the conservation of energy, let us agree to writing the following equivalent statements:

1. F is decreasing on $\langle -\infty, c_1 \rangle$
2. F is increasing on $\langle c_1, c_2 \rangle$
3. F is decreasing on $\langle c_2, c_3 \rangle$
4. F is decreasing on $\langle c_3, c_4 \rangle$
5. F is increasing on $\langle c_4, \infty \rangle$

Note that we do not make the blunder of saying that

"F is decreasing over $\langle c_2, c_4 \rangle$."

Question. After looking at figure 4.1, do you know why it is a blunder to make this statement? Explain.

It may be that, in order to understand why this statement is in error, we should formally define what we mean by these properties. So be it.

DEFINITION. *If F is a function and S is a subset of the domain of F which contains at least two elements, then*

1. *the statement "F is increasing on S" means that if* $x_1 \in S$ *and* $x_2 \in S$, $x_1 < x_2$, *then* $F(x_1) < F(x_2)$

and

2. *the statement "F is decreasing on S" means that if* $x_1 \in S$ *and* $x_2 \in S$, $x_1 < x_2$, *then* $F(x_1) > F(x_2)$.

If you were unable to answer the previous question, try again. Look at the question again. With one eye on the figure (naturally figure 4.1 is the one your eye is on), select a number x_1 a "wee bit" to the left of c_3 and then select a number x_2 a "wee bit" to the right of c_3. Surely $x_1 < x_2$, but note that just as surely $F(x_1) < F(x_2)$, which contradicts the definition.

The student is now encouraged to take these definitions and go back to figure 4.1 and get the import of these properties firmly fixed in his mind.

To further this "fixing in your mind" of the properties of increasing or decreasing, the following examples are given.

Example. The function I (see figure 4.2a) is a good example of a function that is increasing over or on its domain. The function H (see figure 4.2b) is a good example of a function that is decreasing on its domain. In figure 4.2c we show the graph of the function $A(x) = |x|$. This function is decreasing on $\langle -\infty, 0\rangle$ and is increasing on $\langle 0, \infty\rangle$.

The definition that we have recently given and are presently illustrating does present us with an interesting, maybe even puzzling, implication when we apply it to functions whose graphs resemble figure 4.3. This function,

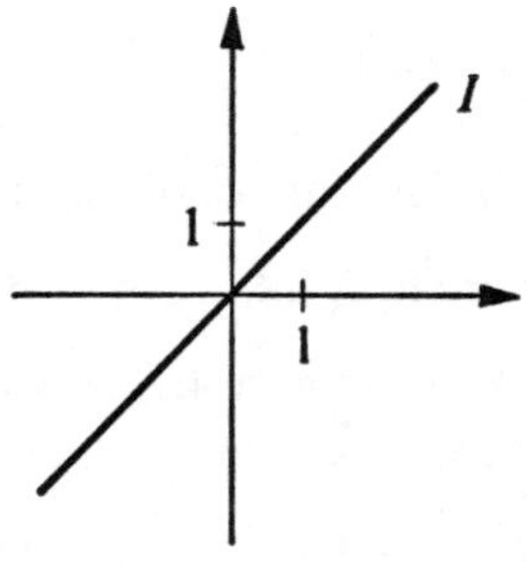

Fig. 4.2a

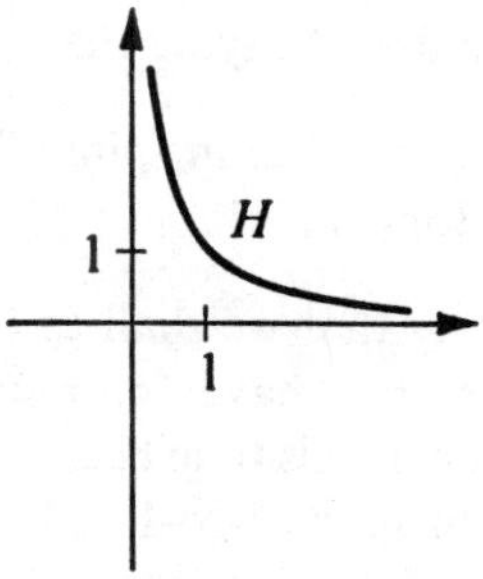

Fig. 4.2b

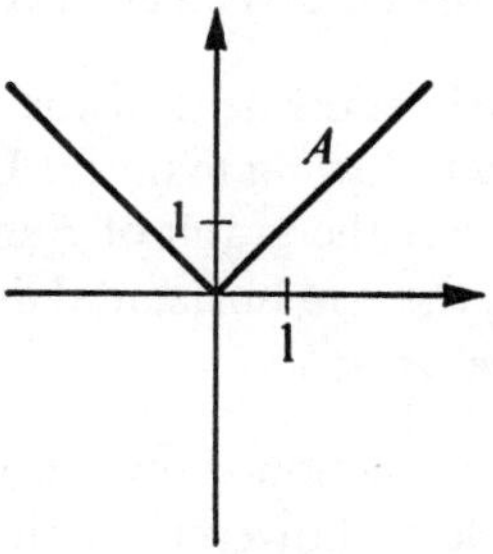

Fig. 4.2c

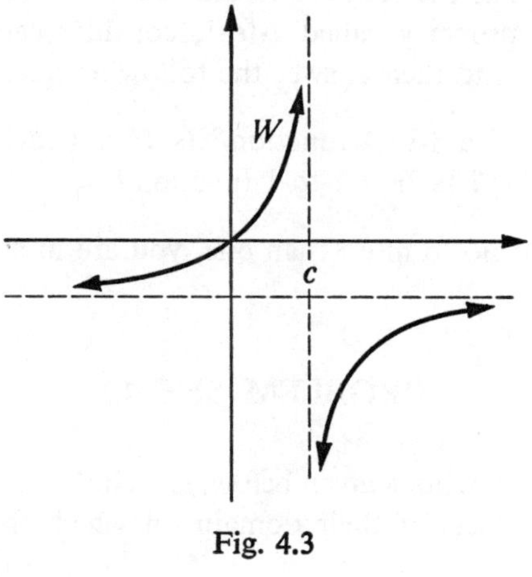

Fig. 4.3

let's call it W, has as its domain the set of all numbers distinct from c. It should be clear from the figure that

W is increasing over $\langle -\infty, c\rangle$

and that

W is increasing over $\langle c, \infty\rangle$.

Yet, we must avoid the blunder of saying that

> W is increasing over its domain in spite of the fact that $\langle -\infty,c\rangle \cup \langle c,\infty\rangle = \text{dom } W$.

One-to-One The property that we shall call *1-to-1* is a property that a function either has or does not have (nothing remarkable about that statement), and we bring it up at this time because we hope to see that there is a connection between the property 1-to-1 and the property just discussed.

It will be recalled that, if F is a function, then no two points in F have the same abscissa. If F also has the property that no two points in F have the same *ordinate*, then we shall call such a function a *1-to-1 function.*

> *Remark.* The geometric equivalent of the preceding paragraph can be paraphrased in the following manner: If F is a function, then no vertical line will intersect the graph of F in more than one point. If F is a 1-to-1 function, then no horizontal line will intersect the graph of F at more than one point.

Clearly, the graph of the function shown in figure 4.1 is not that of a 1-to-1 function. Equally clear, however, is the notion that any linear function whose slope is not zero is a 1-to-1 function.

In order to hint at the connection alluded to in the beginning of this discussion of the property called 1-to-1, consider each of figures 4.2a, 4.2b, 4.2c, and 4.3 and then answer the following questions.

> *Questions.* Is I a 1-to-1 function? Is H a 1-to-1 function? Is A a 1-to-1 function? Is W a 1-to-1 function?

If you answered no to more than one, you are in error.

PROBLEM SET 4.1

1. For each of the functions given below, sketch the graph of the function and give the subsets of their domain on which they are increasing; decreasing.

a. $F(x) = 3x - 5$	**b.** $G(x) = 1 - 2x$
c. $f(x) = (x - 2)\sqrt{x}$	**d.** $g(x) = \lvert 2x - 3\rvert$
e. $u(x) = x + \lvert x - 1\rvert$	**f.** $U(x) = \dfrac{x}{x - 2}$
g. $W(x) = x^2 - x + 6$	**h.** $P(x) = x^2 - 4$

2. Suppose that F is a function whose domain is $[a,b]$.

   a. If F is increasing on $[a,b]$, show that F is a 1-to-1 function. *Hint:* Assume the contrary.
   b. If F is decreasing on $[a,b]$, show that F is a 1-to-1 function.

3. For each of the functions given in problem 1 whose graph you were asked to construct, answer the following question: Is the function a 1-to-1 function?

4. Same directions as were given in problem 1 above.

   a. $f(x) = \begin{cases} 2 & \text{if } x \leq 0 \\ 1 - x & \text{if } x > 0 \end{cases}$ **b.** $g(x) = \begin{cases} \sqrt{(1 - x)} & \text{if } x \leq 1 \\ x + 2 & \text{if } x > 1 \end{cases}$

   c. $F(x) = \begin{cases} |x - 1| & \text{if } x < 1 \\ x^2 - x & \text{if } x \geq 1 \end{cases}$ **d.** $G(x) = \begin{cases} x^2 & \text{if } x \leq 0 \\ 2 & \text{if } x > 0 \end{cases}$

   Remark. It is unfortunate (though deliberate) that most of the functions you have considered so far have had graphs that were some "smooth" curve or line conjectured from, and drawn through, some finite collection of points. Under the "better-late-than-never" philosophy, you are now given a few problems in which the graph of the function is just a collection of isolated points. In these problems, the domain of the function is the key factor.

5. Suppose that $F(x) = x - 1$ and that dom $F = \{x \mid x \in N\}$.

   a. Obtain the graph of F by plotting at least five points.
   b. Is F increasing (decreasing) over its domain? If yes, which?
   c. Is F a 1-to-1 function?

6. Suppose that $F(x) = \dfrac{2x - 1}{x}$ and that dom $F = \{x \mid x \in N\}$.

   a. Obtain the graph of F by plotting at least five points.
   b. Is F increasing (decreasing) over its domain? If yes, which?
   c. Is F a 1-to-1 function?

7. Suppose that $F(x) = (-1)^x$ and that dom $F = \{x \mid x \in N\}$.

   a. Obtain the graph of F by plotting at least five points.
   b. Is F increasing (decreasing) over its domain? If yes, which?
   c. Is F a 1-to-1 function?

Section 2—Inverse Functions Given a function F, we shall now consider the task of constructing from F another set of points F^*, and we shall also try to see what property or properties the given function must have in order that F^* be a function (you should take this as a warning that F^* may not always be a function).

In figure 4.4, another use is found for the function I and another idea is advanced for your consideration. To understand this idea we have only to answer the question, "What are the coordinates of the point P?"

Answer. $P = (m,n)$

An understanding of this simple coordinate geometry is all that is required to anticipate the following definition and to understand how, in figure 4.5, we have constructed the graph of F^* from the given graph of F.

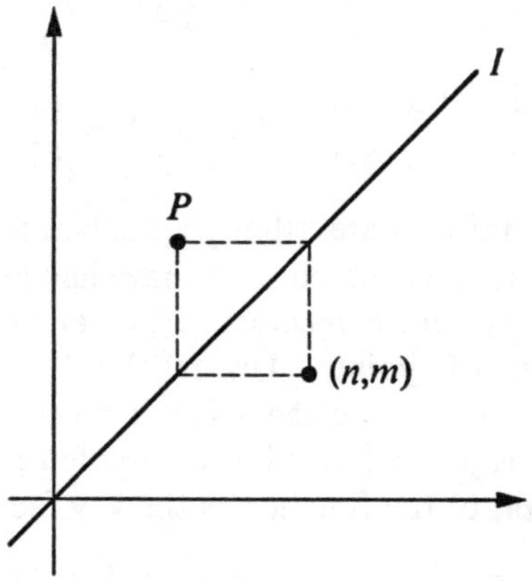

Fig. 4.4

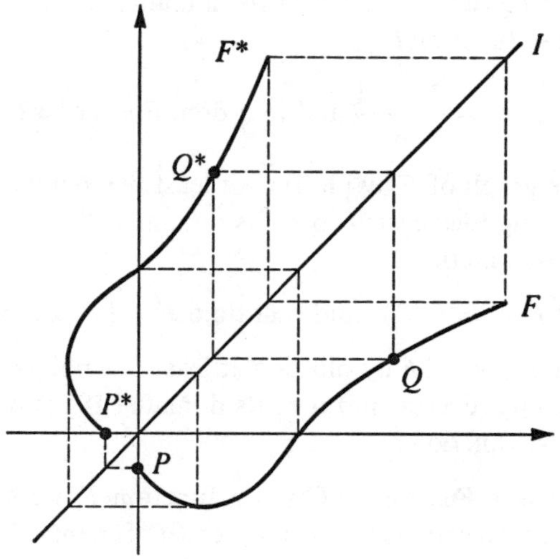

Fig. 4.5

DEFINITION. *If F is a function, then F* is the point set to which the point (a,b) belongs* iff *(b,a) belongs to F.*

Remarks. Notice that in the definition, we very carefully avoided saying that F^* was or was not a function. F^* is called the *inverse of* F and sometimes F^* is a function. When it is a function, then we should have no trouble in seeing that dom $F^* = \text{rng } F$ and that rng $F^* = \text{dom } F$.

Example. If $F(x) = \dfrac{2x}{x-1}$, then, setting $y = F(x)$, we can read the definition and describe the inverse of F with the rule

$$F^* = \left\{(y,x) \;\middle|\; y = \frac{2x}{x-1}\right\}.$$

For those who wish the description of F to inform them whether it is a function or not, more must be done. In this event, we solve the equation $y = \dfrac{2x}{x-1}$ for x and then write

$$F^* = \left\{(y,x) \;\middle|\; x = \frac{y}{y-2}\right\}$$

but it is not always as easy to do this as it was in this example. Since it can be confusing to some to use y as the abscissa of a point, we should avoid this by first solving for x (as before) and then replace x with y and vice versa. *If all this is done, and it should be when we can,* then we would have

$$F^* = \left\{(x,y) \;\middle|\; y = \frac{x}{x-2}\right\}.$$

We can then easily conclude that F^* is a function whose rule of pairing is the equation $F^*(x) = \dfrac{x}{x-2}$ and that

$$\text{dom } F^* = \{x \mid x \neq 2\}, \text{ rng } F^* = \{y \mid y \neq 2\}.$$

Problem. This problem has to do with the previous example. Look at the function F in the example. Find the domain and range of F. Compare these number sets with the domain and range of F^*. Does this comparison present any surprises?

Example. If $F(x) = x^2 - 1$, then we can choose one of the following to describe F^*.

$$F^* = \{(y,x) \mid y = x^2 - 1\}$$

for this is the most natural consequence of the definition; *or*, by solving the equation $y = x^2 - 1$ for x and then replacing x with y and vice versa, we can write

$$F^* = \{(x,y) \mid y = \pm\sqrt{(x + 1)}.$$

Though this is the most repulsive, it does tell us (at least it should) that F^* is not a function. This being true, it is probably just as well to describe F^* as follows:

$$F^* = \{(x,y) \mid y^2 = x + 1\}.$$

For Student Eyes Only There are some deliberate similarities between the following example and the previous one. Purpose, gathering of evidence to reach a conviction.

Example. If $F(x) = \sqrt{(x + 1)}$, then $\text{dom } F = \{x \mid x \geq -1\}$ and $\text{rng } F = \{y \mid y \geq 0\}$. In order to obtain the point set F^*, we go through the following three-step ritual. (Watch that second step, it can be mean.) (1) Set $F(x) = y$, (2) solve for x in terms of y, and (3) replace x with y and vice versa. Taking the first step, we write $y = \sqrt{(x + 1)}$, so (good, the second step is easy) $y^2 = x + 1$, hence $x = y^2 - 1$. The third and final step is trivial, for $y = x^2 - 1$. Now we know that F^* is a function. Therefore we state that

$$F^*(x) = x^2 - 1, \text{dom } F^* = \{x \mid x \geq 0\}, \text{rng } F^* = \{y \mid y \geq -1\}.$$

Because it might offer certain important clues, we now sketch the graphs of the functions F, F^*, and I on the same set of coordinate axes. Look at figure 4.6 and wonder briefly why the point P belongs to both F and F^*.

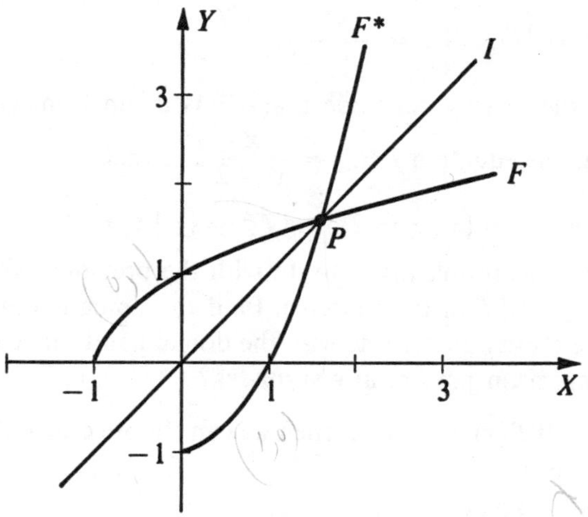

Fig. 4.6

PROBLEM SET 4.2

1. If $F = \{(1,0), (2,3), (3,1), (4,2)\}$, give F^*. Is F^* a function?

In each of problems 2–5, sketch the graph of the function F and on the same coordinate axis, the graph of F^*.

2. $F(x) = \sqrt{(x-1)}$

3. $F(x) = \frac{3}{2}x - 3$

4. $F(x) = 2 - x$

5. $F(x) = x^2 - x - 2$, dom $D = [1/2,3]$

6. In each of problems 2–5, F^* is a function, so give the domain and range of F^* in (a) problem 2, and (b) problem 5.

[7.] **a.** Consider the two graphs shown in figures 4.7 and 4.8 and explain why f^* and g^* are not functions.

b. On the basis of the evidence presented thus far, copy and complete the following statement: "If F is a function, then F^* is a function iff. . . ."

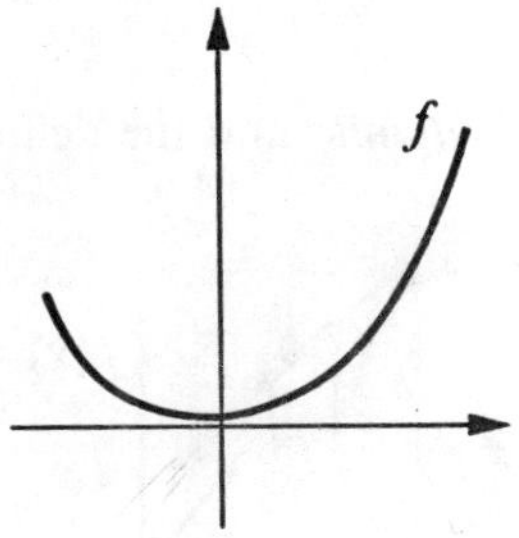

Fig. 4.7

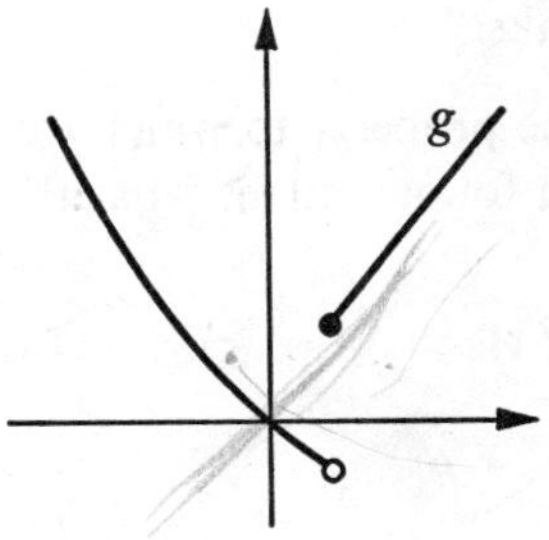

Fig. 4.8

In each of problems 8–14, find a rule describing F^* and if F^* is a function, give its rule of pairing, domain, and range.

8. $F(x) = 2x + 1$

9. $F(x) = \frac{1}{\sqrt{x}}$

10. $F(x) = x^2 - 2x$

11. $F(x) = x^2$

12. $F(x) = \frac{x+1}{x-3}$

13. $F(x) = \frac{x}{x+3}$

14. If $P(x) = x^2 - 1$, dom $P = \{x \mid x \leq 0\}$, then P^* is a function, so give its rule of pairing, domain, and range. Also sketch the graph of P^*.

15. a. If $F(x) = x^2 + x$, describe F^* and state whether or not F^* is a function.

b. If $G(x) = x^2 + x$, dom $G = \{x \mid x \leq -1/2\}$, describe G^* and state whether or not G^* is a function.

16. a. Same as 15a for $F(x) = x^2 - 1$.

b. Same as 15b for $G(x) = x^2 - 1$, dom $G = \{x \mid x \geq 0\}$.

17. Suppose that each of F and F^* is a function. Find $[F \circ F^*](x)$ and $[F^* \circ F](x)$.

Section 3—Continuity Examine each of figures 4.9 through 4.14. Each figure represents the graph of a function which contains a point labeled P.

The functions shown in figures 4.9, 4.10, 4.11, and 4.14 have a certain property at the point P, while those shown in figures 4.12 and 4.13 do not have this property at P. Before explaining what this property is, consider the fact that we have been assuming that many of our functions have this property at each of their points and that we have used this assumption to help sketch graphs of functions. As a matter of fact, we may not even have realized that we were making this assumption. Yet, 'tis many the time that we drew a graph of a function all in "one piece" as it were, and without "breaks."

The property to which we allude is called *continuity* and the definition which follows states what this property is.

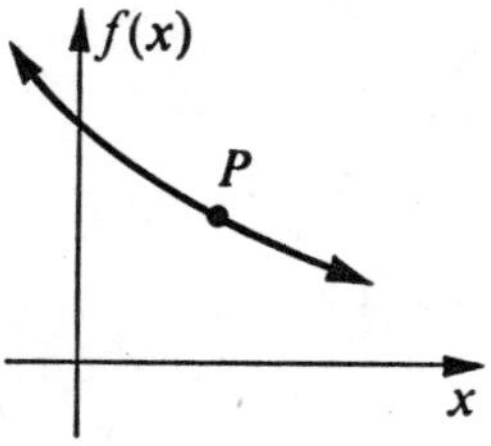

Fig. 4.9

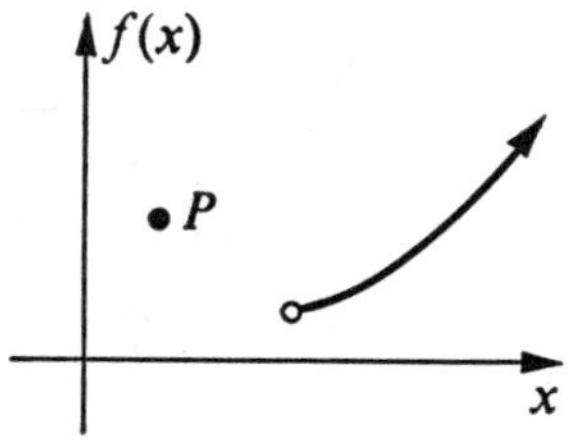

Fig. 4.10

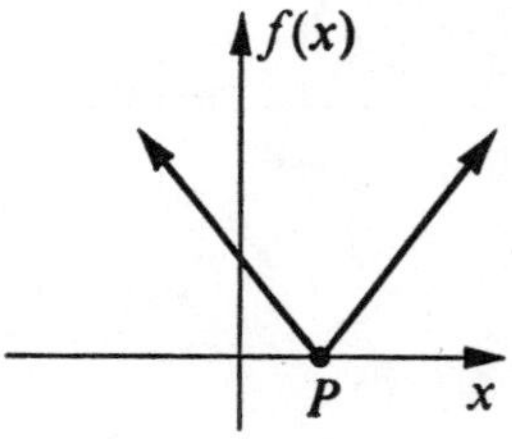

Fig. 4.11

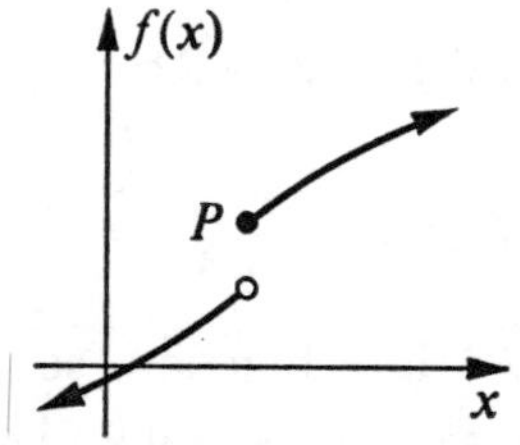

Fig. 4.12

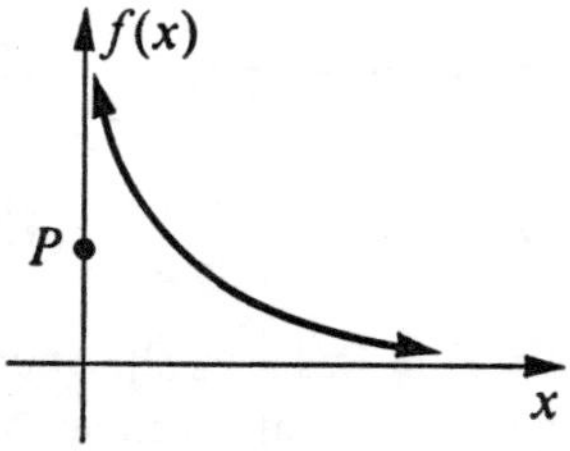

Fig. 4.13

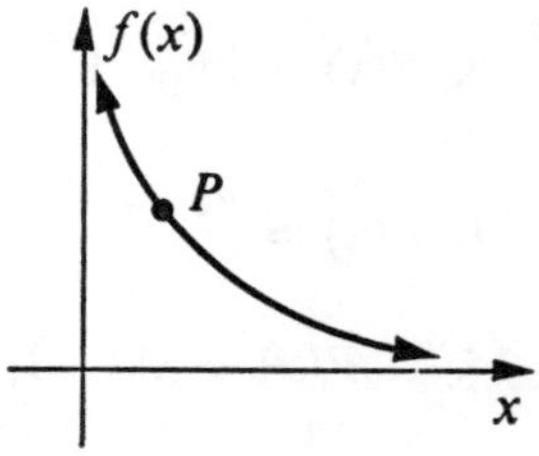

Fig. 4.14

DEFINITION. *If f is a function and P is a point in f, the statement that "f is continuous at P" means that for each pair of horizontal lines h_1 and h_2 with P between them, there is a pair of vertical lines v_1 and v_2 with P between them such that each point of f that is between v_1 and v_2 is between h_1 and h_2.*

The student is encouraged to observe (by drawing) that, in each of figures 4.9 through 4.11, and 4.14 for any pair of horizontal lines with P between them he can produce a pair of vertical lines which meet the requirements of the definition. He should also observe that in each of figures 4.12 and 4.13, he can draw a pair of horizontal lines with P between them such that for no pair of vertical lines with P between them will the requirements of the definition be met. Let us pause for the purpose of making all of these observations.

I now invite your attention to figure 4.15.

Although it is physically impossible for us to draw each and every pair of horizontal lines having the point $[a,f(a)]$ between them (see figure 4.15),

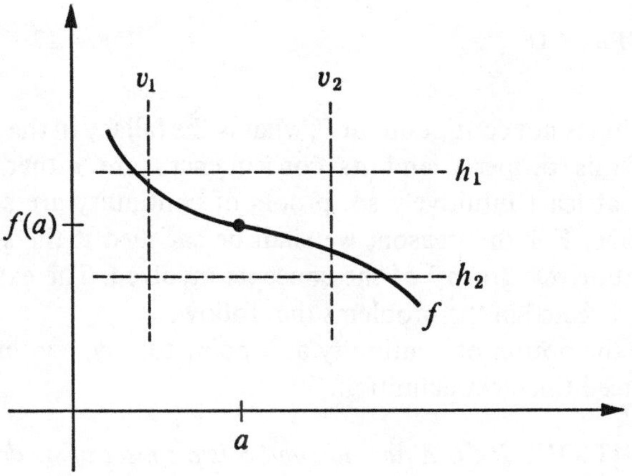

Fig. 4.15. f continuous at $[a,f(a)]$

the student is urged to draw at least one other such pair of horizontal lines closer together than those shown in the figure and to observe that it is still possible to produce the required pair of verticals and so show that f is continuous at the point $[a,f(a)]$. We should marvel at the fact that our minds can do that which we cannot physically accomplish.

In order to secure our grasp more firmly on the property called continuity, consider the following argument attributed to student Karl von

Neinhoffen: "Given the function G whose graph you see in figure 4.16, I maintain that G is continuous at the point P. I argue as follows: surely h_1 and h_2 are a pair of horizontal lines having P between them (see figure 4.17) and just as surely v_1 and v_2 [are] a pair of verticals with P between them. Clearly the requirements of the definition are met, and I have shown that which I have maintained."

End of Karl's argument.

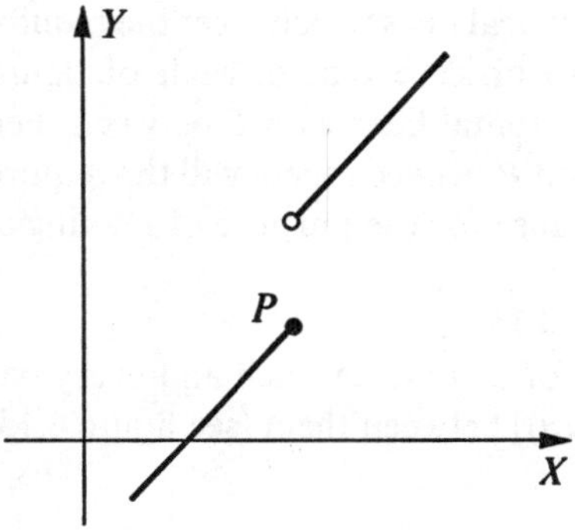

Fig. 4.16

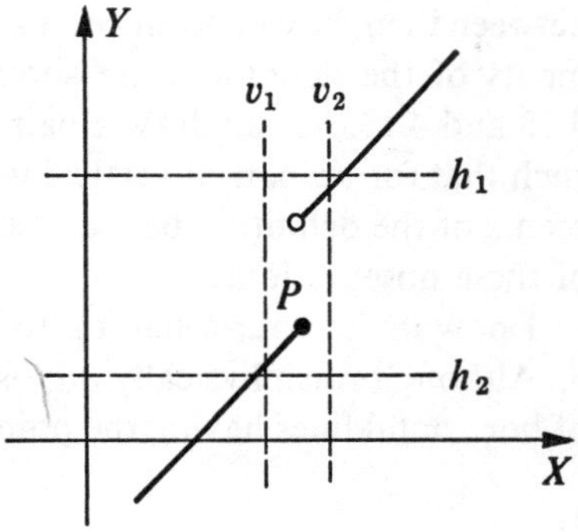

Fig. 4.17

This function is not continuous at P, what is the fallacy in the argument?

Although this property and its consequences are rather simple to comprehend, at least intuitively so, proofs of continuity are surprisingly hard to manage. For this reason, we shall be satisfied if the student can develop an intuitive "grasp" of the concepts involved. The extent of this grasp will be revealed in the problems that follow.

To extend the notion of continuity at a point to, say, continuity on an interval, we need this next definition.

> DEFINITION. *If f is a function and S is a subset of its domain, then the statement that "f is continuous on S" means that f is continuous at each point in S.*

I cannot close this section without calling your attention to at least one of the more important consequences of this property called continuity. This consequence is often called the *intermediate value theorem*, and it is illustrated by figure 4.18.

IVT Suppose f is a function whose domain includes the interval $[a,b]$. If f is continuous on $[a,b]$, $f(a) \neq f(b)$, and *if* t is a number between $f(a)$ and $f(b)$, then there is a number c, $a < c < b$, such that $f(c) = t$.

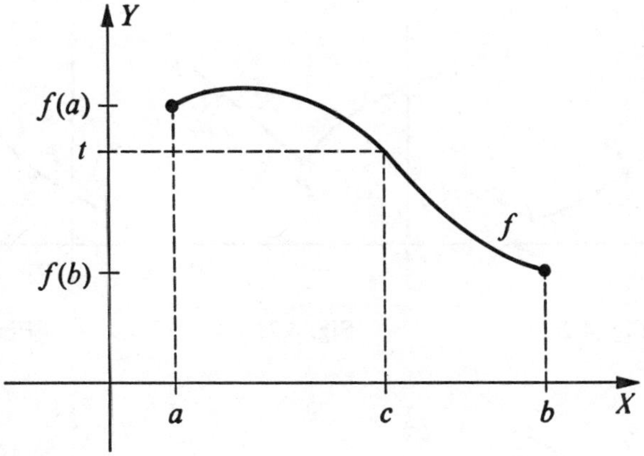

Fig. 4.18. The intermediate value property for continuous functions.

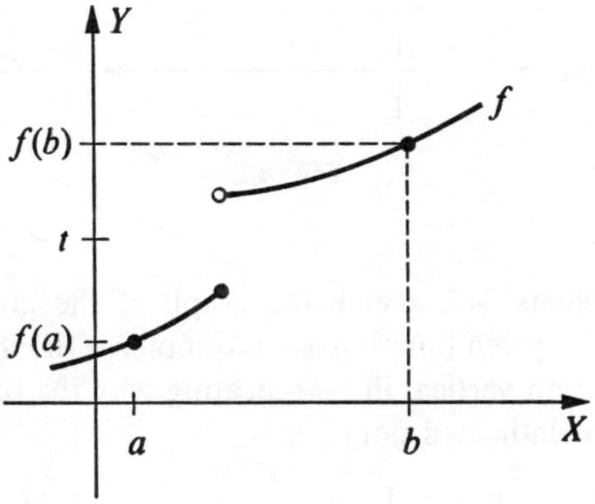

Fig. 4.19

Please note that a function should be continuous on $[a,b]$ in order that the conclusion of the intermediate value theorem be valid. To see this, look at figure 4.19 and observe that, while t is between $f(a)$ and $f(b)$, there is no number c between a and b such that $f(c) = t$.

PROBLEM SET 4.3

1. Indicate, by drawing horizontal and vertical lines with P between them, that F is continuous at P. (See figures 4.20 through 4.22.)
2. Indicate, by drawing horizontal and vertical lines with P between them, that F is not continuous at P. (See figures 4.23 through 4.25).

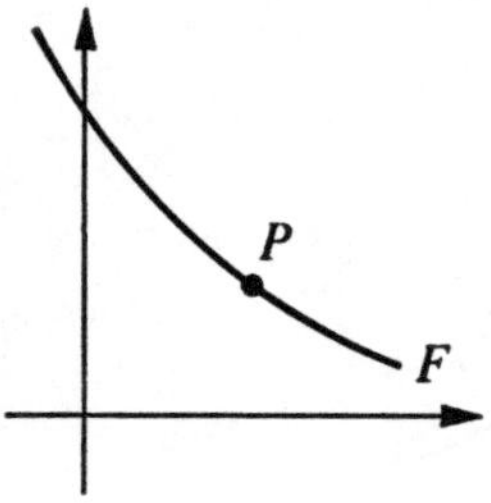

Fig. 4.20

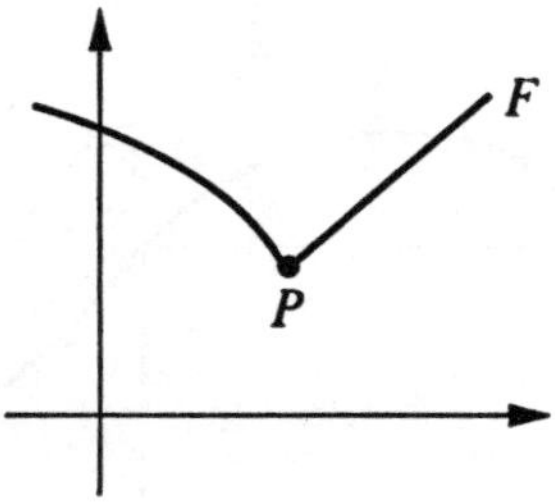

Fig. 4.21

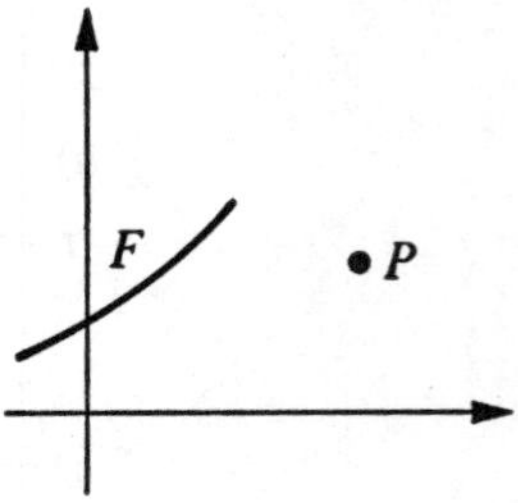

Fig. 4.22

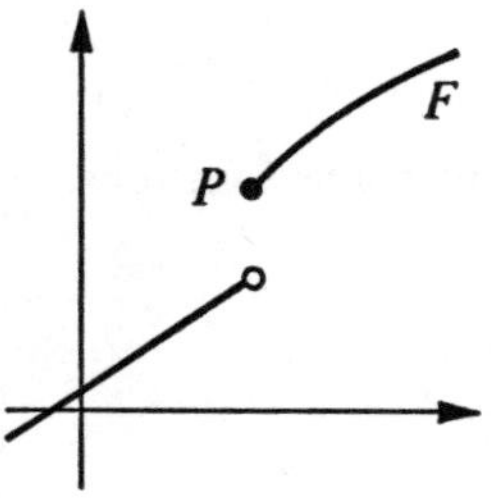

Fig. 4.23

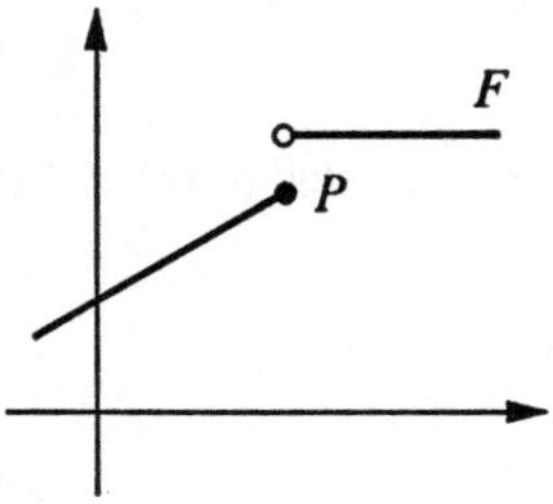

Fig. 4.24

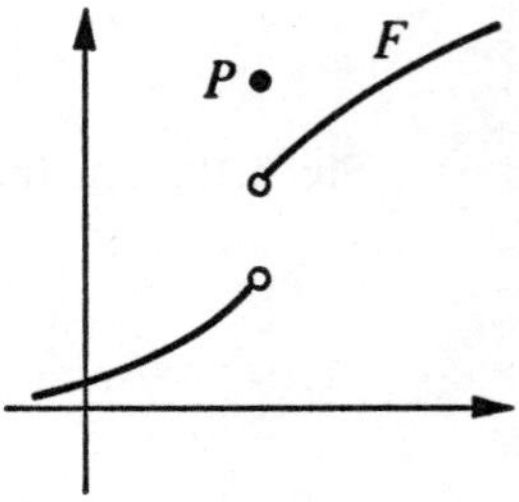

Fig. 4.25

In each of problems 3–7, sketch the graph of the given function, state whether or not the given function is continuous at the indicated point, *and* draw horizontal and vertical lines indicating why the function is or is not continuous at the indicated point.

3. $G(x) = \begin{cases} x - 1 & \text{if } x > 1 \\ x^2 & \text{if } x \leq 1 \end{cases}$

Examine at $[1,G(1)]$.

4. $g(x) = \begin{cases} \sqrt{(x + 2)} & \text{if } x > 2 \\ x & \text{if } x \leq 2 \end{cases}$

Examine at $[2,g(2)]$.

5. $U(x) = \begin{cases} |x + 1| & \text{if } x \leq 0 \\ x - 2 & \text{if } x > 0 \end{cases}$

Examine at $[0,U(0)]$.

6. $W(x) = \begin{cases} \left|\dfrac{x}{x - 2}\right| & \text{if } x \neq 2 \\ 1 & \text{if } x = 2 \end{cases}$

Examine at $[2,W(2)]$.

7. $F(x) = \begin{cases} x + 1 & \text{if } x \leq -1 \\ x/2 & \text{if } x > -1 \end{cases}$

Examine at $(-1,F(-1))$.

8. Suppose that $f(x) = x^2 - x$. If D is the function for which

$$D(x) = \frac{f(x) - f(2)}{x - 2}$$

when $x \neq 2$, sketch the graph of D *UND*, then define $D(2)$ so that D will be continuous at $[2,D(2)]$.

9. Suppose that $g(x) = \dfrac{4}{x+1}$. If D is the function for which

$$D(x) = \frac{g(x) - g(1)}{x - 1}$$

when $x \neq 1$, sketch the graph of D *UND*, then define $D(1)$ so that D will be continuous at $[1,D(1)]$.

10. The graph of a function U is shown in figure 4.26. Its domain is the interval $[a,b]$. Is U continuous on $[a,b]$? In what interval is U increasing?

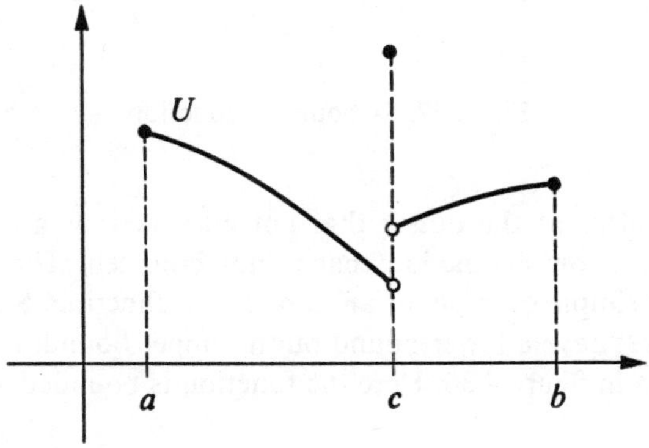

Fig. 4.26

11. Sketch a graph showing that *if* F is continuous on $[a,b]$, $F(a) < 0$, and $F(b) > 0$, then $F(x) = 0$ for at least one x between a and b.

12. Supposing that $f(x) = x^2 + x - 1$ is continuous on $[0,1]$, show that $f(0) < 0 < f(1)$ and then explain why there is a number c, $0 < c < 1$, such that $f(c) = 0$. By making a series of "guesses," obtain c to the nearest tenth.

13. If $F(x) = x^2$, use the intermediate value system to show that $\sqrt{2}$ exists.

14. *Bonus problem.* Suppose that $G(x) = \sqrt{[|x| \cdot (x+1)]}$

a. Determine the domain of G by solving $|x| \cdot (x + 1) \geq 0$.

b. Sketch the graph of G.

c. Discuss the continuity of G at each of the following points: $[0,G(0)]$, $[1,G(1)]$, and $[2,G(2)]$.

Section 4—Bounded Functions The final property that we intend to investigate can be crudely described as follows.

Given the graph of a function, we wish to determine whether there is a pair of horizontal lines by which the graph of the function is bounded, that is, between which the graph is confined or contained. See figure 4.27.

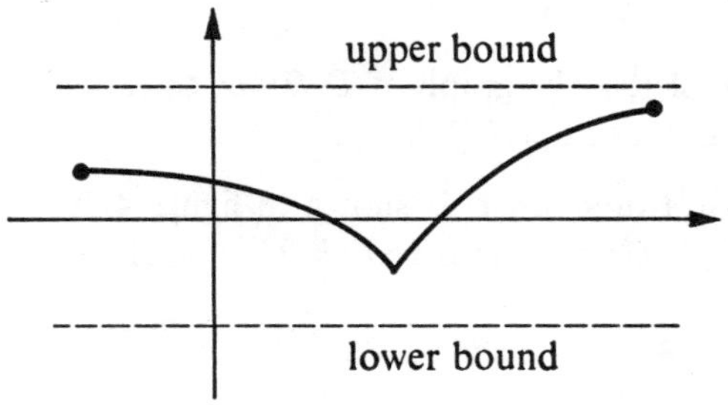

Fig. 4.27. A bounded function.

We make clear at the outset that not all functions are bounded. In figure 4.28 we show a function that is not bounded. The function I is another good simple example of an unbounded function. Some functions (see figure 4.29) have a lower bound but no upper bound. Another possibility is shown in figure 4.30. Here the function is bounded above but not below.

It is probably clear to you by now that in talking about the *bounds* of a function, we are really talking about whether or not the *range* of the function is bounded. Since the range of a function is a number set, then we

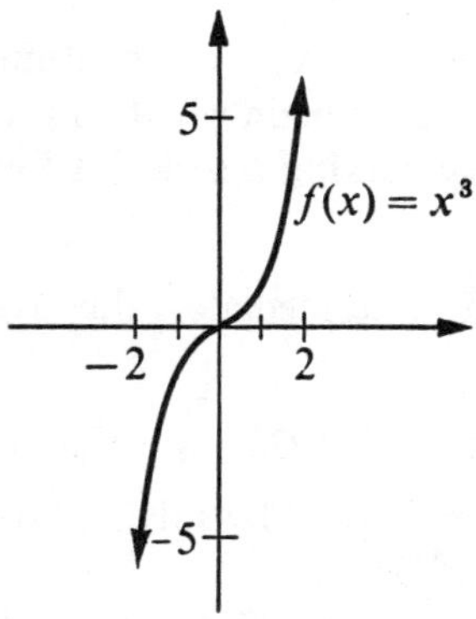

Fig. 4.28. An unbounded function.

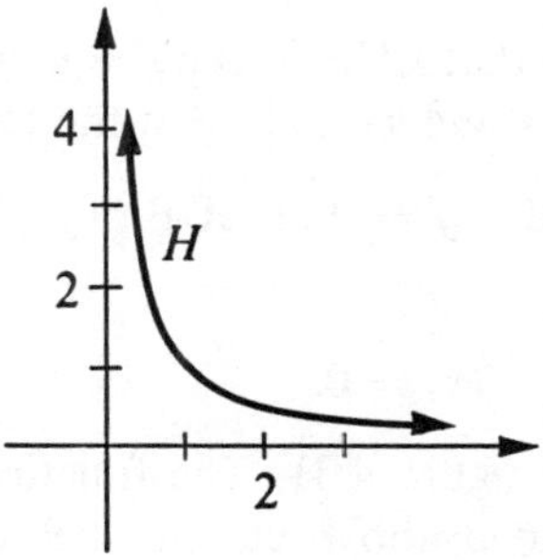

Fig. 4.29. Bounded below but not above.

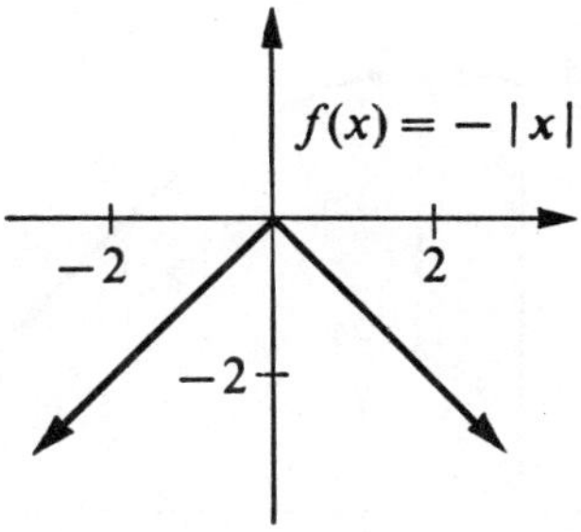

Fig. 4.30. Bounded above but not below.

are talking about a type of problem with which we are not unfamiliar, that is, the problem of bounded number sets (see chapter 1). This triggers fond memories of LUBs and GLBs and of the axiom of completeness. However, let us restrain our enthusiasm long enough to record a definition.

> DEFINITION. *A function F is said to be* bounded above *on a subset S of its domain if there is a number M so that for each x in S,* $F(x) \leq M$.
>
> *A function F is said to be* bounded below *on a subset S of its domain if there is a number m so that for each x in S,* $F(x) \geq m$.
>
> *If F is bounded both above and below on a subset S of its domain, then we shall say that F is* bounded on *S.*

Very closely allied to the notion of the bounds of a function is the notion of the *highest* (maximum) point in the graph of the function and the notion of the *lowest* point in the graph of the function. If we look back at figure 4, we see that the origin is the highest point in the graph of $f(x) = -|x|$.

We can also understand that LUB of rng $f = 0$ and that $0 \in$ rng f. We are authorized to record this understanding in the following manner:

Least upper bound of $f = 0$ or LUB of $f = 0$

and

Maximum value of $f(x) = 0$.

For a contrast, look at figure 4.31. The function g represented by this graph is certainly bounded above but not below. However, *there is no*

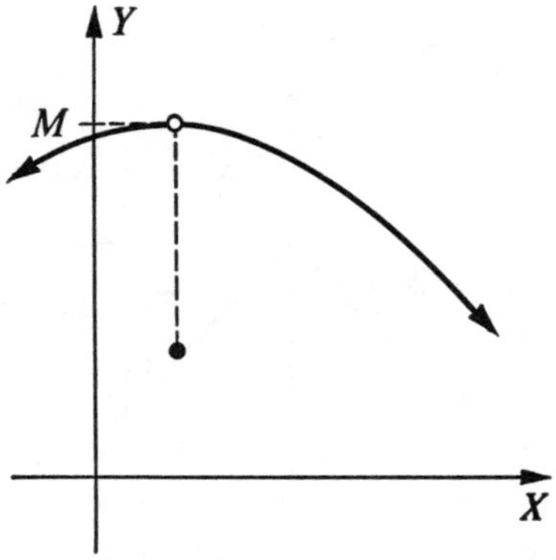

Fig. 4.31

highest point in the graph of g. This is equivalent to saying that the LUB of rng $g = M$ but $M \notin$ rng g. We shall abbreviate this with

LUB of $g = M$.
No Maximum.

After looking at figure 4.28 we shall write that

GLB of $H = 0$.
No minimum.

For additional ideas about the bounds of a function, we give an example.

Example. (See figure 4.32.) If $F(x) = x^2 + x - 2$, we can algebraically determine the boundedness of F by completing the square in x. Let us set $F(x) = y$ and write $y = x^2 + x - 2$ so that $y + 2 = x^2 + x$. After a little thought, we then write $y + 2 + 1/4 = x^2 + x + (1/2)^2$. Hence $y + 9/4 = (x + 1/2)^2$. Thus $y = F(x) =$

$(x + 1/2)^2 - 9/4$. Now if $x = -1/2$, $F(x) = -9/4$, if $x \neq -1/2$, $F(x)$ is equal to some positive number plus $-9/4$, that is, $F(x) > -9/4$. Therefore F is bounded below by $-9/4$ and the point $(-1/2, -9/4)$ is the minimum or lowest point in the graph of F. We record this by writing

GLB of $F = -9/4$
Minimum value of $F(x) = -9/4$.

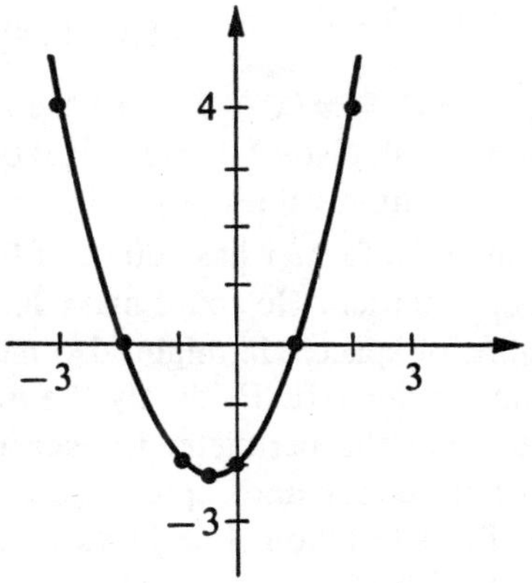

Fig. 4.32

PROBLEM SET 4.4

In each of problems 1–16, sketch the graph of the given function and use it to determine the boundedness of the given function. If the function is bounded below, give its GLB and, if it exists, give its minimum value; if the function is bounded above, give its LUB and, if it exists, give its maximum value.

1. $f(x) = 1 - |x|$
dom $f = R$

2. $F(x) = |x + 1|$
dom $F = R$

3. $G(x) = 1 - \sqrt{(x + 1)}$
dom $G = [-1,4]$

4. $g(x) = \dfrac{2x}{x - 1}$
dom $g = [2,6]$

5. $U(x) = x \cdot |x|$
dom $U = R$

6. $u(x) = x\sqrt{|x|}$
dom $u = R$

7. $w(x) = x^2 - 4x$
dom $w = R$

8. $P(x) = 2x - x^2$
dom $P = R$

9. $Z(x) = \begin{cases} 3 - x & \text{if } x \leq 0 \\ x & \text{if } x > 0 \end{cases}$

10. $W(x) = \begin{cases} x^2 & \text{if } x \leq 0 \\ x + 2 & \text{if } x > 0 \end{cases}$

11. $F(x) = x + |x|$
dom $F = R$

12. $G(x) = x + 1 - |x|$
dom $G = R$

13. $p(x) = 3 + 2x - x^2$
dom $p = R$

14. $f(x) = x^2 + 2x - 3$
dom $f = R$

15. $g(x) = \begin{cases} x^2 & \text{if } 0 \le x \le 2 \\ 6 - 2x & \text{if } 2 < x < 5 \end{cases}$

16. $S(x) = \begin{cases} \sqrt{(x + 2)} & \text{if } x \in \langle -2,2 \rangle \\ 1 & \text{if } x \in [2,4] \end{cases}$

17. Sketch the graph of $p(x) = x^2(x - 1)$ over the interval $\langle -1,1 \rangle$. If p is bounded on this interval, give LUB of p, GLB of p, and, if they exist, the maximum and minimum values of p.

18. Suppose that a gentleman farmer has 100 ft. of fencing with which to enclose a rectangular garden. He could make his garden 10 by 40 ft. giving him 400 sq. ft. of space. He might also make his garden 20 by 30 ft., thus enclosing 500 sq. ft. Evidently the area can vary quite a bit despite the fact that the perimeter in each case is 100 ft. Let x denote the width of his garden and express the area as a function of x. Sketch the graph of this function A and find the value of x that gives a maximum value for $A(x)$.

19. 100 m. (meters) of fencing is to be used to enclose a rectangular plot one side of which lies along a rock wall (see figure 4.33) and will not

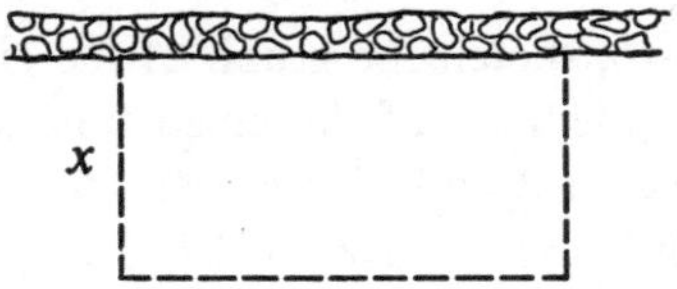

Fig. 4.33

require fencing. We could make this rectangle 10 by 80, thus enclosing 800 sq. m., or we could make it 20 by 60, thereby enclosing 1200 sq. m. Obviously the area enclosed with the same 100 m. of fencing can vary considerably. We wish to enclose the largest possible region. Let x denote the width of the rectangle and express the area of the rectangle in terms of x, that is, as a function of x. Either with or without sketching the graph of the function A, find the dimensions of the largest possible rectangular plot that can be enclosed with the 100 m. of fencing.

20. *Bonus problem.* A battery has an electromotive force of 12 volts and an internal resistance of 1 ohm. When this battery is connected to a load having a resistance r, the current I_a (amperes) flowing through r is given by $I_a = \frac{12}{r+1}$. The power P (watts) delivered to the load is given by $P = I_a^2 \cdot r$. Sketch the graph of the function P and find the value of r which maximizes the power.

21. *Bonus problem.* A rectangle is inscribed under the graph of the function $G(x) = \frac{1}{x^2+1}$ (see figure 4.34). Express the area A of this rectangle as a function of x. Sketch the graph of this function A and find the maximum value of $A(x)$.

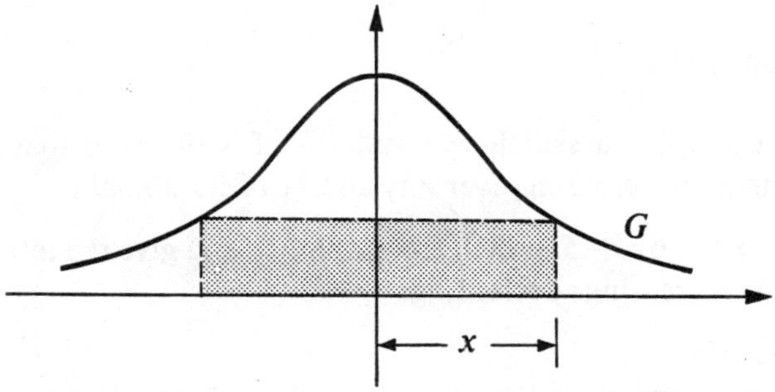

Fig. 4.34

SUMMARY

A picture serves better than a multitude of words

Ah Sho, 690 A.D.

The representation of a function by its graph is a useful means of giving both a quantative and a qualitative picture of the function. While it may be true that we need to study calculus in order to have adequate tools available for sketching the graphs of many functions, we have noted that even our amateurish graphs have served to reveal some important qualities (properties) of functions.

The intervals over which the graph rises (falls) specifies where the function is increasing (decreasing).

If no horizontal line intersects the graph of a function in more than one point, then the function is called a 1-to-1 function.

We have even noted that the properties we discussed are interrelated.

For example, if a function is increasing over its domain, then it is a 1-to-1 function. It was further noted that the inverse of f, f^*, is a function only if f is 1-to-1.

The most sophisticated property that we discussed in this chapter was continuity. Our grasp of this concept will be important in the calculus. The intermediate value theorem for continuous functions was illustrated, and we shall have occasion to use it again in the very next chapter.

The last property that we investigated (aided by our chief witness, the graph) was that of boundedness. We saw that this was equivalent to the boundedness of the number set called the range. If the LUB of the range was an element of the range, then it was labeled as the maximum value of the function; if the GLB of the range was an element of the range, it was labeled as the minimum value.

Review Problems

1 Give an example—a sketch will suffice—of a function that is neither increasing nor decreasing over any subset of its domain.

2. If $f(x) = x^2 - 6x + 5$, sketch the graph of f and give the intervals over which it is increasing; decreasing.

3. a. Define F^*.
   b. What property must F have in order that F^* be a function?
   c. If F^* is a function, give the relationship between the following number sets: dom F, rng F, dom F^*, and rng F^*.
   d. If F is a function, show that $[F \circ F^*](x) = [F^* \circ F](x)$.

4. If $F(x) = \dfrac{x-1}{x+2}$, find $F^*(x)$.

5. Find, if they exist, the maximum or minimum value of the function f given in problem 2.

6. Sketch the graph of a function whose domain is [0,2], that is bounded on [0,2], and is continuous everywhere except at $x = 1$. Sketch the graph of another function whose domain is [0,2], that is unbounded on [0,2], and is continuous everywhere except at $x = 1$.

CHAPTER 5
Polynomial Functions

Section 1—Basic Ideas We first consider the idea that, given a pair of rather simple functions, we can construct a class of functions called *polynomial functions*. Let us begin by going back to the idea that the sum and product of two given functions is a function whose domain is the intersection of the domains of the given functions. Let us continue by letting C denote any linear function whose graph is a horizontal line. Each such linear function C is often called a constant function and, if c is a number, has as a rule of pairing the equation $C(x) = c$. One useful feature of such a function is that its domain is the set of all numbers. Now, let us choose the identity function I and a function C and indicate the nature of our constructions.

Example. If C is a constant function and

$$P = [C + C \cdot I + I \cdot I]$$

then P is the function whose domain is the set of all numbers for which

$$P(x) = C(x) + C(x) \cdot I(x) + I(x) \cdot I(x)$$

that is,

$$P(x) = c + cx + x^2.$$

Example. Suppose $P = [c_0 \cdot I + c_0 \cdot I^2 + c_1 \cdot I^3]$. Here $I^2 = I \cdot I$ and $I^3 = I \cdot I^2$. Hence, $P(x) = c_0 x + c_0 x^2 + c_1 x^3$. So, if $c_0 = 1$ and $c_1 = 2$, $P(x) = x + x^2 + 2x^3$.

Example. Let $P = [C_0 + C_1 \cdot I^2 + C_2 \cdot I^4]$. If $C_0(x) = -3$, $C_1(x) = 2$, and $C_2(x) = 1$, then $P(x) = -3 + 2x^2 + x^4$ which, it seems, is customarily written as

$$P(x) = x^4 + 2x^2 - 3.$$

In each of the examples given, we have tried to indicate that, if we start with the identity function I and the constant functions C, we can, by a finite sequence of the operations of (a) multiplying and (b) adding these functions, obtain a function P (called a polynomial function) of the form

$$P = c_0 + c_1 I + c_2 I^2 + \cdots + c_n I^n$$

where n is a counting number and each of $c_0, c_1, c_2, \ldots, c_n$ is a number. Hence, a polynomial function P is a function whose domain is R, whose range is a subset of R, and whose rule of pairing is the equation

$$P(x) = c_0 + c_1 x + c_2 x^2 + \cdots + c_n x^n.$$

Remark. The numbers $c_0, c_1, c_2, \ldots, c_n$ are called *coefficients* and the function P above is quite often called a *polynomial function of degree n*. Thus the functions in the previous examples are said to be polynomial functions of degrees 2, 3, and 4 respectively. Using other texts as criteria, it also appears to be customary to write the polynomial so that the exponents are in a decreasing order—that is, to write

$$P(x) = c_n x^n + \cdots + c_2 x^2 + c_1 x + c_0.$$

The interest by the mathematician and the student of mathematics in this type of function may not necessarily be caused by the simplicity of their construction, but it is surely aroused by the ease with which their derivatives and integrals are found. This interest might also stem from the practice of using polynomial functions to approximate other, more esoteric, functions for the simple (and quite practical) reason that the polynomial function is easier to evaluate. This practice may be illustrated by using a formula given in calculus texts to write the statement

$$E(x) \approx 1 + x + \frac{x^2}{2!} + \frac{x^3}{3!} + \frac{x^4}{4!} + \frac{x^5}{5!}.$$

In this statement, read $\approx$ as "is approximated by" and note that the right-hand side is a polynomial of degree 5. I might add that polynomial functions are of interest simply because they are everywhere continuous.

Our interest in polynomial functions will begin with the task of sketching their graphs. This task is facilitated by finding $P(0)$ and the zeros of P if there are any. Our ability to factor will also prove invaluable.

Example. If $P(x) = x^3 + x^2 - 6x$, it should be easy to see that $P(0) = 0$ and that $P(x) = x(x - 2)(x + 3)$. Thus $P(x) = 0$ if $x = 0$, $x = 2$, or $x = -3$. Additional points in P can be obtained by computations such as $P(-4) = -24$, $P(-2) = 8$, and $P(1) = -4$. The graph of this function is shown in figure 5.1.

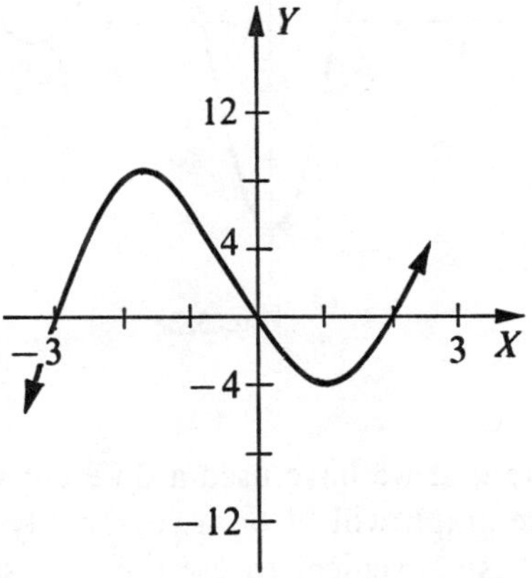

Fig. 5.1

Example. $P(x) = x^4 - 7x^3 + 11x^2 + 7x - 12$. It is obvious that $P(0) = -12$. Let us assume that we are clever enought to factor $P(x)$ and get

$$\begin{aligned} x^4 - 7x^3 + 11x^2 + 7x - 12 &= x^4 + 11x^2 - 12 - 7x^3 + 7x \\ &= (x^2 + 12)(x^2 - 1) - 7x(x^2 - 1) \\ &= (x^2 - 1)(x^2 - 7x + 12) \\ &= (x - 1)(x + 1)(x - 4)(x - 3) \end{aligned}$$

for then we would know that $P(x) = 0$ if $x = 1$, $x = -1$, $x = 3$, or $x = 4$. These and other results are shown in table 5.1.

TABLE 5.1

x	-1	0	1	2	3	4	5
$P(x)$	0	-12	0	6	0	0	48

The graph of this polynomial function is shown in figure 5.2.

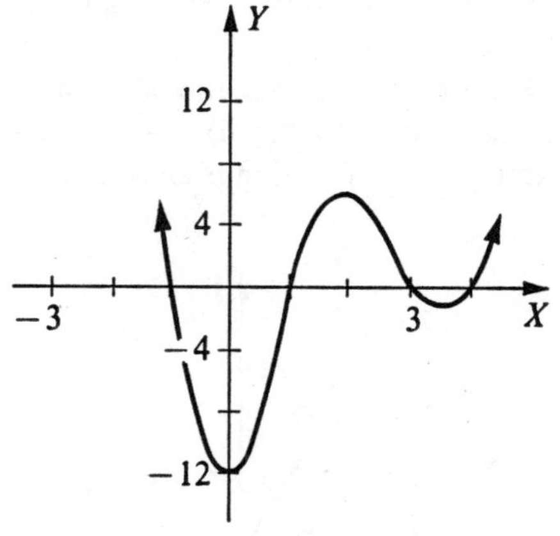

Fig. 5.2

Remark. Note that we have used a different scale on the Y axis in order that the graph will fit on a reasonably small piece of paper. Frequently, it is convenient to use different scales of measurement of the X and Y axes. This convenience usually arises for one of two reasons: (1) in order to get the graph on a reasonably sized piece of paper as was the case in this example, and (2) in applications of mathematics, the physical significance of the numbers on the two axes may be different. In such cases the units of the measurements (time, distance, cost, etc.) are not comparable, and it is useful to select suitable scales for the two axes.

The student may quickly discern that the task of finding the zeros of a polynomial function may be a formidable one because the factoring may not be so easily done as it was in the examples. Indeed, there may not be any zeros of P. This is indicated in the graph of $P(x) = x^2 + x + 1$ shown in figure 5.3. While additional information on how to find the zeros of a polynomial function will be given later, for the present I direct your attention to the idea that if x_1 and x_2 are numbers and $P(x_1) < 0$ but $P(x_2) > 0$, then the continuity of P assures us that there is a number z between x_1 and x_2 for which $P(z) = 0$. This idea is reflected in the next example and is illustrated by figure 5.4.

Example. $P(x) = x^3 + x^2 - 8x - 6$. Our computations are shown in table 5.2. From this table, we see that -3 is a zero of P and that there is a zero of P between -1 and 0 and another between 2 and 3. Remember the intermediate value theorem?

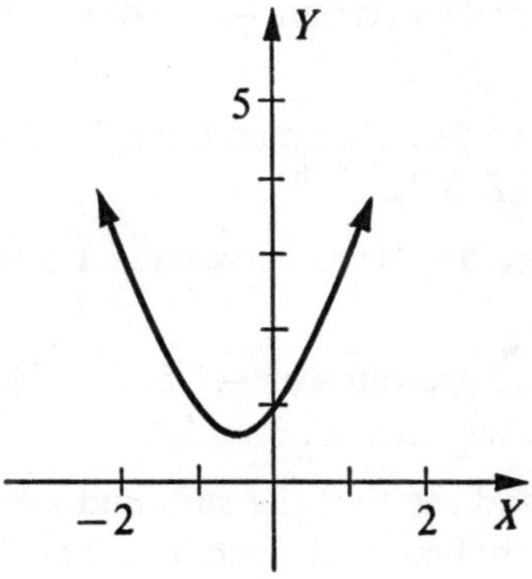

Fig. 5.3

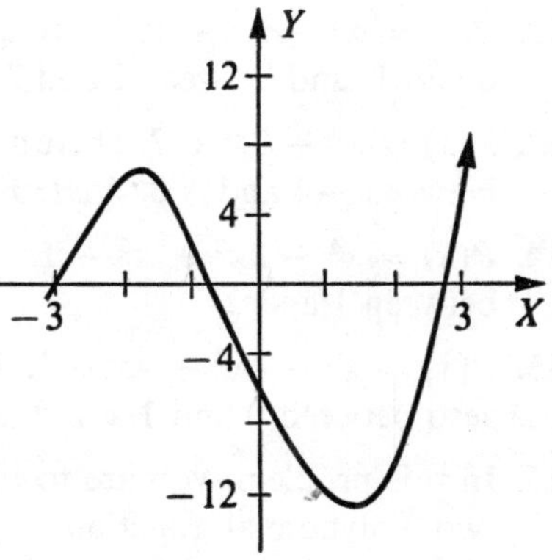

Fig. 5.4

TABLE 5.2

x	-4	-3	-2	-1	0	1	2	3
$P(x)$	-22	0	6	2	-6	-12	-10	6

PROBLEM SET 5.1

1. Assuming that $C_0(x) = 2$, $C_1(x) = -1$, and $C_2(x) = 3$, give the rule of pairing and the degree of each of the following polynomial functions.

   a. $P = C_0 \cdot I + C_1 \cdot I^2$ **b.** $P = C_0 \cdot I^3 + C_1 \cdot I + C_2$
   c. $P = C_2 \cdot I^3 + I^2 + C_1$ **d.** $P = I^5 + C_1 \cdot I^2 + C_2 \cdot I$

In problems 2–11, make use of your knowledge of factoring to give the zeros of the given function P and then sketch its graph.

2. $P(x) = x^2 - x - 2$
3. $P(x) = x^2 - 4x + 3$
4. $P(x) = 2x^2 + 3x - 2$
5. $P(x) = x^2 - 2x + 1$
6. $P(x) = x^3 - 5x^2 + 6x$
7. $P(x) = x^4 - 5x^2 + 4$
8. $P(x) = x^4 + 4x^3 - 4x^2 - 16x$
9. $P(x) = x^3 - 3x^2 - x + 3$
10. $P(x) = x^3 - 4x^2 - x + 4$
11. $P(x) = x^4 + 2x^3 - 15x^2$

In problems 12–15, verify the given statements about the zeros and sketch the graph.

12. $P(x) = x^3 - 3x + 1$. There are zeros between -2 and -1, between 0 and 1, and between 1 and 2.

13. $P(x) = x^3 - 7x + 7$. There are two zeros between 1 and 2 and another between -4 and -3. *Hint:* Find $P(3/2)$.

14. $P(x) = x^4 - 2x^3 + x^2 - 1$. There are zeros between -1 and 0 and between 1 and 2.

15. $P(x) = x^4 - 4x^3 + 4x - 1$. There are zeros at -1 and 1. There is a zero between 0 and 1/2 and another between 3 and 4.

16. In this problem, you are to try and see that the sum and product of two polynomial functions is a polynomial function. Let $P_1(x) = x^2 + x - 1$ and $P_2(x) = x^3 - x + 3$ and give $P(x)$ where **(a)** $P = P_1 + P_2$ and **(b)** $P = P_1 \cdot P_2$.

In the next three problems, we shall try to see that we know a great deal about second degree polynomial functions which are sometimes called *quadratic functions.*

17. Suppose that $P(x) = ax^2 + bx + c$ where each of a, b, and c is a number and $a \neq 0$. Explain how we can determine whether **(a)** P has two zeros, **(b)** P has only one zero, or **(c)** P has no zeros.

18. Suppose that $P(x) = ax^2 + bx + c$ where $a > 0$. Give the details (find z) of an argument which supports the following statement: "There is a number z such that $P(x) \geq z$ for every number x." Geometrically, this means that the graph of P does not contain a point below the horizontal line z.

19. Suppose that $P(x) = ax^2 + bx + c$ where $a < 0$. Give the details (find z) of an argument which supports the following statement: "There is a number z such that $P(x) \leq z$ for every number x." Geometrically, this means that the graph of P does not contain a point above the horizontal line z.

20. In this problem, we call attention to a specific reason for the mathematicians interest in polynomial functions. The *sine function* has the set of all numbers as its domain. The ordinate of the point in this function whose abscissa is x is usually denoted by sin (x) where

$$\sin (x) \approx x - \frac{x^3}{6} + \frac{x^5}{120}.$$

Use this formula to give two decimal place approximations to **(a)** sin (1), **(b)** sin (-1), and **(c)** sin (1/2).

Section 2—Properties of Polynomials It is rather easy to see that the sum and product of two polynomial functions is a polynomial function but, through the process of selecting specific polynomial functions P_1 and P_2, we can quickly observe that the quotient of two polynomials need not be a polynomial. For example, $\dfrac{x^3 + x^2 - 2}{x^2 + x - 1} = x + \dfrac{x - 2}{x^2 + x - 1}$ is not a polynomial and the difficulty is not like that encountered in the real numbers in trying to divide by 0. The difficulty is more analogous to that which arises when we divide one integer by another, the result may not be an integer. However, the process of "long division" encountered in elementary algebra permits us to conjecture the following:

> If P is a polynomial function of degree n and P_1 is a polynomial function of degree m, where $m < n$, then there is a polynomial function P_2 and a polynomial or a constant function r, called the remainder, such that $P(x) = P_1(x) \cdot P_2(x) + r(x)$ where the degree of r is less than the degree of P_1.

For our purposes, the most useful application of the above statement occurs when $P_1(x) = x - a$, that is, where P_1 is a polynomial of degree 1. In this event, you see, r will be a constant function and we can write

2.1 $P(x) = (x - a) \cdot P_2(x) + r(x)$.

Several interesting and useful ideas can be conjectured from this last statement. For example, it is obvious that

$$P(a) = (a - a) \cdot P_2(a) + r(a)$$

thus $P(a) = r(a)$. This implies that if we divide $P(x)$ by $x - a$, then the constant $r(a)$ is the value of P at a. Hence, if $r(a)$ is 0, then $P(x) = (x - a) \cdot P_2(x)$ so that $x - a$ is a factor of $P(x)$ and $P(a) = 0$ so that a is a zero of P. We illustrate these ideas in the following examples.

> *Example.* If $P(x) = x^3 + 3x^2 + x - 2$ and $P_1(x) = x + 2$, then from the "long division" shown below, we find that $P_2(x)$ is $x^2 + x - 1$ and $r(x) = 0$, hence $P(x) = (x + 2)(x^2 + x - 1)$. Therefore $(x + 2)$ is a factor of $P(x)$ and $P(-2) = 0$ or -2 is a zero of P.

$$\begin{array}{rl}
\underline{x + 2(} & x^3 + 3x^2 + x - 2 \quad \underline{)x^2 + x - 1} \\
 & \underline{x^3 + 2x^2} \\
 & \qquad x^2 + x - 2 \\
 & \qquad \underline{x^2 + 2x} \\
 & \qquad\qquad -x - 2 \\
 & \qquad\qquad \underline{-x - 2} \\
 & \qquad\qquad\quad 0
\end{array}$$

Example. If $P(x) = x^4 - x^3 - 17x^2 + 21x + 36$ and $P_1(x) = x + 2$, then the long division would look like this

$$
\begin{array}{l}
\underline{x + 2(}\ x^4 - x^3 - 17x^2 + 21x + 36 \quad \underline{)x^3 - 3x^2 - 11x + 43} \\
\qquad\quad \underline{x^4 + 2x^3} \\
\qquad\qquad\quad -3x^3 - 17x^2 + 21x + 36 \\
\qquad\qquad\quad \underline{-3x^3 - 6x^2} \\
\qquad\qquad\qquad\quad -11x^2 + 21x + 36 \\
\qquad\qquad\qquad\quad \underline{-11x^2 - 22x} \\
\qquad\qquad\qquad\qquad\quad 43x + 36 \\
\qquad\qquad\qquad\qquad\quad \underline{43x + 86} \\
\qquad\qquad\qquad\qquad\qquad\quad -50
\end{array}
$$

revealing that $P(-2) = -50$ as it should be in keeping with statement 2.1.

Admittedly, all this long division is tedious and rather laborious, but there is a method, called *synthetic division*, by which this tedious labor can be simplified. To illustrate this method, we consider the general cubic

$$P(x) = mx^3 + nx^2 + kx + c$$

which is to be divided by $x - a$. The work is exhibited below where r is the remainder.

$$
\begin{array}{l}
\underline{x - a(}\ mx^3 + nx^2 + kx + c \quad \underline{)mx^2 + (n + am)x + (k + an + a^2m)} \\
\qquad\quad \underline{mx^3 - amx^2} \\
\qquad\qquad\quad (n + am)x^2 + kx \\
\qquad\qquad\quad \underline{(n + am)x^2 - (an + a^2m)x} \\
\qquad\qquad\qquad\quad (k + an + a^2m)x + c \\
\qquad\qquad\qquad\quad \underline{(k + an + a^2m)x - (ak + a^2n + a^3m)} \\
\qquad\qquad\qquad\qquad\quad c + ak + a^2n + a^3m = r^*
\end{array}
$$

But surely, we have written down more details than we actually need. In the following, we omit every x and we also omit the second writing of m, $(n + am)$, and $(k + an + a^2m)$ inasmuch as they vanish by subtraction anyway.

$$
\begin{array}{l}
\underline{-a(}\ m + n + k + c \qquad \underline{)m + (n + am) + (k + an + a^2m)} \\
\qquad \underline{\quad -am} \\
\qquad\qquad (n + am) + k \\
\qquad\qquad \underline{\qquad\qquad -(an + a^2m)} \\
\qquad\qquad\qquad (k + an + a^2m) + c \\
\qquad\qquad\qquad \underline{\qquad\qquad\qquad -(ak + a^2n + a^3m)} \\
\qquad\qquad\qquad\qquad c + ak + a^2n + a^3m = r^*
\end{array}
$$

* Note that $r = P(a)$.

However, we can still reduce the details by changing $-a$ to a in the divisor and hence change the subtractions to additions. Also, there is no need to write the quotient in the little box to the right since every term there is to be found in the work below which we have finally condensed to just three lines.

$$\begin{array}{l|llll} a & m + & n & + k & + c \\ & & am & an + a^2m & ak + a^2n + a^3m \\ \hline & m + & (n + am) + & (k + an + a^2m) & + (c + ak + a^2n + a^3m) = r \end{array}$$

Although we have condensed the work shown, we can still extract the necessary details. The following examples will further illustrate the technique of synthetic division.

Example. Using synthetic division, divide $x^4 - 3x^3 + x + 3$ by $x - 2$.

Solution.
$$\begin{array}{l|lllll} 2 & 1 & -3 & +0 & +1 & +3 \\ & & 2 & -2 & -4 & -6 \\ \hline & 1 & -1 & -2 & -3 & +(-3) = r \end{array}$$

hence, the quotient is $x^3 - x^2 - 2x - 3$ and the remainder is -3. By direct substitution, we can also show that $P(2) = -3$.

Example. If $P(x) = 2x^3 + x^2 - 2x - 1$ is divided by $x + 1$, using synthetic division, we write

$$\begin{array}{l|llll} -1 & 2 & +1 & -2 & -1 \\ & & -2 & 1 & 1 \\ \hline & 2 & -1 & -1 & +0 = r \end{array}$$

since the remainder is 0, $x + 1$ is a factor of $P(x)$ and -1 is a zero of P. Thus $P(x) = (x + 1)(2x^2 - x - 1)$. Realizing that we can easily handle quadratic equations, we can conclude that $P(x) = (x + 1)(x - 1)(2x + 1)$, thus the zeros of P are -1, 1, and $-1/2$.

There is one word of caution about the use of the method of synthetic division, for one must notice that if one of the powers of x does not appear in the polynomial, then that power of x has zero coefficient and the synthetic division must have a zero in the corresponding position. To be sure that this caution is understood, consider the following example.

Example. Divide $P(x) = x^4 - 2x - 4$ by $x - 2$.

Solution. Using synthetic division.

$$\begin{array}{r|l} \underline{2} & 1 + 0 + 0 - 2 - 4 \\ & \quad\;\; 2 \quad 4 \quad 8 \quad 12 \\ \hline & 1 + 2 + 4 + 6 + 8 = r \end{array}$$

This means, of course, that $x - 2$ is not a factor of $P(x)$ and hence $P(2) \neq 0$. As a matter of fact, the above shows that $P(2) = 8$. I also ask you to note that the last line of the synthetic division is a row of positive numbers. Since this is the case, numbers larger than 2 can lead only to larger remainders. Thus we have eliminated each number $x \geq 2$ as a zero of P, that is, we have established an *upper bound* of the zeros of P.

PROBLEM SET 5.2

[1] If $P(x) = 3x^3 + x^2 - x + 2$ and $P_1(x) = x - 2/3$, then:

a. by direct substitution, find $P(2/3)$,
b. use long division to divide $P(x)$ by $P_1(x)$, and
c. use synthetic division to divide $P(x)$ by $P_1(x)$.

In problems 2–8, use synthetic division and give the quotient and the remainder when $P(x)$ is divided by $P_1(x)$.

2. $P(x) = 3x^3 + 2x^2 - 6x - 1$, $P_1(x) = x + 2$

3. $P(x) = x^3 - 2x^2 + 4x - 12$, $P_1(x) = x + 3$

4. $P(x) = x^4 + 5x^2 + x - 3$, $P_1(x) = x + 3$

5. $P(x) = x^5 - x^3 + 2x^2 + 1$, $P_1(x) = x - 1$

6. $P(x) = 5x^5 + 2x^4 - 15x^3 - x + 1$, $P_1(x) = x - 3$

7. $P(x) = 4x^4 + 3x^2 - 2x + 5$, $P_1(x) = x + 3$

8. $P(x) = x^4 - 5x + 10$, $P_1(x) = x - 5$

9. If $P(x) = 4x^3 - 7x^2 + 15x - 9$, use synthetic division to show that 3/4 either is or is not a zero of P.

10. If $P(x) = 4x^3 - 3x + 1$, use synthetic division to show that 1/2 either is or is not a zero of P.

11. If $P(x) = x^3 - 4x^2 - 8x - 50$, use synthetic division to find **(a)** $P(2)$, **(b)** $P(3)$, and **(c)** $P(-3)$.

12. If $P(x) = 2x^5 - x^2 + x - 5$, use synthetic division to find **(a)** $P(-1)$, **(b)** $P(1)$, and **(c)** $P(3)$.

13. If $P(x) = x^3 + 2x^2 - 7x - 8$, find an upper bound to the set of zeros of P.

14. If $P(x) = x^4 - 5x^2 + 6x - 9$, find an upper bound to the set of zeros of P.

The following problems are intended to shed more light on the reasons for interest in polynomial functions.

15. **a.** If $P(x) = 2x^3 + x + 3$, give $P(10)$.
b. If $P(x) = x^3 + 3x^2 + 0 \cdot x + 1$, give $P(10)$.
c. If $P(x) = 2x^3 + x^2 + 5x + 4$, give $P(10)$.
d. If $P(x) = x^4 + 2x^2 + 5$, give $P(10)$.
e. Study the above and then write a polynomial function P with non-negative integers as coefficients such that $P(10) = 1{,}027$.

16. From problem 15, you can realize that each counting number can be expressed as the value at $x = 10$ of some polynomial function whose coefficients are the digits, supplying 0 as the coefficient of any missing power of x. This, of course, is the basis of the decimal (base 10) system. For some purposes, it is advantageous to use the binary (base 2) system. In this system, each counting number is the value at $x = 2$ of a polynomial whose coefficients are all zeros or ones. For example, the number 37 in the decimal system is

$$1 \cdot 2^5 + 0 \cdot 2^4 + 0 \cdot 2^3 + 1 \cdot 2^2 + 0 \cdot 2 + 1 = P(2),$$

where $P(x) = x^5 + x^2 + 1$. Hence 37 is denoted by 100101 in the binary system.
a. Express the number 14 in binary notation.
b. Express 20 in binary notation.
c. Suppose the numbers n_1 and n_2 are given in the binary notation as 101 and 1010. Find the sum and product of n_1 and n_2 in the binary notation. *Hint:* Write as polynomials.

Section 3—Bounds of the Zeros The promised additional information on how to find the zeros of P will be postponed until we have considered certain relevant questions.

Questions. If we are given a polynomial function P.

1. how many zeros of P can we expect?
2. is there a number b which is not exceeded by any of the zeros of P?

Before we try to indicate the answers to our questions, we remind ourselves that the domain of P is the set of real numbers and that its range is a subset of the reals. So it is not necessary to consider complex numbers.

The answer to our first question is complicated by the fact that P need not have a zero. However, if z_1 is a zero of P, then we know that $P(x) = (x - z_1) \cdot Q_1(x)$, where Q_1 is a polynomial (or a constant) function of degree 1 less than the degree of P. Now, if z_2 is a zero of Q_1 and $z_2 \neq z_1$, then z_2 is a zero of P and we have $P(x) = (x - z_1)(x - z_2) \cdot Q_2(x)$. If this same sort of argument could be repeated n times, we should have $P(x) = (x - z_1)(x - z_2) \cdots (x - z_n) \cdot Q_n(x)$ where Q_n would be of degree 0, hence Q_n is a constant, say k, and

$$P(x) = k(x - z_1)(x - z_2) \cdots (x - z_n).$$

Thus P would have n distinct zeros. Now, if t is a number distinct from $z_1, z_2, \ldots, z_n$, then when we substitute t for x in the last equation, we obtain

$$P(t) = k(t - z_1)(t - z_2) \cdots (t - z_n) \neq 0$$

since each factor is distinct from 0.

Thus the formal reply to question (1) is given by the statement below.

> If P is a polynomial function of degree n, then P has no more than n distinct zeros.

In the argument which precedes this statement, if we relax the restriction that $z_1 \neq z_2$ and permit $z_1 = z_2$, then we would have written $P(x) = (x - z_1)^2 \cdot Q_2(x)$ and we would have used the terminology "z_1 is a zero of multiplicity two."

> *Example.* If $P(x) = (x - 2)^2(x + 5)^3$, then P is of degree 5 and we can expect to hear that 2 is a zero of multiplicity two and -5 is a zero of multiplicity three. But P has only two distinct zeros. It is enough to say that P has two zeros.

We address ourselves to our second question because in the next section we shall comment upon a method for finding the rational zeros. Happily, this method reduces the number of trails to a finite number. Unhappily, this finite number of trials may be unpleasantly large. We hope to cut down the necessary work to a reasonable minimum. So let us begin.

Suppose $P(x) = c_n x^n + \cdots + c_1 x + c_0$, where $c_n > 0$. Let $b > 0$. Suppose further that in the synthetic division of $P(x)$ by $x - b$, the third line contains no negative numbers, that is, the coefficients of the quotient Q and the remainder r are nonnegative numbers. Since all the coefficients of $Q(x)$ are nonnegative numbers, one of which, c_n, is a positive number, then $Q(x) > 0$ for each $x > 0$. Let us choose $x_1 > b$. Surely $x_1 > 0$.

Hence $Q(x_1) > 0$. Substituting x_1 for x in $P(x) = (x - b)\cdot Q(x) + r$, we have

$$P(x_1) = (x_1 - b)\cdot Q(x_1) + r$$

and, recalling that $r \geq 0$, it follows that $P(x_1) > 0$. Therefore we answer our second question as follows:

> In the synthetic division of $P(x)$ by $x - b$, where each of c_n and b are positive numbers, if all of the numbers in the third row are nonnegative, then there is no zero of P which exceeds b—that is, b is an upper bound of the zeros.

Example.

$$\begin{array}{r|l} \underline{2(} & 1 + 1 - 6 + 2 - 3 \\ & \quad\ \ 2 \quad 6 \quad 0 \quad 4 \\ \hline & 1 + 3 + 0 + 2 + 1 \end{array}$$

Since all the numbers in the third row are nonnegative, 2 is an upper bound of the zeros of $P(x) = x^4 + x^3 - 6x^2 + 2x - 3$.

We can extend the usefulness of synthetic division by observing that if $b > 0$ is an upper bound of the zeros for $P(-x)$, then $-b$ is a lower bound of the zeros for $P(x)$.

Example. If $P(x) = x^4 - 2x^3 + 3x + 5$, then $P(-x) = x^4 + 2x^3 - 3x + 5$.

Choosing $b = 3$, we divide $P(-x)$ by $x - 3$.

$$\begin{array}{r|l} \underline{3(} & 1 + 2 + 0 - \ \ 3 + \ \ 5 \\ & \quad\ \ 3 \quad 5 \quad 15 \quad 36 \\ \hline & 1 + 5 + 5 + 12 + 41 \end{array}$$

which shows that 3 is an upper bound of the zeros for $P(-x)$. Hence, -3 is a lower bound of the zeros of $P(x)$.

We can improve this, however. Reconsider the last example and study the following synthetic division of $P(x)$ by $x - (-3)$, that is, by $x + 3$.

$$\begin{array}{r|l} \underline{-3(} & 1 - 2 + \ \ 0 - \ \ 3 + \ \ \ 5 \\ & \quad -3 \quad 15 \ -45 \quad 144 \\ \hline & 1 - 5 + 15 - 48 + 149 \end{array}$$

So we improve our previous statement to:

> In the synthetic division of $P(x)$ by $x - b$, where $c_n > 0$ and $b < 0$, if the numbers in the third row alternate in sign, then b is a lower bound of the zeros of P.

Let us summarize all of our remarks about the bounds of the zeros in the following recapitulation:

> Suppose $P(x) = c_nx^n + \cdots + c_2x^2 + c_1x + c_0$ and $c_n > 0$. In the synthetic division of $P(x)$ by $x - b$, if:
>
> 1. $b > 0$ and all of the numbers in the third row are nonnegative, then b is an upper bound of the zeros.
> 2. $b < 0$ and the numbers in the third row alternate in sign, then b is a lower bound of the zeros.

PROBLEM SET 5.3

Give the maximum number of distinct zeros for each of the following polynomial functions.

1. $P(x) = 3x^3 + x - 1$ **2.** $P(x) = x^5 + x^2 - 7x + 2$

3. $P(x) = x^5 - 1$ **4.** $P(x) = 2x^4 + x^3 - x^2 - 3$

Find, by inspection, the zeros of the following polynomial functions and give the multiplicity when applicable.

5. $P(x) = (x - 2)(x - 3)^2(x + 4)^3$

6. $P(x) = (x + 1)^4(x - 2)^2$

7. $P(x) = x^2 - 4x + 4$

8. $P(x) = (x^2 + 3x - 10)(x + 5)^2$

Give an upper and a lower bound to the zeros of the given polynomial functions.

9. $P(x) = x^3 + 2x^2 - 7x - 8$

10. $P(x) = x^4 + x^3 - 6x^2 - 8x - 16$

11. $P(x) = 2x^4 + 2x^3 + x^2 - 9x + 20$

12. $P(x) = x^4 + x^3 - 2x - 4$

13. $P(x) = x^3 - 8x + 5$

14. $P(x) = x^5 - 3x^3 + 12$

14. Use synthetic division to evaluate $P(3/2)$ if

$$P(x) = 2x^3 - 2x^2 + x + 6.$$

16. Use synthetic division to evaluate $P(-1/2)$ if

$$P(x) = 2x^4 + x^3 - 6x + 1.$$

Section 4—Rational Zeros of *P* Looking back on the method of synthetic division introduced in the last section, we have some indication that we could use this process to aid us in finding the zeros of a polynomial function if there are any. We have also tried to see how this same process could yield an upper bound to the set of zeros of P. In this context, there is at least one valid question about all this, namely:

> In using synthetic division, is there some method by which a is selected in dividing $P(x)$ by $x - a$ or was a randomly chosen?

It was not randomly chosen as the next statement will indicate. The proof of this statement, however, will not be given in this text.

THEOREM. *If the rational number p/q, in lowest terms, is a zero of the polynomial function P,*

$$P(x) = c_n x^n + \cdots + c_2 x^2 + c_1 x + c_0$$

where each of the coefficients is an integer, then p is a factor of c_0 and q is a factor of c_n.

Remarks. These remarks pertain to the above theorem. Suppose that p/q is a rational number, in lowest terms, and that p/q is a zero of P whose coefficients are integers. Since $P(p/q)$ is 0, we have that

$$c_n(p/q)^n + \cdots + c_2(p/q)^2 + c_1(p/q) + c_0 = 0.$$

If we multiply both sides of this equation by q^n and then add $-c_0 q^n$ to both sides, we would find that

$$c_n p^n + \cdots + c_2 p^2 q^{n-2} + c_1 p q^{n-1} = -c_0 q^n$$

so if we factor p from the left-hand side, we have

$$p \cdot (c_n p^{n-1} + \cdots + c_2 p q^{n-2} + c_1 q^{n-1}) = -c_0 q^n$$

thus p is a factor of $c_0 q^n$. Since p/q is in lowest terms, p is not a factor of q^n, so it must be true that p is a factor of c_0. If we were to begin with the equation

$$c_{n1} p^{n-1} q + \cdots + c_2 p^2 q^{n-2} + c_1 p q^{n-1} + c_0 q^n = -c_n p^n$$

by factoring q from the left-hand side, we could similarly argue that q is a factor of c_n. These remarks are far from a proof, but they do offer evidence that the theorem is true, and we shall assume that this is the case.

A special case of this theorem is worthy of mention. It is, as a matter of fact, the case encountered in many of the problems given in the exercises. Suppose that $c_n = 1$, then, if there are any rational zeros of the polynomial $P(x) = x^n + \cdots + c_2 x^2 + c_1 x + c_0$, they would have to be factors of c_0 and, since c_0 is an integer, the zeros would have to be integers. This special case of our theorem is illustrated in the next example.

Example. Suppose $P(x) = x^3 + 4x^2 + 9x + 10$. Here $c_0 = 10$ so the integral factors of c_0 are ± 1, ± 2, ± 5, and ± 10. Since $c_n = 1$, its integral factors are ± 1 only, so if there is a rational zero of P, it must be an element of the number set $\{-10, -5, -2, -1, 1, 2, 5, 10\}$. We may now determine if any of these numbers is a zero of P by either direct substitution or by appealing to statement 2.1 and synthetic division.

In the previous example, most students would probably prefer direct substitution in spite of their knowledge of synthetic division. Let me point out, however, that synthetic division has the advantage when either n is large or a is other than a small integer. It may also show us that some of the computations are unnecessary by giving us an upper bound to the zeros of P.

If $c_n \neq 1$, things can get a bit sticky. For example, if $P(x) = 2x^4 + x^3 - 2x^2 - 4x - 3$, then the integral factors of c_0 are ± 1 and ± 3, while the integral factors of c_n are ± 1 and ± 2. Therefore, if there are any rational zeros of P they are elements of the number set

$$\{-3, -3/2, -1, -1/2, 1/2, 1, 3/2, 3\}.$$

Before this goes any further, let me emphasize that none of these numbers need be a zero of P. Indeed, there may not be a rational number x such that $P(x) = 0$. Now I shall continue, using synthetic division.

$$\begin{array}{r|rrrrr} \underline{1(} & 2 & +1 & -2 & -4 & -3 \\ & & 2 & 3 & 1 & -3 \\ \hline & 2 & +3 & +1 & -3 & -6 \end{array}$$

which shows that 1 is not a zero of P. It also shows that 1 is not an upper bound to the zeros of P. It also shows that $P(1) = -6$, which is useful in sketching the graph of P.

$$\begin{array}{r|rrrrr} \underline{3/2(} & 2 & +1 & -2 & -4 & -3 \\ & & 3 & 6 & 6 & 3 \\ \hline & 2 & +4 & +4 & +2 & +0 \end{array}$$

which shows that 3/2 is a zero of P and that

$$P(x) = (x - 3/2)(2x^3 + 4x^2 + 4x + 2).$$

Example. Find all the rational zeros of the polynomial function $P(x) = 2x^4 + 7x^3 + 4x^2 - 7x - 6$.

Solution. The integral factors of -6 are ± 6, ± 3, ± 2, and ± 1. The integral factors of 2 and ± 2 and ± 1. Thus the possible rational zeros of P are

$$\{-6, -3, -2, -3/2, -1, -1/2, 1/2, 1, 3/2, 2, 3, 6\}$$

We shall try 2.

$$\begin{array}{r|rrrrr} \underline{2} & 2+ & 7+ & 4- & 7- & 6 \\ & & 4 & 22 & 52 & 90 \\ \hline & 2+ & 11+ & 26+ & 45+ & 84=r \end{array}$$

Hence $P(2) = 84$ and 2 is an upper bound of the zeros of P.

$$\begin{array}{r|rrrrr} \underline{1} & 2+ & 7+ & 4- & 7- & 6 \\ & & 2 & 9 & 13 & 6 \\ \hline & 2+ & 9+ & 13+ & 6+ & 0=r \end{array}$$

hence $P(1) = 0$. The synthetic division above shows that $P(x) = (x - 1)Q(x)$ where $Q(x) = 2x^3 + 9x^2 + 13x + 6$, so that the remaining zeros of P are zeros of Q. The possible rational zeros of Q are $\{-3, -2, -3/2, -1, +1/2, 1/2, 1, 3/2\}$. We try -1.

$$\begin{array}{r|rrrr} \underline{-1} & 2+ & 9+ & 13+ & 6 \\ & & -2 & -7 & -6 \\ \hline & 2+ & 7+ & 6+ & 0=r \end{array}$$

hence $P(-1) = 0$. We also now know that

$$P(x) = (x - 1)(x + 1)(2x^2 + 7x + 6).$$

The quadratic equation $2x^2 + 7x + 6 = 0$ is easy to solve, thus the rational zeros of P are -2, $-3/2$, -1, and 1.

In addition to the problem of finding all of the rational zeros (if there are any) of a given polynomial function P, there is another task that is of interest to us—the graph of P. In the next example, we shall consider both.

Example. If $P(x) = 2x^3 - x - 3$, find all rational zeros of P and sketch its graph.

Solution. $P(0) = -3$ and the possible rational zeros are $\pm \{1/2, 1, 3/2, 3\}$.

$$\begin{array}{r|rrrr} \underline{1} & 2+ & 0- & 1- & 3 \\ & & 2 & 2 & 1 \\ \hline & 2+ & 2+ & 1- & 2=r \end{array} \qquad \begin{array}{r|rrrr} \underline{3/2} & 2+ & 0- & 1- & 3 \\ & & 3 & 9/2 & 21/4 \\ \hline & 2+ & 3+ & 7/2+ & 9/4=r \end{array}$$

The above shows that $P(1) = -2$, $P(3/2) = 9/4$, that 3/2 is an upper bound of the zeros, and that there is a zero of P between 1 and 3/2, but it is not a rational number. Why?

$$\begin{array}{r|rrrr} \underline{-1/2} & 2+ & 0- & 1- & 3 \\ & & -1 & 1/2 & 1/4 \\ \hline & 2- & 1- & 1/2- & 11/4 \end{array} \qquad \begin{array}{r|rrrr} \underline{-1} & 2+ & 0- & 1- & 3 \\ & & -2 & 2 & -1 \\ \hline & 2- & 2+ & 1- & 4 \end{array}$$

which shows that $P(-1/2) = -11/4$, $P(-1) = -4$, and that -1 is a lower bound of the zeros. A similar use of synthetic division would show that $P(1/2) = -13/4$ hence, *there are no rational zeros of P.* Why? Though direct substitution can be used, I shall obtain additional points in P by using synthetic division.

$$\begin{array}{l} \underline{2(}\ 2 + 0 - 1 - \ 3 \\ \qquad\ \ 4 \quad 8 \quad 14 \\ \hline 2 + 4 + 7 + 11 \end{array} \qquad \begin{array}{l} \underline{-3/2(}\ 2 + 0 - \ 1 - \ \ 3 \\ \qquad\qquad -3 \quad 9/2 \quad 21/4 \\ \hline 2 - 3 + 7/2 - 33/4 \end{array}$$

The graph of this function is shown in figure 5.5.

Example. If $P(x) = x^4 + x^3 - 7x^2 - 13x - 6$, find all the rational zeros of P and sketch its graph.

Solution. Noting that $c_n = 1$, the possible rational zeros of P are $\pm\{1, 2, 3, 6\}$.

$$\begin{array}{l} \underline{1(}\ 1 + 1 - 7 - 13 - \ 6 \\ \qquad\ \ 1 \quad 2 \quad -5 \quad -18 \\ \hline 1 + 2 - 5 - 18 - 24 \end{array} \qquad \begin{array}{l} \underline{-1(}\ 1 + 1 - 7 - 13 - 6 \\ \qquad\ -1 \quad 0 \quad 7 \quad 6 \\ \hline 1 + 0 - 7 - \ 6 + 0 \end{array}$$

thus we have that $P(1) = -24$ and $P(-1) = 0$. We also see that

$$P(x) = (x + 1)(x^3 - 7x - 6).$$

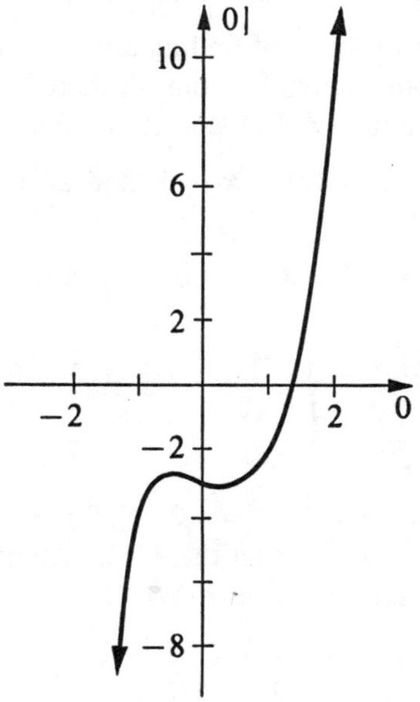

Fig. 5.5

We continue with the polynomial $Q(x) = x^3 - 7x - 6$ for the zeros of Q are also zeros of P.

$$\begin{array}{r|rrrr} -2 & 1 & +0 & -7 & -6 \\ & & -2 & 4 & 6 \\ \hline & 1 & -2 & -3 & +0 \end{array}$$

hence $P(-2) = 0$, and now we have that

$$P(x) = (x + 1)(x + 2)(x^2 - 2x - 3).$$

Factoring the quadratic, we have that

$$\begin{aligned} P(x) &= (x + 1)(x + 2)(x - 3)(x + 1) \\ &= (x + 1)^2(x + 2)(x - 3). \end{aligned}$$

Hence, the rational zeros of P are -2, -1, and 3. The graph of P is shown in figure 5.6.

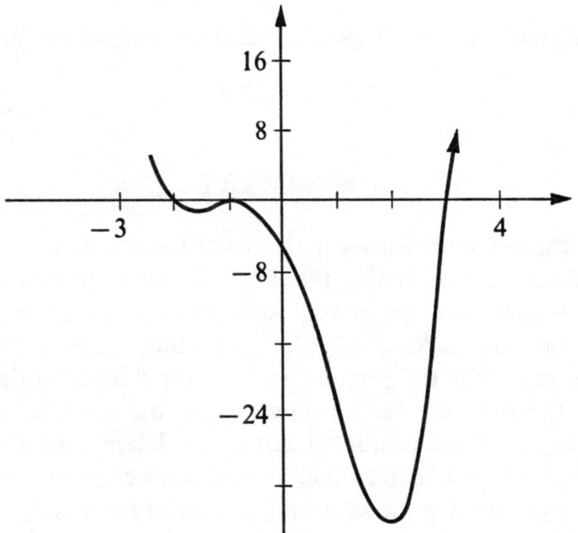

Fig. 5.6

PROBLEM SET 5.4

Find all the rational zeros of the following polynomial functions.

1. $P(x) = x^3 + x^2 - 3x - 4$
2. $P(x) = x^4 - 2x^3 + 3x^2 + x + 6$

3. $P(x) = 2x^4 + 5x^3 - 11x^2 - 20x + 12$

4. $P(x) = 2x^3 - 3x^2 - 11x + 6$

5. $P(x) = x^3 - 6x^2 + 11x - 6$

6. $P(x) = x^4 + x^3 - 11x^2 - 9x + 18$

7. $P(x) = x^4 + 2x^3 - 3x^2 - 4x + 4$

8. $P(x) = x^4 - 2x^3 + x^2 - 1$

9. $P(x) = x^4 - 4x^3 - 4x + 12$

10. $P(x) = 3x^3 + 2x^2 - 3x - 1$

11. $P(x) = 3x^3 + 8x^2 - x - 20$

12. $P(x) = 6x^3 + 11x^2 - 4x - 4$

13. $P(x) = 2x^4 + 5x^3 - x^2 + 5x - 3$

Sketch the graphs of the polynomial functions assigned in problems 1–13 above.

SUMMARY

In this chapter we discussed polynomial functions. The major notion considered was that of finding the zeros of such functions. In so doing, considerable use was made of your ability to factor certain polynomials. Some use was also made of your skill in handling quadratic equations. It is also the case that the primary reasons for discussing the factor and remainder theorems are for learning how to use synthetic division; for determining the possible rational zeros; for determining the bounds of the zeros—all of these matters had a direct connection to the central topic of this chapter—finding the zeros of polynomial functions.

Review Problems

1. If P is a certain polynomial function, then the synthetic division illustrated below shows that $P(-2) = 0$, hence that -2 is a zero of P.

$$\begin{array}{r|l} \underline{-2} & 1 + 2 + 1 + 0 - 4 \\ & \quad\ -2\ -0\ -2\ +4 \\ \hline & 1 + 0 + 1 - 2 + 0 \end{array}$$

It also shows other things. So answer the following questions.

a. What is $P(x)$?
b. What is $Q(x)$?
c. What does this show about the factors of $P(x)$?

2. Find the zeros of Q as given in the problem above. Are the zeros of Q also zeros of P?

3. Use the information obtained in problems 1 and 2 to sketch the graph of the polynomial function P.

4. Give the zeros of P if $P(x) = 4x^2 - 24x - 3$.

5. List all possible rational zeros of P if

$$P(x) = 2x^3 - 8x^2 + 2x + 12.$$

In each of problems 6–8, use synthetic division to answer the given question.

6. Is $x - 3$ a factor of $P(x) = 3x^5 - 2x^4 + 5x^2 + x - 11$?

7. What is the number c if -2 is a zero of

$$P(x) = x^4 - 2x^3 - x^2 + cx - 6?$$

8. If $P(x) = x^4 - 2x^2 + 7$, what is $P(2)$?

CHAPTER 6
Functions L and E

H(x) = 1/x

Section 1—Area of a Region Bounded by H We shall borrow from the set of all points the notion of a rectangular region and we shall assume that we can find the area of a rectangle. We also need to remember that the point (x,y) belongs to the special function H only if x is a positive number and y is the reciprocal of x. We then go on to the following: Let S denote the region of the plane, represented by the shaded portion of figure 6.1, to which the point (u,z) belongs only if u belongs to the interval $[a,b]$ of positive numbers and $z = 0$, $z = 1/u$, or z is between 0 and $1/u$.

> *Question.* We can define the area of a rectangular region, but how shall we define a number suitable to be called *the area of region* S?

Since the region S shown in figure 6.1 is not rectangular, nor triangular, not circular, nor any other of the regions whose areas we learned how to compute in plane geometry, it must be admitted that it may not be easy for us to answer the question posed above at this time. Therefore, we shall simply bypass the question for now and simply *assume* that the region S has area and that there is some positive number which we can assign to the area of this region. Let us denote this number by the symbol

$$\int_a^b H.$$

Care must be taken to observe that the symbol we have just given denotes nothing more nor less than a positive number that we assume is the area

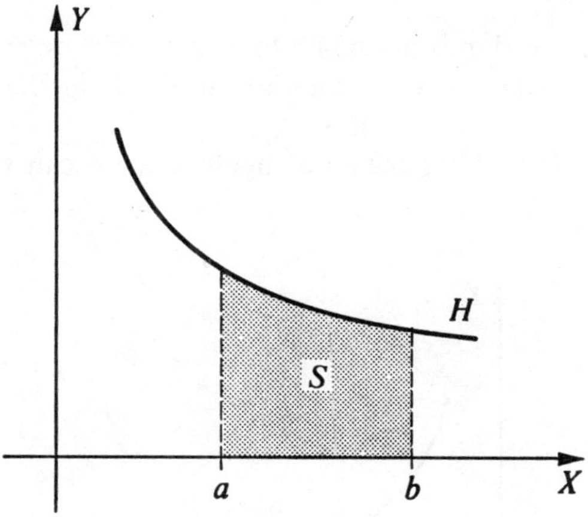

Fig. 6.1

of region *S*. It cannot and is not claimed that this symbol gives us some magic formula for computing the area of region *S*. It can be claimed, however, that we will denote the area of the shaded region in figure 6.2 by the notation

$$\int_1^x H.$$

We shall read this as "the area of the region bounded by the graph of *H* on the interval [1,*x*]."

Now, given $0 < a < b$, we are concerned, and rightly so, with the magnitude of this number, that is, we are concerned with the question, What is the area of the region *S* ? Patience. . . . We have noted above that

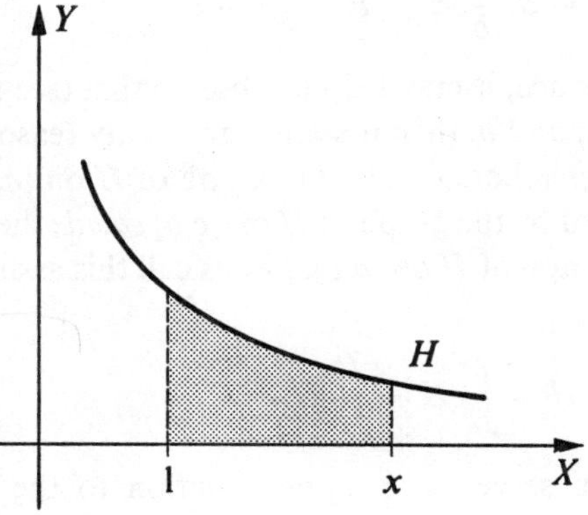

Fig. 6.2

answering this question is not going to be easy, but we shall, immediately if not sooner, make some attempt to answer it in the comments that follow.

In the first place, by glancing at figure 6.3, we can see that it seems

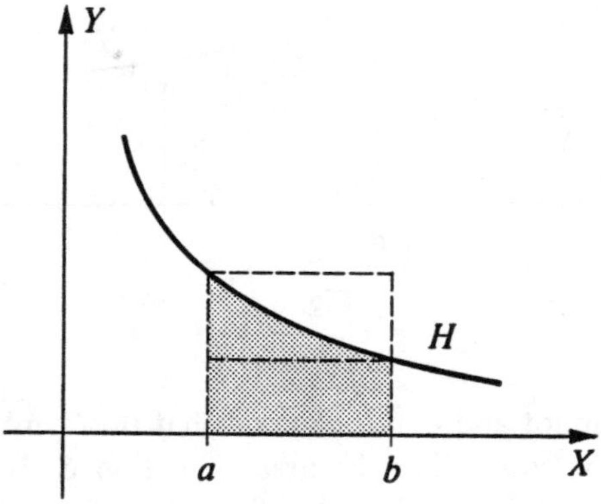

Fig. 6.3

reasonable to assert that the area of the region S is a positive number between the areas of the two rectangles shown in this figure, one of which is included in S and the other including S.

Question. What are the areas of these two rectangles?

When this question has been correctly answered, we can see that (for reference purposes call this property 1)

Property 1 $$(b - a)\cdot\frac{1}{b} < \int_b^a H < (b - a)\cdot\frac{1}{a}.$$

In the second place, it may help to observe that (see figure 6.4), if c is a number between a and b, then it seems innocently reasonable to state that *the area of the region* bounded by the graph of H on $[a,c]$ *plus* the area of the region bounded by the graph of H on $[c,b]$ *equals* the area of the region bounded by the graph of H on $[a,b]$. Let us call this assumption property 2 and write it as

Property 2 $$\int_a^b H = \int_a^c H + \int_c^b H.$$

Property 1 can serve as an approximation to the number $\int_a^b H$, for example, if $[a,b] = [1,4]$, then this property states that $3/4 < \int_1^4 H < 3$.

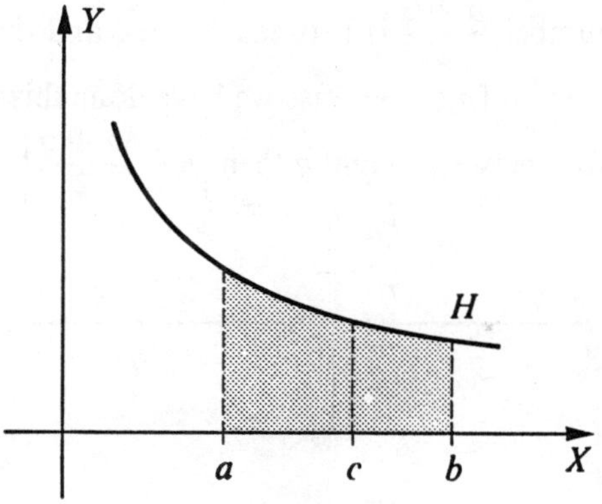

Fig. 6.4

This hardly seems acceptable as a reasonably accurate estimate, however. We might also note, as suggested by student O'Leary, that the area of the trapezoid shown in figure 6.5 could be used to estimate the area of the region S. In connection with this suggestion, there are two pertinent questions which must be asked.

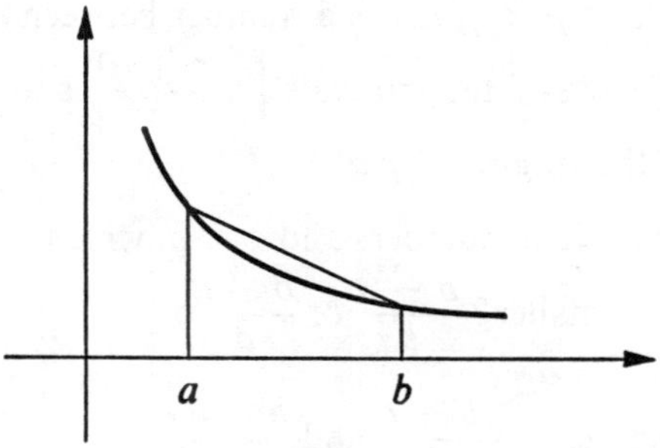

Fig. 6.5

Question. What is the area of this trapezoid? *and*, equally important,

Question. How closely does the area of this trapezoid approximate the area of the region S?

On the basis of these last few comments, if you have concluded that we are seeking some suitable means of approximating the number $\int_a^b H$, you are correct. It seems like an honorable undertaking. To this end then, let us conclude this day's effort by recalling that if p and q are numbers,

$p < q$, then the number $\frac{p+q}{2}$ is between p and q and that it is called the midpoint of the interval $[p,q]$. Because we have done this before, we know that if r is a number between p and q then $\left| r - \frac{p+q}{2} \right| < \frac{q-p}{2}$.

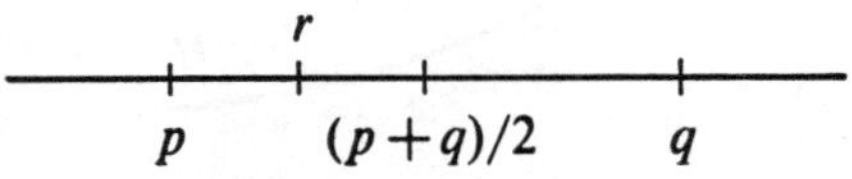

Fig. 6.6

PROBLEM SET 6.1

Problem Set on Intervals

1. Give a geometric interpretation of property 2.

2. Suppose that p and q are numbers and that $p < q$. Find the lengths of the intervals $\left[p, \frac{p+q}{2}\right]$ and $\left[\frac{p+q}{2}, q\right]$.

3. Suppose that $0 < p < q$. If r is a number between p and q verify that the length of each of the intervals $\left[r, \frac{p+q}{2}\right]$ and $\left[\frac{p+q}{2}, r\right]$ is less that one-half the length of $[p,q]$.

4. If a and b are positive numbers and $a < b$, which of the following two numbers is the smaller? $\frac{b-a}{b}$ or $\frac{b-a}{a}$.

5. Find the "average" of $\frac{b-a}{b}$ and $\frac{b-a}{a}$.

6. If z is the number found in problem 5, find the length of each of the intervals $\left[z, \frac{b-a}{a}\right]$ and $\left[\frac{b-a}{b}, z\right]$.

7. Solve the inequality $\left| x - \frac{b^2 - a^2}{2ab} \right| < \frac{(b-a)^2}{2ab}$.

[8] Begin this problem by drawing horizontal lines through the points $(1/a, a)$ and $(1,1)$ in figure 6.7 and then note that the region bounded by these horizontal lines, the Y axis, and the graph of H is congruent to the region S and should therefore have the same area. Now then,

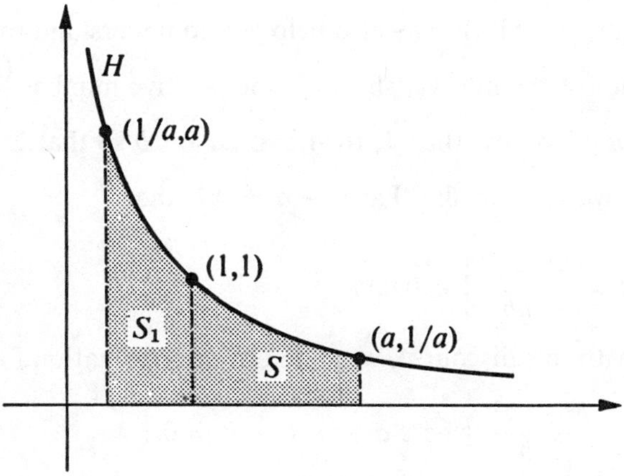

Fig. 6.7

go to work and show that the area of region S_1 equals the area of region S.

9. Give a geometric representation of $\int_1^3 H$.

[10] Give the area of the trapezoid shown in figure 6.5 of this section.

Section 2—Approximation Formula for $\int_a^b H$ While we have not defined the area of the region S shown in figure 6.1 of the previous section, we have assumed that it exists and have attributed to it several properties. Because of our work in the preceding section, we are now in a position to obtain an approximation to the positive number $\int_a^b H$, that is, the area of region S.

Property 1 asserts that $\int_a^b H$ is between $\frac{b-a}{b}$ and $\frac{b-a}{a}$, so that, by using the ideas of the last section, we can reason that $\int_b^a H$ differs from the average of $\frac{b-a}{b}$ and $\frac{b-a}{a}$ by less than $\frac{(b-a)^2}{2ab}$. But the average of the two numbers above is the positive number $\frac{b^2-a^2}{2ab}$, hence we can reason that $\int_a^b H$ is approximated by $\frac{b^2-a^2}{2ab}$ and that the error of this approximation is less that $\frac{(b-a)^2}{2ab}$. We recap this discussion by the following statement:

$$\left|\int_v^b H - \frac{b^2-a^2}{2ab}\right| < \frac{(b-a)^2}{2ab}$$

and give figure 6.8 which may also help you to understand this discussion, for the length of the interval shown is the positive number $\dfrac{(b-a)^2}{ab}$.

Now, if a is not less than 1, then $2 \leq 2a < 2b$ so that $2 < 2ab$, hence $\dfrac{1}{2ab} < \dfrac{1}{2}$. If we also require that $b - a = 0.1$, then

$$\left| \int_a^b H - \frac{b^2 - a^2}{2ab} \right| < 0.005.$$

Armed with this discourse, we state our approximation formula

$$\int_a^b H = \frac{b^2 - a^2}{2ab}, \text{ where } a \geq 1,\ b - a = 0.1.$$

The error in this approximation is less than 0.005, that is, two decimal places, if we require that $a \geq 1$ and that $b - a = 0.1$. Therefore we will require these things.

With no further delay, we turn our attention to the computations below.

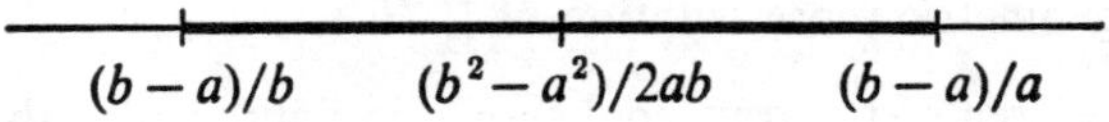

Fig. 6.8

Example.

$$\int_1^{1.1} H = \frac{(1.1)^2 - 1^2}{2.2} = \frac{0.21}{2.2}.$$

We pause to reach an agreement. In the numerical work of simplifying the quotient on the right-hand side, you are expected to round off the third decimal place in accordance with the customary rule for doing these things. Thus,

$$\int_1^{1.1} H = 0.095.$$

Example. Study this example rather carefully.

$$\int_{1.1}^{1.2} H = \frac{(1.2)^2 - (1.1)^2}{2.64} = \frac{0.23}{2.64} = 0.087$$

and so, by property 2 of the preceding section,

$$\int_1^{1.2} H = \int_1^{1.1} H + \int_{1.1}^{1.2} H = 0.095 + 0.087 = 0.182.$$

Example. This will be our last example, for by now the path is clearly marked.

$$\int_1^{1.3} H = \int_1^{1.2} H + \int_{1.2}^{1.3} H \text{ by property 2.}$$

Our approximating formula yields (check this)

$$\int_{1.2}^{1.3} H = 0.080$$

hence

$$\int_{1.2}^{1.3} H = 0.182 + 0.080 = 0.262.$$

PROBLEM SET 6.2

Exercise Complete this table of approximations (see table 6.1).

TABLE 6.1

x	$\int_1^x H$	x	$\int_1^x H$
1.1	0.095	1.6	
1.2	0.182	1.7	
1.3	0.262	1.8	
1.4		1.9	
1.5		2.0	

Problem H-1 Find *any* evidence you can to support the statement "if $1 < x$, then $\int_1^x H = \int_{1/x}^1 H$".

Problem H-2 Find *any* evidence you can to support the statement "if $0 < c$, then $\int_a^b H = \int_{ca}^{cb} H$". *Hint:* The areas of the trapezoids in figure 6.9 might help.

Section 3—Function L Using our work in the last section as a guide, it is very easy for us to understand that, if $x_1 > 1$, $x_2 > 1$, and $x_1 \neq x_2$, then

$$\int_1^{x_1} H \neq \int_1^{x_2} H.$$

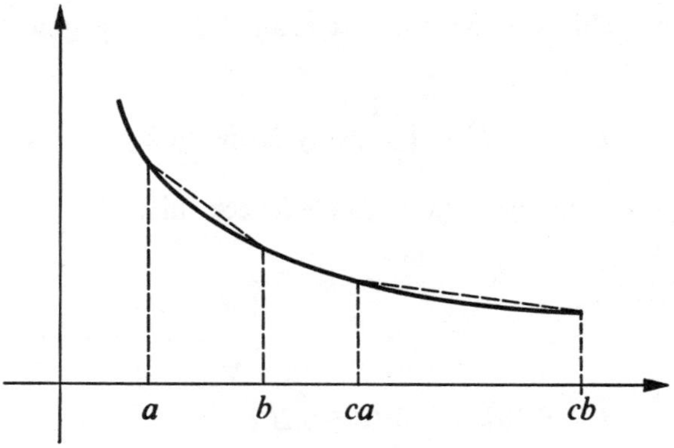

Fig. 6.9

Remark. Read this as "the area of the region bounded by the graph of H on $[1,x_1] \neq$ the area of the region bounded by the graph of H on $[1,x_2]$".

To state the obvious, then, the area of a region bounded by the graph of H on $[1,x]$ depends upon the number x, that is, there is a collection of ordered number pairs $(x, \int_1^x H)$ in which no two pairs have the same first element. Does this sound familiar?

We now define another special function which we will denote by L.

DEFINITION. *L denotes the function whose domain is the set of positive numbers and for which*

$$L(x) = \begin{cases} -\int_x^1 H & \text{when } 0 < x < 1 \\ 0 & \text{when } x = 1 \\ \int_1^x H & \text{when } x > 1 \end{cases}$$

Our immediate objective now is to discover or develop some of the properties of this function L. In order to do this, we shall continue to use the properties given in section 1, and we shall feel free to use any or all of the problems in problem set 6.2.

We resume our investigation of the number $\int_a^b H$ by pointing out that there are five cases (but no more) for us to consider. (See figures 6.10 through 6.14.) Let us now consider case 1. The definition of L tells us

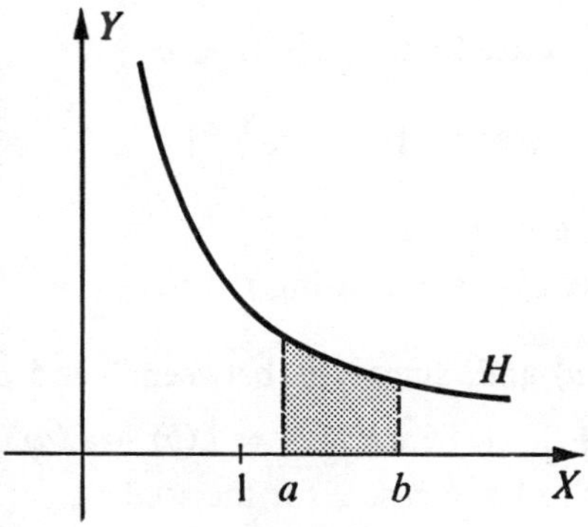

Fig. 6.10

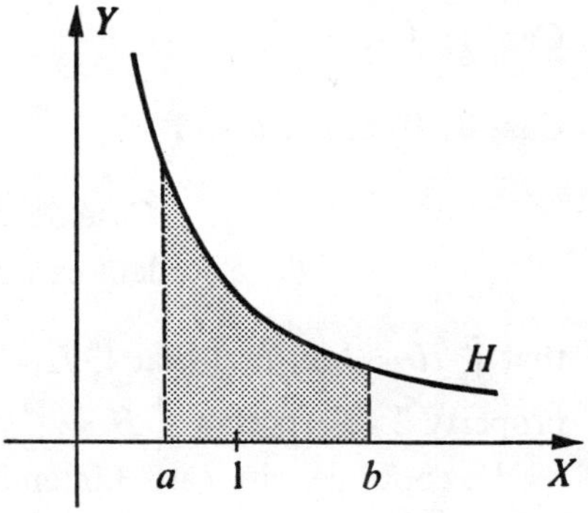

Fig. 6.11

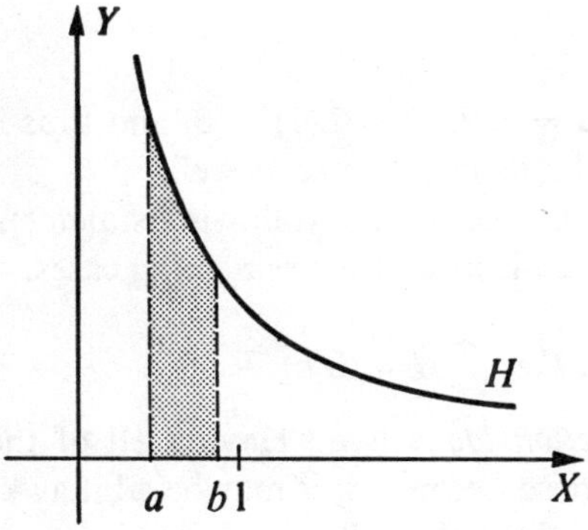

Fig. 6.12

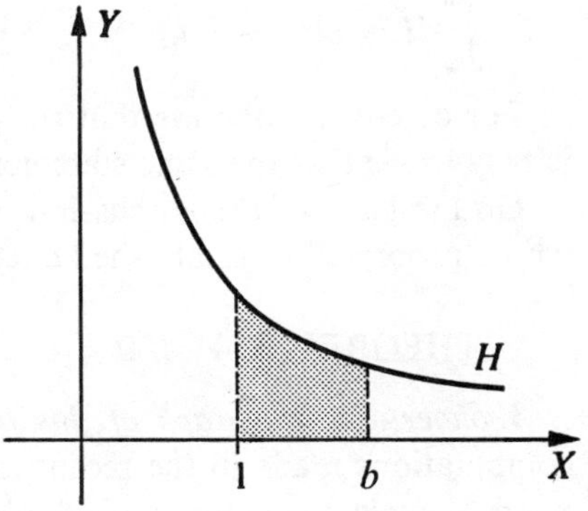

Fig. 6.13

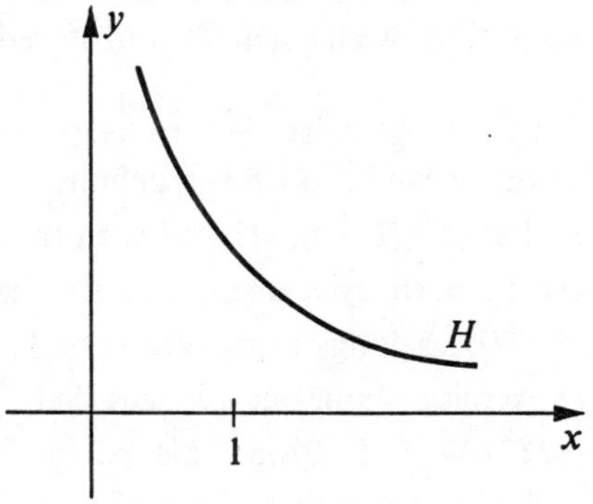

Fig. 6.14

Case 1: $1 < a < b$ Case 2: $0 < a < 1 < b$

Case 3: $0 < a < b < 1$ Case 4: $1 = a < b$

Case 5: $0 < a < b = 1$

(The student can make his own drawing.)

that $\int_1^b H = L(b)$ and that $\int_1^a H = L(a)$ and, since a is between 1 and b, property 2 asserts that $\int_1^b H = \int_1^a H + \int_a^b H$, hence $\int_a^b H = L(b) - L(a)$.

We continue with case 3 intending to leave case 2 for the student.

In case 3, $a < b < 1$ so that $\int_a^1 H = -L(a)$ and $\int_b^1 H = -L(b)$. We also note that $\int_a^1 H = \int_a^b H + \int_b^1 H = \int_a^b H - L(b)$, hence $-L(a) = \int_a^b H - L(b)$, so we again conclude that

$$\int_a^b H = L(b) - L(a).$$

For case 4, we observe that since $a = 1$, $L(a) = L(1) = 0$, and thus it is easy to see that the same statement holds in this case as well.

On the basis of this discussion, we conjecture the following statement whose proof will be established once we dispose of the remaining cases.

THEOREM L-1: *If* $0 < a < b$, *then* $\int_a^b H = L(b) - L(a)$.

Wonder what the graph of this function looks like? Having all of the computations made in the recent exercise before us, it may be a bit awkward to explain why we didn't think of this before. But hindsight is generally closer to twenty-twenty than any other kind.

A bit chagrined perhaps, we address ourselves without further delay to the task of securing some knowledge or facts relative to the nature of the graph of this special function. In doing this, we will make use, quite naturally, of the computations we have made and stored in a little table of approximations.

We begin the task by asking the student to keep one eye fixed on this table and to keep the other eye fixed on the definition of the function L. With one eye he sees that $\int_1^{1.5} H = 0.405$ and with the other he sees that $L(1.5) = \int_1^{1.5} H$. So, with both eyes open, he sees that $L(1.5) = 0.405$. Hence, the point (1.5,0.405) belongs to the graph of L.

Repeating this eye exercise again, one eye sees that $\int_1^2 H = 0.693$ while the other sees that $L(2) = \int_1^2 H$. From this pair of observations, we conclude that $L(2) = 0.693$, hence that (2,0.693) is another point on the graph of L.

A repetition of this procedure will produce other points on the graph of L.

However, by just looking at the definition alone we can see that $L(1) = 0$ and hence the point (1,0) belongs to the graph of L.

Perhaps the most "eye popping" observation of all comes when we look at problem H-1 and, using $x = 2$, see it tells us that $\int_{1/2}^{1} H = \int_{1}^{2} H$. For if we look at our tables, we see $\int_{1}^{2} H = 0.693$, and when we look at the definition, we see that $L(1/2) = -\int_{1/2}^{1} H$. Thus $L(1/2) = -0.693$ and so we have another point on the graph of L, namely, $(1/2, -0.693)$.

Plotting all of these observations on a piece of graph paper, we get something like figure 6.15a. Hoping that we are not making any mistake in doing so, we see that if we join our points in some "smooth" manner we have something like figure 6.15b as a graph (as least a part of one) of the function L.

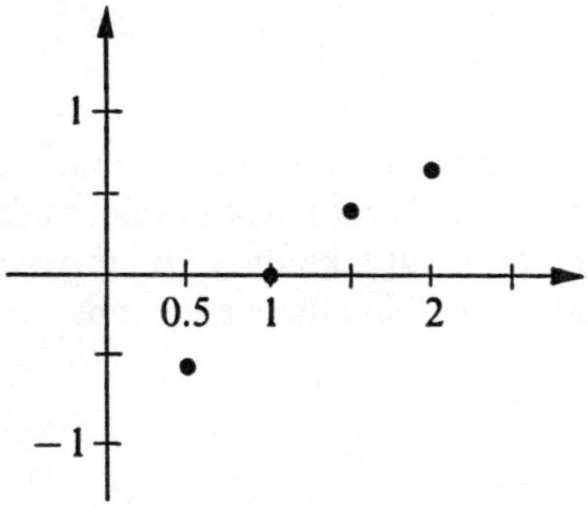

Fig. 6.15a

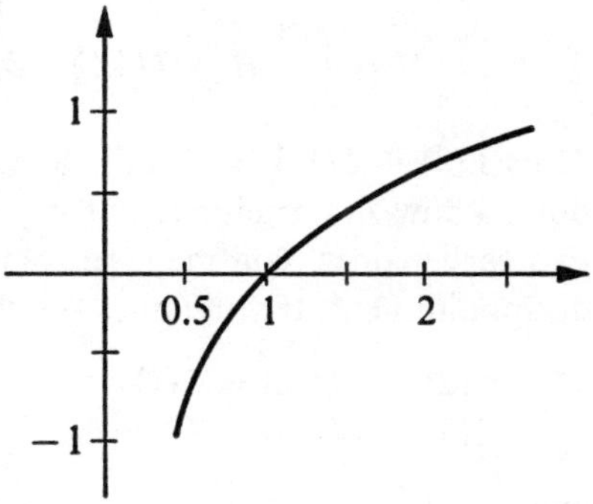

Fig. 6.15b

Pleased with our pictures, we let our eyes fondly wander back over our marvelous observations. After some minutes, two things attract our attention. They are called problems H-1 and H-2.

$$\int_{1/x}^{1} H = \int_{1}^{x} H \quad \text{and} \quad \int_{a}^{b} H = \int_{ca}^{cb} H \qquad \text{for } c > 0$$

Problem H-1 seems to be crying out to us to see beyond what we have already noted in our recent eye exercises, $L(1/2) = -L(2)$. Alarmed at this, we look at this another time

$$\int_{1/3}^{1} H = \int_{1}^{3} H \quad \text{so} \quad L(1/3) = -L(3)$$

and in spite of the fact that our computations have not progressed to the point where we know $L(3)$, we think we may have something here!

With a great deal of determination, we rewrite our last observation in the form $L(3) + L(1/3) = 0 = L(1)$. Likewise, we write our earlier observation in the form

$$L(2) + L(1/2) = L(1).$$

In a continued effort to establish what seems to be another startling observation, we select from our tables $L(1.5)$ not only because we have it listed, but mainly because $1.5 = 3/2$. In rapid succession we write

$$L(1.5) = -L(2/3) \qquad \text{(our first discovery)}$$

adding $L(3)$ to both sides* we get

$$L(3) + L(1.5) = L(3) - L(2/3).$$

By theorem L – 1, the right side is the number $\int_{2/3}^{3} H$ and if we use problem H-2 with $c = 3/2$, we find that

$$\int_{2/3}^{3} H = \int_{1}^{9/2} H = L(4.5) \quad \text{since} \quad 9/2 = 4.5.$$

Thus $L(3) + L(1.5) = L(4.5)$. Some of us may be disappointed that we don't know $L(4.5)$ either, but if you place this latest equation alongside two earlier ones, you may see something that will take the edge off your disappointment. If you fail to see the pattern in these three equations

$$L(2) + L(1/2) = L(1)$$
$$L(3) + L(1/3) = L(1)$$

and

$$L(3) + L(1.5) = L(4.5)$$

we will continue in the exercise until you do.

PROBLEM SET 6.3

1. What general statement would summarize our observations that $L(1/2) = -L(2)$ and that $L(1/3) = -L(3)$?

2. Let's begin again with the statement that

$$L(1.5) = -L(2/3).$$

* If you are wondering why $L(3)$ was added, don't. In the exercise you will see that we could have added any number to both sides.

This time, add $L(5)$ to both sides instead of $L(3)$. Apply theorem $L - 1$ and then problem H-2, using $c = 3/2$ as before. Follow through in the fashion indicated until you have a statement of the form $L(x) + L(z) = L(u)$.

3. Begin again with the statement $L(1.5) = -L(2/3)$. Add $L(4)$ to both sides. Follow the same procedure again using $c = 3/2$. Finish when you have a statement of the form $L(x) + L(z) = L(u)$.

4. Repeat the same process, this time adding $L(2)$ to both sides. Finish when you have the same type of statement as in the previous problems, that is, in the form $L(x) + L(z) = L(u)$.

5. This time we begin with a different opening. We have $L(1.8)$ and we know that $1.8 = 9/5$. Start then with

$$L(1.8) = -L(5/9).$$

Add the same number to both sides, say $L(2)$. Apply theorem L-1 and then problem H-2, using $c = 9/5$, and follow the procedure outlined in the text and in the problems above until you have obtained a statement in the form $L(x) + L(z) = L(u)$.

6. Take all of the problems 2–6 that you have worked; place the statements of the desired form that you have obtained alongside the three given in the text. Study these and then *conjecture* a statement that seems to reflect what you have noted as a common conclusion.

7. If you have worked problem 4 of this problem set, go back to it and notice that you can now compute $L(3)$. *Do so.*

8. Prove theorem L-1 for the cases not covered in the text.

9. *Small bonus problem.* Start with the assumption that $x > 1$ and assume also that $L(x) = -L(1/x)$. Add $L(x)$ to both sides and follow through, choosing your own c.

Section 4—"Big" L Theorem Sometimes one cannot see the forest because the trees are in the way. Thus it may be that we were unable to see any general pattern emerge from our efforts in the last problem set. Because it is imperative that the student knows this result, we cannot refrain from stating the following theorem.

THEOREM L-2. *If each of x and z is a positive number, then:*

1. $L(xz) = L(x) + L(z)$,
2. $L(1/z) = -L(z)$, and
3. $L(x/z) = L(x) - L(z)$.

We must stress the fact that this is an important and useful property of the function L. So, even if the student has conjectured this theorem as a result of his recent endeavors, there is some need to see if this conjectured pattern can be established.

It follows, therefore, that we are unable to say no more about the theorem and we shall begin at least to try and show that it is a true statement.

Suppose that $z < 1$, then $xz < x$ so that $\int_{xz}^{x} H = L(x) - L(xz)$ by theorem L-1. Using problem H-2 with $c = 1/x$, we find $\int_{xz}^{x} H = \int_{z}^{1} H$. Since $z < 1$, $\int_{z}^{1} H = -L(z)$ is the definition of L. It now follows that $L(x) - L(xz) = -L(z)$, that is, $L(xz) = L(x) + L(z)$.

If we suppose that $1 < z$, then $x < xz$ so that $\int_{x}^{xz} H = L(xz) - L(x)$ by theorem L-1. Again using H-2 with $c = 1/x$, we obtain $\int_{x}^{xz} H = \int_{1}^{z} H$. Since $1 < z$, $\int_{1}^{z} H = L(z)$ by the definition of L. Therefore we have that $L(xz) - L(x) = L(z)$, that is, $L(xz) = L(x) + L(z)$ as before.

Finally, if we suppose that $z = 1$, we can see that since $L(1) = 0$, the statement $L(xz) = L(x) + L(z)$ is true. This establishes the first part of our theorem.

We shall now use this part of the theorem to help establish the other parts of the theorem.

Thus $L(z \cdot 1/z) = L(z) + L(1/z)$. But $z \cdot 1/z = 1$ and $L(1) = 0$, hence $L(z) + L(1/z) = 0$ and the second part of the theorem is established.

From the idea that $L(x/z) = L(x \cdot 1/z)$, the last part of the theorem can be shown.

We shall conclude this section by demonstrating the usefulness of this theorem. Using the computations already made, we may obtain approximations to $L(x)$ for other values of x.

Example. $L(3) = L(2 \cdot 1.5)$, hence $L(3) = L(2) + L(1.5)$. Thus, consulting our table, $L(3) = 0.693 + 0.405 = 1.098$.

This example suggests that you were able to show that $L(2) = 0.693$.

Example.
$$\begin{aligned} L(4.2) &= L(3 \cdot 1.4) = L(3) + L(1.4) \\ &= 1.098 + 0.336 \\ &= 1.434. \end{aligned}$$

Example. $L(1/3) = -L(3) = -1.098$. We could also look at this example in the following manner.

$$L(1/3) = L(1) - L(3) \quad \text{for} \quad L(1) = 0$$

and we get the same result.

Since the next assignment will require that we have correctly approximated $L(x)$ for the values given in the table of problem set 6.2, each student should check his table with the instructor before going on.

PROBLEM SET 6.4

1. Let problem H-1 help you to evaluate $\int_{5/6}^{1} H$.

2. Prove the last part of theorem L-2. *Hint:* $\frac{x}{z} = x \cdot \frac{1}{z}$.

In each of problems 3–6, use theorem L-2 and the computations made and recorded in a previous table to find approximations to the given numbers.

3. $L(2.2)$, $L(3,6)$, $L(4)$, and $L(4/3)$.

4. $L(3.4)$, $L(1/4)$, $L(12)$, and $L(3/4)$.

5. $L(6)$, $L(9)$, $L(3.2)$, and $L(2/3)$.

6. $L(3.9)$, $L(8)$, $L(3/8)$, and $L(27)$.

[7] Find $L(5)$. This is critical. Try to see how you can use problem 1 above to help. When finished, record your result in the table of approximations.

8. Use the results of problem 7 to find $L(10)$ and $L(1/5)$.

9. Find $L(0,3)$, $L(0,6)$, and $L(13)$.

[10] Find approximations to the numbers $L(2.7)$ and $L(2.8)$. Later we shall use these approximations to establish a result that in some respects is remarkable (also simple), hence is remarkably simple or perhaps simply remarkable.

11. Suppose that $x > 0$.

a. Show that $L(x^2) = 2 \cdot L(x)$ and that $L(x^3) = 3 \cdot L(x)$.
b. If $n \in N$, use part *a* to *conjecture* a similar result for $L(x^n)$.
c. Show that $L(\sqrt{x}) = \frac{L(x)}{2}$.

Special problem. Using some of the previous computations, sketch the graph of the function L on the interval $\langle 0,5]$. (See figure 6.16.)

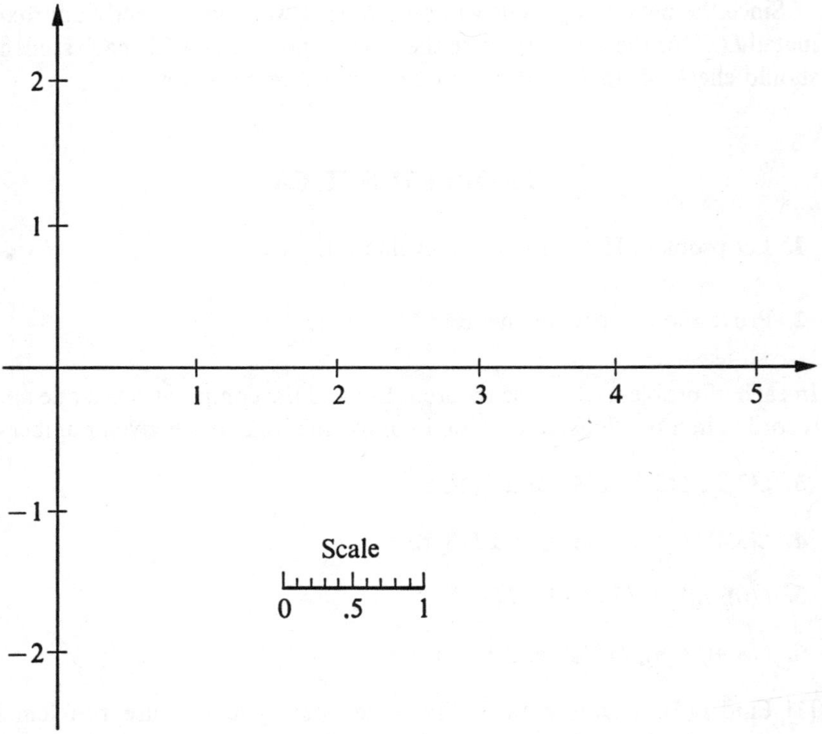

Fig. 6.16

Section 5—More Properties of L The ideas and computations in problem set L are so vital to us that this may be a good time to pause and catch our breath before plunging on. While resting, we might note that the computations we have made and the graphs we have drawn have certainly suggested that the function L is a strictly increasing function. Let us nail down this idea by hammering home the fact that this property of L is a consequence of some of the properties already presented. For example, $\frac{b-a}{b} < \int_a^b H < \frac{b-a}{a}$, $\int_a^b H = L(b) - L(a)$, and $L(xz) = L(x) + L(z)$.

If x and z are positive numbers and $x < z$, then

$$0 < \frac{z-x}{z} < \int_x^z H = L(z) - L(x).$$

Therefore $L(x) < L(z)$ and we have shown that L is an increasing function. Glory be! We feel like celebrating, for 'tis a fine thing that we have done, on the one hand. But on the other hand, and there usually is another, you know, what we have done raises a sobering thought (and they are the worst kind): Does this rascal (meaning the function L to be sure) recognize no bounds to its behavior? Or is it like some of those sneaky number sets we were studying earlier, putting on a good show of ascending into infinity only to have the lid (upper bound to the more learned of ye) slammed down on it. For, there is nothing to be seen that would prevent there being a number h such that the horizontal line h, speaking quite sternly to L, might say, "this far but no farther." If such a number h does exist, then the graph of L, still keeping all of its nice properties, mind you, might have the appearance suggested by figure 6.17. Clearly $h > 1$, for we have already shown that $L(3) > 1$. But still, there is nothing to be seen that would "give the lie" to all this. But then, one hears of a lad who was blind in one eye and couldn't see out of the other.

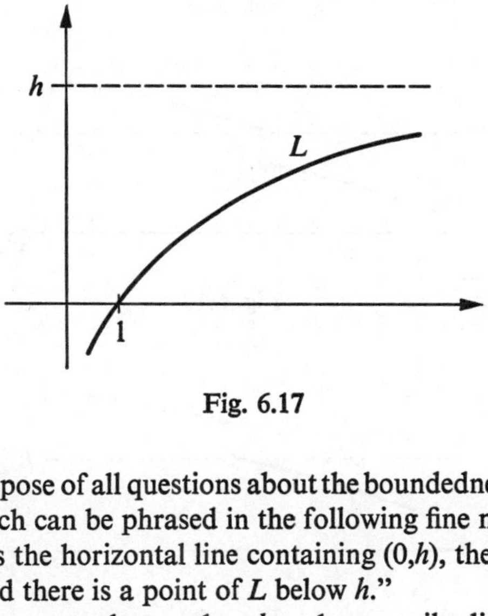

Fig. 6.17

Well, let's dispose of all questions about the boundedness of L by arguing a statement which can be phrased in the following fine manner: "If h is a number and h is the horizontal line containing $(0,h)$, then there is a point of L above h and there is a point of L below h."

Note that the cases where $-1 < h < 1$ are easily disposed of, and so we let our argument be directed to the case where $h > 1$. If $h > 1$, there is a counting number m such that $m \leq h < m + 1$. Surely $3^{m+1} > 0$ for $3^{m+1} = 3 \cdot 3 \cdot 3 \cdot \cdots (m + 1)$ times. Now $L(3^{m+1}) = (m + 1) \cdot L(3)$ and, since $L(3) > 1$, hence $L(3^{m+1}) > m + 1 > h$. This gives us a point of L above h and $(1,0)$ is a point of L below h. Now, if $h < -1$, then $(1,0)$ will serve as a point above h. If we then select an integer $j > h$, then $-j$ will be a counting number. Let us set $-j = k$. Thus, in a manner similar to the above, we can obtain a point of L below h.

Hints. $L(3^k) = k \cdot L(3)$ and $k \cdot L(3) > k > -h$.

Fatigued we may be but finished we are not. For the above suggests the question: Does h contain a point of L? We know that what is suggested by figure 6.18 is impossible because the domain of L is the set of all positive numbers. We may think that this does it and just fill in the dot. But maybe the graph of L looks like either figure 6.19 or figure 6.20. "Oh no!" we

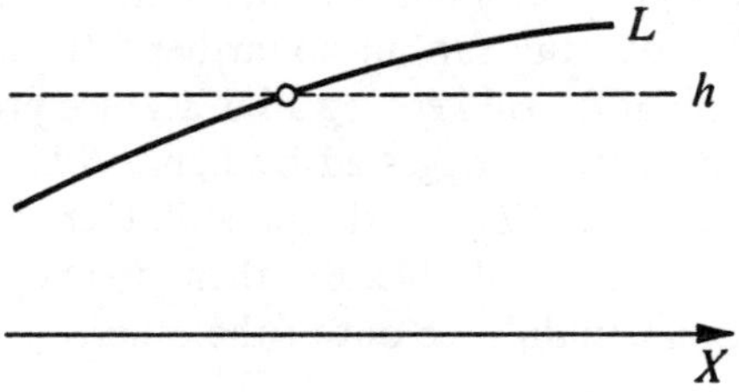

Fig. 6.18

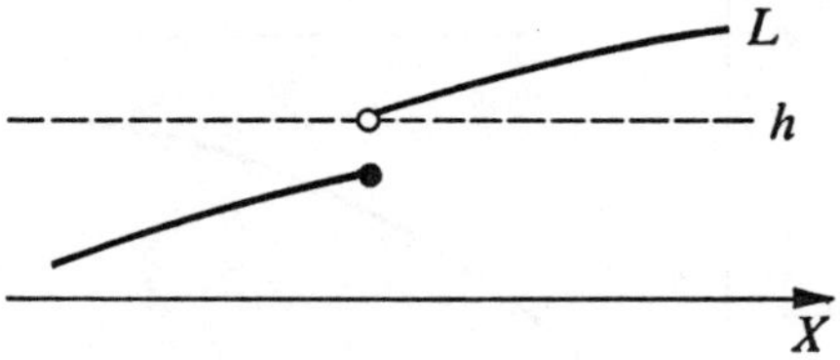

Fig. 6.19

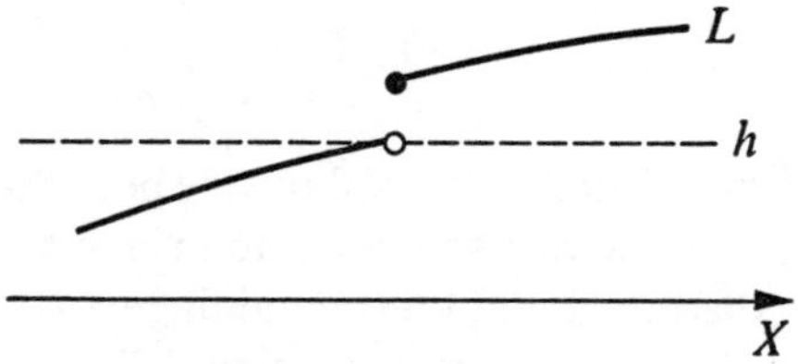

Fig. 6.20

wail, "for L is continuous." Is it now? Look back at your graph of L and try to understand an earlier remark about knowingly or unknowingly assuming that our functions have this property. I submit that this assumption about the function L is only a hope, fathered, perhaps, by a trust in the "decency" of the function. Be of good cheer, for the following statement can be proven.

"If h is a number, then h contains a point of L." Given this statement, it follows that L is everywhere continuous. The proof is difficult, however, and is relegated to the appendix. Not to be banished to the appendix, though, is an important corollary to this statement, namely,

The range of L is the set of real numbers.

In order that our note will be complete, the results of the first argument in this section, to the effect that L is a strictly increasing function, will be formally summarized by the following:

THEOREM L-3. *If x and z are positive numbers and $x < z$, then* $L(x) < L(z)$.

PROBLEM SET 6.5

1. a. Define the function L. Give its domain and range.
b. List the properties of the function L.

2. Assuming that the following statement is true,

"If h is a number and h is the horizontal line containing the point $(0,h)$, then h contains a point of L"

show that h does not contain two points of L. *Hint:* Suppose that h did contain two points of L and examine the consequences of this supposition.

3. Briefly explain why the statement that $L(s) = L(t)$ implies that $s = t$.

4. Is L^*, the inverse of L, a function? Explain your answer.

5. If $(p,q) \in L$, then $(q,p) \in L^*$. This implies that: $L(p) =$ ______, $L^*(q) =$ ______, hence $[L \circ L^*](q) =$ ______, and $[L^* \circ L](p) =$ ______.

6. a. Give a geometric interpretation of the number $\int_1^{8.1} H$.
b. Use the approximating formula to find $\int_8^{8.1} H$ to three decimal places.
c. If $\int_1^8 H = 2.079$, use property 2 to approximate $\int_1^{8.1} H$.
d. Now $L(8.1) = \int_1^{8.1} H$. Using $L(3) = 1.098$ and $L(10) = 2.303$, approximate $L(8.1)$ by another method.

7. a. Give a geometric interpretation of theorem L-1.
b. Use theorems L-1 and L-2 to approximate the number $\int_{1/27}^{27} H$ if $L(3) = 1.098$.

8. If $L(z) = 2.152$, $L(x) = 1.282$, and $L(e) = 1$, give:

a. $L(xz)$ **b.** $L(x/z)$ **c.** $L(1/x)$ **d.** $L(xe^2)$
e. $L(z^3)$ **f.** $L(e/z)$ **g.** $L(1/e)$ **h.** $L(ex/z)$
i. give the order relation for the number set z, x, e, e^2, e^3, that is, _____ < _____ < _____ < _____ < _____.

9. Give a brief explanation or demonstration that each of the following is true:

a. $L(\sqrt{x}) = \dfrac{L(x)}{2}$ **b.** $L(\sqrt[3]{x}) = \dfrac{L(x)}{3}$ **c.** $L(x\sqrt{x}) = \dfrac{3L(x)}{2}$

10. Suppose that c is a number and that $1 < c$. Let f be the line containing the point $(c,1/c)$ whose slope is $-1/c^2$. Give the point-slope equation of f.

11. Refer to figure 6.20. If f is the line in problem 10, RT is a horizontal line, and the line through P, Q, and R is vertical, find the following:

a. the coordinates of the points P, Q, R, and T.
b. the area of the trapezoid with vertices at P, T, $(1,0)$, and $(c,0)$.
c. the area of the trapezoid with vertices at Q, T, $(1,0)$, and $(c,0)$.
d. the area of the rectangle with vertices at R, T, $(1,0)$, and $(c,0)$.

12. Refer to figure 6.21. Give the following: (**a**) area of the right triangle PRT, (**b**) area of the right triangle QRT, and (**c**) area of the triangle PQT (not a right triangle).

We are now ready to obtain some formulas for $\int_1^c H$.

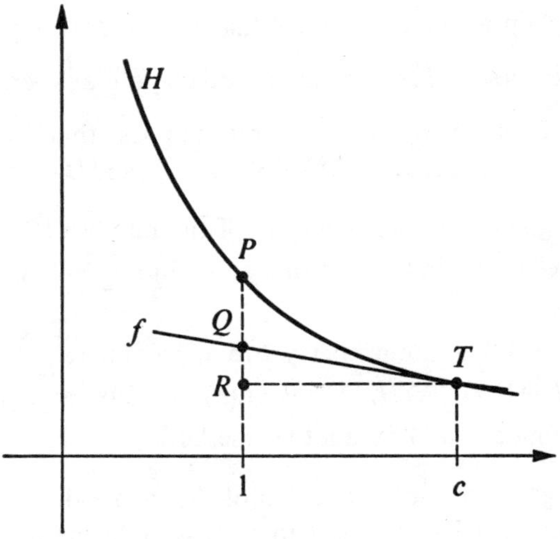

Fig. 6.21

Bonus Problems

13. It should seem reasonable (because it is true) that $\int_1^c H$ equals the area of the rectangle in problem 11d plus a positive number k_1 where k_1 is less than the area of the triangle in problem 12a. Write an equation and an inequality which expresses these ideas.

14. $\int_1^c H$ equals the area of the rectangle in problem 11d plus the area of the triangle in problem 12b plus a positive number k_2 where k_2 is less than the area of the triangle in problem 12c. Do the same as in problem 13.

15. *Do not* try to envisage other triangles, and so forth, but *do look* at what you have written in problems 13 and 14 and then *conjecture* another formula for $\int_1^c H$ using k_3.

16. Return to the line in problem 10 and show that there is no point of f above graph of H. *Hint:* Assume that there is a point of f above the graph of H. Call this point $[t,f(t)]$. It follows that $t \neq c$ and that $f(t) > H(t)$, that is, $f(t) > 1/t$. Why? Substitute this point into the equation of f and show that (give details) $\frac{2c - t}{c^2} > 1/t$. Using this inequality, go on to obtain a contradiction.

Section 6—Function E In the preceding sections of this chapter we established the ideas that L was an increasing function with domain the set of positive numbers and range the set of all numbers. Therefore the point set to which the point (a,b) belongs only if (b,a) belongs to L is a function (see inverse function). Let us christen this function "E."

> DEFINITION. *The function E is the set of points to which (a,b) belongs only if (b,a) belongs to L.*
>
> *Remarks.* From this definition and some other ideas, the student should see (this is important, also easy) that the domain of E is the set of all numbers and that the range of E is the set of positive numbers, that is, if x is a number, then $E(x) > 0$. The graphical implication of this inequality is that the graph of the function E will not, repeat not, intersect the X axis. It will be confined to the first and second quadrants.

Using the definition of the function E and the properties of the function L, we can establish several useful results. For example:

1. From the fact that the function L is increasing over its domain (is a 1-to-1 function), we can argue that the function E is an

increasing function over its domain. Thus if $x < z$, then $E(x) < E(z)$.

2. We may also argue that if x is a number, then $E(x)$ is some positive number, let's call it p. Letting $E(x) = p$ means, of course, that $(x,p) \in E$, hence $(p,x) \in L$; but this, in turn, means that $L(p) = x$. Thus, under composition, L and E behave like we expect them to, namely, $L(E(x)) = x$. If we let x be a positive number (to be sure it's in the domain of L), then we can also show that $E(L(x)) = x$.

Before we get into any more arguments, let us formally record what we have already argued as follows:

THEOREM E-1. *If x and z are numbers and $x < z$, then $E(x) < E(z)$.*

THEOREM E-2. *If x is a number, then $[L \circ E](x) = x$.* And *if $x > 0$, then $[E \circ L](x) = x$.*

Now, let's have a go at it again.

3. While you may not have proven this, you may nevertheless grasp the idea that if h is a number, then the horizontal line containing the point $(0,h)$ will contain a point on the graph of L. Since 1 is a number, then the horizontal line containing $(0,1)$ will intersect the graph of L at some point whose ordinate is 1. We can look back at either our computations or the graph of the function L that we have sketched, and we can readily see that the point in L whose ordinate is 1 will have an abscissa between 2 and 3. Thus there is some number x such that $L(x) = 1$. So we make the following definition.

DEFINITION. *e is the number such that $L(e) = 1$.*

Remark. If we go' back to some computations and the "simply remarkable" idea referred to, we can make a "fine" estimate of this number e. So let's call a time out to look this up. . . . We see that we were able to show that $L(2.7) < 1$ while $L(2.8) > 1$, hence we can conclude that $2.7 < e < 2.8$.

In order to prepare for at least part of the following problem set, let us drag out some of the computations that we have previously made and place them in a convenient place, namely in table 6.2.

TABLE 6.2

x	$L(x)$	x	$L(x)$
1	0	e	1
1.5	0.405	1/2	−0.693
2	0.693	$1/e$	−1
3	1.098	5	1.609
4	1.386	10	2.302

PROBLEM SET 6.6

1. Suppose that x and z are numbers and that $x < z$.

a. If $E(x) = E(z) = y$, then (x,y) and (z,y) belong to E, which would mean that (y,x) and (y,z) belong to L. But this is impossible. Why?
b. If we let $E(x) = y$ and $E(z) = w$, then (x,y) and (z,w) belong to E, which would mean that (y,x) and (w,z) would belong to L. So if we were to suppose that $E(x) > E(z)$, that is, that $y > w$, we would have a contradiction. Explain the contradiction.
c. From parts *a* and *b*, what can we conclude?

2. Suppose that $x > 0$ and let $L(x) = p$. This implies that $(x,p) \in L$, which means that $(p,x) \in E$, hence $E[L(x)] =$ ______. This should add to your understanding of theorem E-2.

3. Find $E(0)$. Give some explanation.
Find $E(1)$. Give some explanation.

4. Use the result $2.7 < e < 2.8$ and a little arithmetic to find a and b where $a < 1/e < b$.

5. **a.** If $L(2) = 0.693$, then $E(x) = 2$ for what number x?
b. If $L(5) = 1.609$, then $E(x) = 5$ for what number x?
c. If $L(1/3) = -1.098$, then $E(x) = 1/3$ for what number x?
d. $E(x) = 0.5$ for what number x?

6. Make use of most if not all of the results given in the brief table recently constructed to sketch the graphs of the functions L and E. Put both graphs on one set of coordinate axes.

7. Is the function E bounded above or below? If you answer yes to either, give LUB of E or GLB of E?

Section 7—More on E and the Number b^x We have recently shown that the function E has the set of all numbers for its domain and the set of positive numbers as its range. Thus the runction E is not bounded above, but it is bounded below by zero. However, $E(x)$ has no minimum value. We have also noted that this function is an increasing function over its domain. To advance our knowledge of this special function, consider the following.

Suppose each of t and c is a number, then there is a point in L whose ordinate is t and there is a point in L whose ordinate is c, that is, there is a positive number x_1 such that $L(x_1) = t$ and there is a positive number x_2 such that $L(x_2) = c$. Now theorem L-2 assures us that $L(x_1x_2) = L(x_1) + L(x_2) = t + c$, therefore, $(x_1x_2, t + c)$ is a point in L. By consulting the definition, we then know that $(t + c, x_1x_2) \in E$, hence, $E(t + c) = x_1x_2$. Likewise, since $(x_1,t) \in L$, we have $(t,x_1) \in E$, hence, $E(t) = x_1$. All of this means that the following statement is a theorem.

THEOREM E-3. *If each of t and c is a number, then*

$$E(t + c) = E(t)\cdot E(c).$$

The pair of statements $E(t + c) = E(t)\cdot E(c)$ and $L(xz) = L(x) + L(z)$ are called addition formulas, and they play an important role in many applications of the functions L and E.

One immediate consequence of this theorem is the following:

THEOREM E-4. *If x is a number, then* $E(-x) = 1/E(x)$.

The proof of this theorem will be left to the student.

Now, but not before, we give this definition.

DEFINITION. *If b is a positive number and x is a number, then b^x (read as "b to the x") is the number $E(x\cdot L(b))$, that is, $b^x = E[x\cdot L(b)]$.*

There are several interesting consequences of this definition, two of which will be shown immediately.

1. By this definition, the number $e^x = E[x\cdot L(e)]$. Since $L(e) = 1$, however, this simplifies to

$$E(x) = e^x.$$

This consequence is interesting in that it supplies a rule of pairing for the function E, something which the definition of this function did not mention.

2. By this definition the number $e^{L(x)} = E[L(x) \cdot L(e)]$. Since $L(e) = 1$ we have that $e^{L(x)} = E[L(x)]$. Now, we have shown before (see theorem E-2) that $E[L(x)] = x$, hence

$$e^{L(x)} = x.$$

Remark. This second consequence, unlike the first, requires that $x > 0$ in order that $L(x)$ be a number.

This consequence is extremely interesting for reasons that we shall now try to see. In order to be sure that our sight is unimpaired, you are informed that we will make use of a table (an extension of those that you have made) that appears at the end of this section.

The results given in these examples are an approximation taken directly from the aforementioned table. The first example illustrates the basic plan.

Example. Find $e^{2.041}$.

A glance at the table shows that $L(7.7) = 2.041$, thus $e^{2.041} = e^{L(7.7)}$. Since we have shown that $e^{L(x)} = x$, it follows that

$$e^{2.041} = 7.7.$$

The next example illustrates how we handle an evaluation that falls between two entries in the table. The arithmetic process used is commonly called *interpolation.*

Example. Find $e^{1.50}$.

From the tables we see that the number x for which $L(x) = 1.50$ falls between $L(4.4)$ and $L(4.5)$. Therefore we make the following computations:

$$L(x) - L(4.4) = 1.50 - 1.482 = 0.018$$
$$L(4.5) - L(4.4) = 1.504 - 1.482 = 0.022$$

The ratio $\dfrac{0.018}{0.022} = 0.8$ to one decimal place, which is the only accuracy we can expect. Thus

$$x = 4.48.$$

So now we have that $e^{1.50} = e^{L(4.48)}$ and hence

$$e^{1.50} = 4.48.$$

The next example will illustrate the procedure when we fall outside the table.

Example. Find $e^{6.78}$.

Notice first that the table is not extensive enough to show the number x for which $L(x) = 6.78$. But then note underneath the table that $2 \cdot L(10) = L(100) = 4.605$. With a little arithmetic, then, we have $6.78 = 4.605 + 2.175$, hence $e^{6.78} = e^{4.605+2.175} = e^{4.605} \cdot e^{2.175}$. From the tables we now see that $L(8.8) = 2.175$, hence

$$e^{6.78} = e^{L(8.8)} \cdot e^{L(100)} \text{ so that}$$
$$e^{6.78} = 8.8(100) \text{ or } 880.$$

For our final example we illustrate how we first must use the definition of b^x in order to change the form of the problem into a more familiar form.

Example. Find $\sqrt[3]{6.5}$, that is, find $(6.5)^{1/3}$.

By the definition of b^x we know that $(6.5)^{1/3} = E(1/3 \cdot L(6.5))$. From the tables we find $L(6.5) = 1.872$, thus $1/3 \cdot L(6.5) = 0.624$. Thus we have progressed to the point of having

$$\sqrt[3]{6.5} = E(0.624) = e^{0.624} \qquad (\text{Recall } E(x) = e^x).$$

So now we have our problem in a familiar form, and we go on to find $e^{0.624}$. We can interpolate to find that the number x for which $L(x) = 0.624$ is 1.87 (see below). Thus

$$\sqrt[3]{6.5} = e^{L(1.87)} = 1.87$$

$$0.624 - 0.588 = 0.036; \; 0.642 - 0.588 = 0.054 \text{ and } \frac{0.036}{0.054} = 0.7.$$

TABLE 6.3

A TABLE FOR $L(x)$

x	0.0	0.1	0.2	0.3	0.4	0.5	0.6	0.7	0.8	0.9
1	0.000	0.095	0.182	0.262	0.336	0.405	0.470	0.531	0.588	0.642
2	0.693	0.742	0.788	0.833	0.875	0.916	0.956	0.993	1.030	1.065
3	1.099	1.131	1.163	1.194	1.224	1.253	1.281	1.308	1.335	1.361
4	1.386	1.411	1.435	1.459	1.482	1.504	1.526	1.548	1.569	1.589
5	1.609	1.629	1.649	1.668	1.686	1.705	1.723	1.740	1.758	1.775
6	1.792	1.808	1.825	1.841	1.856	1.872	1.887	1.902	1.917	1.932
7	1.946	1.960	1.974	1.988	2.001	2.015	2.028	2.041	2.054	2.067
8	2.079	2.092	2.104	2.116	2.128	2.140	2.152	2.163	2.175	2.186
9	2.197	2.208	2.219	2.230	2.241	2.251	2.262	2.272	2.282	2.293
10	2.303	2.313	2.322	2.332	2.342	2.351	2.361	2.370	2.380	2.389

$L(0.1) = -2.303$ $\quad 2 \cdot L(10) = 4.604$ $\quad 4 \cdot L(10) = 9.210$

$L(10) = 2.303$ $\quad 3 \cdot L(10) = 6.908$ $\quad 5 \cdot L(10) = 11.513$

Note: Because $e^{L(x)} = x$, table 6.3 can be utilized to give approximations to e^x.

PROBLEM SET 6.7

Find approximations to each of the following:

1. $e^{1.723}$ **2.** $e^{2.23}$ **3.** $e^{1.253}$

4. $e^{2.351}$ **5.** $e^{1.15}$ **6.** $e^{0.42}$

7. $3^{1.8}$ **8.** $\sqrt{2}$ **9.** $\sqrt{6}$

10. $10^{2.4}$

Hint for problems 11 and 12: $E(-x) = 1/E(x)$.

11. $e^{-0.788}$ **12.** $e^{-4.205}$

Section 8—Exponential Functions The definition of b^x has several interesting consequences. For example, if z is a number, then $1^z = E[z \cdot L(1)]$. But $L(1) = 0$ and $E(0)$ is 1, hence 1^z is 1. Also, note that e^x is $E[x \cdot L(e)]$ and, since $L(e) = 1$, we find that $E(x) = e^x$ an equation which can serve as a rule of pairing for the function E. Furthermore, since the "laws of exponents" are usually encountered in high school algebra, we may reasonably suppose that the student knows them. It may be of interest, however, to see that these laws are rather easily established consequences of this definition even when the exponents are not integers. Nevertheless, the primary concern of this section is the application of the definition of *b to the x* to a class of functions labeled *exponential functions.* The primary task shall be that of sketching the graphs of these functions.

If F is a function for which $F(x) = b^x$, then F is an exponential function. The special function E is certainly in this class of functions and, perhaps to stress its importance, it is frequently called *the* exponential function. Granted that E is an important exponential function, it seems proper to review some of our recent achievements. Since $L(2) = 0.693$ and $L(5) = 1.609$, we know that $(0.693,2)$ and $(1.609,5)$ are points in E. Also, $L(1/e) = -1$, thus $(-1,1/e)$ belongs to E. Furthermore, since $L(0.3) = -1.204$ and $L(0.4) = -0.916$, we can deduce that $0.3 < 1/e < 0.4$. Now the statement that $2.7 < e < 2.8$ yields the finer result $0.35 < 1/e < 0.38$. We shall now consider examples which reflect the primary task of this section, sketching graphs of exponential functions.

Examples. If F is the exponential function such that $F(x) = 2^x$, then, by making use of the laws of exponents, we can show that $F(-2) = 1/4$, $F(-1) = 1/2$, $F(0) = 1$, $F(1) = 2$, and $F(2) = 4$. The graph of F appears in figure 6.22.

In figure 6.23, we exhibit the graph of $G(x) = 3^{1-x}$. We find that $G(-1) = 9$, $G(0) = 3$, $G(1) = 1$, and $G(2) = 1/3$.

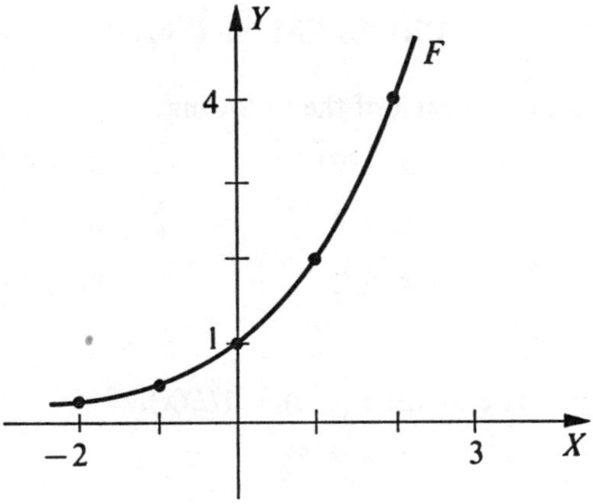

Fig. 6.22

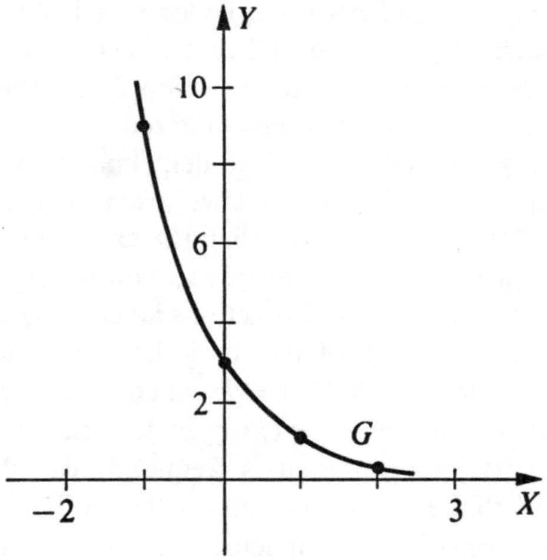

Fig. 6.23

If x is a number, then $E(-x)$ is $1/E(x)$ by theorem E-6. So, if we let $1/E$ denote the exponential function to which the point (x,y) belongs only if $(-x,y)$ belongs to E, we can sketch its graph by examining the information given just prior to these examples. If you examine this information, you will see that the following points belong to the function $1/E$: $(1,1/e)$, $(0,1)$, $(-0.693,2)$, $(-1,e)$, and $(-1.609,5)$. The graph of the function $1/E$ is shown in figure 6.24.

Example. Here we will look once again at composite functions. In spite of the loud groans, suppose $F(x) = x + 1$, then

$$[E \circ F](x) = e^{x+1}.$$

Hence,

$$[E \circ F](-2) = 1/e, \qquad [E \circ F](-1) = 1,$$
$$[F \circ E](0) = e, \qquad [E \circ F](1) = e^2,$$

and (check this) $[E \circ F](0.609) = 5$. The graph of this composite is shown in figure 6.25.

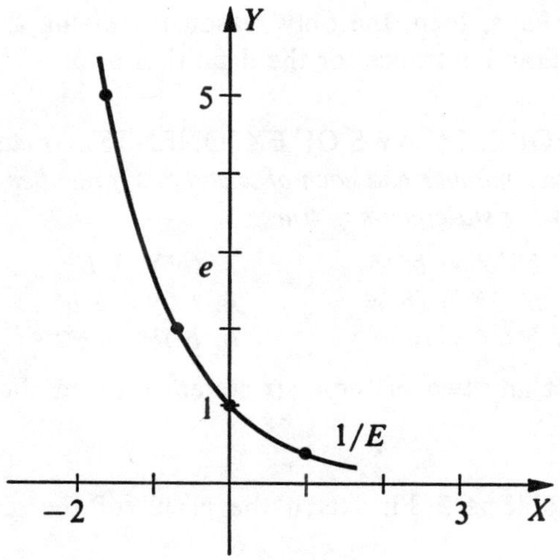

Fig. 6.24

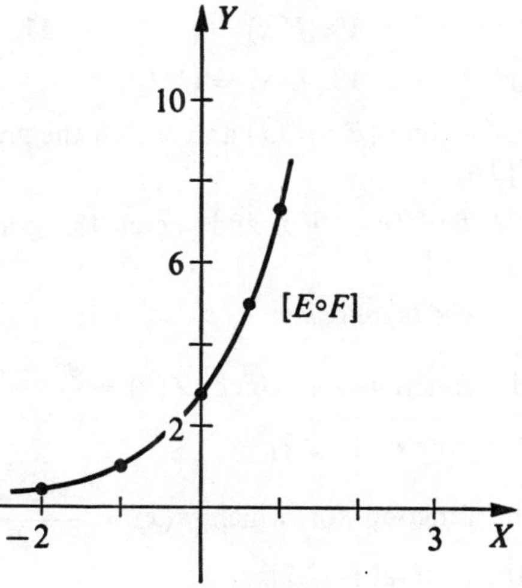

Fig. 6.25

PROBLEM SET 6.8

1. We can use the result $2.7 < e < 2.8$ to give an estimate for e^2. However, a finer result can be obtained by starting with $\int_1^{7.3} H = 1.988$, which implies that $L(7.3) < 2$. If you will now compute $L(7.4)$, you can then give an estimate for e^2.

2. It is quite likely that the student already knows the theorem stated below. Perhaps, then, the only reason for giving the theorem is to provide a grand entrance for the definition of b^x.

> THEOREM (LAWS OF EXPONENTS). *If each of b and c is a positive number and each of x and z is a number, then each of the following statements is true:*
>
> 1. $b^x \cdot b^z = b^{x+z}$ 2. $(b^x)^z = b^{xz}$
> 3. $b^x \cdot c^x = (bc)^x$ 4. $b^{-x} = 1/b^x$
> 5. $b^x/c^x = (b/c)^x$ 6. $b^x/b^z = b^{x-z}$
>
> Select any two of these six statements and show that they are true.

In each of problems 3–13, sketch the graph of the given exponential function.

[3] E **4.** $F(x) = 2^{x-1}$ **5.** $F(x) = 2^{x+1}$

6. $G(x) = 3^{x+1}$ **7.** $G(x) = 3^{x-2}$ **8.** $W(x) = 3^{-x}$

9. $W(x) = 4^{1-x}$ **10.** $F(x) = e^{1-x}$ **11.** $F(x) = e^{-x}$

12. $G(x) = (1/2)^{x-1}$ **13.** $G(x) = (1/2)^{x+1}$

14. If $G(x) = x - 2$, find $[E \circ G](x)$ and sketch the graph of $[E \circ G]$ over the interval $[0,4]$.

15. If $F(x) = x/2$, find $[E \circ F](x)$ and sketch the graph of $[E \circ F]$ over the interval $[-2,4]$.

And now, einige gescheite aufgaben

16. Let $\mathscr{S}$ be the function for which $\mathscr{S}(x) = \dfrac{e^x - e^{-x}}{2}$ and sketch its graph over the interval $[-2,2]$.

17. Let $\mathscr{C}$ be the function for which $\mathscr{C}(x) = \dfrac{e^x + e^{-x}}{2}$ and sketch its graph over the interval $[-2,2]$.

18. Show that $\mathscr{S}(t + a) = \mathscr{S}(t)\cdot\mathscr{C}(a) + \mathscr{C}(t)\cdot\mathscr{S}(a)$.

SUMMARY

The summary of this chapter could be simply the one sentence: In this chapter we considered two important functions L and E. But after all the work we put in, that would be a rank injustice.

This chapter began with a consideration of the area of the region bounded by the graph of H on $[a,b]$. The symbol $\int_a^b H$ was introduced to denote the area of this region. It was then observed that the area of the region bounded by the graph of the function H on $[1,x]$ was a function of the number x. This led to the definition of the function L. We then proceeded rather vigorously, sometimes even rigorously, to develop the properties of this function. It was determined that this function was continuous, increasing, and unbounded over its domain, the set of positive numbers. It was also shown that $L(xz) = L(x) + L(z)$ which is very useful. The number e was defined by the statement $L(e) = 1$ and we were able to show that $2.7 < e < 2.8$.

The inverse of L is denoted by E and was shown to have the property that $E(t + c) = E(t) \cdot E(c)$. The number b^x was defined and from it, two little statements tying together the functions L and E and the number e, to wit $E(x) = e^x$ and $e^{L(x)} = x$.

The chapter concluded with a study of a class of functions called exponential and the observation that the function E is an important member of this class.

Review Problems

1. Give the domain and range of the function L; same for the function E.

2. On the same set of coordinate axes, sketch the graphs of the functions L and E.

3. Give the following numbers:

a. $L(1)$ **b.** $L(1/e)$ **c.** $E(1)$ **d.** $L(e^2)$ **e.** $E(-1)$
f. $L(e)$ **g.** $E(0)$ **h.** $e^{L(3)}$ **i.** $L(\sqrt{e})$ **j.** $e^{-L(2)}$
k. $e^{2L(3)}$

4. If $L(s) = 0.72$ and $L(t) = 1.10$ give:

a. $L(st)$ **b.** $L(s/t)$ **c.** $L(1/s)$ **d.** $L(s^2)$ **e.** $L(es)$
f. $L(\sqrt{t})$ **g.** $L(1/e)$ **h.** $L(e^{1.1})$ **i.** $L(e^2/t)$
j. order the following numbers: s, 1, e, $1/t$, 0, s^2, t, and $1/e$.

5. If $E(a) = c$ and $L(k) = m$ give:

a. $L(c)$ **b.** $E(m)$ **c.** $[L \circ E](a)$ **d.** $[E \circ L](k)$
e. $L(e^c)$ **f.** $E(-a)$ **g.** $L(1/k)$ **h.** $E(a + m)$

6. Give the definition of the number b^x.

7. On the same set of coordinate axes, sketch the graphs of the functions $f(x) = 2^x$ and $g(x) = 2^{-x}$.

CHAPTER 7
Logarithm

Section 1—Base *b* Logarithm Our intentions now are to point out an application that is often made of the function L. This being the case, perhaps it would be advisable for the student to review the L theorems before considering the following definition.

DEFINITION. *If* b *is a positive number,* $b \neq 1$, *and* x *is a positive number, the* base b logarithm of x, *denoted by* $\log_b (x)$, *is the number* $L(x)/L(b)$, *that is,*

$$\log_b (x) = \frac{L(x)}{L(b)}.$$

Remarks. Base 10 logarithms are called *common logarithms* and while they may be approximated with the above definition and our earlier computations, we are content simply to call attention to the fact that tables of common logarithms can be found almost everywhere, that is, such tables are quite common. If one uses such a table, it should be a matter of personal satisfaction to note that

$$\log_{10} (2) = \frac{L(2)}{L(10)} = 0.301$$

and that

$$\log_{10} (5) = \frac{L(5)}{L(10)} = 0.699.$$

It is very important that we be able to see that $\log_b(1)$ is 0, that $\log_b(b)$ is 1, and that $\log_b(1/b)$ is -1, so be sure that you see why each of these statements is true.

If b is a positive number distinct from 1 and if each of x and z is a positive number, then $\log_b(xz) = \frac{L(xz)}{L(b)}$. Since $L(xz) = L(x) + L(z)$ by theorem L-2, it follows that

$$(1)\quad \log_b(xz) = \frac{L(x)}{L(b)} + \frac{L(z)}{L(b)} = \log_b(x) + \log_b(z).$$

In a similar manner, we can show that

$$(2)\quad \begin{aligned} &\log_b(1/z) = -\log_b(z), \\ &\log_b(x/z) = \log_b(x) - \log_b(z), \end{aligned}$$

and, if c is a positive number different from 1, then

$$(3)\quad \log_b(c)\cdot\log_c(b) = 1.$$

To the above list of statements, frequently called the properties of logarithms, we add the following: If p is a number, then

$$(4)\quad \log_b(x^p) = p\cdot\log_b(x).$$

To establish this last property, we first consider the definition of b^x to learn that $x^p = E[p\cdot L(x)]$, hence

$$\log_b(x^p) = \frac{\{L(E[p\cdot L(x)])\}}{L(b)}.$$

But theorem E-2 asserts that $L(E[p\cdot L(x)]) = p\cdot L(x)$, therefore

$$\log_b(x^p) = p\cdot\frac{L(x)}{L(b)}$$

and the property is established.

To illustrate the use that can be made of these properties of logarithms, consider the following.

Example. Find a number x, it must be positive, such that $\log_b(x) - \log_b(2) = -1$.

Solution. $\log_b(x) - \log_b(2)$ is $\log_b(x/2)$ and -1 is $\log_b(1/b)$, hence $\log_b(x/2) = \log_b(1/b)$. It follows from this that $x/2 = 1/b$ and so $x = 2/b$.

PROBLEM SET 7.1

If each of x and z is a positive number and each of b and c is a positive number different from 1, show that each of statements 1 through 3 is true.

1. $\log_b(1/z) = -\log_b(z)$.

2. $\log_b(x/z) = \log_b(x) - \log_b(z)$.

3. $\log_b(c)\cdot\log_c(b) = 1$.

4. If $\log_3(x) - \log_3(t) = 1$, express x in terms of t.

5. Find a number x such that $\log_5(x) - \log_5(x+4) = -1$.

6. If $0 < \log_{10}(x) - \log_{10}(z) < 1$, can you conclude that $1 < x - z < 10$? Explain.

7. If m is a counting number and x is a positive number, then $\log_{10}(10^m\cdot x) = m + \log_{10}(x)$. Verify this statement for $m = 1$, $m = 2$, and $m = 3$.

[8] The "base e logarithm of x" may be denoted by $\log_e(x)$, but traditionally it is denoted by $\text{Ln}(x)$ and is called the *natural logarithm of x*. Show that $\text{Ln}(x) = L(x)$.

9. Suppose that p is a number and that each of x and b is a positive number, $b \neq 1$. Using the definition of b^x and theorem E-2, show that if $b^p = x$, then $\log_b(x) = p$. This problem is often used as the definition of the base b logarithm, but not here.

10. Assuming that x is a positive number and that p is a number, use a definition and theorem E-2 to verify that $\log_b(x^p) = p\cdot\log_b(x)$.

11. Give approximations to $\log_{10}(1.4)$ and $\log_{10}(14)$.

[12] If z is a positive number, find $e^{\text{Ln}(z)}$.

13. If a is a positive number and x is a number, show that $L(a^x) = x\cdot L(a)$.

14. If n is a counting number and b is a positive number, show that $(b^{1/n})^n = b$. *Hint:* See the definition of b^x.

15. **a.** Find $\text{Ln}(10)\cdot\log_{10}(e)$.
b. Use the above to give an approximation to $\log_{10}(e)$.

Section 2—Logarithmic Equations Our primary purpose here is to indicate how the properties of logarithms can be utilized for the practical purpose of solving certain equations. It will be important to recall that $\log_b(1)$ is 0, $\log_b(b)$ is 1, and that $\log_b(1/b) = -1$.

Example. Find a number x such that

$$2 \cdot \log_{10}(x) - \log_{10}(x + 6) = 0.$$

Solution. Since $\log_b(x) - \log_b(z) = \log_b(x/z)$ and $\log_b(x^2) = 2 \cdot \log_b(x)$, we can start our solution to the given problem by writing the equation

$$\log_{10}\left[\frac{x^2}{x + 6}\right] = \log_{10}(1)$$

from which it imediately follows that $x^2 = x + 6$ or $x^2 - x - 6 = 0$. So we solve the quadratic equation to find that $x = 3$ or $x = -2$. However, $\log_{10}(-2)$ is not defined and so the only solution to our example is given by $x = 3$. Prior to the next example problem, you are reminded that $\log_e(x)$ is usually denoted by $\text{Ln}(x)$ and is called the natural logarithm of x. We can also recall that $\text{Ln}(x) = L(x)$.

Example. Find a number x such that $\text{Ln}(2) - \text{Ln}(x - 1) = -1$.

Solution. The given equation is equivalent to

$$\text{Ln}\left[\frac{2}{x - 1}\right] = \text{Ln}(1/e)$$

hence we should have no trouble in deducing that

$$\frac{2}{x - 1} = \frac{1}{e}$$

therefore $x = 2e + 1$.

Example. Solve $\log_3(x + 1) + \log_3(x - 1) = 3$.

Solution. $\log_3(3^3) = 3 \cdot \log_3(3) = 3$ by property 4 of section 1, so let us begin with the equation

$$\log_3(x^2 - 1) = \log_3(27),$$

for then we have $x^2 - 1 = 27$ and therefore $x = 2\sqrt{7}$.

For our next example, we select an equation known as an exponential equation.

Example. Solve $2^{x+1} = 3^x$.

Solution. (see definition of b^x) $2^{x+1} = E[(x + 1) \cdot L(2)]$ and $3^x = E[x \cdot L(3)]$, therefore $(x + 1) \cdot L(2) = x \cdot L(3)$. From this we get $x \cdot [L(3) - L(2)] = L(2)$, that is, $x \cdot L(3/2) = L(2)$. Thus the solution of our equation is the number $\dfrac{L(2)}{L(1.5)}$ and we are not surprised to see this written as $x = \dfrac{\text{Ln}(2)}{\text{Ln}(1.5)}$.

For those practical people who desire reasonably good approximations, we can consult our tables (see p. 251) and go on to give

$$x = \frac{0.693}{0.405} = 1.71.$$

Example. Solve $e^{2x} = 7$.

Solution. $e^{2x} = E[2x \cdot L(e)] = E(2x)$, thus $E(2x) = 7$. Hence $L[E(2x)]$ is $L(7)$, which implies that $2x = L(7)$ or $x = \frac{L(7)}{2}$. To give an approximation to our solution, we consult the tables already mentioned and write

$$x = \frac{1.946}{2} = 0.973.$$

PROBLEM SET 7.2

[1] Verify each of the following statements:

a. $\log_b (b)$ is 1
b. $\log_b (1)$ is 0
c. $\log_b (1/b)$ is -1
d. $\log_b (b^n)$ is n

Solve the following equations:

2. $\log_{10} (x + 3) + \log_{10} (x - 2) = \log_{10} (6)$.

3. $\log_{10} (x + 1) - \log_{10} (3) = 1$.

4. $\log_3 (x + 2) - \log_3 (x - 6) = 2$.

5. $\log_8 (16) + \log_8 (x - 2) = 2$.

6. $\log_2 (x^2 - 3x + 6) - \log_2 (x - 1) = 2$.

7. $\log_{10} (x + 6) + \log_{10} (5) - \log_{10}(x) = 2 - \log_{10} (x + 3)$.

8. $\text{Ln} (x) - \text{Ln} (3) = 1$.

9. $\text{Ln} (x) + \text{Ln} (2) = -1$.

10. $\text{Ln} (x + 2) - \text{Ln} (4x + 3) = \text{Ln} (1/x)$.

11. $\text{Ln} (x + 3) - \text{Ln} (2x - 1) = 1$.

12. $\text{Ln} (x) = \text{Ln} (e - 1) + \text{Ln} (e + 1) - \text{Lo}\left[\frac{e^2 - 1}{e}\right]$.

For each of the following exponential equations, give the solution and then give an approximation to the solution.

13. $3^{x-1} = 5$ **14.** $2^{3-x} = 5^x$ **15.** $e^{x-1} = 5$

16. $e^{x+1} = 10$ **17.** $e^{x-1} = 2$ **18.** $e^{2x-1} = 5$

19. $4^{2x} = 3^{x+3}$ **20.** $3^{1-2x} = 2^{x+5}$

Section 3—Applications We now direct a small amount of attention to some applications of the functions L and E. It may well be that, because many natural phenomena can be represented by models based upon this pair of functions, it was only natural that they be called *the natural logarithm function* and *the exponential function.*

One is warned (he may also be irked) that, since some of us deem the following problems to be practical, he should be prepared to give practical answers, that is, reasonable approximations.

Example. The number of milligrams M of radium present at the end of t years is given by the formula

$$M = M_1 \cdot 10^{-0.000174t}$$

where M_1 is the number of milligrams present at $t = 0$.

Problem. How long will it take for the given sample M_1 to be reduced to one-half its original size?

Analysis. The problem can be translated into the equation

$$1/2 \cdot M_1 = M_1 \cdot 10^{-0.000174t}$$

and, assuming that $M_1 \neq 0$ (a practical assumption), we solve the equation $1/2 = 10^{-0.000174t}$ for t. There are several ways of doing this depending upon the tools available. Perhaps the first shown is the easiest.

Solution 1. If base 10 logarithm tables are common, we can start with $\log_{10}(1/2) = -0.000174t$ (see prop. 4, sect. 1) or $\log_{10}(2) = 1.74 \cdot 10^{-4}t$ (see prop. 1, sect. 1). From the tables we find that $\log_{10}(2) = 0.301$, and so we find that $t = \dfrac{0.301}{1.74} \cdot 10^4 = 1730$ years.

Solution 2. Again we start with the equation

$$1/2 = 10^{-0.000174t}$$

and go on to $L(1/2) = -1.74 \cdot 10^{-4}t \cdot L(10)$. But $L(1/2) = -L(2)$, thus $L(2) = 1.74 \cdot 10^{-4}t \cdot L(10)$. From this last equation it follows that $1.74 \cdot 10^{-4}t = \dfrac{L(2)}{L(10)}$, hence $t = \dfrac{L(2) \cdot 10^4}{1.74 \cdot L(10)}$. Consulting our tables, we find that $L(10) = 2.303$ and that $L(2) = 0.693$ so substitution gives us $t = \frac{6.93}{4.01} \cdot 10^3 = 1730$. We charge the relatively small variances in the solutions to experimental error, but we must realize that our computations assure us of only three digit (slide rule) accuracy.

Example. The intensity B of light x feet below the surface of a body of water is given by the formula $B = e^{-0.86x}$. (It should harm no one to consider that B is a function whose domain is the set of non-negative numbers and whose rule of pairing is the equation $B(x) = e^{-0.86x}$.)

Problem. Find (a) the intensity at the surface, and (b) the intensity 10 feet below the surface.

Solution. (a) here $x = 0$ so $B = 1$. (*b*) since $x = 10$, we start with the equation $B = e^{-8.6}$ which is the solution. If suitable tables for e^{-z} are handy, we can quickly give the approximation desired as $B = 0.00019$. If no such tables are available, we can consider another method of solution which utilizes the tables for $L(x)$ which are available. Alternate solution to part (b): start, as before, with $B = e^{-8.6}$. By definition, $e^{-8.6} = E[-8.6 \cdot L(e)]$. Since $L(e)$ is 1, we have that $L(B) = L[E(-8.6)]$, but $L[E(-8.6)] = -8.6$ by theorem E-2, thus $L(B) = -8.6$. Theorem L-2 assures us that $L(1/B) = -L(B)$ and so we write $L(1/B) = 8.6$. Keeping one eye on the table, we see that (approximately)

$$L(1/B) = 6.908 + 1.692 = 3 \cdot L(10) + L(5.4), \text{ hence}$$
$$1/B = 5400 \text{ and } B = 0.00019.$$

Perhaps a practical summary of this is that it would be pretty dark as well as wet 10 feet below the surface.

PROBLEM SET 7.3

1. The pH of a solution is defined by the equation

$$\text{pH} = \log_{10}(1/C)$$

where C is the concentration of hydrogen ions in the solution. Further we are told, if pH < 7, the solution is acidic; if pH > 7, the solution is alkaline. Accepting all of this, determine the nature of a solution in which

$$C = 1.6 \cdot 10^{-3}.$$

2. The radioactive isotope C^{11} decomposes rapidly. If, at one instant, there is 1 microgram of this isotope present, then t minutes later there will be M micrograms present where $M = (0.5)^{t/2}$. Find the amount of C^{11} present at the end of 10 minutes; 20 minutes.

3. A special case of Newtons Law for the rate at which a hot body cools is given by the equation $50 = 75 \cdot e^{-20r}$. Find r.

4. The charge Q on an electrical condenser at any time t is given by the equation $Q = Q_1 \cdot e^{-kt}$ where Q_1 is the charge at $t = 0$ and k is a constant, depending on the make or style of the condenser. Find Q when $t = 4$, $k = 1/2$, and $Q_1 = 10$.

5. In a series circuit containing a coil of inductance K, we modify Ohms law to read

$$A = \frac{v}{r}(1 - e^{rx/k}).$$

Using natural logarithms, solve for x.

6. Sketch the graph of the function $F(x) = xe^x$.

7. In the study of radioactivity, the function $G(t) = 1 - e^{-t/k}$, with domain the set of nonnegative numbers, determines the probability that a particle will be emitted from a radioactive substance before time t. In this equation, k is a positive number depending upon the nature of the substance. Assuming that $k = 1$, sketch the graph of G on the interval [0,3]. Use a large unit distance for better accuracy.

8. The U.S. Army found that the cost C in dollars of replacing a *single* M-1 rifle with the M-14 rifle over a period of x years was given by the equation

$$C = 125 \cdot e^{-0.1x}.$$

Find the approximate cost of reequipping 1000 men over a period of two years.

9. An approximation to the pressure p in millimeters of mercury at a height h kilometers above mean sea level is given by $p = 760 \cdot e^{-0.144h}$. After noting that the sea level pressure is 760, find the height at which the pressure is 1/2 that at sea level.

10. In this problem, no tables are needed. If x is a positive number such that $\text{Ln}\,[x + \sqrt{(x^2 + 1)}] = 2$, try to show that $x = \dfrac{e^2 - e^{-2}}{2}$.

TABLE 7.1

A TABLE FOR Ln (x)

x	0.0	0.1	0.2	0.3	0.4	0.5	0.6	0.7	0.8	0.9
1	0.000	0.095	0.182	0.262	0.336	0.405	0.470	0.531	0.588	0.642
2	0.693	0.742	0.788	0.833	0.875	0.916	0.956	0.993	1.030	1.065
3	1.099	1.131	1.163	1.194	1.224	1.253	1.281	1.308	1.335	1.361
4	1.386	1.411	1.435	1.459	1.482	1.504	1.526	1.548	1.569	1.589
5	1.609	1.629	1.649	1.668	1.686	1.705	1.723	1.740	1.758	1.775
6	1.792	1.808	1.825	1.841	1.856	1.872	1.887	1.902	1.917	1.932
7	1.946	1.960	1.974	1.988	2.001	2.015	2.028	2.041	2.054	2.067
8	2.079	2.092	2.104	2.116	2.128	2.140	2.152	2.163	2.175	2.186
9	2.197	2.208	2.219	2.230	2.241	2.251	2.262	2.272	2.282	2.293
10	2.303	2.313	2.322	2.332	2.342	2.351	2.361	2.370	2.380	2.389

Ln 0.1 = −2.303 2·Ln 10 = 4.605 4·Ln 10 = 9.210

Ln 10 = 2.303 3·Ln 10 = 6.908 5·Ln 10 = 11.513

Notes. Because $e^{\mathrm{Ln}\,x} = x$, table 7.1 can be utilized to give approximations to e^x.

Examples. $e^{0.53} = 1.7$ $e^{1.50} = 4.5$, and $e^2 = 7.4$. Since $e^{-x} = 1/e^x$, a little additional arithmetic yields the approximations $e^{-0.53} = 0.59$, $e^{-1.50} = 0.22$, and $e^{-2} = 0.14$.

Section 4—*L* Class Functions Taking for a starting point the definition of the base b logarithm of x and feeling free to use the properties already developed, we now consider a class of functions to which a function F belongs if $F(x) = \log_b(x)$. Quickly we can note that $F(x) = \dfrac{L(x)}{L(b)}$, that is, $F(x) = \dfrac{1}{L(b)}\cdot L(x)$, but this can be interpreted as meaning that $F(x) = K\cdot L(x)$, where K is some constant obviously dependent upon b. Thus the graph of F can be constructed from the graph of L.

Example. (See figure 7.1.) To sketch the graph of $F(x) = \log_2(x)$, we already know that $F(2) = 1$, $F(1) = 0$, $F(1/2) = -1$, $F(4) = 2$, and $F(1/4) = -2$. We may also observe that $F(x) = \dfrac{1}{L(2)}\cdot L(x)$ and that $L(2) = 0.693$, hence $F(x) = 1.44\cdot L(x)$, this last being an approximation suitable for graphical purposes. From this we can approximate $F(3) = (1.44)(1.099)$, that is, $F(3) = 1.58$. In like manner, $F(1.5)$ is 0.60. Note that dom $F = \{x \mid x > 0\}$.

Example. (See figure 7.2.) If $g(x) = \log_2(x - 2)$, then dom $g = \{x \mid x > 2\}$. It is easy to see that $g(6) = 2$, $g(4) = 1$, $g(3) = 0$, and

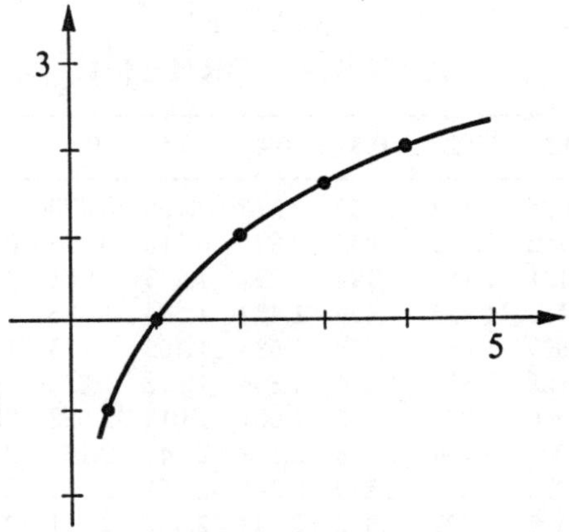

Fig. 7.1

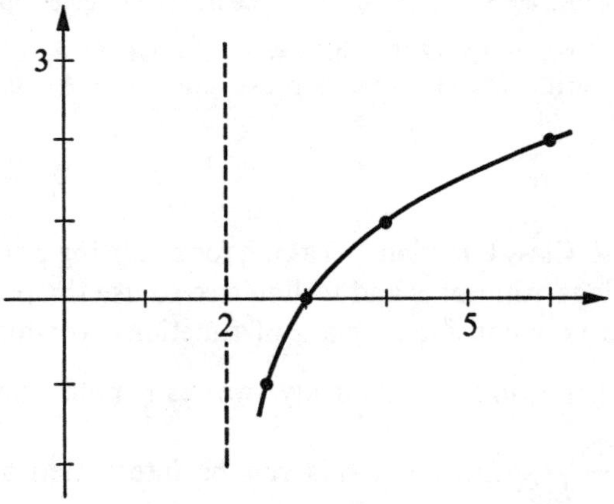

Fig. 7.2

that $g(2.5) = -1$. If these calculations do not provide us with enough points to indicate the graph of g, we can surely note that $g(5) = 1.44 \cdot L(4) = 1.99$ approximately.

I believe that these two examples are enough on this and consider in the next example another use of the old idea of the composite function.

Example. Suppose that G is a function, then if $[L \circ G]$ is a function we know that $[L \circ G](x) = L[G(x)]$ and we need to remember that if $G(x)$ is to be in the domain of L, then $G(x)$ must be a positive number. To be specific, let G be the function such that $G(x) = x + 1$, the domain of G is the set of all numbers. We can then show that $[L \circ G](x) = \text{Ln}(x + 1)$. We can also show that $\text{dom}[L \circ G] = \{x \mid x > -1\}$. The graph of this composite function is shown in figure 7.3. In constructing this graph, we made use of the following: $[L \circ G](-1/2) = \text{Ln}(1/2)$, $[L \circ G](0) = \text{Ln}(1)$, $[L \circ G](e - 1) = \text{Ln}(e)$, and $[L \circ G](5) = \text{Ln}(6)$.

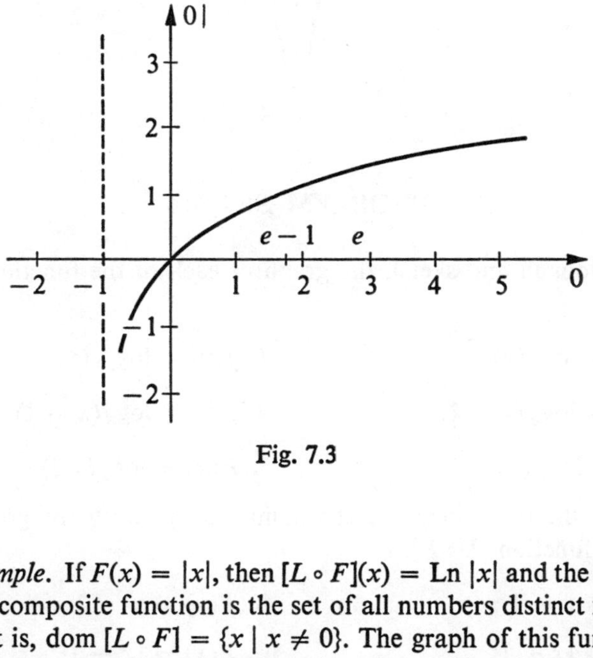

Fig. 7.3

Example. If $F(x) = |x|$, then $[L \circ F](x) = \text{Ln}|x|$ and the domain of this composite function is the set of all numbers distinct from zero. That is, $\text{dom}[L \circ F] = \{x \mid x \neq 0\}$. The graph of this function can be constructed by observing that $[L \circ F](1) = [L \circ F](-1) = 0$, $[L \circ F](e) = [L \circ F](-e) = 1$, and that $[L \circ F](1/e) = [L \circ F](-1/e) = -1$. As a matter of fact, one can easily note that if x is not zero, then we have that $[L \circ F](x) = [L \circ F](-x)$. If additional points are needed, we can surely produce them. For example,

$$[L \circ F](5) = [L \circ F](-5) = 1.609.$$

The graph of this composite function is indicated in figure 7.4.

Remark. Note that the graph is Y-symmetrical.

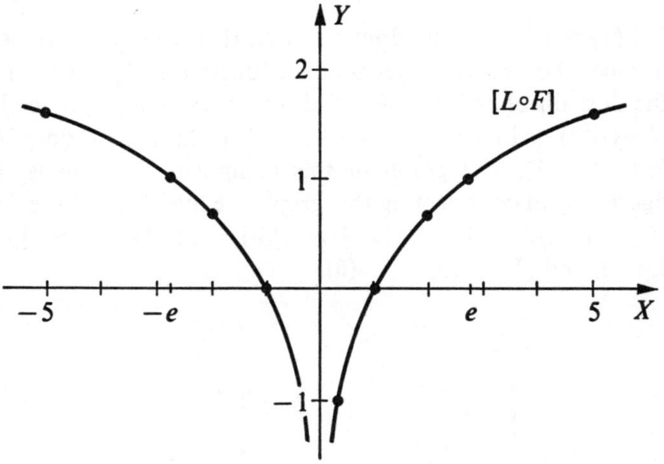

Fig. 7.4

PROBLEM SET 7.4

Give the domain and sketch the graph of each of the functions defined below.

1. $F(x) = \log_3 (x)$

2. $f(x) = \log_{10} (x)$

3. $G(x) = \log_2 (x + 1)$

4. $g(x) = \log_2 (x - 1)$

5. $p(x) = \log_4 (2x)$

6. $P(x) = \log_3 (x/3)$

In each of the following, give the domain and sketch the graph of the composite function $[L \circ F]$.

7. $F(x) = x - 1$

8. $F(x) = x - 2$

9. $F(x) = |x + 1|$

10. $F(x) = |x - 1|$

11. $F(x) = ex$

12. $F(x) = x/e$

13. $F(x) = x/2$

14. $F(x) = 2x$

15. We have not considered the slope quotient $\dfrac{L(x) - L(a)}{x - a}$ but, if $F(x) = \log_b (x)$ and a and x are positive numbers, show that

$$\frac{F(x) - F(a)}{x - a} = K \cdot \frac{L(x) - L(a)}{x - a}$$

where K is a constant.

SUMMARY

The definition of a number called *the base b logarithm* perhaps confirmed your suspicions as to why the letter L was used to designate the special function considered in the last chapter. For, from this definition, it followed that $L(x) = \text{Ln}(x) = \log_e(x)$. Base 10 or common logarithms were also introduced.

A few applications served to indicate why the functions L and E are called *the natural logarithm* and *the exponential functions* respectively.

The solution of logarithmic and exponential equations was considered, and we made considerable use of the L and E theorems in finding their solutions.

In the concluding section of this chapter, we looked at a class of functions called *L class* or *log functions.* It is easy to recall the properties of these functions for they are a direct analogy to the properties of the function L.

Review Problems

1. Find each of the following numbers:

a. $\log_b(b)$ **b.** $\log_b(1)$ **c.** $\log_b(1/b)$
d. $\log_b(b^2)$ **e.** $\log_b(e)$

2. Complete each of the following.

a. $\log_b(xz) =$ _______ **b.** $\log_b(1/z) =$ _______
c. $\log_b(x/z) =$ _______ **d.** $\log_b(x^p) =$ _______

3. Rewrite each of the following as a single logarithm.

a. $\log_b(2x) + \log_b(3x) - \log_b(x)$
b. $\text{Ln}(2\pi) - \text{Ln}(\sqrt{2\pi}) - \text{Ln}(e)$
c. $2\log_{10}(\sqrt{2}) + 2\log_{10}(\sqrt{3})$

4. Solve the following equations.

a. $\log_{10}(x + 2) + \log_{10}(x - 1) = 1$
b. $\text{Ln}(x - 1) + \text{Ln}(2) = 2$
c. $\log_5(2x) - \log_5(x + 1) = 0$

5. Sketch the graph of each of the functions.

$f(x) = \log_2(x - 1)$ and $g(x) = \log_3(x/3)$

A REVIEW

It
is time
to pause and
regroup. So the next
few pages are devoted to a
review test in order to see if
there are topics that need reviewing.

A Review Test In order to see what you may need to review, do as many of the following as you can without looking back.

1. If $P = (3, -1)$ and $Q = (1,0)$ are two points, give:

a. an equation of the line containing P and Q.
b. the equation of the circle whose center is P and which contains Q.

2. Describe $U = \{x \mid |x - 3| < 1\}$ as an interval.

3. Give, if they exist, the GLB and LUB of:

$$A = \left\{x \,\middle|\, \frac{x}{x-2} \leq 0\right\} \qquad B = \left\{x \,\middle|\, x = \frac{2n-1}{3n}, n \in N\right\}$$

4. a. The statement that "F is a function" means . . .
b. The statement that "F is a 1-to-1 function" means . . .

5. Give the domain and range of each of the following functions:

a. I **b.** H **c.** L

d. E **e.** $f(x) = |x|$ **f.** $g(x) = \dfrac{x}{x+2}$

g. $p(x) = x^2 - 2x$

6. If $F(x) = \dfrac{x-1}{x}$ give: (a) $F(1)$, (b) $F(-1)$, (c) $F(a+1)$, and (d) $\dfrac{F(x) - F(a)}{x - a}$.

7. If $F(x) = \dfrac{x}{x-1}$ and $G(x) = 2x + 1$, give $[F \circ G](x)$ and $[G \circ F](x)$.

8. **a.** What is meant by the statement "F is continuous at $P = [a,F(a)]$"?
b. Sketch a function F that is not continuous at $P = [a,F(a)]$.
c. Illustrate the intermediate value theorem.

9. **a.** What property must F have in order that F^* is a function?
b. If $F(x) = \sqrt{(x+1)}$, give $F^*(x)$, dom and rng of F^*.

10. If $P(x) = 3x^3 - 2x^2 - 9x + 6$,

a. Give all possible rational zeros (roots) of P.
b. Use synthetic division to find $P(2/3)$.

The answers to these problems are found in the appendix.

CHAPTER 8
Circular Functions

Section 1—Sine and Cosine Functions If P is a point on the circle with center at the origin and with radius 1 unit, then the length of arc, θ, measured *counterclockwise* from the point (1,0) to the point P is called the *arc* of P. In figure 8.1, for example, the arc of A is 1 and the arc of B is $\frac{1}{2}$.

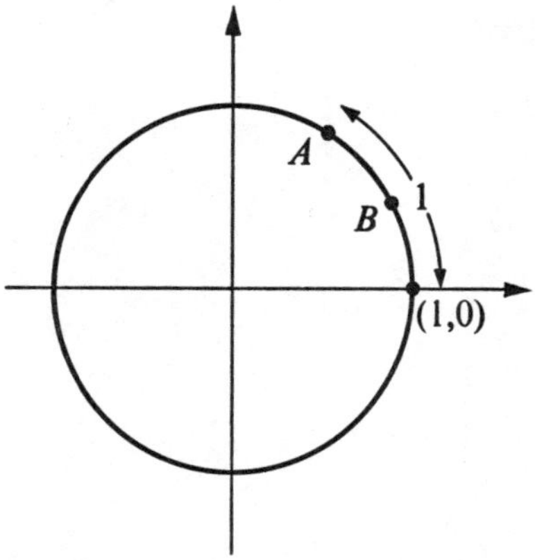

Fig. 8.1. The arc of a point on the unit circle.

If P is the point (x,y) and θ is the arc of P, then the values of x and y are uniquely determined by the number θ. In other words, x and y are functions of θ. The function which determines the horizontal coordinate of P is called the *cosine* function and the function which determines the vertical component of P is called the *sine* function. This is usually abbreviated as $x = \cos\theta$ and $y = \sin\theta$. (See figure 8.2.)

The circumference of the unit circle is denoted by the number 2π where π is approximately equal to 3.14159. Hence the arcs of the points A, B, C,

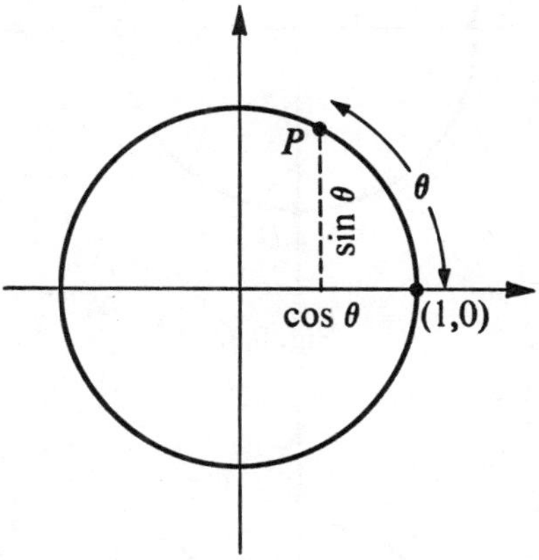

Fig. 8.2. $P = (\cos\theta, \sin\theta)$.

and D in figure 8.3a are 0, $\pi/2$, π, and $3\pi/2$ respectively. Thus the following should be rather obvious:

$$\begin{array}{ll} \cos 0 = 1 & \sin 0 = 0 \\ \cos \pi/2 = 0 & \sin \pi/2 = 1 \\ \cos \pi = -1 & \sin \pi = 0 \\ \cos 3\pi/2 = 0 & \sin 3\pi/2 = -1 \end{array}$$

An arc measured *clockwise* from (1,0) is considered to be negative. Thus the arc of B is $-3\pi/2$ as well as $\pi/2$. The arc of C is $-\pi$ as well as π. If we also consider the points P and Q (see figure 8.3b), then the following fundamental property of the circular functions should easily follow.

FP-1
$$\cos(-\theta) = \cos\theta \quad \text{and} \quad \sin(-\theta) = -\sin\theta$$

If $0 < \theta < \pi/2$, then *the point P on the unit circle whose arc is θ is in the first quadrant*, hence $x > 0$ and $y > 0$, that is, $\cos\theta > 0$ and $\sin\theta > 0$. Because it is easier to say and to write for the statement italicized above, we shall say that "θ is in the first quadrant" and write $\theta \in$ Qd. I.

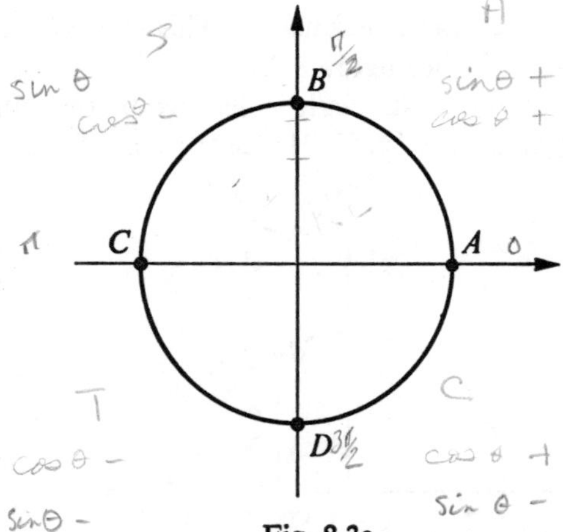

Fig. 8.3a

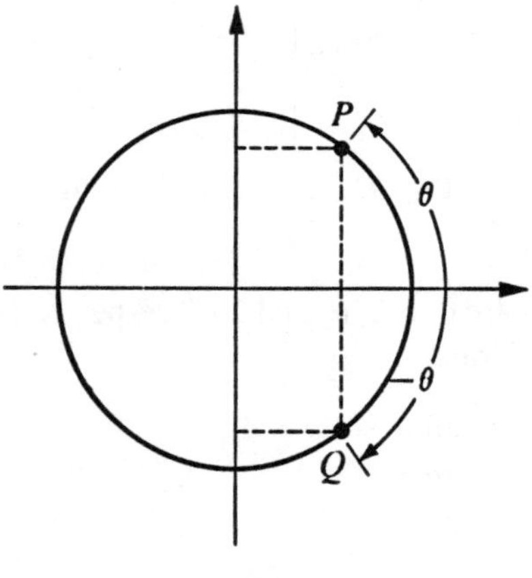

Fig. 8.3b

Example. If $3\pi/2 < \theta < 2\pi$, then $\theta \in$ Qd. IV.

Table 8.1 may help you to determine when the circular functions sine and cosine have positive or negative values.

A very fundamental observation (see figure 8.2) is to note that if $P = (\cos\theta, \sin\theta)$ is a point on the unit circle, then its coordinates must satisfy the equation of the unit circle $x^2 + y^2 = 1$, thus

FP-2 $(\cos\theta)^2 + (\sin\theta)^2 = 1.$

Remark. $\cos^2\theta$ is the usually used notation for $(\cos\theta)^2$ and likewise $\sin^2\theta$ is used for $(\sin\theta)^2$. Because of this established tradition, we shall use this notation.

TABLE 8.1

Quadrant in which θ lies	$\cos \theta$	$\sin \theta$
I	+	+
II	−	+
III	−	−
IV	+	−

In figure 8.4, the point P is one of the points in common to the unit circle and the function I and shows that $\cos \pi/4 = \sin \pi/4$. Armed with this, FP-2 tells us that $2 \cdot \cos^2 (\pi/4) = 1$ or $\cos \pi/4 = \sin \pi/4 = \frac{\sqrt{2}}{2}$.

Further note that if the value of one of the circular functions sine or cosine and the quadrant to which θ belongs is known, then FP-2 can be used to find the value of the other circular function. This is illustrated in the example which follows.

Example. Given: $\cos \theta = 2/3$ and $\theta \in$ Qd. IV. Find $\sin \theta$.

Solution. $4/9 + \sin^2 \theta$ is 1 by FP-2, hence

$$\sin^2 \theta = 5/9.$$

Since $\theta \in$ Qd. IV, we know that $\sin \theta < 0$, therefore $\sin \theta = -\frac{\sqrt{5}}{3}$.

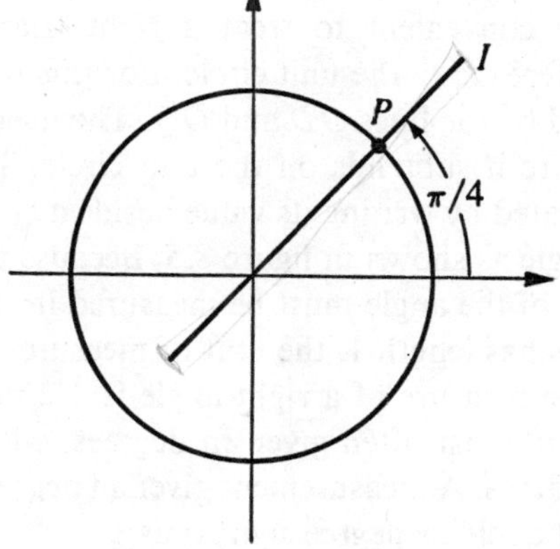

Fig. 8.4. Equality of cos π/4 and sin π/4.

If $Q = (a,b)$ is a point not on the unit circle (see figure 8.5), then the line segment joining Q and the origin O will intersect the unit circle at some point P whose arc is θ. Since the right triangles OCP and ODQ are similar, the corresponding sides are proportional:

$$\frac{DQ}{OQ} = \frac{CP}{OP} \text{ and } \frac{OD}{OQ} = \frac{OC}{OP}$$

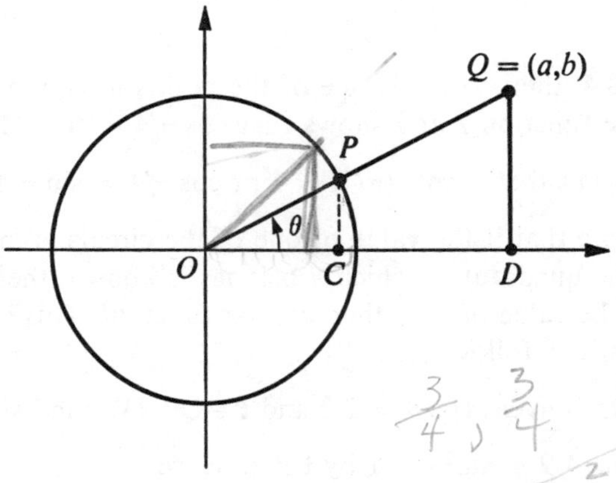

Fig. 8.5

Since $OP = 1$, $CP = \sin\theta$, $OC = \cos\theta$, and $OQ = \sqrt{(a^2 + b^2)}$, then $\sin\theta = \dfrac{b}{\sqrt{(a^2 + b^2)}}$ and $\cos\theta = \dfrac{a}{\sqrt{(a^2 + b^2)}}$.

It is frequently convenient to treat a right triangle such as ODQ without explicit reference to the unit circle. For this reason, we can speak of the *angle* formed by the lines OD and OQ. The measure of this angle is the length of the arc it subtends on the unit circle. The measure of this angle is often indicated by writing its value beside a curved arrow between the sides of the angle as shown in figure 8.5. Because the arc which determines the measure of the angle must be measured in the same metric unit in which the radius has length 1, the unit of measure of an angle is called a *radian*. Hence the measure of a right angle is $\pi/2$ radians. Measures of angles, however, are most often given in degrees, with 90 degrees corresponding to $\pi/2$ radians. A measurement given in degrees is usually written with a small raised circle or degree sign; thus

$$\pi \text{ radians} = 180°.$$

As shown in figure 8.6, an arc whose length is more than 2π (i.e., more than the circumference of the unit circle) can be measured. In this figure, it should be clear that

$$\cos(\theta + 2\pi) \text{ is } \cos\theta$$

and that

$$\sin(\theta + 2\pi) \text{ is } \sin\theta$$

so that, in general, if k is any integer

$$\text{FP-3} \quad \begin{aligned} \cos(\theta + 2k\pi) &= \cos\theta \\ \sin(\theta + 2k\pi) &= \sin\theta \end{aligned}$$

The sine and cosine functions therefore repeat themselves at intervals of 2π and are said to be *periodic functions* with period 2π.

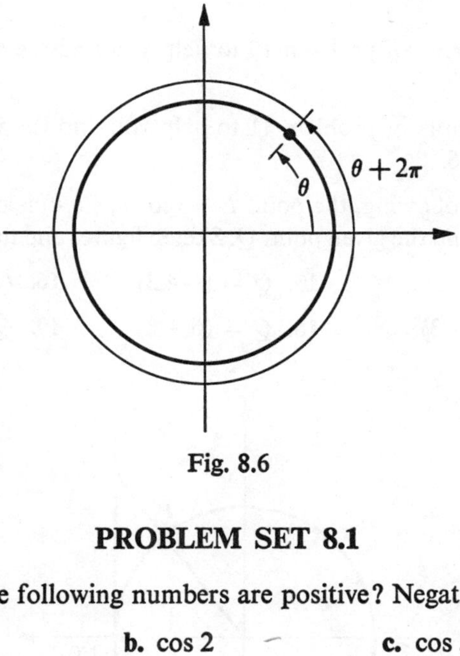

Fig. 8.6

PROBLEM SET 8.1

1. Which of the following numbers are positive? Negative?

a. $\sin 2$ **b.** $\cos 2$ **c.** $\cos 5$
d. $\sin 3.5$ **e.** $\sin 1.6$ **f.** $\cos -1$

2. Determine the quadrant in which θ lies under the given conditions.

a. $\sin\theta > 0$ and $\cos\theta < 0$. **b.** $\cos\theta > 0$ and $\sin\theta > 0$.

c. $\cos\theta < 0$ and $\sin\theta < 0$. **d.** $\dfrac{\sin\theta}{\cos\theta} > 0$ and $\cos\theta < 0$.

3. Determine the values of $\cos t$ and $\sin t$ for each of the following values of t. $t = -\pi/4$, $-\pi/2$, $3\pi/4$, and $7\pi/4$.

4. You can show that the circular functions sine and cosine are bounded functions if you can explain why **(a)** $|\sin \theta| \leq 1$ and **(b)** $|\cos \theta| \leq 1$. *Do so* but be brief.

Find the values of $\cos \theta$ and $\sin \theta$ under the following conditions.

5. $\cos \theta = -\frac{4}{5}$, $\theta \in$ Qd. III

6. $\sin \theta = -\frac{1}{2}$, $\theta \in$ Qd. IV

7. $\cos \theta = \frac{5}{6}$, $\theta \in$ Qd. I

8. $\sin \theta = \frac{2}{3}$, $\theta \notin$ Qd. II

9. $\dfrac{\sin \theta}{\cos \theta} = \frac{3}{4}$, $\theta \in$ Qd. I

10. Use geometric methods to determine the values of $\cos \pi/3$ and $\sin \pi/3$.

11. Use geometric methods to determine the values of $\cos \pi/6$ and $\sin \pi/6$.

12. Use the results of problem 10 to help you find the values of $\cos 2\pi/3$ and $\sin 2\pi/3$.

13. Use the results of problem 11 to help you find the values of $\cos 7\pi/6$ and $\sin 7\pi/6$.

In each of the following, the point $P = (\cos \theta, \sin \theta)$ is on the line joining the origin and the given point Q. Sketch figures and find $\cos \theta$ and $\sin \theta$.

14. $Q = (3,4)$

15. $Q = (-4,3)$

16. $Q = (5,-4)$

17. $Q = (-3,-3)$

18. $Q = (6,-8)$

19. $Q = (\surd(2,2)$

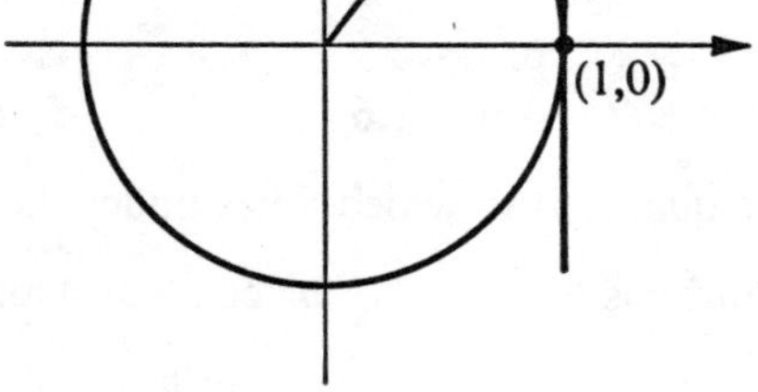

Fig. 8.7

20. Give the degree measure of an angle whose measure is:

a. $\pi/3$ radians **b.** $3\pi/4$ radians **c.** $\pi/5$ radians
d. $\pi/8$ radians **e.** 1 radian

[21] Let $0 < \delta < \pi/2$. In each part of this problem, draw a unit circle and let P be the point whose arc is δ and then plot the point P' whose arc is θ where θ is as given below. After this is done, you are to express $\cos\theta$ and $\sin\theta$ in terms of $\cos\delta$ or $\sin\delta$.

a. $\theta = \pi - \delta$ **b.** $\theta = \pi + \delta$ **c.** $\theta = 2\pi - \delta$

22. In figure 8.7 you are to determine the coordinates of the point T. The point T is the intersection of a vertical line through (1,0) and the line joining the origin and $P = (\cos\theta, \sin\theta)$.

SPECIAL PROBLEM

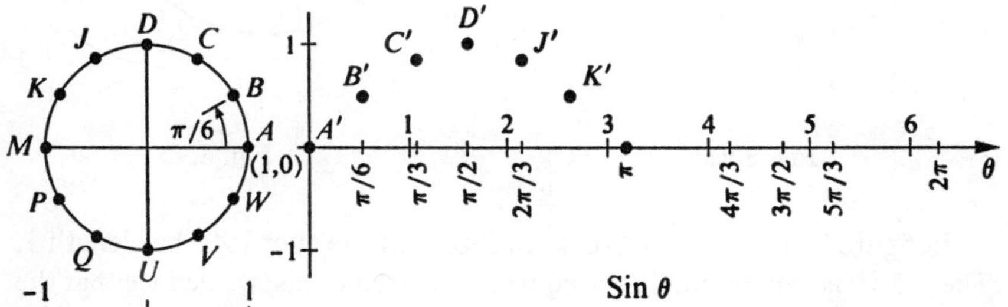

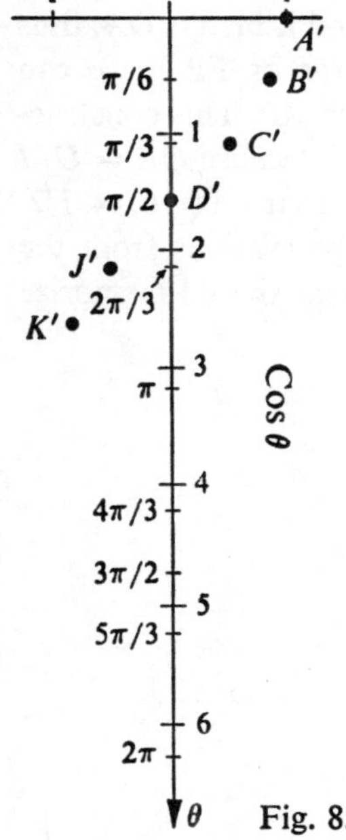

Fig. 8.8

Each of the points A, B, C, . . . are equally spaced on the unit circle. Thus the arc of B is $\pi/6$; the arc of C is $\pi/3$; and so forth.

To obtain the graph of $\sin\theta$, project the ordinate of A horizontally over to intersect the line $\theta = 0$ at A'; project the ordinate of B horizontally over to intersect the line $0 = \pi/6$; and so forth.

To obtain the graph of $\cos\theta$, project the abscissa of A vertically down to intersect the line $\theta = 0$; project the abscissa of B vertically down to intersect the line $\theta = \pi/6$ at B'; and so forth. Proceeding in the manner indicated, draw the graphs of $\sin\theta$ and $\cos\theta$. (See figure 8.8.)

Section 2—Special Values of θ The values of the circular functions sine and cosine for certain special and important values of θ can be found through the use of fairly simple geometric methods. For example, the sin ($\pi/4$) and the cos ($\pi/4$) obtained in the previous section. For additional examples, take a look at figures 8.9a and 8.9b.

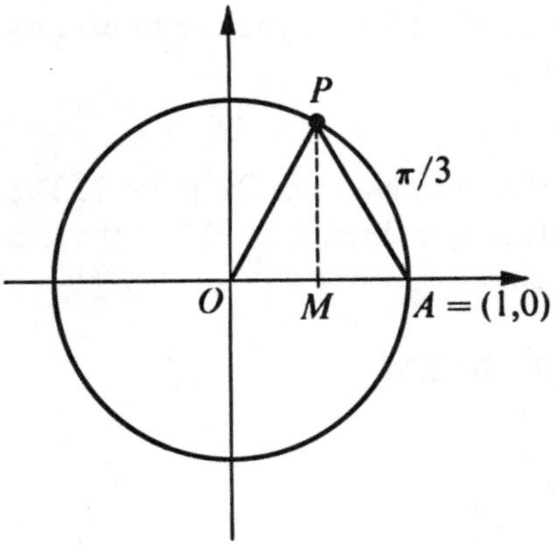

Fig. 8.9a

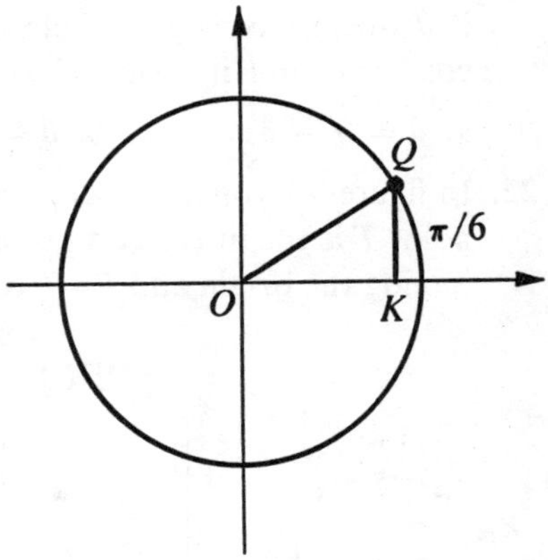

Fig. 8.9b

In figure 8.9a, we locate the point P so that the chord AP has length 1. The ΔOAP is equilateral, hence equiangular. From this, we deduce that the arc of P is $\pi/3$. A perpendicular PM is constructed and it bisects OA, thus $\cos(\pi/3) = 1/2$. By using the theorem of Pythagoras or FP-2, we can show that $\sin(\pi/3) = \sqrt{3}/2$. We next bisect the chord AP. This construction yields the point Q which we show in figure 8.9b. Because $QK = OM$ and $OK = PM$, we know that $\cos(\pi/6) = \sqrt{3}/2$ and that $\sin(\pi/6) = 1/2$.

The above results together with some of our observations from the previous section are collected in table 8.2. The student should memorize these results.

TABLE 8.2

θ	$\sin\theta$	$\cos\theta$
0	0	1
$\pi/6$	$1/2$	$\sqrt{3}/2$
$\pi/4$	$\sqrt{2}/2$	$\sqrt{2}/2$
$\pi/3$	$\sqrt{3}/2$	$1/2$
$\pi/2$	1	0

The usefulness of this table can be extended by making use of our solution to problem 19 in problem set 8.1, one part of which we recapitulate here: The arc of P is δ and the arc of Q is θ, $\theta = \pi + \delta$. The two right triangles shown in figure 8.10 are congruent so that

$$|\cos \theta| = \cos \delta$$

and

$$|\sin \theta| = \sin \delta$$

Since $\theta \in$ Qd. III, the above means that

$$\cos \theta = -\cos \delta$$

and

$$\sin \theta = -\sin \delta.$$

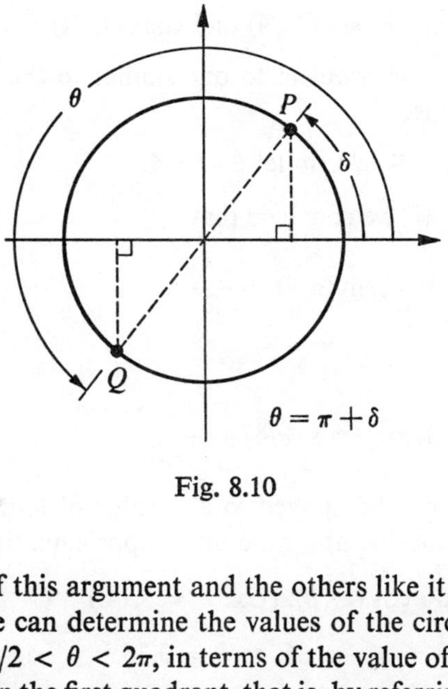

Fig. 8.10

The importance of this argument and the others like it in problem 19 lies in the fact that we can determine the values of the circular functions for any argument θ, $\pi/2 < \theta < 2\pi$, in terms of the value of the same function for some number in the first quadrant, that is, by referring to some number δ, $0 < \delta < \pi/2$, for which one of the following is true: $\theta = \pi + \delta$ or $\theta = \pi - \delta$ *or* $\theta = 2\pi - \delta$.

Example. Find $\sin (5\pi/3)$ and $\cos (5\pi/3)$. Here $\theta = 5\pi/3$.

Solution. $5\pi/3 = 2\pi - \pi/3$ so $\delta = \pi/3$. Thus $|\sin (5\pi/3)| = \sin (\pi/3)$ and $|\cos (5\pi/3)| = \cos (\pi/3)$. Since $\theta \in$ Qd. IV, $\sin \theta < 0$ and

$\cos\theta > 0$, so we write $\sin(5\pi/3) = -\sin(\pi/3)$ and $\cos(5\pi/3) = \cos(\pi/3)$. After a glance at Table 8.2, we conclude that

$$\sin(5\pi/3) = -\frac{\sqrt{3}}{2} \text{ and that } \cos(5\pi/3) = \frac{1}{2}.$$

Remark. The number δ described above is often called a *reference number.*

If $\theta > 2\pi$, then we have to appeal to FP-3 first, that is, we should recall that for any integer k,

$$\sin(\theta + 2k\pi) = \sin\theta \text{ and } \cos(\theta + 2k\pi) = \cos\theta.$$

Example. Find $\sin(11\pi/4)$ and $\cos(11\pi/4)$. Here $\theta = 11\pi/4$.

Solution. Since $11\pi/4 > 2\pi$ and $11\pi/4 = 2\pi + 3\pi/4$, our first step is to write

$$\sin(11\pi/4) = \sin(3\pi/4) \text{ and } \cos(11\pi/4) = \cos(3\pi/4)$$

thus reducing the problem to one similar to the previous example. Let us continue.

$$3\pi/4 = \pi - \pi/4, \text{ so let } \delta = \pi/4.$$

Since $\theta \in$ Qd. II, we conclude that

$$\sin(3\pi/4) = \sin(\pi/4) = \frac{\sqrt{2}}{2}$$

and

$$\cos(3\pi/4) = -\cos(\pi/4) = -\frac{\sqrt{2}}{2}.$$

The same ideas can be applied to any value of θ, though at times we may need to do a considerable amount of approximating.

Example. Find $\cos(2.78)$. (Use $\pi = 3.14$.)

Solution. $2.78 = \pi - 0.36$, so $\delta = 0.36$.

$$|\cos(2.78)| = \cos(0.36)$$

Since $\pi/2 < 2.78 < \pi$, $\theta \in$ Qd. II, therefore,

$$\cos(2.78) = -\cos(0.36).$$

Unlike previous examples, we come to a halt here for we have not determined an evaluation of $\cos(0.36)$ at this time.

PROBLEM SET 8.2

1. Find $\sin\theta$ and $\cos\theta$ for each of the given θ.

   a. $\theta = 5\pi/6$ b. $\theta = 4\pi/3$ c. $\theta = 7\pi/4$

2. Complete table 8.3. (The results obtained in the examples are listed.)

TABLE 8.3

θ	$\sin\theta$	$\cos\theta$	θ	$\sin\theta$	$\cos\theta$
$2\pi/3$			$4\pi/3$		
$3\pi/4$	$\sqrt{2}/2$	$-\sqrt{2}/2$	$3\pi/2$		
$5\pi/6$			$5\pi/3$	$-\sqrt{3}/2$	$1/2$
π			$7\pi/4$		
$7\pi/6$			$11\pi/6$		
$5\pi/4$			2π		

3. In each of problems 3 through 11, find δ, $0 < \delta < \pi/2$, and express each of the following in terms of the same function of δ. (In problems 8–11, use one of the approximates $\pi = 3.14$ or $2\pi = 6.28$).

3. $\sin(8\pi/5)$ 4. $\cos(7\pi/8)$ 5. $\cos(7\pi/12)$

6. $\sin(19\pi/12)$ 7. $\sin(13\pi/10)$ 8. $\cos(5.60)$

9. $\cos(4.24)$ 10. $\sin(2.14)$ 11. $\sin(1.74)$

12. If $\sin t = -\dfrac{3}{\sqrt{13}}$ and $\pi/2 < t < \pi$, give $\cos t$.

13. If $\dfrac{\sin u}{\cos u} = \dfrac{1}{2}$ and $\pi < u < 3\pi/2$, give $\sin u$ and $\cos u$. Verify that $\sin^2 u + \cos^2 u$ is 1.

14. Simplify each of the following statements. *Hint:* They are simple and and can be rapidly done.

   a. $\sin^2(\pi/5) + \cos^2(\pi/5)$ b. $\sin^2(1) + \cos^2(1)$

   c. $(1 - \cos\theta)(1 + \cos\theta)$ d. $\cos\theta - \cos(-\theta)$

   e. $\dfrac{1 - \sin^2 x}{1 - \sin(-x)}$

15. *Bonus problem.* In plane geometry there are many intriguing constructions, one of which is given below. (See figure 8.11.)

To inscribe a regular pentagon in a circle, draw two perpendicular diameters AB and CD. Bisect AO at E. With E as center and ED as radius, draw arc DF. The construction of the pentagon can then be completed using chord DF as the sides. Assuming that the radius of the circle is 1, *find the length of the chord DF.* If this chord is the length of each side of the regular pentagon, then one-half its length is the sin θ. Because each central angle is 360° divided by 5, that is, 72°, then $\theta = 36°$. Thus, using the tables in the appendix, you can at least check approximately whether or not you have correctly found the length of the chord DF as required.

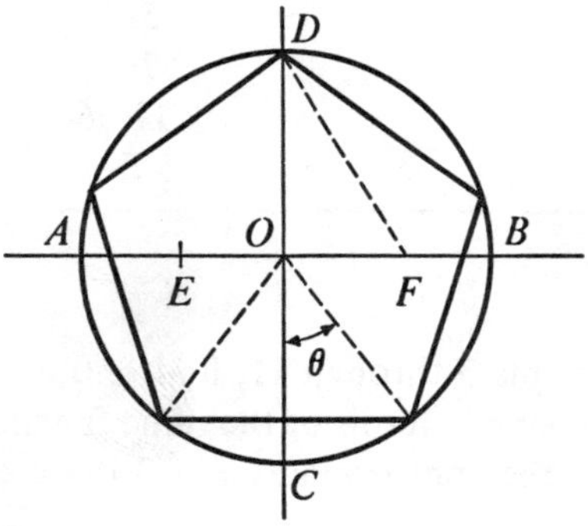

Fig. 8.11

Section 3—Addition Formulas for Cos and Sin If we examine a function f whose domain includes the numbers x, t, and $(x + t)$, we are concerned with whether or not there is any relationship expressing $f(x + t)$ in terms of $f(x)$ and $f(t)$, Such relationships do occur; one simple example is given by the formula $I(x + t) = I(x) + I(t)$. A more sophisticated example was encountered earlier in our study of the function E. Remember?

$$E(x + t) = E(x) \cdot E(t).$$

Such relationships are commonly referred to as *addition formulas*, and we shall attempt to show that the circular functions cosine and sine *do have* such a formula.

We can modestly begin by observing that $\cos(2\pi)$ is 1 and that $\cos(\pi) + \cos(\pi) = -2$. In general, then,

$$\cos(x + t) \neq \cos(x) + \cos(t).$$

In a similar fashion, you should be able to show that

$$\sin(x + t) \neq \sin(x) + \sin(t). \text{ Do so.}$$

Disappointed that the relationship is not this simple, we dig in for the battle, for we need to know the answer to the question:

Can we express $\cos(x + t)$ and $\sin(x + t)$ in terms of $\cos(x)$, $\cos(t)$, $\sin(x)$, and $\sin(t)$?

The attack begins by drawing two unit circles. In figure 8.12, the arc of P is x, the arc of Q is t, and the arc of W is $-t$. Thus the length of the arc from W to P is $(x + t)$. In figure 8.13, we locate the point Z whose arc is $(x + t)$. Since, in equal circles, equal arcs subtend equal chords, this can be done by simply making chord WP equal in length to chord AZ. Our plan of attack is based upon the assertion that these chords are equal and will employ the distance formula so let us call "time out" while we look up this formula. . . .

Because the distance formula requires them, let us list the coordinates of the points P, Q, W, and Z for easy reference.

$$P = (\cos[x], \sin[x]) \qquad Q = (\cos[t], \sin[t])$$
$$W = (\cos[-t], \sin[-t]) \qquad Z = (\cos[x + t], \sin[x + t])$$

Let the battle begin!

Glancing at figure 8.12 and using the distance formula, we find that

$$s^2 = [\cos(x) - \cos(-t)]^2 + [\sin(x) - \sin(-t)]^2.$$

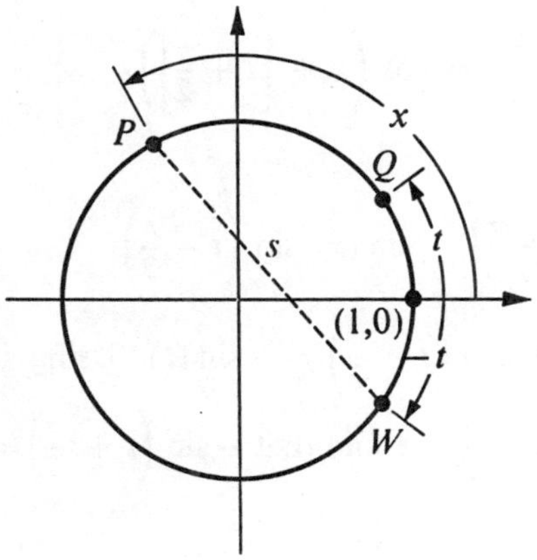

Fig. 8.12

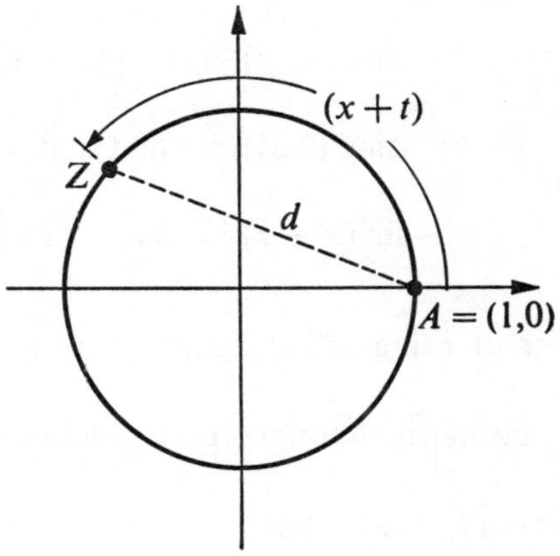

Fig. 8.13

Our memory banks remind us that $\cos(-t)$ is $\cos(t)$ and that $\sin(-t)$ is $-\sin(t)$, so we rewrite the above as

$$s^2 = [\cos(x) - \cos(t)]^2 + [\sin(x) + \sin(t)]^2.$$

A glance at figure 8.13 and the distance formula gives

$$d^2 = [\cos(x+t) - 1]^2 + [\sin(x+t) - 0]^2.$$

If we expand the right-hand side of the last two results and simplify by using the fundamental property

$$\cos^2\theta + \sin^2\theta = 1$$

we will have (*student should verify this*) the equations

$$s^2 = 2 - 2\cos(x)\cos(t) + 2\sin(x)\sin(t)$$

and

$$d^2 = 2 - 2\cos(x+t).$$

Since $s = d$ implies that $s^2 = d^2$, we set the above equal to each other and simplify. This activity produces the equation

$$\cos(x+t) = \cos(x)\cdot\cos(t) - \sin(x)\cdot\sin(t). \tag{8.31}$$

Pleased with our success and anticipating that the student can show (see problem 1 in problem set 8.3) that

$$\cos\left(u + \frac{\pi}{2}\right) = -\sin(u) \tag{8.32}$$

we use this to see that, letting $u = x + t$,

$$-\sin(x+t) = \cos\left([x+t] + \frac{\pi}{2}\right) = \cos\left(x + \left[t + \frac{\pi}{2}\right]\right).$$

So, by using (8.31), we find that

$$-\sin(x+t) = \cos(x)\cdot\cos\left(t + \frac{\pi}{2}\right) - \sin(x)\cdot\sin\left(t + \frac{\pi}{2}\right)$$

and using (8.32) again we find that $\cos\left(t + \frac{\pi}{2}\right) = -\sin(t)$. Using it again (monotonous, isn't it) with $u = t + \frac{\pi}{2}$, we find that $-\sin\left(t + \frac{\pi}{2}\right) = \cos(t + \pi)$. Thus

$$-\sin(x+t) = -\cos(x)\cdot\sin(t) + \sin(x)\cdot\cos(t+\pi).$$

The use of (8.31) shows that $\cos(t + \pi)$ is $-\cos(t)$ so that the above can be written in the following form

$$-\sin(x + t) = -\cos(x)\cdot\sin(t) - \sin(x)\cdot\cos(t) \quad \text{or as}$$

$$\sin(x + t) = \sin(x)\cdot\cos(t) + \cos(x)\cdot\sin(t) \tag{8.33}$$

The formulas given in equations 8.31 and 8.33 are the desired ones, so we collect them below to stress that they are important.

The Addition Formulas for cos *and* sin

$$\begin{aligned}\cos(x + t) &= \cos(x)\cdot\cos(t) - \sin(x)\cdot\sin(t)\\ \sin(x + t) &= \sin(x)\cdot\cos(t) + \cos(x)\cdot\sin(t)\end{aligned}$$

To indicate the utility of the addition formulas, we offer the following examples:

Example 1. This example illustrates how the addition formulas can produce other useful formulas. In the formula for $\cos(x + t)$, let $t = -r$ and see that

$$\begin{aligned}\cos(x - r) &= \cos(x)\cdot\cos(-r) - \sin(x)\cdot\sin(-r)\\ &= \cos(x)\cdot\cos(r) + \sin(x)\cdot\sin(r)\end{aligned}$$

which could be called a *subtraction formula.*

Example 2. A computation example.

Given: $\sin(x) = 2/3$ and $0 < x < \pi/2$
$\sin(t) = 3/5$ and $\pi/2 < t < \pi$

Find: $\cos(x + t)$, $\cos(x - t)$, and $\sin(x + t)$.

Preliminaries. Since $\sin(x) = 2/3$, $\cos^2(x) = 5/9$ and, since $0 < x < \pi/2$, $\cos(x) = \sqrt{5}/3$.

Since $\sin(t) = 3/5$, $\cos^2(t) = 16/25$ and, since $\pi/2 < t < \pi$, $\cos(t) = -4/5$.

Solutions.

1. $$\begin{aligned}\cos(x + t) &= \cos(x)\cdot\cos(t) - \sin(x)\cdot\sin(t)\\ &= \frac{\sqrt{5}}{3}\cdot\frac{-4}{5} - \frac{2}{3}\cdot\frac{3}{5} = \frac{-4\sqrt{5} - 6}{15}\\ &= -\frac{6 + 4\sqrt{5}}{15}.\end{aligned}$$

2. $$\cos(x - t) = \frac{6 - 4\sqrt{5}}{15}.$$

3. $\sin(x + t) = \sin(x)\cdot\cos(t) + \cos(x)\cdot\sin(t)$

$$= \frac{2}{3}\cdot\frac{-4}{5} + \frac{\sqrt{5}}{3}\cdot\frac{3}{5}$$

$$= \frac{-8 + 3\sqrt{5}}{15}.$$

PROBLEM SET 8.3

1. Show that $\cos\left(u + \frac{\pi}{2}\right) = -\sin(u)$.

2. Verify each of the following:

a. $[\cos(x) - \cos(t)]^2 + [\sin(x) + \sin(t)]^2$
$= 2 - 2\cdot\cos(x)\cdot\cos(t) + 2\cdot\sin(x)\cdot\sin(t)$.

b. $[\cos(x + t) - 1]^2 + [\sin^2(x + t)] = 2 - 2\cdot\cos(x + t)$.

[3] Show that $\sin(x - r) = \sin(x)\cdot\cos(r) - \cos(x)\cdot\sin(r)$. *Hint:* Use an addition formula with $t = -r$.

4. Use the addition formulas to show that:

a. $\sin\left(x + \frac{\pi}{2}\right) = \cos(x)$

b. $\cos\left(x + \frac{\pi}{4}\right) = \frac{\sqrt{2}}{2}[\cos(x) - \sin(x)]$

c. $\sin\left(\frac{\pi}{2} - x\right) = \cos(x)$

d. $\cos\left(x - \frac{\pi}{2}\right) = \sin(x)$

5. If $\sin(x) = 4/5$, $0 < x < \pi/2$, and $\cos(t) = 12/13$, $0 < t < \pi/2$, find $\cos(x + t)$ and $\sin(x + t)$.

[6] Show that $\sin(2x) = 2\sin(x)\cos(x)$. *Hint:* Use $\sin(x + x)$.

[7] Verify each of the following statements:

a. $\cos(2x) = \cos^2 x - \sin^2 x$ **b.** $\cos(2x) = 2\cos^2 x - 1$
c. $\cos(2x) = 1 - 2\sin^2 x$

8. Use the results of problem 7 to show that

a. $\cos(t) = 2\cos^2(t/2) - 1$ **b.** $\cos(t) = 1 - 2\sin^2(t/2)$

9. Simplify each of the following:

a. $2 \sin (x/2) \cos (x/2)$ **b.** $[\sin (x/2) - \cos (x/2)]^2$

10. Simplify each of the following:

a. $\cos (75°) \cdot \cos (30°) + \sin (75°) \cdot \sin (30°)$
b. $\sin (20°) \cdot \cos (10°) + \cos (20°) \cdot \sin (10°)$

11. Simplify each of the following:

a. $\cos^2 (\pi/12) + \sin^2 (\pi/12)$ **b.** $\cos^2 (\pi/12) - \sin^2 (\pi/12)$
c. $2 \cdot \sin (\pi/12) \cdot \cos (\pi/12)$

12. If $\sin (u) = 3/5$ and $0 < u < \pi/2$, find

a. $\sin (2u)$ **b.** $\cos (2u)$ **c.** $\cos (u/2)$

13. Since $\frac{\pi}{3} - \frac{\pi}{4} = \frac{\pi}{12}$, use the addition or subtraction formulas to find $\cos (\pi/12)$ and $\sin (\pi/12)$. Leave answers in radical form.

14. Use one of the statements in problem 7, with $x = \pi/8$, to find $\cos (\pi/8)$.

15. Suppose that the line segment joining the origin with the point (3,4) intersects the unit circle at the point P whose arc is t and that the line segment joining the origin with the point (−6,8) intersects the unit circle at the point Q whose arc is x. (Draw a figure.) Find the coordinates of the points P and Q, and use them to find the following numbers:

a. $\cos (x + t)$ **b.** $\sin (x + t)$ **c.** $\cos (2t)$ **d.** $\sin (2x)$

16. This problem is not difficult but it is long. Show that

$$2 \cdot \cos \left(\frac{x}{2} + \frac{t}{2}\right) \cdot \sin \left(\frac{x}{2} - \frac{t}{2}\right) = \sin (x) - \sin (t).$$

Section 4—Graphs of Cos and Sin We have recently noted that the circular functions cosine and sine are bounded, periodic functions having as their domain the set of all real numbers. The periodicity of these functions simplifies the drawing of their graphs.

The list of values of $\cos x$ for certain values of x in the interval $[0,2\pi]$ given in table 8.4, can be used to plot the points shown in figure 8.14.

Since the period of the cosine function is 2π, you can plot the points

$$(x + 2k\pi, \cos [x + 2k\pi])$$

at the same time, for as many values of the integer k as you wish. Figure 8.15 shows two additional blocks of points on the graph of this function, corresponding to the choices $k = -1$ and $k = 1$.

TABLE 8.4

VALUES FOR $\sin x$ AND $\cos x$

x	$\cos x$	$\sin x$
0	1	0
$\frac{\pi}{6}$	$\frac{\sqrt{3}}{2} \approx 0.87$	$\frac{1}{2}$
$\frac{\pi}{4}$	$\frac{\sqrt{2}}{2} \approx 0.71$	$\frac{\sqrt{2}}{2} \approx 0.71$
$\frac{\pi}{3}$	$\frac{1}{2}$	$\frac{\sqrt{3}}{2} \approx 0.87$
$\frac{\pi}{2}$	0	1
$\frac{2\pi}{3}$	$-\frac{1}{2}$	$\frac{\sqrt{3}}{2} \approx 0.87$
$\frac{3\pi}{4}$	$-\frac{\sqrt{2}}{2} \approx -0.71$	$\frac{\sqrt{2}}{2} \approx 0.71$
$\frac{5\pi}{6}$	$-\frac{\sqrt{3}}{2} \approx -0.87$	$\frac{1}{2}$
π	-1	0
$\frac{7\pi}{6}$	$-\frac{\sqrt{3}}{2} \approx -0.87$	$-\frac{1}{2}$
$\frac{5\pi}{4}$	$-\frac{\sqrt{2}}{2} \approx -0.71$	$-\frac{\sqrt{2}}{2} \approx -0.71$
$\frac{4\pi}{3}$	$-\frac{1}{2}$	$-\frac{\sqrt{3}}{2} \approx -0.87$
$\frac{3\pi}{2}$	0	-1
$\frac{5\pi}{3}$	$\frac{1}{2}$	$-\frac{\sqrt{3}}{2} \approx -0.87$
$\frac{7\pi}{4}$	$\frac{\sqrt{2}}{2} \approx 0.71$	$-\frac{\sqrt{2}}{2} \approx -0.71$
$\frac{11\pi}{6}$	$\frac{\sqrt{3}}{2} \approx 0.87$	$-\frac{1}{2}$
2π	1	0

Trusting in the decency of this function, we hope that it is reasonable to assume that the graph of $f(x) = \cos x$ is a smooth unbroken curve, that is, it is a continuous function. Connecting the points in figure 8.15 with a smooth curve produces the graph of the Cosine function as shown in figure 8.16.

Using the table of values, you can construct the graph of the sine function in a similar fashion. In the next pair of illustrations, figure 8.17

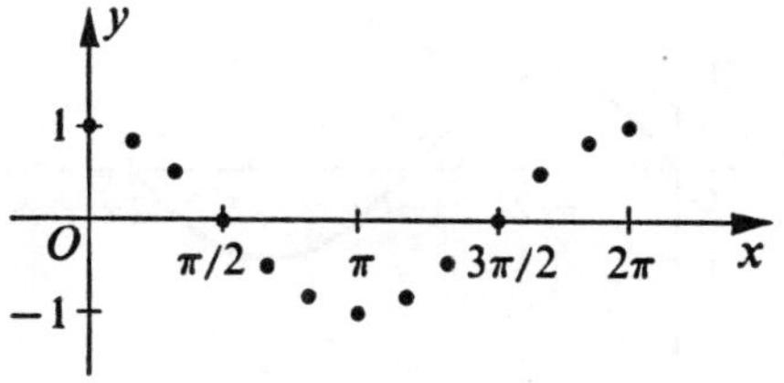

Fig. 8.14

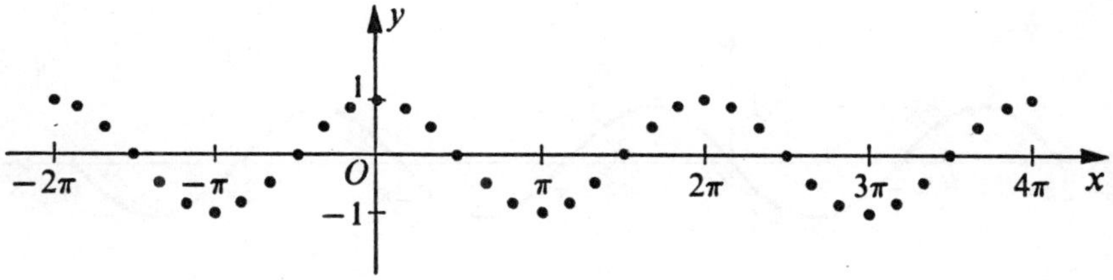

Fig. 8.15

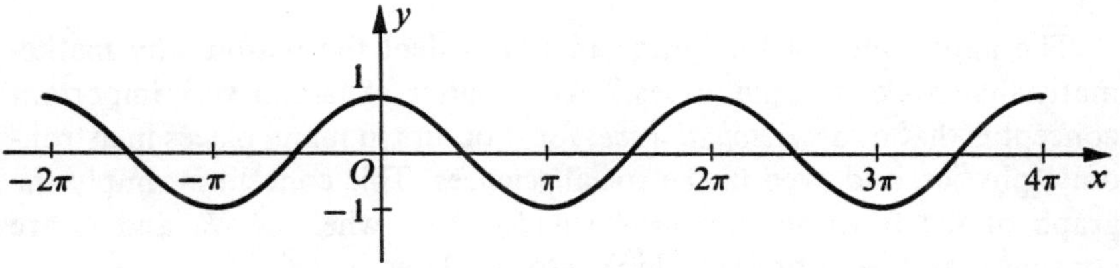

Fig. 8.16. $\{(x,y): y = \cos x, -2\pi \leq x \leq 4\pi\}$.

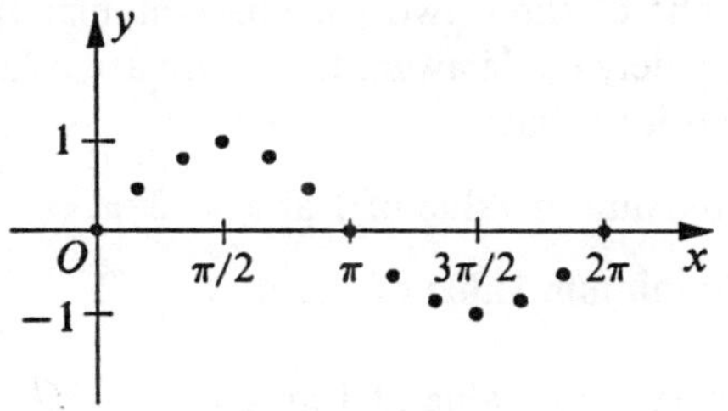

Fig. 8.17

shows points (x, $\sin x$) in the interval $0 \leq x \leq 2\pi$ and, if these points are connected with a smooth curve, you have the graph of $f(x) = \sin x$, which is shown in figure 8.18.

Using the periodicity of the sine function, you can reproduce the above graph over successive intervals of length 2π as shown in figure 8.19.

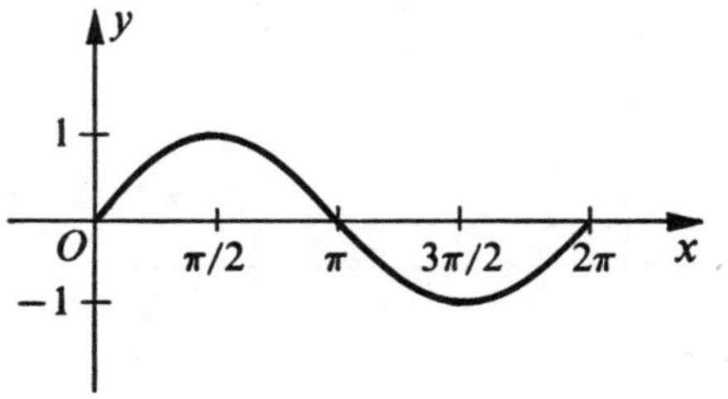

Fig. 8.18. $\{(x,y): y = \sin x, 0 \leq x \leq 2\pi\}$.

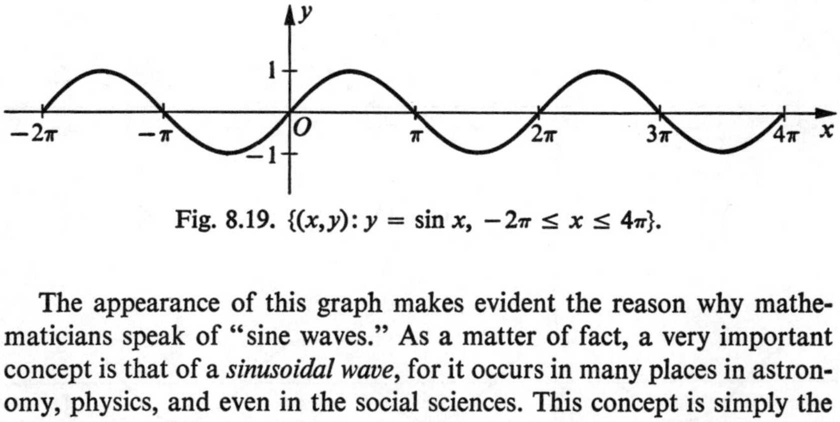

Fig. 8.19. $\{(x,y): y = \sin x, -2\pi \leq x \leq 4\pi\}$.

The appearance of this graph makes evident the reason why mathematicians speak of "sine waves." As a matter of fact, a very important concept is that of a *sinusoidal wave*, for it occurs in many places in astronomy, physics, and even in the social sciences. This concept is simply the graph of the function $F(x) = A \cdot \sin(Bx + C)$ where A, B, and C are constants, and we promise to look into this later.

Because we assume that the cosine and sine functions are continuous functions, the graphs of these two principal circular functions (you can soon expect to be rudely made aware that there are others) also give mute testimony to the notions that:

$\cos x$ has a maximum value of 1 at $x = 2k\pi$,

$\cos x$ has a minimum value of -1 at $x = (2k + 1)\pi$,

$\sin x$ has a maximum value of 1 at $x = \frac{\pi}{2} + 2k\pi$,

$\sin x$ has a minimum value of -1 at $x = \frac{3\pi}{2} + 2k\pi$,

$\cos x$ is 0 at $x = \frac{\pi}{2} + k\pi$,

$\sin x$ is 0 at $x = k\pi$, for each integer k;

and that:

In quadrant I $\begin{cases} \cos x \text{ decreases from 1 to 0,} \\ \sin x \text{ increases from 0 to 1.} \end{cases}$

In quadrant II $\begin{cases} \cos x \text{ decreases from 0 to } -1, \\ \sin x \text{ decreases from 1 to 0.} \end{cases}$

In quadrant III $\begin{cases} \cos x \text{ increases from } -1 \text{ to 0,} \\ \sin x \text{ decreases from 0 to } -1. \end{cases}$

In quadrant IV $\begin{cases} \cos x \text{ increases from 0 to 1,} \\ \sin x \text{ increases from } -1 \text{ to 0.} \end{cases}$

We promised to pay attention to the graph of the function $F(x) = A \cdot \sin(Bx + C)$ in detail at a later time. At this moment, however, there is no good reason why we cannot begin to look into this matter, at least in some small way.

Example 1. $F(x) = 2 \sin x$.

Recall that $\sin x$ has period 2π and that $|\sin x| \leq 1$ so that $-2 \leq 2 \sin x \leq 2$. Hence, the period of F is 2π; its maximum value is 2; its minimum value is -2. Table 8.5 can be constructed and can be used to help sketch the graph of F as indicated in figure 8.20.

TABLE 8.5

x	$F(x)$
0	0
$\pi/4$	$\sqrt{2} \approx 1.41$
$\pi/2$	2
π	0
$3\pi/2$	-2
2π	0

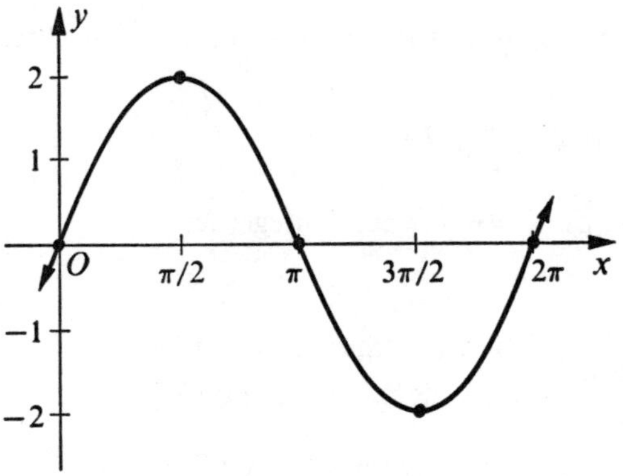

Fig. 8.20

Example 2. $G(x) = \cos 2x$.

Cos x has period 2π, that is, as x goes from 0 to 2π, cos x makes one complete cycle. Thus cos $2x$ makes one complete cycle as $2x$ goes from 0 to 2π or as x goes from 0 to π. (See table 8.6.) Hence, the period of G is π. A formal explanation of this follows. (See figure 8.21.)

TABLE 8.6

x	$2x$	$G(x)$
0	0	1
$\pi/4$	$\pi/2$	0
$\pi/2$	π	-1
$3\pi/4$	$3\pi/2$	0
π	2π	1

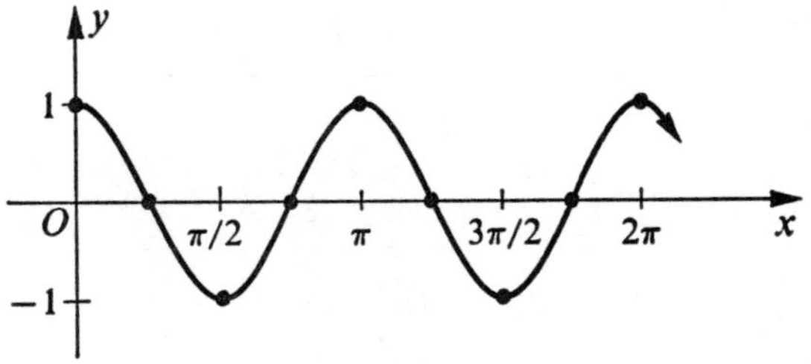

Fig. 8.21

To argue that the function G in the previous example has period π, we begin by noting that

$$\cos (2x + 2\pi) = \cos 2x$$

and then see that, for every number t,

$$G(t + \pi) = \cos (2t + 2\pi) = \cos 2t = G(t)$$

or $G(x + \pi) = G(x)$ for every number x. This establishes the statement about the period of G.

Example 3. $U(x) = \frac{3}{2} \sin (\pi x)$.

Here we try to combine the ideas examined in the first two examples. (See figure 8.22 and table 8.7.) The maximum value of $U(x)$ is 3/2 and the minimum value of $U(x)$ is $-3/2$. Further, since

$$\sin (\pi x + 2\pi) = \sin (\pi x)$$

we can see that, for every number t,

$$U(t + 2) = \frac{3}{2} \sin (\pi[t + 2]) = \frac{3}{2} \sin (\pi t + 2\pi) = \frac{3}{2} \sin (\pi t).$$

Thus

$$U(t + 2) = U(t) \text{ or } U(x + 2) = U(x)$$

which shows that the period of U is 2.

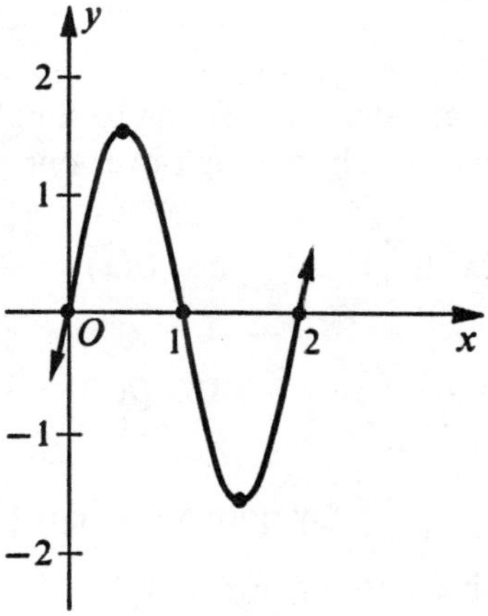

Fig. 8.22

TABLE 8.7

x	πx	$\sin(\pi x)$	$U(x)$
0	0	0	0
1/2	$\pi/2$	1	3/2
1	π	0	0
3/2	$3\pi/2$	-1	$-3/2$
2	2π	0	0

PROBLEM SET 8.4

Sketch the graph of the given functions over one period.

1. $F(x) = 4 \sin x$

2. $f(x) = \frac{5}{2} \sin x$

3. $G(x) = \frac{3}{4} \cos x$

4. $g(x) = 2 \cos x$

5. $u(x) = \pi \sin x$

6. $U(x) = \sqrt{2} \sin x$

7. $v(x) = \cos(2x)$

8. $V(x) = \sin(2x)$

9. $W(x) = \sin(x/2)$

10. $w(x) = \cos(x/2)$

11. $Z(x) = 2 \sin\left(\frac{\pi x}{3}\right)$

12. $z(x) = \frac{1}{2} \sin\left(\frac{\pi x}{2}\right)$

13. $f(x) = \frac{1}{2} \cos(2x)$

14. $F(x) = \frac{1}{2} \sin(2x)$

In problems 15–18, use an addition formula to simplify the rule of pairing of the function and then sketch its graph over one period.

15. $K(x) = -4 \cos\left(2x + \frac{\pi}{2}\right)$

16. $G(x) = -\sqrt{3} \sin\left(\pi x - \frac{\pi}{2}\right)$

17. $Y(x) = \frac{3}{2} \sin\left(\frac{x}{2} + \frac{\pi}{2}\right)$

18. $Q(x) = 3 \cos(3x - \pi)$

19. Suppose $F(x) = \cos\left(\frac{2x}{3}\right)$. By noting that $\cos\left(\frac{2x}{3} + 2\pi\right) = \cos\left(\frac{2x}{3}\right)$ show that, for each number t, $F(t + 3\pi) = F(t)$, hence F has period 3π.

20. Suppose that $G(x) = \sin(\pi x)$. By noting that $\sin(\pi x + 2\pi) = \sin(\pi x)$, show that, for each number t, $G(t + 2) = G(t)$, hence G has period 2.

21. Sketch the graph of $F(x) = \sin(x)$ on $[-\pi/2,\pi/2]$.

22. Sketch the graph of $G(x) = \cos(x)$ on $[0,\pi]$.

23. Sketch the graph of $A(x) = |\sin x|$ on $[0,2\pi]$.

Section 5—Tangent Function There is some truth in the notion that cos and sin are the most important of the circular functions since they are used to define the other circular functions.

> *Remark.* From the above statement, we reach the distressing conclusion that there are more circular functions. This conclusion will be amply verified.

In particular, the *tangent function* (perhaps the most useful of the remaining circular functions) is defined as follows:

$$\tan = \left\{(x,y) \,\middle|\, y = \frac{\sin x}{\cos x}\right\}.$$

Since $\cos x = 0$ for $x = (2k + 1)\cdot\frac{\pi}{2}$ where k is an integer,

$$\text{dom tan} = \left\{x \,\middle|\, x \in R \text{ and } x \neq (2k + 1)\cdot\frac{\pi}{2}, k \in J\right\}$$

rng tan is R.

Thus, unlike the cosine and sine functions, the tangent function is not defined for all x and it is not a bounded function.

The tangent function can be defined in a geometric manner. Begin by drawing a unit circle and then construct a vertical line containing the point $A = (1,0)$. If $x \geq 0$, then there is a point P on the unit circle whose arc is x. Now draw a line segment joining the origin 0 and the point P. Extend this segment until it intersects the vertical line. (If you have chosen $x = \pi/2$, try to see why we have a little trouble in doing this and then choose some other value for x). Label this point of intersection T (see figures 8.23a through 8.23d). No matter which of these figures you concentrate on, it can be seen that ΔOBP is similar to ΔOAT, hence

$$\frac{AT}{OA} = \frac{BP}{OB} \text{ or } \frac{AT}{1} = \frac{\sin x}{\cos x}.$$

We then define the ordinate of the point T to be the value of $\tan x$.

$$\tan x = \frac{\sin x}{\cos x}, \quad \cos x \neq 0.$$

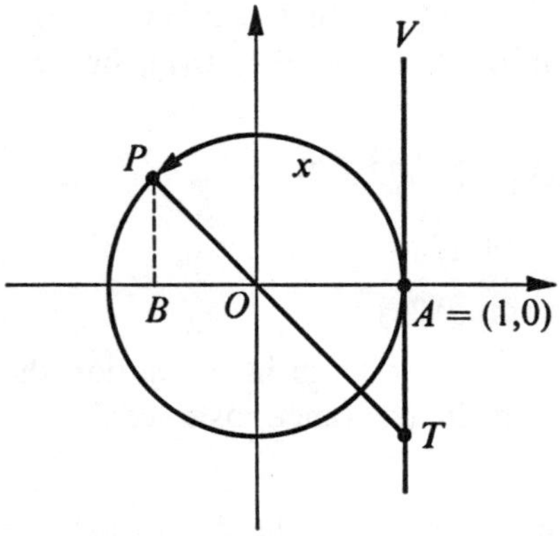

Fig. 8.23a. $\frac{\pi}{2} < x < \frac{\pi}{2}$.

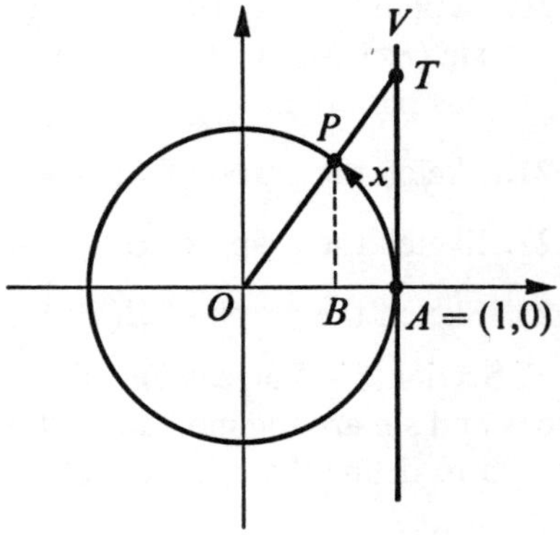

Fig. 8.23b. $0 < x < \frac{\pi}{2}$.

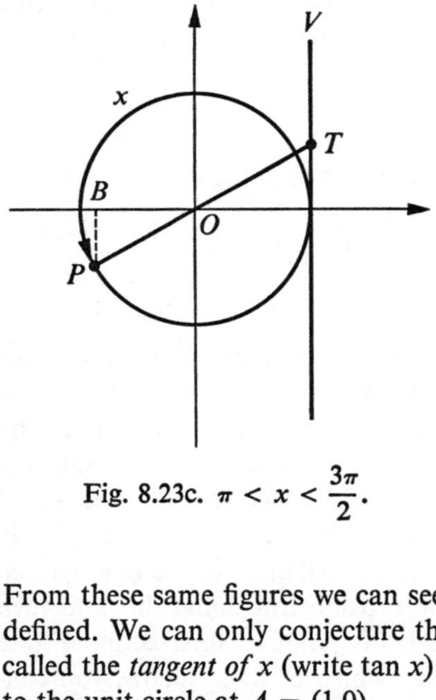

Fig. 8.23c. $\pi < x < \frac{3\pi}{2}$.

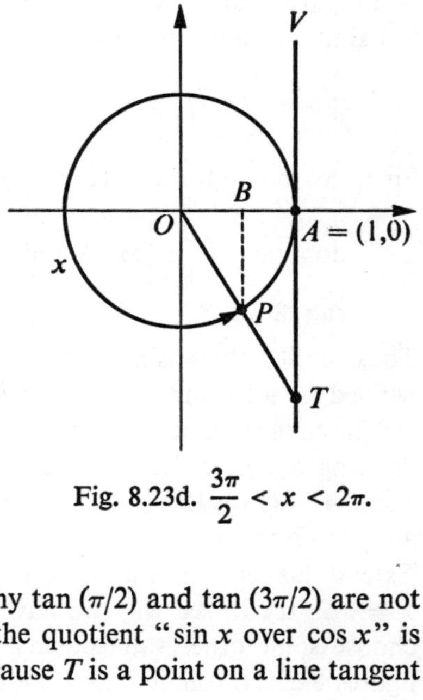

Fig. 8.23d. $\frac{3\pi}{2} < x < 2\pi$.

From these same figures we can see why tan $(\pi/2)$ and tan $(3\pi/2)$ are not defined. We can only conjecture that the quotient "sin x over cos x" is called the *tangent of* x (write tan x) because T is a point on a line tangent to the unit circle at $A = (1,0)$.

In figure 8.24, the arc of P is x and the arc of Q is $(x + \pi)$. Since P, O, and Q are collinear, the extension of the segments OP and OQ will intersect the vertical line V at the point T. This implies that

$$\tan(x) = \tan(x + \pi).$$

So, unlike the cosine and sine functions, the period of the tangent function is π.

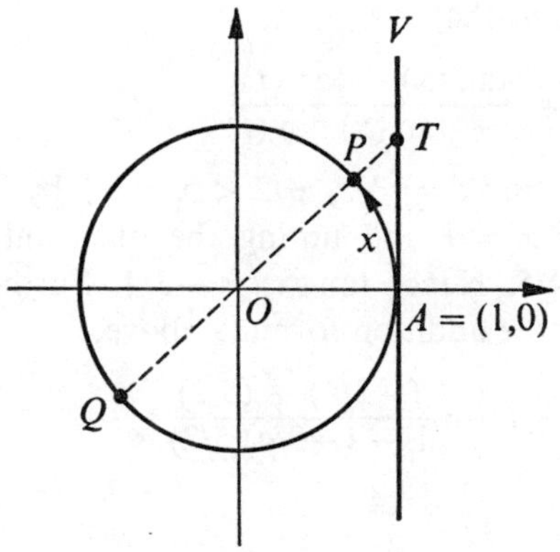

Fig. 8.24. tan (x) = tan $(x + \pi)$.

We can obtain this same idea by using the addition formulas for cos and sin. Thus

$$\tan(x+\pi) = \frac{\sin(x+\pi)}{\cos(x+\pi)} = \frac{\sin(x)\cos(\pi) + \cos(x)\sin(\pi)}{\cos(x)\cos(\pi) - \sin(x)\sin(\pi)}$$

$$= \frac{-\sin(x)}{-\cos(x)} = \frac{\sin(x)}{\cos(x)}$$

$$= \tan(x).$$

As indicated above, by using the definition of the tangent function and the established results for cos (x) and sin (x), you can obtain analogous results for tan (x). For example:

$$\tan(\pi/3) = \frac{\sin(\pi/3)}{\cos(\pi/3)} = \frac{\sqrt{3}/2}{1/2} = \sqrt{3}.$$

$$\tan(-x) = \frac{\sin(-x)}{\cos(-x)} = \frac{-\sin(x)}{\cos(x)} = -\tan(x).$$

One of the more important results that can be found in this way is the addition formula for the tangent function.

$$\tan(x+t) = \frac{\sin(x+t)}{\cos(x+t)} = \frac{\sin(x)\cos(t) + \cos(x)\sin(t)}{\cos(x)\cos(t) - \sin(x)\sin(t)}.$$

Now, by dividing numerator and denominator of the quotient above by cos (x) cos (t), you get

$$\tan(x+t) = \frac{\dfrac{\sin(x)\cos(t)}{\cos(x)\cos(t)} + \dfrac{\cos(x)\sin(t)}{\cos(x)\cos(t)}}{\dfrac{\cos(x)\cos(t)}{\cos(x)\cos(t)} - \dfrac{\sin(x)\sin(t)}{\cos(x)\cos(t)}}$$

from which it follows that

$$\tan(x + t) = \frac{\tan(x) + \tan(t)}{1 - \tan(x)\tan(t)}.$$

Example. If $\sin(x) = 3/5$, $\pi/2 < x < \pi$, by using the statement $\cos^2 x + \sin^2 x = 1$ and noting the quadrant, we can show that $\cos(x) = -4/5$. Hence $\tan(x) = -3/4$. Furthermore, if $\tan(t) = 5/2$, then, by the addition formula above,

$$\tan(x + t) = \frac{(-3/4) + (5/2)}{1 - (-3/4)(5/2)}$$

and so $\tan(x + t) = \dfrac{14}{23}$.

Using the known values of cos x and sin x for certain special values of x, we can construct table 8.8. Since we have already shown that $\tan(-x) = -\tan(x)$, then the last column in this table comes as a surprise to no one except Manfred.

TABLE 8.8

x	$\tan(x)$	$\tan(-x)$
0	0	0
$\frac{\pi}{6}$	$\frac{\sqrt{3}}{3} \approx 0.577$	$-\frac{\sqrt{3}}{3}$
$\frac{\pi}{4}$	1	-1
$\frac{\pi}{3}$	$\sqrt{3} \approx 1.732$	$-\sqrt{3}$
$\frac{\pi}{2}$	*Not defined*	

We can use this table to determine some points in the graph of the tangent function on the interval $\langle -\pi/2, \pi/2 \rangle$. If we then make use of the fact that the period of this function is π, we can plot a few more points as shown in figure 8.25.

The dashed vertical lines in the above figure serve to remind us that $\tan(x)$ is not defined for $x = \dfrac{\pi}{2} + k\pi$ where k is an integer. They may also serve to remind us that the tangent function is unbounded.

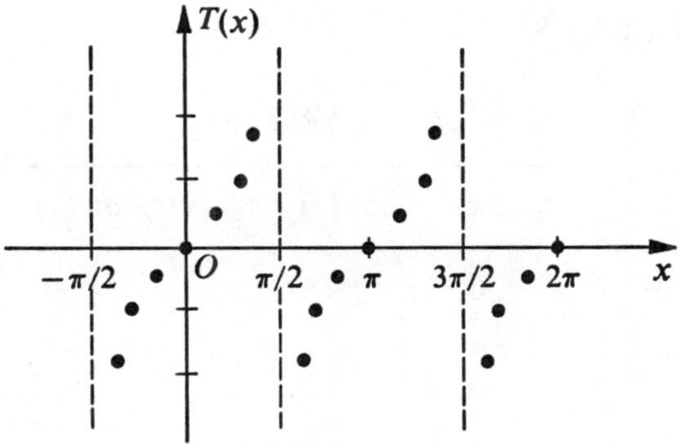

Fig. 8.25

Armed with the points shown in the previous figure, if we assume that the tangent function is continuous over its domain, we can draw a smooth curve through these points obtaining a reasonably good sketch of the graph of this function. (See figure 8.26.)

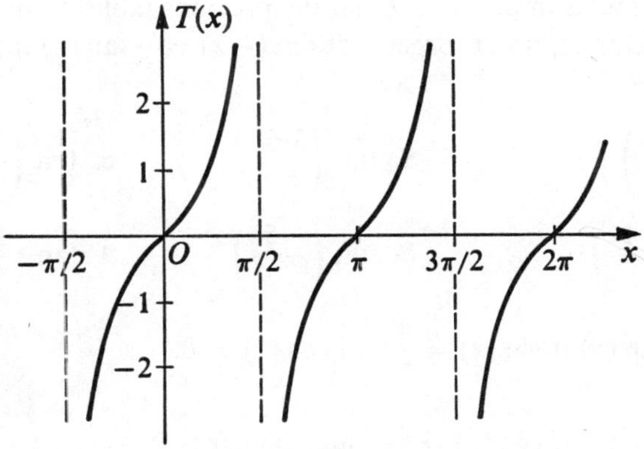

Fig. 8.26

PROBLEM SET 8.5

1. **a.** In which quadrants is $\tan(x) > 0$? $\tan(x) < 0$?
b. For what values of x is $\tan(x)$ undefined?

2. Use the addition formula to show that **(a)** $\tan(x + \pi) = \tan(x)$ and **(b)** $\tan(x - \pi) = \tan(x)$.

[3] Complete table 8.9.

TABLE 8.9

x	$\cos(x)$	$\sin(x)$	$\tan(x)$
0			
$\frac{\pi}{6}$			
$\frac{\pi}{4}$			
$\frac{\pi}{3}$			
$\frac{\pi}{2}$			

Use the formulas in problem 2 and the previous tables to find each of the following. Oh yes, don't forget that $\tan(-x) = -\tan(x)$ for you can use that also.

4. $\tan\left(\frac{3\pi}{4}\right)$ **5.** $\tan\left(\frac{5\pi}{6}\right)$ **6.** $\tan\left(\frac{2\pi}{3}\right)$

7. $\tan\left(-\frac{4\pi}{3}\right)$ **8.** $\tan\left(-\frac{5\pi}{4}\right)$ **9.** $\tan\left(\frac{17\pi}{6}\right)$

10. Find $\tan(x)$ if $\sin(x) = \frac{4}{5}$ and $\cos(x) > 0$.

11. Find $\tan(x)$ if $\cos(x) = -\frac{3}{5}$ and $\sin(x) > 0$.

12. Find $\tan(x)$ if $\cos(x) = -\frac{5}{13}$ and $\sin(x) > 0$.

13. Find $\tan(x)$ if $\sin(x) = \frac{5}{13}$ and $\cos(x) > 0$.

14. Find $\tan(x + t)$ if $\sin(x) = \frac{7}{25}$, $\sin(t) = \frac{3}{5}$, $0 < x < \frac{\pi}{2}$, and $0 < t < \frac{\pi}{2}$.

15. Find $\tan(x+t)$ if $\sin(x) = \frac{8}{17}$, $\sin(t) = \frac{4}{5}$, $0 < x < \frac{\pi}{2}$, and $0 < t < \frac{\pi}{2}$.

[16] Obtain a formula for $\tan(2x)$ in terms of $\tan(x)$.

Prove each of the following statements:

17. $\tan(x-t) = \dfrac{\tan(x) - \tan(t)}{1 + \tan(x)\cdot\tan(t)}$. *Hint:* $x - t$ is $x + (-t)$.

18. $\tan\left(x + \dfrac{\pi}{4}\right) = \dfrac{1+\tan(x)}{1-\tan(x)}$.

19. $\dfrac{2\tan(x)}{1+\tan^2 x} = \sin(2x)$.

20. $\dfrac{1-\cos(x)}{\sin(x)} = \tan\left(\dfrac{x}{2}\right)$. *Hint:* let $x = 2t$.

In each of the following, give the period p of the function and then sketch its graph over the interval

$\langle -p/2, p/2\rangle \cup \langle p/2, 3p/2\rangle$.

Example. $F(x) = \tan(3x)$.

First note that $\tan(3x+\pi) = \tan(3x)$. Why? Then, factoring the left member, we can write

$$\tan\left(3\left[x + \frac{\pi}{3}\right]\right) = \tan(3x).$$

Hence, $F\left(x+\dfrac{\pi}{3}\right) = \tan\left(3\left[x+\dfrac{\pi}{3}\right]\right) = \tan(3x) = F(x)$ so the period of F is $\dfrac{\pi}{3}$ or $p = \dfrac{\pi}{3}$. (See figure 8.27.)

21. $F(x) = \tan(2x)$

22. $F(x) = \tan(4x)$

23. $F(x) = \tan\left(\dfrac{x}{3}\right)$

24. $F(x) = \tan\left(\dfrac{x}{2}\right)$

25. $F(x) = \tan\left(\dfrac{\pi}{2}x\right)$

26. $F(x) = \tan(\pi x)$

27. In this problem, you are to think about the graph of the tangent function and to let T be the function for which $T(x) = \tan(x)$. Give the domain of T if 0 must be in the domain and T must be 1-to-1 on its domain. This is a simple task so do not complicate it.

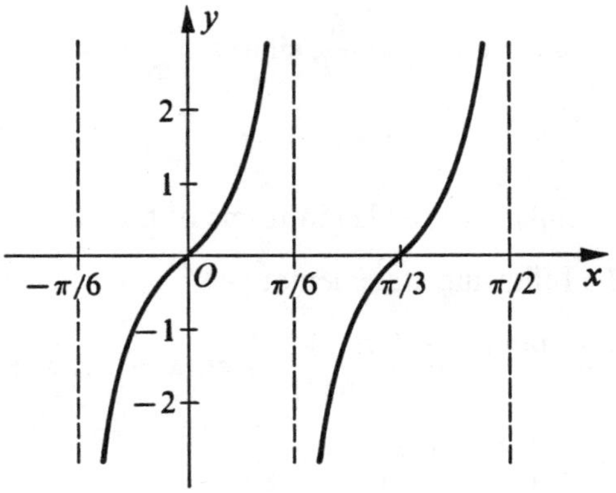

Fig. 8.27

Section 6—Functions C^*, S^*, and T^* You are now asked to recall that if F is a function, then F^* is called the inverse of F and that F^* is the set of ordered pairs to which (a,b) belongs only if (b,a) belongs to F. You have already observed that F^* is not always a function in spite of the fact that F is a function. Further, you know that F^* is a function only if F is a 1-to-1 function. It doesn't take much time to observe that the functions cosine, sine, and tangent are certainly not 1-to-1. Hence, their inverses are not functions. But, if we define three functions C, S, and T by imposing some restrictions on the domains of the three principal circular functions cosine, sine, and tangent, then the inverses of these three functions, C^*, S^*, and T^*, will be functions. (See figures 8.28a through 8.28c.)

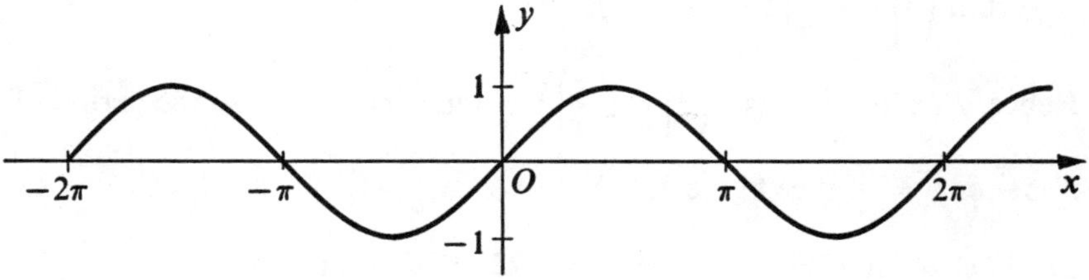

Fig. 8.28a. $\{(x,y): y = \sin x\}$.

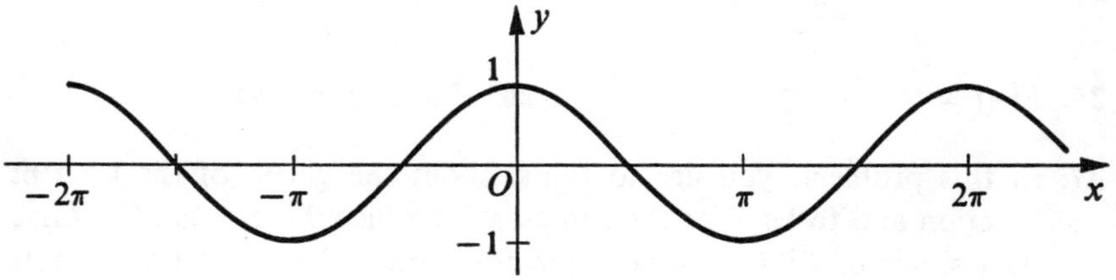

Fig. 8.28b. $\{(x,y): y = \cos x\}$.

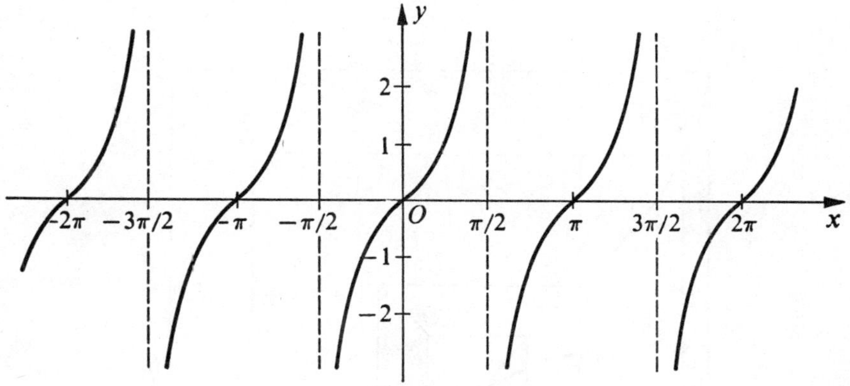

Fig. 8.28c. $\{(x,y): y = \tan x\}$.

Remark. The restrictions to be imposed on the domains can be different from those that we will choose here. However, our choices are the ones commonly used.

Example. (definition): Let C be the function for which

$$C(x) = \cos(x) \mid \text{dom } C = [0,\pi].$$

The graph of this function is shown in figure 8.29a, and you can see that this function is univalent, hence its inverse is a function.

In figure 8.29b, the graph of the function I has been drawn and, by "running around some rectangles," the graph of the inverse of C has been constructed.

Now, the dom $C = [0,\pi]$ and rng $C = [-1\ 1]$, thus dom $C^* = [-1,1]$ and rng $C^* = [0,\pi]$. To elaborate, for $0 \leq t \leq \pi$, let $\cos(t) = b$, thus

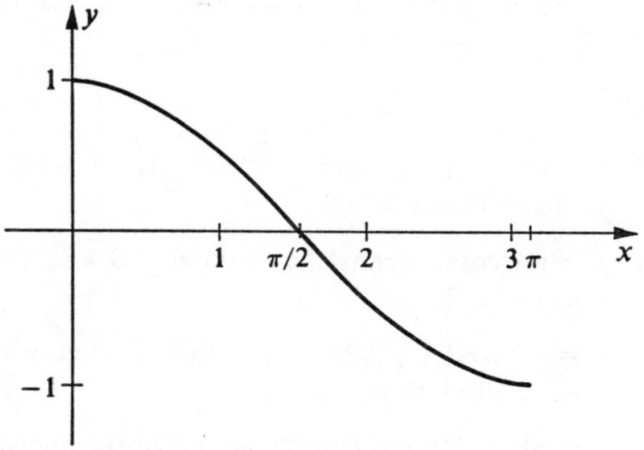

Fig. 8.29a. The graph of the function C.

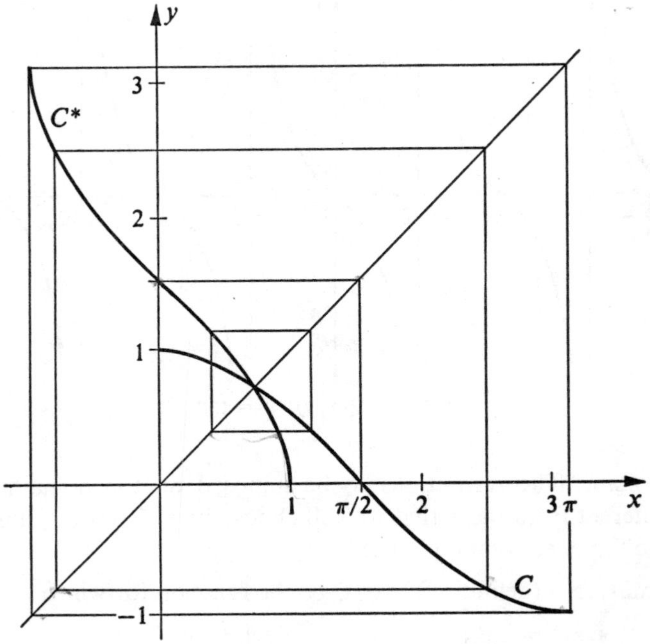

Fig. 8.29b. The graph of C^*.

$(t,b) \in C$ which implies that $(b,t) \in C^*$. So, if we agree to use traditional notation and to denote our function C^* by

$$C^*(x) = \text{arc cos}\,(x), \text{ that is, } C^* = \{(x,y) \mid y = \text{arc cos}\, x\}$$
$$\text{dom } C^* = [-1,1]$$

we must agree that the statement $y = \text{arc cos}\, x$, $-1 \leq x \leq 1$, means that (1) $\cos(y) = x$ and that (2) $0 \leq y \leq \pi$.

Examples.

1. $y = \text{arc cos}\,(1/2)$ means that $\cos(y) = 1/2$ and that $y \in [0,\pi]$, thus $y = \pi/3$.
2. $y = \text{arc cos}\, 0$ means that $\cos(y) = 0$ and that $y \in [0,\pi]$, thus $y = \pi/2$.
3. $y = \text{arc cos}\,(-\sqrt{2}/2)$ means that $\cos(y) = -\sqrt{2}/2$ and that $y \in [0,\pi]$, thus $y = 3\pi/4$.

From the graph of C^* and from table A.2 of the appendix, we can follow this next example.

4. $y = \text{arc cos}\,(0.765)$ means that $\cos(y) = 0.765$ and that $y \in [0,\pi]$, thus $y \approx 0.70$.

From the above examples, we can make some useful generalizations relative to the meaning of the statement

$Y = \text{arc cos}\,X.$

a. If $0 \le X \le 1$, then $\cos(Y) = X$ and $Y \in [0,\pi/2]$.
b. If $-1 \le X \le 0$, then $\cos(Y) = X$ and $Y \in [\pi/2,\pi]$

and

c. there is one one solution to the statement.

In a similar manner, we let S be the function given by

$S(x) = \sin(x) \mid \text{dom}\,S = [-\pi/2,\pi/2].$

The graph of this function is shown in figure 8.30 and it is a 1-to-1 function.

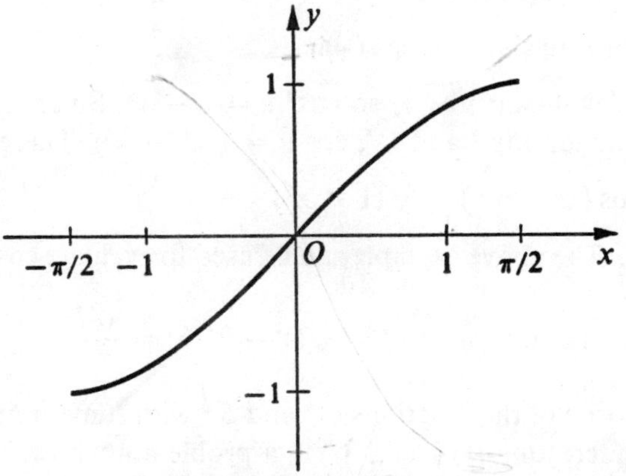

Fig. 8.30. The function S.

Its inverse, therefore, is the function that we will denote as follows:

$S^*(x) = \text{arc sin}\,(x)$
$\text{dom}\,S^* = [-1,1]\ \text{rng}\,S^* = [-\pi/2,\pi/2]$

The graph of this function will be left as an exercise for the student.

I think that it will be sufficient to point out that the statement $Y = \text{arc sin}\,X$ means that $\sin(Y) = X$ and that $Y \in [-\pi/2,\pi/2]$. Hence,

a. if $-1 \le X \le 0$, then $\sin(Y) = X$ and $Y \in [-\pi/2,0]$,
b. if $0 \le X \le 1$, then $\sin(Y) = X$ and $Y \in [0,\pi/2]$,

and

c. for each such number X between -1 and 1, there is only one number Y.

Examples.

1. $y = \text{arc sin}(-1)$ means that $\sin y = -1$ and that $y \in [-\pi/2, 0]$, thus $y = -\pi/2$.

2. $y = \text{arc sin}(0.84)$ means that $\sin y = 0.84$ and that $y \in [0, \pi/2]$. Thus, by using table A.2, $y \approx 1.0$.

3. Find $\cos(\text{arc sin } 1/2)$.

Solution. Let $\text{arc sin } 1/2 = y$. This means that $\sin y = 1/2$ and that $y \in [0, \pi/2]$, thus $y = \pi/6$. Hence,

$$\cos(\text{arc sin } 1/2) = \cos(y) = \cos(\pi/6) = \sqrt{3}/2.$$

4. Show that $\cos(\text{arc sin } x) = \sqrt{(1 - x)^2}$.

Proof. Let $y = \text{arc sin } x$. It follows that $\sin y = x$ and that $y \in [-\pi/2, \pi/2]$. Hence,

$$\cos(\text{arc sin } x) = \cos(y)$$

But $\cos^2 y + \sin^2 y = 1$, so $\cos^2 y = 1 - x^2$. Since $\cos(y)$ is positive in quadrants I and IV, $\cos y = \sqrt{(1 - x^2)}$. Therefore

$$\cos(\text{arc sin } x) = \sqrt{(1 - x^2)}.$$

5. The above example can be used to evaluate $\cos(\text{arc sin } 3/4)$.

Solution. $\cos(\text{arc sin } 3/4) = \sqrt{(1 - 9/16)} = \dfrac{\sqrt{7}}{4}.$

The composite of the functions C^* and S^* with some simple function F presents an interesting problem. Well, a problem anyhow.

Example. Suppose that $F(x) = 2x$, then $[S^* \circ F]$ is the function whose rule of pairing is the equation

$$[S^* \circ F](x) = \text{arc sin}(2x).$$

To obtain the domain of this composite function, we go back to the definition and go through the ritual given below (we have gone through this ritual before).

$x \in \text{dom } F$ implies that $x \in R$
$F(x) \in \text{dom } S$ implies that $2x \in [-1, 1]$, that is, $x \in [-1/2, 1/2]$.

Therefore $\text{dom } [S^* \circ F] = [-1/2, 1/2]$.

By selecting certain numbers in the domain of this function, we can compute the corresponding ordinate and tabulate these results in some table such as shown below. Plot these points as indicated in figure 8.31. If we then trust that this function is continuous over its domain, we can sketch its graph by drawing a "smooth curve" through the points that we have plotted.

Our final effort in behalf of the inverses of the circular functions begins by defining a function T with

$$T(x) = \tan x \mid \text{dom } T = \langle -\pi/2, \pi/2 \rangle$$

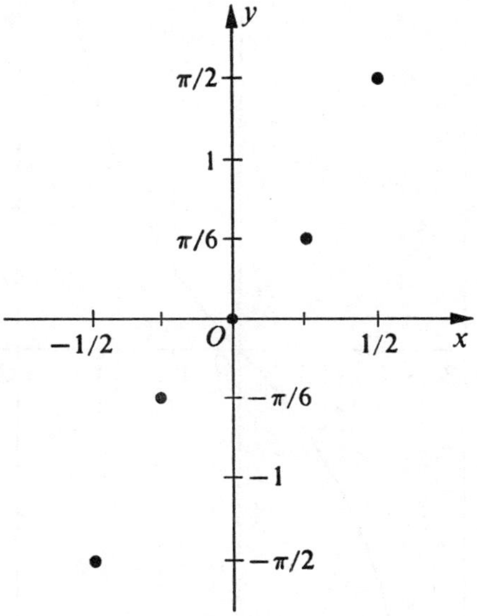

Fig. 8.31

x	$2x$	arc sin $(2x)$
$-\frac{1}{2}$	-1	$-\frac{\pi}{2}$
$-\frac{1}{4}$	$-\frac{1}{2}$	$-\frac{\pi}{6}$
0	0	0
$\frac{1}{4}$	$\frac{1}{2}$	$\frac{\pi}{6}$
$\frac{1}{2}$	1	$\frac{\pi}{2}$

In figure 8.32a we see that T is a 1-to-1 function so that its inverse T^* is a function.

We can now construct the graph of the inverse function T^* by some familiar geometric methods (see figure 8.32b).

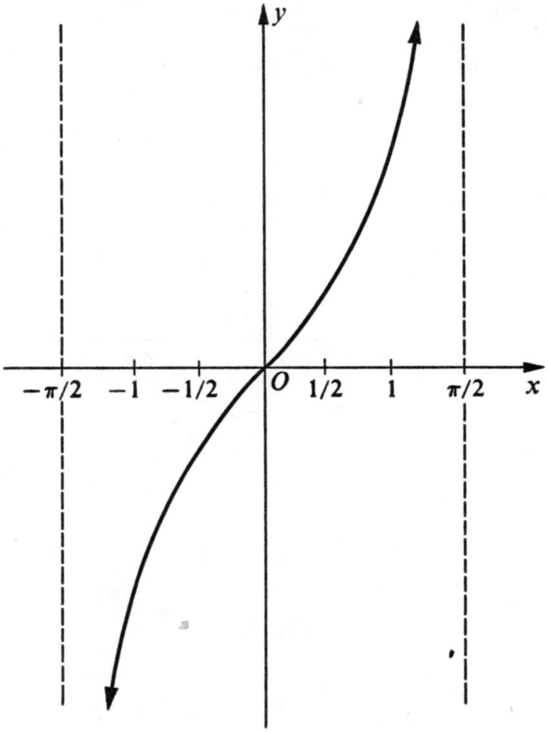

Fig. 8.32a. The function *T*.

Using the common notation, we define T^* as follows:

$$T^*(x) = \text{arc tan } x$$
$$\text{dom } T^* = R \text{ and rng } T^* = \langle -\pi/2, \pi/2 \rangle$$

and we hasten to point out that the statement

$$Y = \text{arc tan } x$$

means that (1) $\tan Y = X$ and that (2) $Y \in \langle -\pi/2, \pi/2 \rangle$.

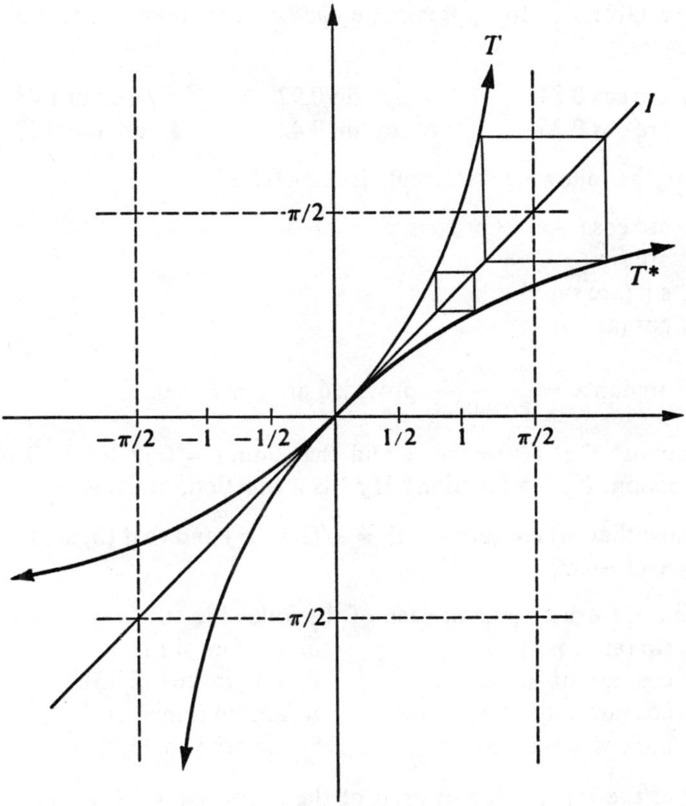

Fig. 8.32b

PROBLEM SET 8.6

1. Construct the graph of the function S^*.

2. Find each of the following numbers.

a. arc cos $-1/2$ **b.** arc sin $1/2$ **c.** arc cos $\sqrt{3}/2$
d. arc sin $\sqrt{3}/2$ **e.** arc cos $-\sqrt{2}/2$ **f.** arc sin $-1/2$
g. arc sin 0 **h.** arc cos 0 **i.** arc tan 0
j. arc tan 1 **k.** arc tan -1 **l.** arc sin 1

3. Use table A.2 to approximate each of the following.

a. arc cos 0.38 **b.** arc sin 0.948 **c.** arc tan 0.36
d. arc cos -0.32 **e.** arc sin 0.991 **f.** arc cos 0.03
g. arc sin 0.783 **h.** arc tan 1.96

4. Use table A.3 to approximate each of the following to the nearest degree.

a. arc cos 0.84 **b.** arc sin 0.92 **c.** arc tan 1.48
d. arc cos 0.25 **e.** arc sin 0.42 **f.** arc tan 0.57

5. Are the following statements true or false?

a. arc cos $(-x)$ = arc cos x
b. arc sin $(-x)$ = $-$arc sin x
c. sin (arc sin x) = x
d. cos (arc cos x) = x

e. $\text{arc tan } x = \dfrac{\text{arc sin } x}{\text{arc cos } x}$ provided arc cos $x \neq 0$.

6. Suppose that $f(x) = \sin x$ and that dom $f = [\pi/2, 3\pi/2]$. Graph this function. If f^* a function? If f^* is a function, describe it.

7. Show that **(a)** cos (arc sin x) = $\sqrt{(1 - x^2)}$ and that **(b)** sin (arc cos x) = $\sqrt{(1 - x^2)}$.

8. Find the exact value of each of the following.

a. sin (arc cos 3/4) **b.** cos (arc sin 1)
c. cos (arc sin $\sqrt{2}/2$) **d.** sin (arc cos 12/13)
e. cos (arc sin $-3/5$) **f.** arc sin (tan $\pi/4$)
g. arc cos (cos $2\pi/3$) **h.** arc tan (sin $3\pi/2$)

9. Find the exact value of each of the following. **(a)** cos (arc sin 3/5 + arc sin 4/5). *Hint:* Let y = arc sin 3/5 and z = arc sin 4/5 for then the above is equivalent to cos $(y + z)$, and we have an addition formula that will help. **(b)** sin (arc sin $-3/5$ + arc sin 4/5). *Hint:* same as above.

10. Suppose that $F(x) = \dfrac{x}{2}$. In each of the following, give the rule of pairing, the domain, the range, and sketch the graph of the given composite function.

a. $[S^* \circ F]$ **b.** $[C^* \circ F]$ **c.** $[T^* \circ F]$

Express each of the following in terms of x.

11. cos (2 arc cos x). *Hint:* Let y = arc cos x and then use a formula for cos $(2y)$.

12. cos (2 arc sin x). *Hint:* Let y = arc sin x and then use a formula for cos $(2y)$.

13. $\sin(2 \arccos x)$. *Hint:* Let $y = \arccos x$ and then use the formula for $\sin(2y)$.

Section 7—Other Circular Functions The cosine and sine are the basic circular functions in the sense that they are used to define the other circular functions. As a matter of fact, you have already witnessed one example of this, for the tangent function was defined as the quotient of these two basic functions. Thus there is some merit in the notion that mastery of the cosine and sine functions is the primary, if not the only, objective in our study of the circular functions. However, in order that the student be at least exposed to the other circular functions that he will read about in any further study of mathematics, we bow to tradition and use the cosine and sine functions to define three other functions.

DEFINITION. *The functions cotangent, secant, and cosecant are defined as follows:*

$$\text{cotangent} = \left\{(x,y) \,\middle|\, y = \frac{\cos x}{\sin x}, \sin x \neq 0\right\}$$

$$\text{secant} = \left\{(x,y) \,\middle|\, y = \frac{1}{\cos x}, \cos x \neq 0\right\}$$

$$\text{cosecant} = \left\{(x,y) \,\middle|\, y = \frac{1}{\sin x}, \sin x \neq 0\right\}$$

or, more briefly,

$$\cot x = \frac{\cos x}{\sin x}, \quad \sin x \neq 0$$

$$\sec x = \frac{1}{\cos x}, \quad \cos x \neq 0$$

$$\csc x = \frac{1}{\sin x}, \quad \sin x \neq 0$$

Since the definition expresses each of the functions cot, sec, and csc in terms of the sin or cos, many of their properties can be determined by using the known properties of these two basic functions.

A Digression Before we make any investigation into the properties of these three new functions, I would like to call to your attention two statements that are important because they are frequently encountered.

Satz 1. $1 + \tan^2 x = \sec^2 x$.
Satz 2. $1 + \cot^2 x = \csc^2 x$.

To prove these two statements is easy, so get a small scrap of paper and begin. To prove Satz 1, write $1 + \tan^2 x = 1 + \dfrac{\sin^2 x}{\cos^2 x}$ and simplify this. To prove Satz 2, write $1 + \cot^2 x = 1 + \dfrac{\cos^2 x}{\sin^2 x}$ and simplify.

End of digression.

Properties of cot, sec, and csc

Since $\cot x = \dfrac{\cos x}{\sin x}$, it can be argued that dom cot $= \{x \mid \sin x \neq 0\}$. We can improve on this by noting that $\sin x$ is 0 when $x = k\pi$, where k is an integer. Hence,

$$\text{dom cot} = \{x \mid x \neq k\pi, k \in J\}.$$

Since the domain of the function cosecant is the same as the domain of the cotangent (*Why?*), then

$$\text{dom csc} = \{x \mid x \neq k\pi, k \in J\}.$$

The domain of the secant function is the same as the domain of the tangent. For in each case, it is required that $\cos x \neq 0$. Since $\cos x$ is 0 for $x = \frac{\pi}{2}(2k + 1) = \frac{\pi}{2} + k\pi$ where k is an integer,

$$\text{dom sec} = \left\{x \;\middle|\; x \neq \frac{\pi}{2} + k\pi, k \in J\right\}.$$

The ranges of the sec and csc are not difficult to find. We know that $\sec x = \dfrac{1}{\cos x}$ and that $|\cos x| \leq 1$, hence $\dfrac{1}{|\cos x|} \geq 1$. Thus $|\sec x| \geq 1$, that is, the range of the secant is given by

$$\text{rng sec} = \{y \mid y \geq 1 \text{ or } y \leq -1\}.$$

In a similar fashion,

$$\text{rng csc} = \{y \mid y \geq 1 \text{ or } y \leq -1\}.$$

The significance of this to the graphs of these two functions should not be overlooked. It implies that no point in the graph of either the secant or cosecant will lie between the horizontal lines $y = 1$ and $y = -1$.

It is not so easy to determine the range of the cot, but we can use table A.2 in the appendix as evidence to support the claim that as x increases from 0 to $\pi/2$, cot x decreases through all the positive real numbers, from "infinity" to 0. It is also true that

$$\cot(-x) = \frac{\cos(-x)}{\sin(-x)} = -\cot x$$

hence, we can conclude that rng cot $= R$.

It is easy to determine the signs of the cot, sec, and csc by using what we know about the signs of the sin and cos. Table 8.10 indicates the signs of these function values.

TABLE 8.10

x in quadrant:	I	II	III	IV
cos x	+	−	−	+
sin x	+	+	−	−
cot x	+	−	+	−
sec x	+	−	−	+
csc x	+	+	−	−

We shall leave to the student (see problems 5–7) the task of verifying the following statements:

cotangent has period π.
secant has period 2π.
cosecant has period 2π.

Graphs To sketch the graph of the function $f(x) = \cot x$, it helps to recall that $\cot x = 0$ when $\cos x$ is 0, that is, for $x \in \left\{\frac{\pi}{2} + k\pi, k \in J\right\}$. Also recall that

$$\text{dom cot} = \{x \mid x \neq k \ , k \in J\}.$$

By selecting some values of x in the interval from 0 to π, we can plot a few points of the graph of f. See figure 8.33.

We then connect these few points with a "smooth" curve. The dashed vertical lines serve only to remind us that the cot 0 and cot π are not defined. Since this function has period π, we can repeat this procedure over successive intervals of length π to obtain the graph of the cotangent function as shown in figure 8.34.

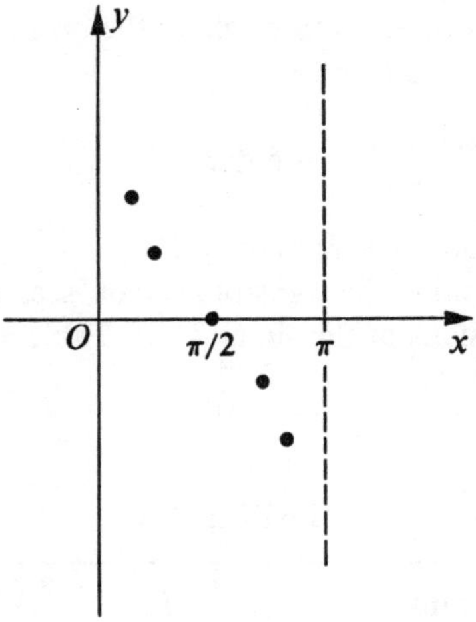

Fig. 8.33

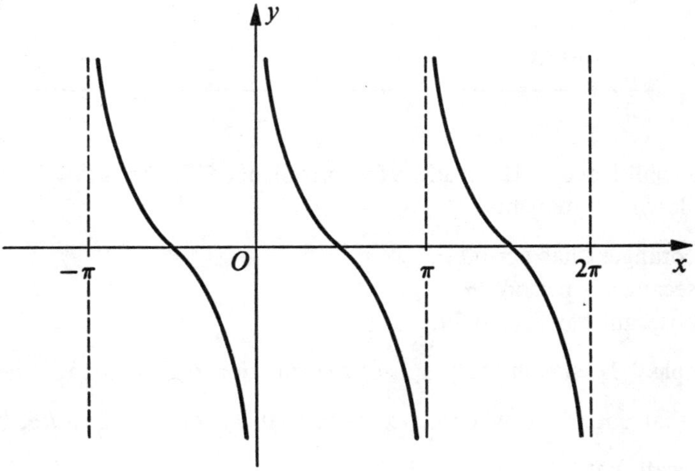

Fig. 8.34

PROBLEM SET 8.7

1. Sketch the graph of the function $f(x) = \sec x$ on the interval $[0,2\pi]$.

2. Sketch the graph of the function $f(x) = \csc x$ on the interval $[0,2\pi]$.

3. Use the graphs to complete table 8.11.

TABLE 8.11

as x increases from __ to __	$0 \quad \frac{\pi}{2}$	$\frac{\pi}{2} \quad \pi$	$\pi \quad \frac{3\pi}{2}$	$\frac{3\pi}{2} \quad 2\pi$
$\sin x$	Increases	Decreases		
$\cos x$	Decreases			
$\tan x$	Increases		Increases	
$\cot x$				Decreases
$\sec x$				
$\csc x$				

4. Verify each of the following statements.

a. $\cot x = \dfrac{1}{\tan x}$, $\tan x \neq 0$

b. $\cot(-x) = -\cot x$
c. $\sec(-x) = \sec x$
d. $\csc(-x) = -\csc x$

5. Using the fact that $\cot(x + \pi) = \dfrac{\cos(x + \pi)}{\sin(x + \pi)}$, show that $\cot(x + \pi) = \cot x$.

6. Using the addition formula for the cosine function to show that $\sec(x + 2\pi) = \sec x$.

7. Use the addition formula for the sine function to show that $\csc(x + 2\pi) = \csc x$.

8. Give the exact value of each of the following:

a. $\cot 0$ **b.** $\cot \pi/6$ **c.** $\cot \pi/4$
d. $\cot \pi/3$ **e.** $\cot(-\pi/4)$ **f.** $\cot(-3\pi/4)$

9. Give the exact value of each of the following:

a. $\sec 0$ **b.** $\sec \pi/6$ **c.** $\sec \pi/4$
d. $\sec \pi/3$ **e.** $\sec(-\pi/6)$ **f.** $\sec(7\pi/6)$

10. Give the exact value of each of the following:

a. $\csc \pi/6$ b. $\csc \pi/3$ c. $\csc \pi/2$
d. $\csc(-\pi/3)$ e. $\csc(5\pi/3)$ f. $\csc(3\pi/2)$

11. Verify each of the following statements.

a. $\sec^2 x(1 - \sin^2 x) = 1$
b. $\cos x \cdot \csc x = \cot x$
c. $\sin x(\sec x - \csc x) = \tan x - 1$
d. $\cos^2 x(1 + \tan^2 x) = 1$
e. $\cot(x + t) = \dfrac{\cot x \cot t - 1}{\cot x + \cot t}$
f. $\cot(2x) = \dfrac{\cot^2 x - 1}{2 \cot x}$

12. If $\sin x = 3/5$ and $0 < x < \pi/2$, give $\cos x$, $\tan x$, $\cot x$, $\sec x$, and $\csc x$.

13. If $\sin x = 3/5$ and $\pi/2 < x < \pi$, give $\cos x$, $\tan x$, $\cot x$, $\sec x$, and $\csc x$.

14. Using figure 8.35 argue that

$d(\overline{OB}) = \cot x$, $d(\overline{OT}) = \sec x$, $d(\overline{OC}) = \csc x$.

15. Let $\dfrac{x}{2} = \tan z$ and show that $4 + x^2 = 4 \sec^2 z$.

16. Let $\dfrac{x}{3} = \tan z$ and show that $9 + x^2 = 9 \sec^2 z$.

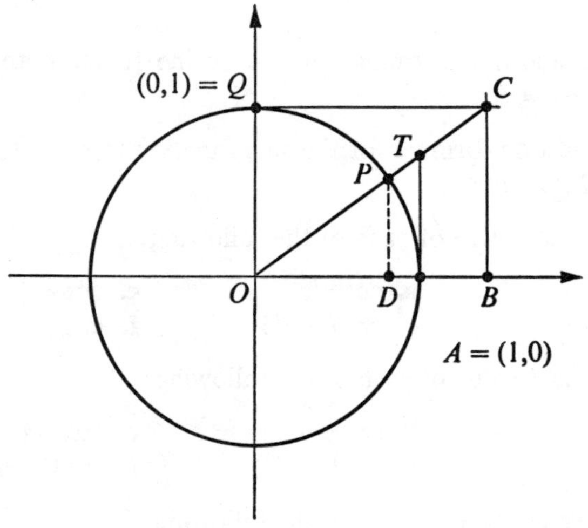

Fig. 8.35

17. Let $\frac{x}{3} = \sec z$ and show that $x^2 - 9 = 9\tan^2 z$.

18. Let $\frac{x}{2} = \sec z$ and show that $x^2 - 4 = 4\tan^2 z$.

Section 8—Equations In this section we consider equations of the form $Z(x) = 0$, where Z is an algebraic combination of trigonometric (circular) functions. $Z(x)$ will be a polynomial in one of the trigonometric functions or the product of two or more polynomials, each involving only one of the trigonometric functions, or $Z(x)$ will be reducible to one of the above.

The task of finding the solution set S of such equations can range from "difficult" to "unbelievably hard." The only cheerful feature of solving such equations lies in the fact that, since the trigonometric functions have period π or 2π, it is sufficient to find the solutions in the interval $[0,2\pi\rangle$ and to note that all other solutions are obtained from the periodicity condition. The solutions that are in the interval $[0,2\pi\rangle$ are called the *principal solutions*. We shall denote the set of principal solutions by S_p.

A Recap of Some Useful Information (See Tables 8.12 and 8.13)

TABLE 8.12

x	$\sin x$	$\cos x$	$\tan x$
0	0	1	0
$\pi/6$	$1/2$	$\sqrt{3}/2$	$\sqrt{3}/3$
$\pi/4$	$\sqrt{2}/2$	$\sqrt{2}/2$	1
$\pi/3$	$\sqrt{3}/2$	$1/2$	$\sqrt{3}$
$\pi/2$	1	0	∞
π	0	-1	0
$3\pi/2$	-1	0	$-\infty$

$$\sin^2 x + \cos^2 x = 1$$

$$1 + \tan^2 x = \sec^2 x$$

$$\cos 2x = 2\cos^2 x - 1 = 1 - 2\sin^2 x$$

$$\sin 2x = 2\sin x \cos x$$

TABLE 8.13

	Quadrant				
	I	II	III	IV	Period
sin	+	+	−	−	2π
cos	+	−	−	+	2π
tan	+	−	+	−	π

$$\tan 2x = \frac{2\tan x}{1-\tan^2 x}$$

Example 1. Find all solutions to the equation

$$\sin(3x) = -\sqrt{3}/2.$$

Solution. I think that it might help if we let $3x = t$ and solve the equation $\sin t = -\sqrt{3}/2$.

Now $\sin t < 0$ when $t \in$ Qd. III or Qd. IV. We also know that $\sin(\pi/3) = \sqrt{3}/2$. Using $\pi/3$ as a reference number,

$$t = \pi + \pi/3 = \frac{4\pi}{3} \text{ (Qd. III soln.)}$$

and

$$t = 2\pi - \pi/3 = \frac{5\pi}{3} \text{ (Qd. IV soln.).}$$

Since the period of the sine function is 2π, we then have

$$t = \frac{4\pi}{3} + 2k\pi \quad \text{or} \quad t = \frac{5\pi}{3} + 2k\pi.$$

If we now recall that $t = 3x$, we can see that

$$x = \frac{4\pi}{9} + \frac{2k\pi}{3} \quad \text{or} \quad x = \frac{5\pi}{9} + \frac{2k\pi}{3}.$$

Hence,

$$S_p = \left\{\frac{4\pi}{9}, \frac{5\pi}{9}\right\}$$

therefore

$$S = \left\{x \,\middle|\, x = \frac{4\pi}{9} + \frac{2k\pi}{3}\right\} \cup \left\{x \,\middle|\, x = \frac{5\pi}{9} + \frac{2k\pi}{3}\right\}.$$

Example 2. Find all solutions to the equation

$$2\cdot\cos^2 x - \cos x = 0.$$

Solution.

$2\cdot\cos^2 x - \cos x = 0$ implies that
$\cos x\,(2\cos x - 1) = 0$, hence
$\cos x = 0$ or $\cos x = 1/2$.

To find the principal solutions, we know that $\cos x = 0$ for $x = \pi/2$ and $x = 3\pi/2$. Since $\cos x > 0$ only in Qd. I and Qd. IV, we know that $\cos x = 1/2$ for $x = \pi/3$ and $x = 2\pi - \pi/3 = \frac{5\pi}{3}$. (Note the use made of $\pi/3$ as a reference number.)

Thus the set of principal solutions is given by

$$S_p = \{\pi/2, \pi/3, 3\pi/2, 5\pi/3\}.$$

Since the period of the cosine is 2π,

$$S = \left\{x \,\middle|\, x = \frac{\pi}{2} + 2k\pi\right\} \cup \left\{x \,\middle|\, x = \frac{3\pi}{2} + 2k\pi\right\}$$
$$\cup \left\{x \,\middle|\, x = \frac{\pi}{3} + 2k\pi\right\} \cup \left\{x \,\middle|\, x = \frac{5\pi}{3} + 2k\pi\right\}.$$

This looks long (which it is) and impressively complex (which it isn't).

Example 3. Find all solutions to the equation

$$\cos(2x) + \sin^2 x = 1.$$

Solution. Now $\cos(2x) = 1 - 2\sin^2 x$, thus the given equation is equivalent to the equation

$$1 - \sin^2 x = 1$$

which can be simplified to

$$\sin^2 x = 0.$$

Therefore $S_p = \{0,\pi\}$ and so

$$S = \{x \mid x = 0 + 2k\pi\} \cup \{x \mid x = \pi + 2k\pi\}.$$

What follows is not always easy to do but, when it is easy to do, we take great delight in doing it. By selecting a few integers k, we can easily see that

$$S = \{\ldots, -2\pi, 0, 2\pi, 4\pi, \ldots\} \cup \{\ldots, -\pi, \pi, 3\pi, 5\pi, \ldots\}.$$

So, at least in this example, it is "no sweat" to observe that the solution set can be compactly written as

$$S = \{x \mid x = k\pi\}.$$

Note: This "compactness" of S is not always so easy. Look at example 2. I think some sort of a medal should be given to anyone who can put the S of this example in a compact form.

Example 4. This is the first example of an equation involving two distinct trig functions which cannot be simplified (as in example 3) to an equivalent equation in only one such function.

Find all solutions to the equation

$$\tan x \cdot \sin x - \sin x = 0.$$

Solution. By factoring, we can write the given equation as $\sin x (\tan x - 1) = 0$. Setting each factor equal to zero, that is,

$$\sin x = 0, \tan x = 1$$

we see that $\sin x = 0$ if $x = 0$ or $x = \pi$ while $\tan x = 1$ if $x = \pi/4$ or $x = 5\pi/4$, hence

$$S_p = \{0, \pi/4, \pi, 5\pi/4\}.$$

Now, all of the solutions to the equation $\sin x = 0$ can be compactly written as

$$\{x \mid x = 0 + 2k\pi\} \cup \{x \mid x = \pi + 2k\pi\} = \{x \mid x = k\pi\}$$

while the solutions to the equation $\tan x = 1$ are given by (recall that the period of the tangent function is π)

$$\{x \mid x = \pi/4 + k\pi\} \cup \{x \mid x = 5\pi/4 + k\pi\} = \{x \mid x = \pi/4 + k\pi\}.$$

It should now be reasonably clear that

$$S = \{x \mid x = k\pi\} \cup \{x \mid x = \pi/4 + k\pi\}.$$

Example 5. Find the principal solutions only of the equation

$$\sin x + \cos x = 1.$$

Solution. It may at first appear that this equation is not in the form that we announced we would consider, since it involves the sum of two circular functions. However, we can write

$$\sin x = 1 - \cos x$$

and, squaring both sides, obtain

$$\sin^2 x = 1 - 2\cos x + \cos^2 x.$$

Since $\sin^2 x = 1 - \cos^2 x$, the above equation can be put in the form

$$2\cos^2 x - 2\cos x = 0$$

or, by factoring,

$$\cos x(\cos x - 1) = 0.$$

Setting each factor equal to zero

$$\begin{array}{lcl} \cos x = 0 & \textit{or} & \cos x = 1 \\ x = \pi/2,\ 3\pi/2 & & x = 0 \end{array}$$

By substituting in the original equation, we can see that $3\pi/2$ is not a solution. This extraneous solution is a result of our squaring. Thus

$$S_p = \{0, \pi/2\}.$$

In the next example, we try to indicate how we shall use table A.2 in the appendix to approximate the solutions when they are not the "nice special" values we are so fond of.

Example 6. Find the principal solutions to the equation

$$\sin x = 0.64.$$

Solution. First note that this equation has solutions in the first and second quadrants only. By looking at table A.2, we see that

$$\sin 0.70 = 0.64422$$

and that

$$\sin 0.69 = 0.63654.$$

Since we must interpolate, we write

$$0.01 \left\{ \begin{array}{l} \sin 0.70 = 0.6442 \\ \sin x \quad\ = 0.64000 \\ \sin 0.69 = 0.63654 \end{array} \right. \quad 0.00346 \longrightarrow 0.00768$$

and set up the proportionality

$$\frac{\Delta}{0.01} = \frac{0.00346}{0.00768}$$

hence $\Delta = \dfrac{346}{768} \cdot (0.01) \approx 0.004$. So one solution is approximated by

$0.69 + 0.004 = 0.694$. To obtain the second quadrant solution, we use $\pi \approx 3.142$ and write

$$3.142 - 0.694 = 2.448.$$

Therefore

$$S_p = \{0.694, 2.448\}.$$

PROBLEM SET 8.8

Note Whenever it is necessary to approximate the required solutions, use table A.2 and let $\pi = 3.142$ and $2\pi = 6.283$. In each of the following, give S_p.

1. $\cos x = 1/2$

2. $\tan x = \sqrt{3}$

3. $\tan 2x = \sqrt{3}$

4. $\cos 2x = \sqrt{2}/2$

5. $\sin x = -0.7033$

6. $\cos x = 0.6675$

7. $\tan x = 1.5640$

8. $\tan x = -2.8350$

9. $\cos 2x + \cos x = 0$

10. $\cos 2x - \sin x = 0$

11. $\cos^2 x - 2 \cos x = 0$

12. $2 \sin^2 x - 3 \sin x + 1 = 0$

13. $3 \tan x = 2 \cos x$. *Hint:* $\tan x = \dfrac{\sin x}{\cos x}$

14. $4 \tan^2 x = 3 \sec^2 x$. *Hint:* $\sec^2 x = 1 + \tan^2 x$.

In each of the following, give S.

15. $2 \cos^2 x + \cos x = 0$

16. $\sin^2 x - \sin x = 0$

17. $\sin 2x + \cos x = 0$

18. $\sin 2x - \sin x = 0$

19. $\cos^2 x + 2 \cos x + 1 = 0$

20. $\sin^2 x + 2 \sin x - 3 = 0$

21. $2 \cos^2 x - \sin x = 1$

22. $2 - \sin x = 2 \cos^2 x$

23. $2 \sin^2 x - \sin x = 1$

24. $2 \cos^2 x - 3 \cos x + 1 = 0$

25. Suppose $0 < x < \pi/2$ and $0 < y < \pi/2$. Find x and y so that $\tan (x - y) = 1$ and $\cos (x + y) = 0$.

Section 9—$A \cdot \sin (Bx + C)$ This section calls attention to the concept of a "sinusoidal wave," that is, the graph of the function $W(x) = A \cdot \sin (Bx + C)$ where A, B, and C are constants. The main objective of the discussion shall be to indicate the relationships between these constants and the properties of the function W called *amplitude*, *period*, and *phase shift*. The discussion will be limited in scope by the assumption that $A > 0$ and that $B > 0$.

Earlier (section 4 of this chapter), we at least hinted at two of these relationships. It may prove helpful to repeat this information.

Amplitude Since $|\sin \theta| \leq 1$, $A \cdot |\sin (Bx + C)| \leq 1 \cdot A$ from which it follows that

$$-A \leq \sin (Bx + C) \leq A.$$

Hence, the function W has a maximum value of A and a minimum value of $-A$. The constant A is called the *amplitude* of the sinusoidal wave (sine wave).

Period It is easy to see that

$$A \cdot \sin (Bx + C) = A \cdot \sin [(Bx + C) + 2\pi].$$

It requires only a slightly better vision to note that

$$\begin{aligned} A \cdot \sin \left[B\left(x + \frac{2\pi}{B}\right) + C\right] &= A \cdot \sin [(Bx + 2\pi) + C] \\ &= A \cdot \sin [(Bx + C) + 2\pi] \\ &= A \cdot \sin (B + C) \end{aligned}$$

This means that $W\left(x + \frac{2\pi}{B}\right) = W(x)$, and so the period of W is $\frac{2\pi}{B}$.

It may prove helpful to illustrate these two notions by figure 8.36. In this figure, I let $C = 0$ for simplicity.

The remainder of our discussion will be devoted to determining the effect of the constant C on the graph of W.

Phase Shift Consider the graph of $f(x) = \sin (x + C)$. When $x + C = 0$, $x = -C$, and when $x + C = 2\pi$, $x = 2\pi - C$. The graph (indicated in figure 8.37) is therefore a sine wave shifted to the *left* a distance C when $C > 0$. (If $C < 0$, the shift will be the *right*.)

The constant $-C$ is called the phase shift of f.

If $g(x) = \sin (Bx + C)$, we note that when $Bx + C = 0$, $x = -\frac{C}{B}$ and when $Bx + C = 2\pi$, $x = \frac{2\pi}{B} - \frac{C}{B}$. So the graph will be shifted to the *left* by an amount $\frac{C}{B}$ when $C > 0$. The shift will be to the *right* when $C < 0$. (See figure 8.38.)

The constant $-\frac{C}{B}$ is called the phase shift of g.

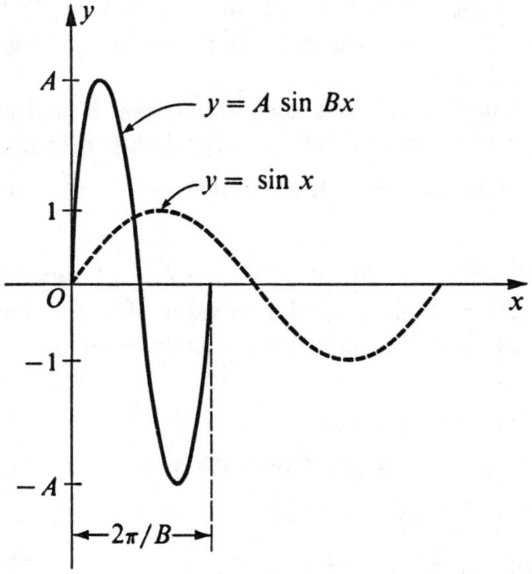

Fig. 8.36

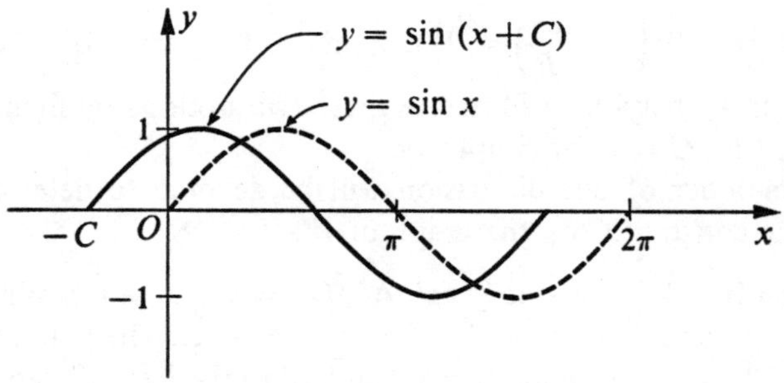

Fig. 8.37

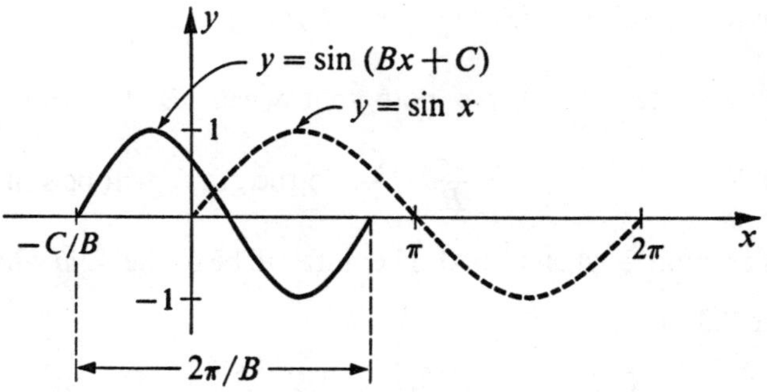

Fig. 8.38

Finally, we combine all these ideas in the representation of the general sine wave

$$W(x) = A \cdot \sin(Bx + C).$$

The amplitude is A.
The period is $2\pi/B$.
The phase shift is $-C/B$. To the left when $C > 0$. To the right when $C < 0$. (See figure 8.39.)

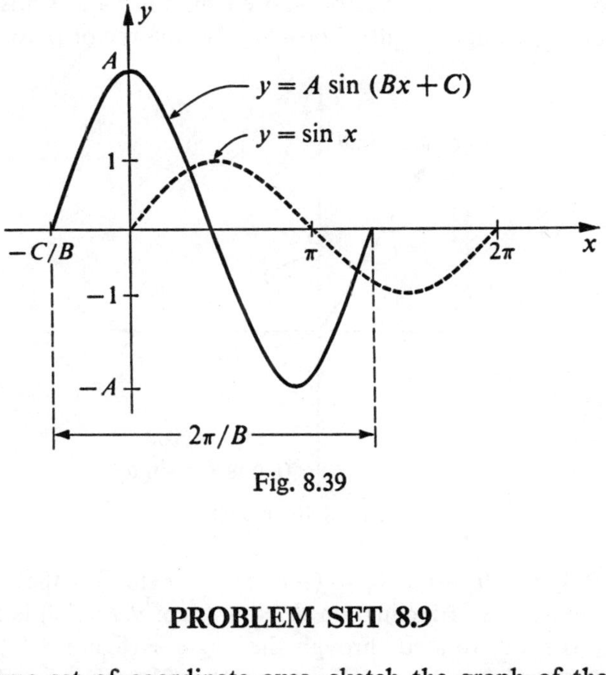

Fig. 8.39

PROBLEM SET 8.9

On the same set of coordinate axes, sketch the graph of the function $F(x) = \sin x$ and of the given function W. Also give the amplitude, period, and phase shift of W.

1. $W(x) = 2 \sin \left(x - \frac{\pi}{4}\right)$

2. $W(x) = 2 \sin \left(x + \frac{\pi}{4}\right)$

3. $W(x) = 3 \sin \left(x + \frac{\pi}{3}\right)$

4. $W(x) = 3 \sin \left(x - \frac{\pi}{3}\right)$

5. $W(x) = \frac{1}{2} \sin (3x)$

6. $W(x) = \frac{1}{2} \sin (2x)$

7. $W(x) = 2 \sin \left(x + \frac{\pi}{2}\right)$

8. $W(x) = 2 \sin \left(x - \frac{\pi}{2}\right)$

9. $W(x) = \frac{3}{2} \sin (2x + \pi)$

10. $W(x) = 3 \sin \left(\frac{x}{2} + \frac{\pi}{2}\right)$

11. $W(x) = 2 \sin \left(\pi x + \frac{\pi}{2}\right)$

12. $W(x) = \frac{3}{2} \sin \left(\pi x - \frac{\pi}{2}\right)$

13. $W(x) = \sin \left(\frac{\pi}{2} [x + 1]\right)$

14. $W(x) = \sin \left(\frac{\pi}{2} [x - 1]\right)$

Section 10—Polar Coordinates A two dimensional vector is uniquely determined by its length and its angle of inclination (direction). Thus, for $r > 0$ $(r \cos \theta, r \sin \theta)$ is the vector whose length is r and whose angle of inclination is θ (see figure 8.40). Consider also the vector $(r \cos \theta, r \sin \theta)$

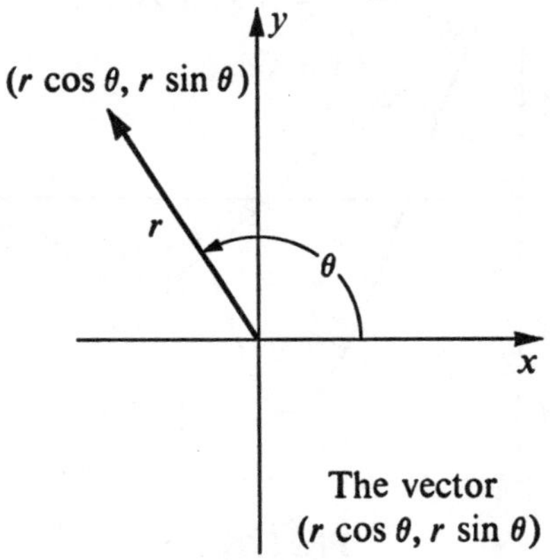

The vector
$(r \cos \theta, r \sin \theta)$

Fig. 8.40. $r > 0$.

when $r < 0$. If $r < 0$, $-r > 0$, so $(-r \cos \theta, -r \sin \theta)$ is the vector with length $-r$ and angle of inclination θ. Now $(r \cos \theta, r \sin \theta)$ is the vector $(-r \cos \theta, -r \sin \theta)$ rotated through the angle π (figure 8.41). That is, when $r < 0$, then $(r \cos \theta, r \sin \theta)$ is the vector of length $-r$ and direction $\theta + \pi$.

We can summarize the above by the following statements:

Given real numbers r and θ, $r \neq 0$:

1. If $r > 0$, then $(r \cos \theta, r \sin \theta)$ is the vector of length r and angle of inclination θ.
2. If $r < 0$, then $(r \cos \theta, r \sin \theta)$ is the vector of length $-r$ and angle inclination $\theta + \pi$.
3. If r is 0, then we call the vector $(r \cos \theta, r \sin \theta)$ the zero vector or *pole*.

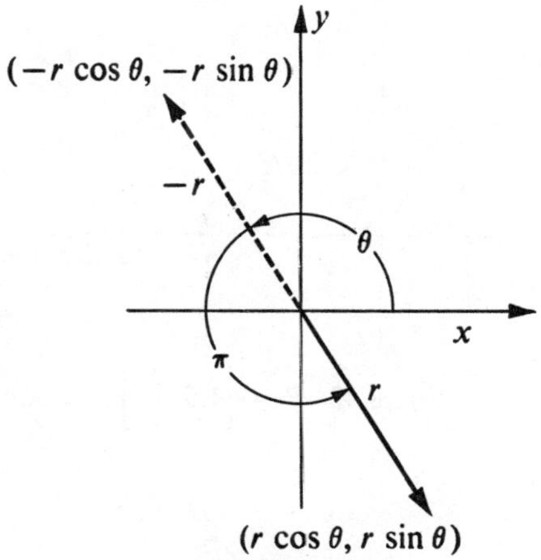

Fig. 8.41. $r < 0$.

Since we are interested in thinking of the vector $(r\cos\theta, r\sin\theta)$ as a point in the plane, the real numbers r and θ are called the *polar coordinates* of the point $(x,y) = (r\cos\theta, r\sin\theta)$ and we shall write these coordinates in the form $[r,\theta]$. It may also help to know that the origin is called the *pole* and the nonnegative X axis is called the *polar axis*. (See figures 8.42 through 8.48.)

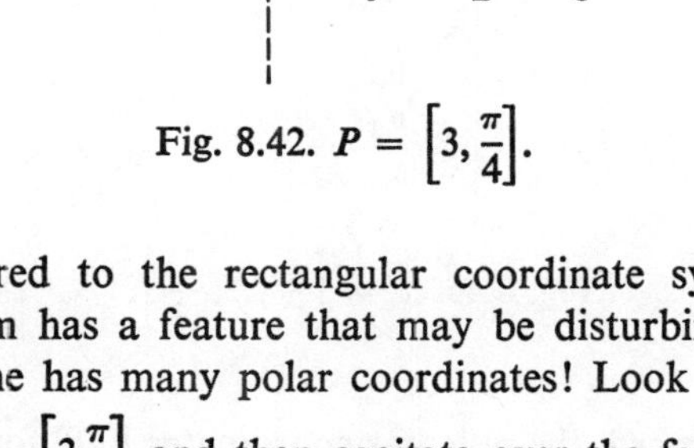

Fig. 8.42. $P = \left[3, \frac{\pi}{4}\right]$.

When compared to the rectangular coordinate system, the polar coordinate system has a feature that may be disturbing. To wit, each point in the plane has many polar coordinates! Look back at the first figure's point $P = \left[3, \frac{\pi}{4}\right]$ and then cogitate over the fact that this same point also has the following polar coordinates:

$$\left[-3, \frac{5\pi}{4}\right], \left[3, \frac{9\pi}{4}\right], \text{ and } \left[-3, -\frac{3\pi}{4}\right].$$

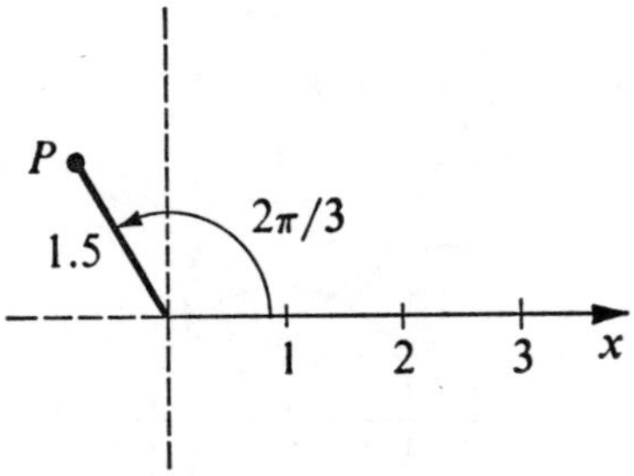

Fig. 8.43. $P = \left[1.5, \frac{2\pi}{3}\right]$.

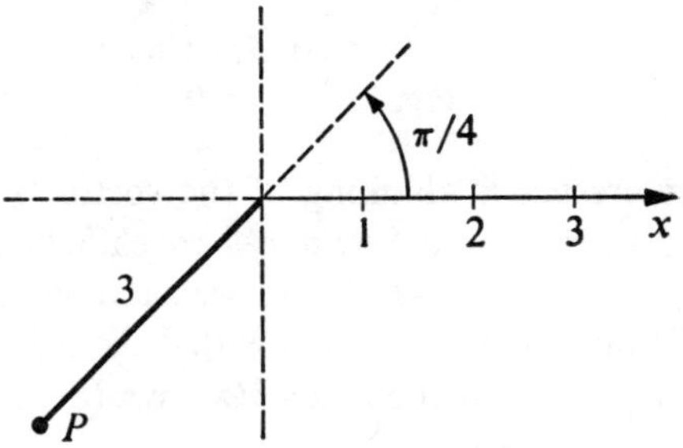

Fig. 8.44. $P = \left[-3, \frac{\pi}{4}\right]$.

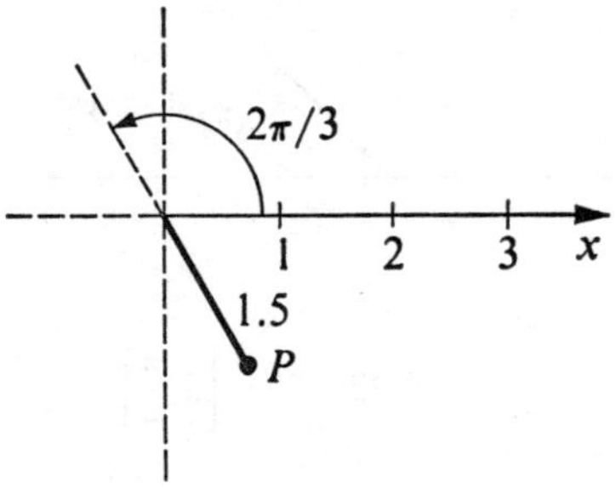

Fig. 8.45. $P = \left[-1.5, \frac{2\pi}{3}\right]$.

As a matter of record, the pole has coordinates $[0,\theta]$ for every number θ.

Question. After, or while, looking at the figure where $P = \left[-1.5, \frac{2\pi}{3}\right]$ can you give another polar coordinate of P?

Answer. Yes.

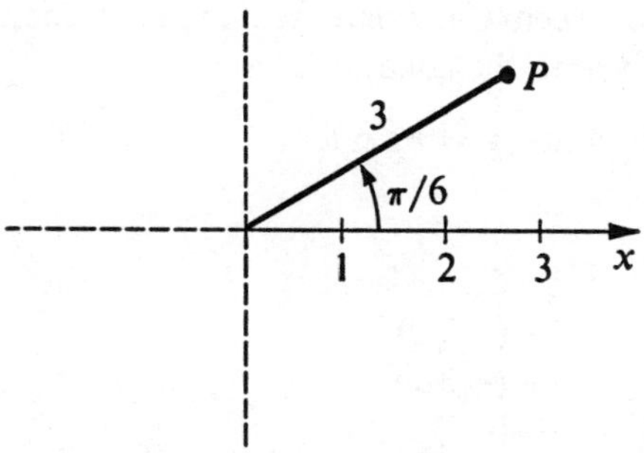

Fig. 8.46. $P = [3,30°]$

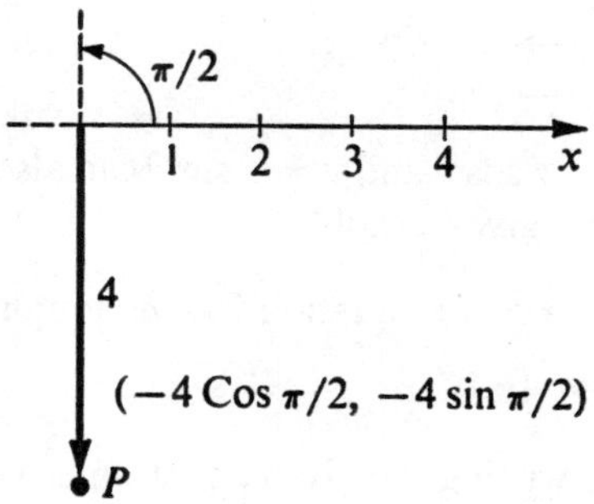

Fig. 8.47. $P = \left[-4, \frac{\pi}{2}\right]$.

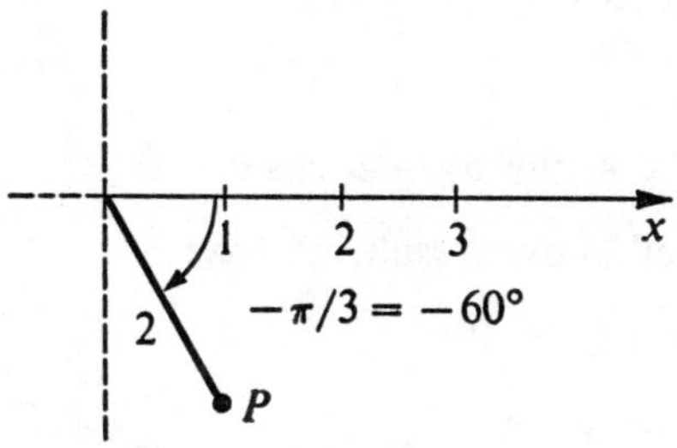

Fig. 8.48 $P = [2, -60°]$.

Question. Well?

Answer. $P = \left[1.5, -\frac{\pi}{3}\right]$.

At this time, you are asked to look back at the summary given on the first page of this section.

The relation between Cartesian (rectangular) coordinates and polar coordinates are given in the equations

$$x = r\cos\theta \quad \text{and} \quad y = r\sin\theta$$

Example.

Polar	to Cartesian
$[2,\pi]$	$\to(-2,0)$
$[2,\pi/6]$	$\to(\sqrt{3},1)$
$[0,\pi/4]$	$\to(0,0)$
$[-3,\pi/2]$	$\to(0,-3)$
$[\sqrt{2},3\pi/4]$	$\to(-1,1)$
$[-2,\pi/2]$	$\to(0,-2)$
$[2,3\pi/2]$	$\to(0,-2)$
$[5,0]$	$\to(5,0)$

The equations $x = r\cos\theta$ and $y = r\sin\theta$ can also be used to determine a polar equation for a given graph.

Example. Find a polar equation for the graph of

$$x^2 + y^2 - 4x = 0.$$

Remark. After writing the given equation as $(x - 2)^2 + y^2 = 4$, we know that this is a circle center (2,0) and radius 2.

Solution. Substituting, we get

$$r^2\cos^2\theta + r^2\sin^2\theta - 4r\cos\theta = 0$$

and factoring

$$r^2(\cos^2\theta + \sin^2\theta) - 4r\cos\theta = 0$$

or, using a well known result, we have

$$r^2 - 4r\cos\theta = 0$$

which is equivalent to

$$r(r - 4\cos\theta) = 0.$$

The graph of the equation $x^2 + y^2 - 4x = 0$, then, is the union of the graphs of

(1) $r = 0$

and

(2) $r = 4\cos\theta.$

But the graph of (1) is the pole, nothing more. Hence, if the graph of (2) contains the pole, then (1) contributes nothing additional to the graph and so the graph of (2) will be the same as that of the given equation.

Clearly, $4 \cos (\pi/2) = 0$ so that $[0,\pi/2]$ satisfies (2). But this is the pole. Therefore $r = 4 \cos \theta$ is the desired equation.

In the example above, it is worthwhile to note that $x^2 + y^2 = r^2$. If we had observed this before, we could have shortened our work in the example. Maybe figure 8.49 will help us to remember this.

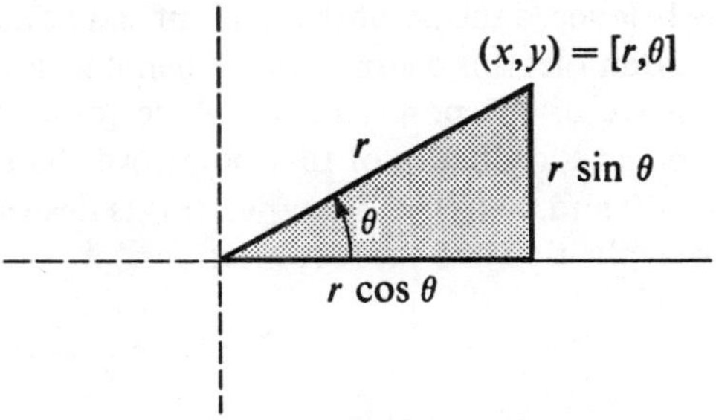

Fig. 8.49

Example. Find a polar equation for the graph of $x - y = 3$.

Solution. $r \cos \theta - r \sin \theta = 3$.

Yes, we are through.

Example. Find a polar equation for $4y = x^2$.

Solution.

$$4r \sin \theta = r^2 \cos^2 \theta$$
$$r^2 \cos^2 \theta - 4r \sin \theta = 0$$
$$r(r \cos^2 \theta - 4 \sin \theta) = 0$$

Since the pole satisfies the equation $r \cos^2 \theta - 4 \sin \theta = 0$, the desired polar equation is given by

$$r \cos^2 \theta - 4 \sin \theta = 0.$$

Example. ATTENTION! This is different. Give the Cartesian equation for the polar equation $r = 2 \cos \theta$.

Solution. Since $r \cos \theta$ is x, it follows that $\cos \theta = \frac{x}{r}$ provided $r \neq 0$. Substitute this into the given polar equation and get the result

$$r = \frac{2x}{r} \quad \text{or} \quad r^2 = 2x.$$

Recalling that $x^2 + y^2 = r^2$ (I told you this would be useful), we can obtain the equation

$$x^2 + y^2 = 2x.$$

While we have belabored the point that a point can be assigned different polar coordinates, don't let this disturb you. When it is desirable to have a 1-to-1 correspondence of the points in the plane (other than the origin or pole) and the polar coordinates of that point, we shall achieve this by the restrictions $r > 0$ and $0 \leq \theta < 2\pi$. When this is desirable, we shall be decent enough to explicitly state these restrictions.

PROBLEM SET 8.10

Plot each of the following polar coordinates *and* give the corresponding Cartesian coordinates.

1. $[1,\pi]$ **2.** $[5,\pi/6]$ **3.** $[4,2\pi/3]$

4. $[-3,3\pi/2]$ **5.** $[6,5\pi/6]$ **6.** $[1,90°]$

7. $[3\sqrt{2},45°]$ **8.** $[-3,0°]$ **9.** $[-2,3\pi/4]$

Before you proceed to problems 10–15, take a look at the following examples.

Example. Given $(-3,-3\sqrt{3})$, find $[r,\theta]$.

Solution. Since $x = r \cos \theta$, $y = r \sin \theta$, and $r = \pm\sqrt{(x^2 + y^2)}$, so if we assume that $r > 0$, then $r = \sqrt{(9 + 27)} = 6$, hence $\cos \theta = -1/2$. Now we must stop and realize that the given point is in the third quadrant, for it then follows that $\theta = \frac{\pi}{3} + \pi = \frac{4\pi}{3}$ or 240°. Therefore

$$[r,\theta] = [6,4\pi/3].$$

Note: if we had assumed $r < 0$, then we could have answered $[r,\theta] = [-6,\pi/3]$, but we prefer the simple things in life.

Example. Given (3,7) find $[r,\theta]$.

Solution. Assuming $r > 0$, $r = \sqrt{(9 + 49)} = \sqrt{58}$ and $\cos\theta = \dfrac{3}{\sqrt{58}}$. Since this is not one of the familar values of θ, we shall resort to approximations using tables in the appendix. First we show that $\cos\theta$ 0.394 and then find (to the nearest degree) that θ 67°. Therefore

$$[r,\theta] = [7.616,67°].$$

Find polar coordinates for each of the following.

10. (2,2) **11.** $(-3,3)$ **12.** $(\sqrt{3}/2,-1/2)$

13. (3,4) **14.** $\left(\dfrac{3\sqrt{3}}{2},\dfrac{3}{2}\right)$ **15.** $(-4\sqrt{2},4\sqrt{2})$

Give a polar equation for each of the given Cartesian equations.

16. $x^2 + y^2 = 9$ **17.** $x^2 + y^2 - 6y = 0$

18. $x^2 + y^2 + 4y = 0$ **19.** $y = x^2$

20. $x^2 - y^2 = 4$ **21.** $2x - y = 2$

22. $(x^2 - y^2)^2 = x^2y$ [Bifolium]

23. $(x^2 + y^2)^2 = x^2 - y^2$ [Lemniscate of Bernoulli]

Give the Cartesian equation for each of the given polar equations.

24. $r = 5$ **25.** $r = 3\sin\theta$

26. $r = 2\cos\theta$ **27.** $r(1 - \sin\theta) = 1$

28. $r^2\cos(2\theta) = 9$. *Hint:* $\cos(2\theta) = \cos^2\theta - \sin^2\theta$

Section 11—Polar Graphs It is true that the graph of a polar equation can be drawn by finding the graph of its corresponding Cartesian equation. This is certainly the best procedure for the polar equation

$$2r\cos\theta + r\sin\theta = 1$$

for, if we substitute $x = r\cos\theta$ and $y = r\sin\theta$, we have the Cartesian equation

$$2x + y = 1.$$

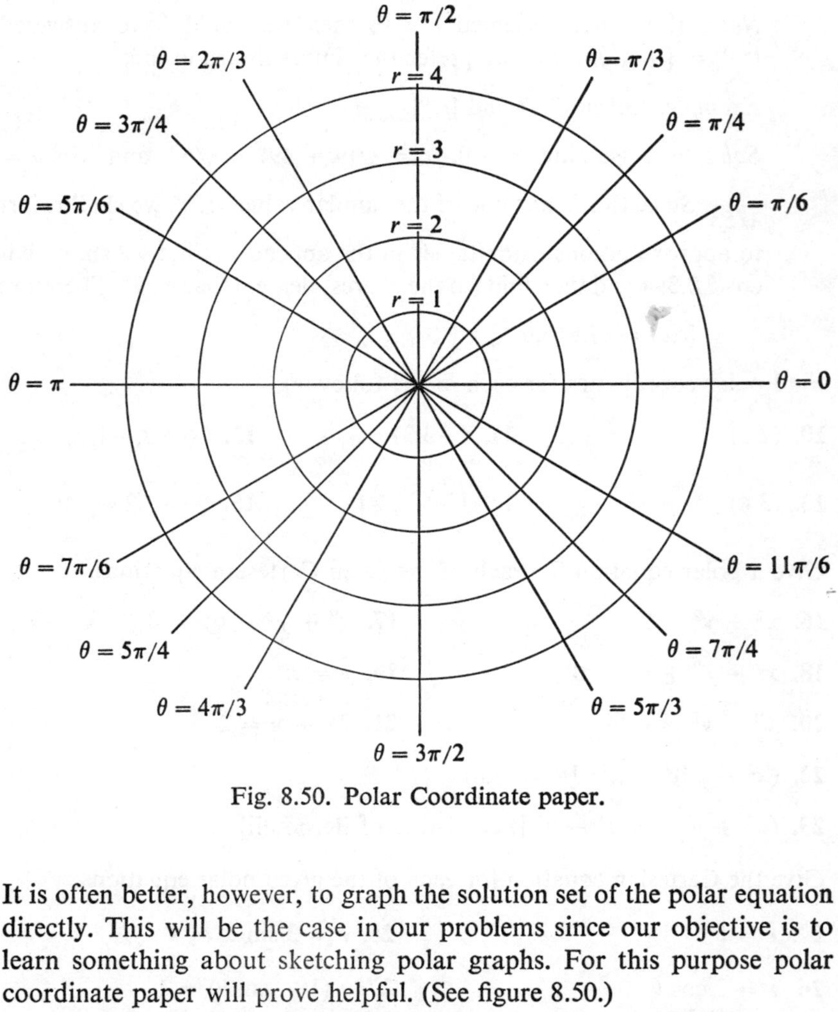

Fig. 8.50. Polar Coordinate paper.

It is often better, however, to graph the solution set of the polar equation directly. This will be the case in our problems since our objective is to learn something about sketching polar graphs. For this purpose polar coordinate paper will prove helpful. (See figure 8.50.)

One way of sketching the graph of a polar equation is simply to plot a reasonable number of points on the graph and then to make a good guess as to what it looks like.

Example 1. Sketch the graph of $r = 2 \cos \theta$.

Begin by making a table of ordered pairs $[r,\theta]$ that satisfies this equation. For convenience, use familiar values of θ and compute the associated values of r. (See table 8.14.) If we were to connect these points with a smooth curve, the graph appears to be a circle center at (1,0) and radius 1. We can easily verify this by obtaining the rectangular equation. Let us pause to do this. (See figure 8.51.) Simply plotting points is not always the most

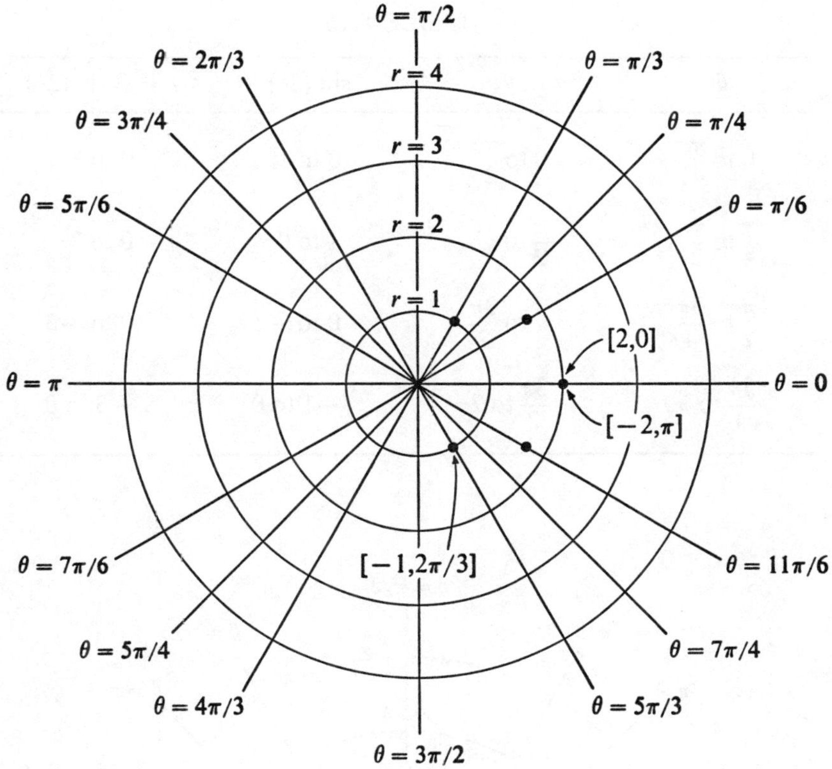

Fig. 8.51

TABLE 8.14

θ	0	$\frac{\pi}{6}$	$\frac{\pi}{3}$	$\frac{\pi}{2}$	$\frac{2\pi}{3}$	$\frac{5\pi}{6}$	π	$\frac{7\pi}{6}$	$\frac{4\pi}{3}$	$\frac{3\pi}{2}$
r	2	$\sqrt{3}$	1	0	-1	$-\sqrt{3}$	-2	$-\sqrt{3}$	-1	0

efficient way to sketch the polar graph. Sometimes it helps to examine how r changes as θ increases.

Example 2. Sketch the polar graph of $r = 3\cdot\sin(2\theta)$.

Before starting the graph, see table 8.15 showing how r varies with θ. By making the additional computation $\theta = \frac{\pi}{6}$, $r \approx 2.6$ we can make the

TABLE 8.15

θ	2θ	$\sin(2\theta)$	$r = 3\sin(2\theta)$
0 to $\frac{\pi}{4}$	0 to $\frac{\pi}{2}$	0 to 1	0 to 3
$\frac{\pi}{4}$ to $\frac{\pi}{2}$	$\frac{\pi}{2}$ to π	1 to 0	0 to 3
$\frac{\pi}{2}$ to $\frac{3\pi}{4}$	π to $\frac{3\pi}{2}$	0 to -1	0 to -3
$\frac{3\pi}{4}$ to π	$\frac{3\pi}{2}$ to 2π	-1 to 0	-3 to 0

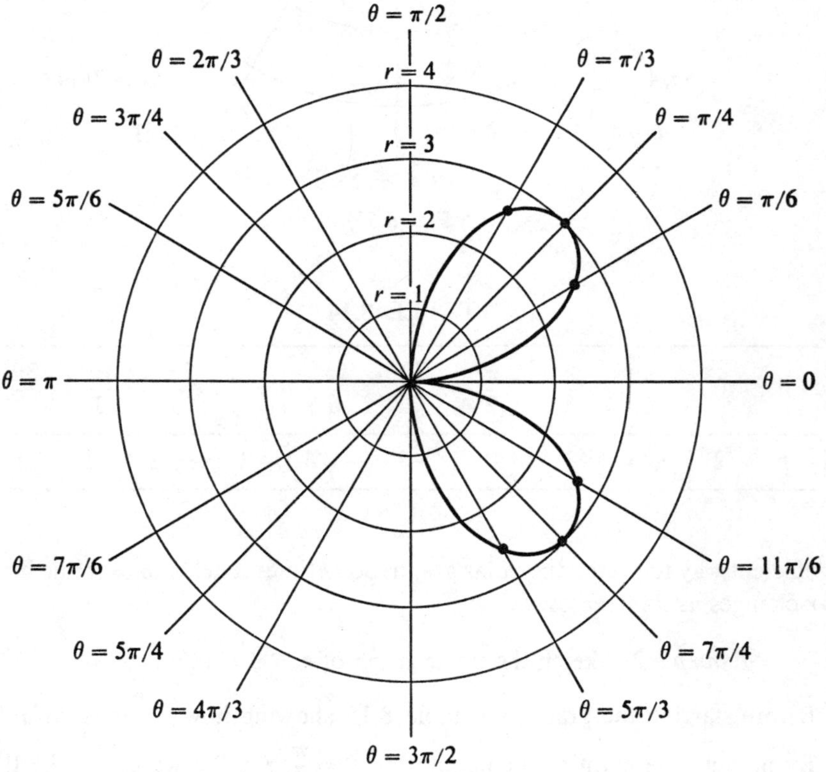

Fig. 8.52

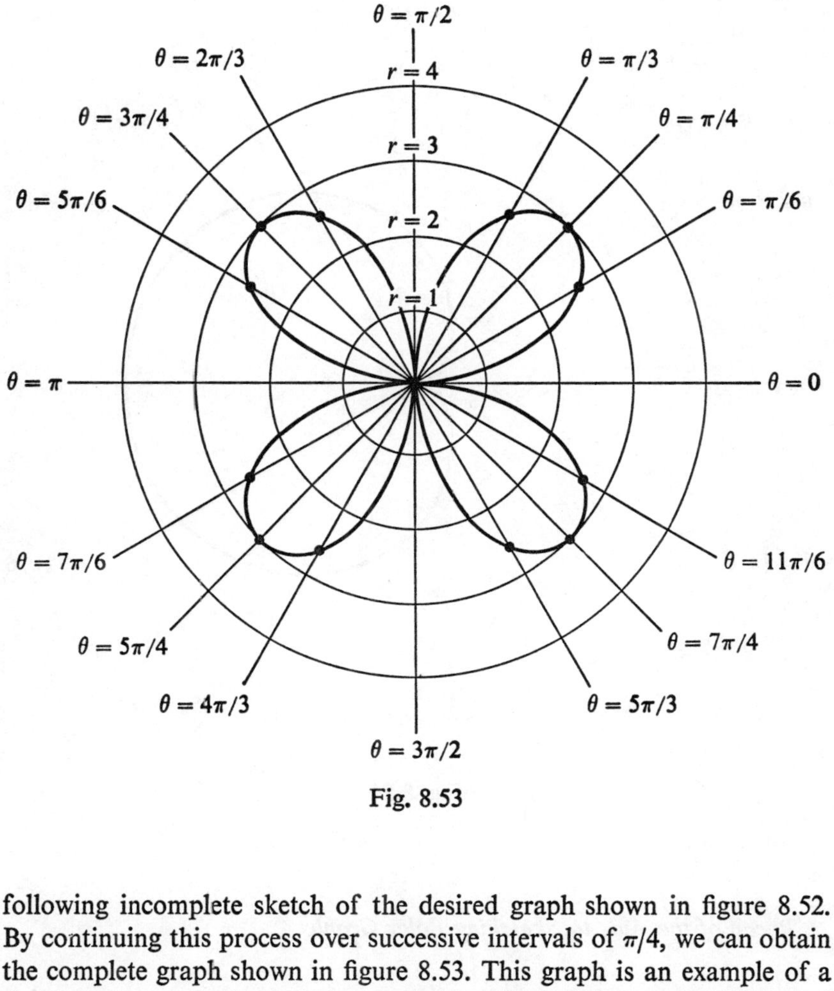

Fig. 8.53

following incomplete sketch of the desired graph shown in figure 8.52. By continuing this process over successive intervals of $\pi/4$, we can obtain the complete graph shown in figure 8.53. This graph is an example of a *four-leaved rose*. It also helps to know that if the replacement of θ by $-\theta$ does not alter the equation, then the graph is symmetric to the line containing the polar axis (the X axis if you please).

Example 3. Sketch the graph of $r = 2(1 + \cos\theta)$. (See figure 8.54.) Since $\cos(-\theta)$ is $\cos(\theta)$, the graph is symmetrical as described above. Let us also show how r varies with θ.

$$\begin{array}{ll} \theta & 0 \to \pi/6 \to \pi/3 \to \pi/2 \to 2\pi/3 \to 5\pi/6 \to \pi \\ r & 4 \to 3.7 \to 3 \to 2 \to 1 \to 0.3 \to 0 \end{array}$$

(Read $\to$ as "goes from to")

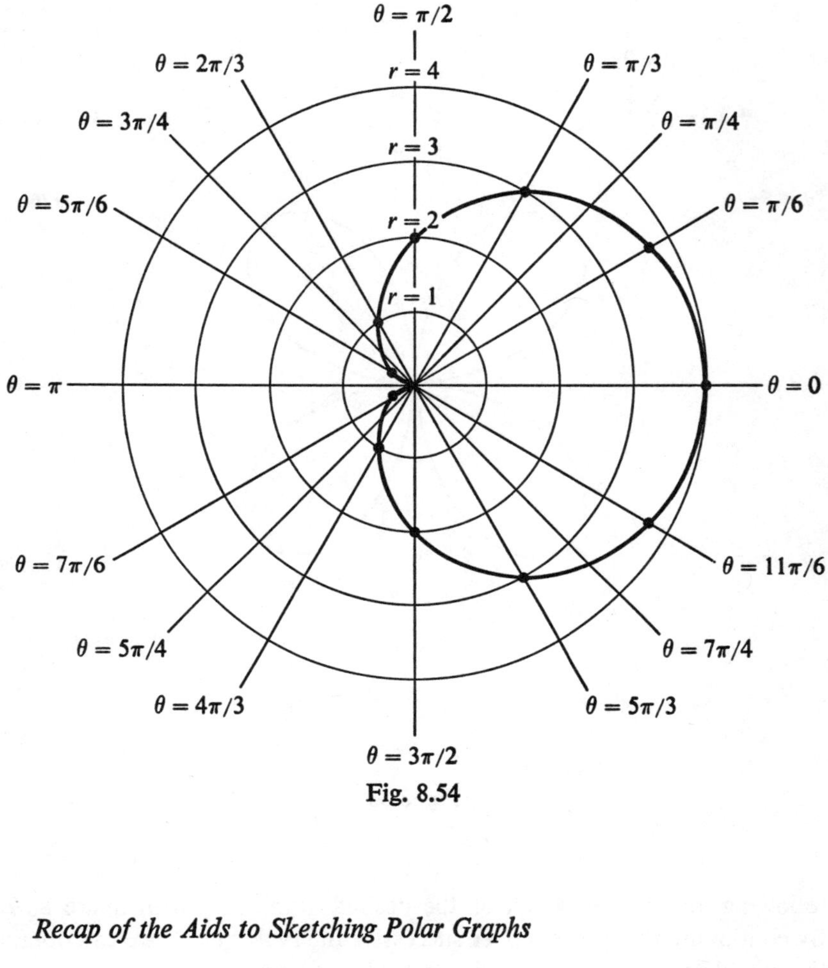

Fig. 8.54

Recap of the Aids to Sketching Polar Graphs

1. Examine how r varies with θ.
2. If the replacement of θ by $-\theta$ does not alter the equation, then the graph is symmetric to the X axis.
3. If the replacement of r by $-r$ does not alter the equation, then the graph is symmetric to the pole.

The equation $r^2 = 4\cos(2\theta)$ is an example of (3) above since $(-r)^2 = r^2$. Also note that $\cos(-2\theta) = \cos(2\theta)$ so the graph is also symmetric to the X axis. The graph is shown in figure 8.55.

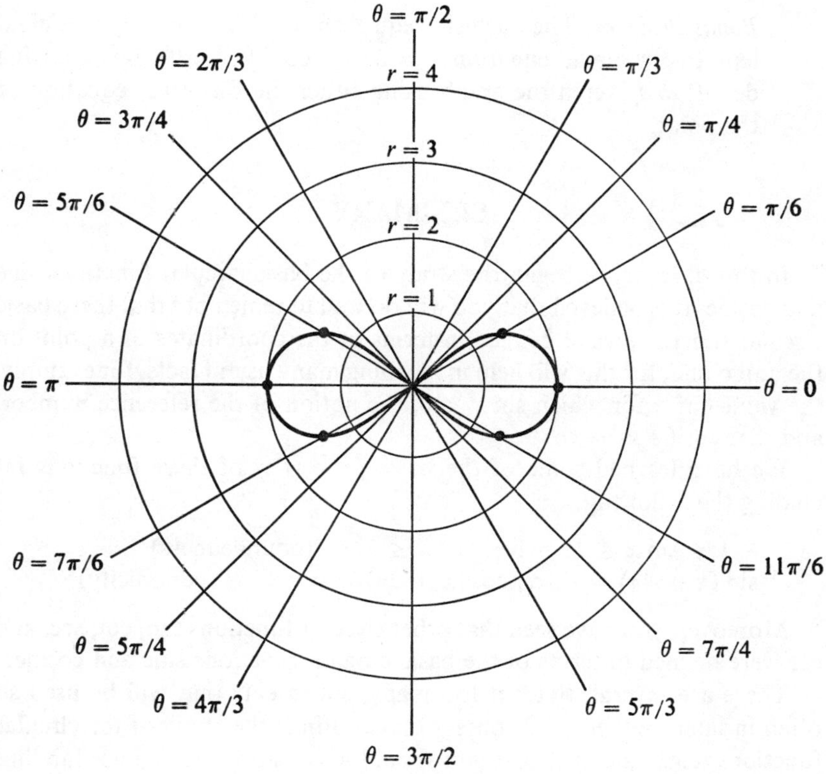

Fig. 8.55

PROBLEM SET 8.11

Sketch the graphs of the given polar equations.

1. $r = 3\cos\theta$

2. $r = 2\sin\theta$

3. $r = 2\sin(2\theta)$

4. $r = 3\cos(2\theta)$

5. $r = 2(1 + \sin\theta)$

6. $r = 2(1 - \cos\theta)$

7. $r = 1 - 2\sin\theta$

8. $r = 2 + \sin\theta$

9. $r = \sin(3\theta)$ [Three-leaved rose]

10. $r^2 = 9\cos(2\theta)$ [Lemniscate]

11. $r = 1 + 2\cos\theta$ [Limacon]

12. $r = \theta$ [Spiral]

Bonus Problem. The Cartesian equation $(x^2 + y^2)^2 = 2x^2y$ is equivalent to the polar equation $r = 2 \sin \theta \cos^2 \theta$. Verify this showing details *and* sketch the graph using either the Cartesian equation or the polar.

SUMMARY

In this chapter we began the study of the basic circular functions sine and cosine. It is believed that you will do well to remember that these basic circular functions were defined in terms of the coordinates of a point on the unit circle; for this will help in recalling many useful facts, for example, the values of x for which $\sin x < 0$, the notion of the reference number, and that $\cos(-x) = \cos x$.

We have learned some of the basic properties of these functions including the following:

$$-1 \leq \sin x \leq 1; \ -1 \leq \cos x \leq 1 \qquad \text{(boundedness)}$$
$$\sin(x + 2\pi) = \sin x; \ \cos(x + 2\pi) = \cos x \qquad \text{(periodicity)}$$

Moreover, you have seen that other circular functions tan, cot, sec, and csc were defined in terms of the basic circular functions sine and cosine.

There are several, but not too many, statements that will be used so often in later mathematics courses that continue the study of the circular functions (e.g., calculus) that you should make sure that you are familiar with them. Namely,

$$\sin^2 x + \cos^2 x = 1$$
$$1 + \tan^2 x = \sec^2 x$$
$$\sin(x + t) = \sin x \cos t + \cos x \sin t$$
$$\cos(x + t) = \cos x \cos t - \sin x \sin t$$
$$\tan x = \frac{\sin x}{\cos x}.$$

Review Problems

1. Determine *all* values of θ that satisfy:

a. $\sin \theta = 0$ **b.** $\cos \theta = 0$ **c.** $\cos \theta = 1$
d. $\sin \theta = 1/2$ **e.** $\sin \theta = 1$ **f.** $\tan \theta = 1$
g. $\cos \theta = 1/2$ **h.** $\cos \theta = \sqrt{3}/2$ **i.** $\sin \theta = \sqrt{2}/2$

2. Define tan, cot, sec, and csc functions in terms of the basic circular functions sine and cosine.

3. Give the periods of the functions cos and tan.

4. Sketch the graphs of the functions sin, cos, and tan over one period.

5. Give the set of principal solutions (S_p) for the following equation: $2 \sin^2 x + \sin x - 1 = 0$.

6. Find polar coordinates for each of the following points: **(a)** $(-1,0)$ **(b)** $(3,-2)$ **(c)** $(1,1)$.

7. Discuss the polar graph of $r = 1 + \cos \theta$. This discussion should include how r varies with θ and symmetry.

Ein Moment zur Ruhe aber nicht zum Schlafen.

It cannot be denied that our vigorous if not rigorous pursuit of the elementary functions has been both exhilarating and demanding. In the interest of avoiding a cardiac arrest, let us pause for a moment and consider an idea that requires nothing more complicated than the arithmetic of the nonnegative integers. I confess that there is another reason for this respite—we will soon need the factorial notation.

***n*-Factorial** If n is a counting number, then $n!$, read as *n factorial*, is the counting number $n \cdot (n - 1)!$ and, for convenience, $0!$ is 1. That is,

$$n! = n \cdot (n - 1)!$$

and

$$0! = 1.$$

Example. To evaluate 4!, we begin as follows:

$$4! = 4 \cdot 3!$$

so we can find 4! if we can find 3!

$$3! = 3.2!$$

and now we need to know 2!

$$2! = 2 \cdot 1!.$$

Relentlessly we go on with

$$1! = 1 \cdot 0!$$

and finally, this is part of the convenience mentioned, we come to rest with

$$0! = 1.$$

Retracing our efforts, starting with the above, we have

$$1! = 1$$
$$2! = 1\cdot 2$$
$$3! = 1\cdot 2\cdot 3$$
$$4! = 1\cdot 2\cdot 3\cdot 4 = 24.$$

From this example we can learn that

$$5! = 1\cdot 2\cdot 3\cdot 4\cdot 5 = 120$$

or, more generally, that

$$n! = 1\cdot 2\cdot 3\cdot \cdots \cdot (n-1)\cdot n.$$

Example. To simplify $\frac{7!}{5!}$ we write

$$\frac{7!}{5!} = \frac{7\cdot 6!}{5!} = \frac{7\cdot 6\cdot 5!}{5!}$$

and, after "cancelling," get

$$\frac{7!}{5!} = 42.$$

Example. To simplify $6! - 5!$ we note that

$$\begin{aligned} 6! - 5! &= 6\cdot 5! - 5! \text{ and factoring} \\ &= 5!\cdot(6-1) \\ &= 5!\cdot 5 \text{ or } 600. \end{aligned}$$

Example (of a general nature)

$$\begin{aligned} (n+1)! + n! &= n!\cdot(n+1) + n! \\ &= n!\cdot[(n+1)+1] \end{aligned}$$
$$n!\cdot(n+2).$$

Moral. Sometimes it is practical to use

$$n! = (n-1)!\cdot n$$

while on other occasions we may wish to use

$$n! = 1\cdot 2\cdot 3\cdot \cdots (n-1)\cdot n.$$

PROBLEM SET $n!$

Evaluate each of the following:

1. $6!$ **2.** $\frac{9!}{8!}$ **3.** $\frac{5!\cdot 7!}{8!}$

4. $\frac{12!}{10!\cdot 2!}$ **5.** $\frac{8!}{5!\cdot 3!}$ **6.** $\frac{4! + 3!}{4!}$

7. $\dfrac{5! + 3!}{3!}$ **8.** $\dfrac{6!}{7! - 6!}$ **9.** $\dfrac{4! + 3!}{4! - 3!}$

10. $1 + \dfrac{1}{1!} + \dfrac{1}{2!} + \dfrac{1}{3!}$

Simplify each of the following:

11. $\dfrac{(n + 1)!}{n + 1}$ **12.** $\dfrac{(n + 1)!}{(n + 2)!}$ **13.** $\dfrac{n}{n!}$

14. $\dfrac{(n - 1)!}{n!}$ **15.** $(n + 1)! - n!$ **16.** $\dfrac{1}{n!} - \dfrac{1}{(n + 1)!}$

17. $\dfrac{n!}{(n - 3)! \cdot 3!}$ Assume $n > 3$.

CHAPTER 9
Sequence Function

Section 1—Sequences, Basic Ideas The primary basic idea of this chapter is to have a clear notion as to what it is we are going to talk (yea, even argue) about. Verily, it seems obvious that we should start with the following.

> DEFINITION. *The statement that "f is a sequence" means that f is a function whose domain is the set of counting numbers.*

One should easily see that every sequence is a function but not every function is a sequence.

If f is a sequence, then the number $f(1)$ is called the *first term of the sequence*; $f(2)$ is called *the second term of the sequence*; and so forth, so that, if n is a counting number, then the number $f(n)$ is called the *n-th term of the sequence*. In some books, $f(n)$ may be called the general term of the sequence. While we realize that a sequence is a set of ordered number pairs (points), it is nevertheless traditional to indicate a sequence by just giving its terms. Hence, you can expect to see a sequence indicated as a number set. For example

$$\{f(1), f(2), f(3), \ldots, f(n), \ldots\}.$$

Let us now consider some rather simple tasks, namely, giving the terms of a sequence. We shall do this in two equally simple ways.

Example 1. Suppose that f is the sequence for which $f(n) = 2n - 3$. We easily find that $f(1) = -1$, $f(2) = 1$, $f(3) = 3$, and even that

$f(7) = 11$. Furthermore, if k is a counting number larger than 1, then $f(k - 1) = 2k - 5$.

Example 2. If f is the sequence for which $f(n) = 1 - 1/n$, then $f(1) = 0, f(2) = 1/2, f(3) = 2/3$, and $f(4) = 3/4$. It is not unusual to see this sequence indicated by the following:

$$f = \left\{0, 1/2, 2/3, 3/4, \ldots, \frac{n-1}{n}, \ldots\right\}.$$

For the next example, I choose to show you another commonly used method of defining sequences referred to as the *recursive method.* For reasons that are obscure to me but may be clear to you, some students call this the *cursed method.*

Example 3. Suppose f is the sequence for which $f(1) = 2$ and, for $n > 1, f(n) = 2 - f(n - 1)$. Clearly, the first term of this sequence is 2. Try to see just as clearly that

since $2 > 1, f(2) = 2 - f(1)$, hence $f(2) = 0$
since $3 > 1, f(3) = 2 - f(2)$, hence $f(3) = 2$
since $4 > 1, f(4) = 2 - f(3)$, hence $f(4) = 0$, and so forth.

Example 4. Suppose that f is the sequence such that $f(1) = 1$, and, for $n > 1$, $f(n) = n - f(n - 1)$. It follows that $f(2) = 2 - f(1)$, hence $f(2) = 1$; $f(3) = 3 - f(2)$, hence $f(3) = 2$; $f(4) = 4 - f(3)$, hence $f(4) = 2$. Compare the task of finding $f(7)$ with the same task in example 1.

While additional examples might be helpful, we ought not conclude this section on the basic ideas of sequences until we have some notion of their graphs of such functions and a discussion of their properties as described in chapter 4. We shall do this at least for the four sequences given in the previous examples.

1. Graphs Using the computations made in these examples, we see that the graphs of these four functions resemble figures 9.1 through 9.4. Perhaps the easiest mistake to make when sketching the graph of a sequence is to forget momentarily its domain and to join the points plotted by some line or curve. For example, looking at figure 9.1, there might be a great temptation to draw a line through the points shown. But this would be a mistake.

Question. Why would this be a mistake?

Answer. A line drawn through these points would surely include a point whose abscissa is 5/2 and 5/2 is not a counting number, that is, $5/2 \notin \text{dom } F$.

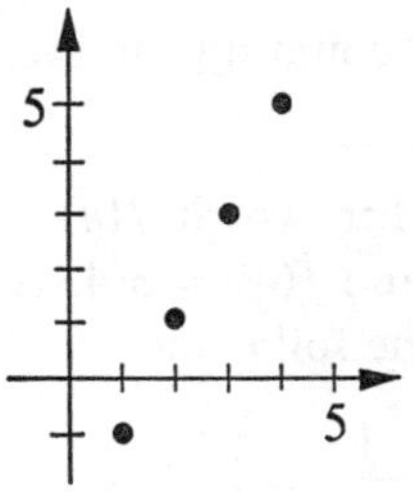

Fig. 9.1. $F(n) = 2n - 3$.

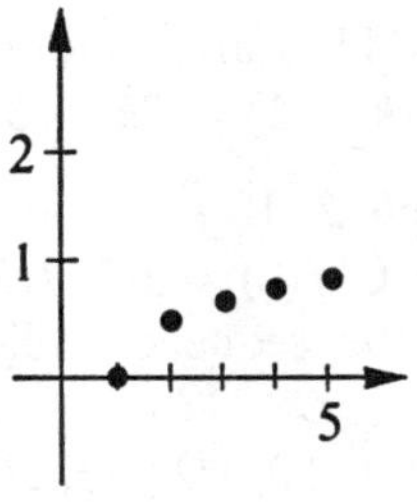

Fig. 9.2. $f(n) = \dfrac{n-1}{n}$.

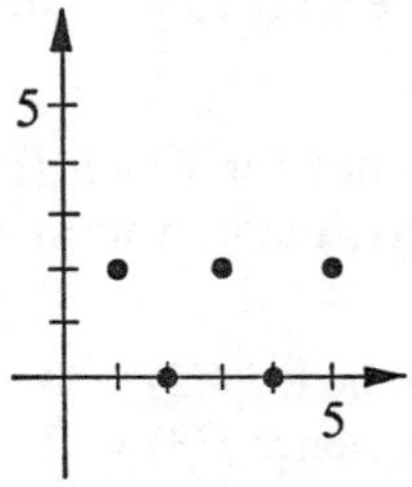

Fig. 9.3. $f(1) = 2$, and $f(n) = 2 - f(n - 1)$, $n > 1$.

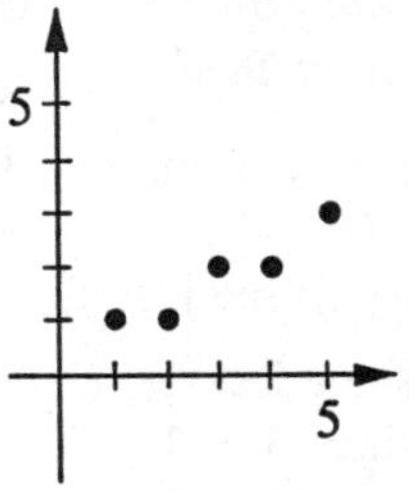

Fig. 9.4. $f(1) = 1$, and $f(n) = n - f(n - 1)$, $n > 1$.

2. Some Other Properties If we again look at the graphs shown above, we can easily conjecture that the first and second sequences are increasing over their domains and are therefore 1-to-1 functions. Not so for the other two, however. Moreover, it is not too difficult to see that each of the functions shown in these four figures is bounded below, but those shown in figures 9.1 and 9.4 are not bounded above. The sequence represented by figure 9.2 is interesting. Clearly it is an increasing function. But just as clearly we recall that the problem of determining the boundedness of a function is the problem of finding the bounds of a number set called the range of the function. Since this is not a problem unknown to us, it is with a good deal of joy, maybe even pleasure, that we note that $\dfrac{n-1}{n} = 1 - 1/n < 1$ for every counting number n. Hence, for figure 9.2, f is bounded above. To recap,

figure 9.1: GLB of $F = -1$, no upper bound;
figure 9.2: LUB of $f = 1$, GLB of $f = 0$;
figure 9.3: LUB of $f = 2$, GLB of $f = 0$;
figure 9.4: GLB of $f = 1$, no upper bound.

3. Continuity It is quite satisfying to observe that, at each point in the graph of a sequence, we can go through the horizontal and vertical line routine (see figure 9.5) and thereby establish that such functions are continuous at each point, that is, are continuous over their domain.

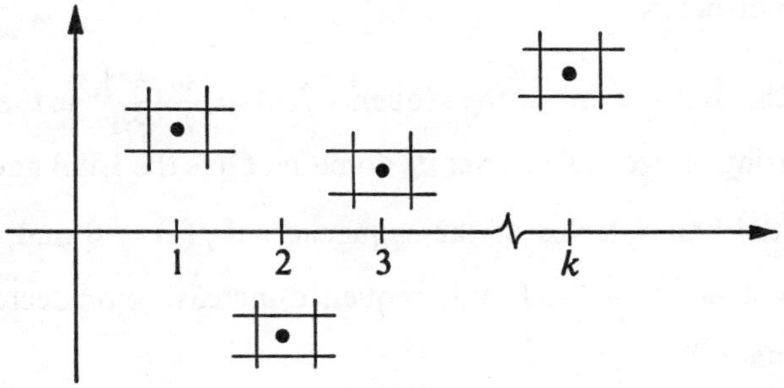

Fig. 9.5

PROBLEM SET 9.1

1. Give the first 5 and the $(n-1)$st terms of each of the sequences:

a. $F(n) = 5 - 3n$ **b.** $f(n) = \dfrac{n}{n+1}$ **c.** $g(n) = n! - n.$

2. Same as problem 1 for the following sequences:

a. $F(n) = 2n + 1$ **b.** $f(n) = n^2$ **c.** $G(n) = \dfrac{2^n}{n!}.$

3. Give the first 5 terms of the sequence F if $F(1) = a$ and, for $n > 1$, $F(n) = F(n-1) + d$. This sequence is known as the *arithmetic sequence.*

4. Give the first 4 terms of the sequence F if $F(1) = 1$ and, for $n > 1$,
$$F(n) = \frac{F(n-1)}{2} + \frac{1}{F(n-1)}.$$

5. If $F(n) = 7 - 2n$, find the first 5 terms of F, sketch its graph, and discuss its boundedness being sure to give, if they exist, the LUB or GLB of F.

6. If $F(n) = \dfrac{3n+1}{2n-1}$, find the first 5 terms of F, sketch its graph, and discuss its boundedness being sure to give, if they exist, the LUB or GLB of F.

7. If $f(n) = -|n - 3|$, sketch the graph of f, give the subset of its domain over which it is increasing, and determine whether it is bounded above or below.

8. If $f(n) = n^2 - 4n$, sketch the graph of f, give the subset of its domain over which it is decreasing, and determine whether it is bounded above or below.

9. Give the first 4 terms of the sequence $f(n) = \frac{2n + 1}{n + 1}$. Is the sequence f increasing or decreasing over its domain? Give the LUB and GLB of f.

10. Give the first 4 terms of the sequence f if $f(1) = 0$ and, for $n > 1$, $f(n) = 1 + \frac{f(n - 1)}{2}$. Is this sequence increasing or decreasing over its domain?

11. In this problem, you are to try to guess the n-th term of the sequence whose first 5 terms are:

a. 1, 3, 5, 7, and 9 **b.** 1, 4, 7, 10, and 13
c. 1, 2, 6, 24, and 120

12. Examine the idea that each term of the sequence F given in problem 4 is a better approximation of $\sqrt{2}$ than the preceding term by showing that $|2 - [F(2)]^2| > |2 - [F(3)]^2|$.

13. Suppose that f is the sequence such that $f(1) = 2$ and, for $n > 1$, $f(n) = f(n - 1) - \frac{1}{n(n - 1)}$. Give the first 5 terms of this sequence *and then*, on the basis of this evidence, conjecture a nonrecursive rule for $f(n)$.

14. If $f(n) = \frac{2n - 1}{n + 1}$, find the *smallest* counting number m such that $|F(m) - 2| < 0.01$.

15. If $F(n) = \frac{3n + 1}{2n - 1}$, find the *smallest* counting number m such that $\left|F(m) - \frac{3}{2}\right| < 0.01$.

16. If each of a and x is a number distinct from 0, and G is a sequence such that $G(1) = a$ and, for $n > 1$, $G(n) = x \cdot G(n - 1)$, give the first 4 terms of G and then, on the basis of this evidence, conjecture a nonrecursive rule for $G(n)$.

17. *Small bonus problem*: Suppose that f is a sequence such that, for $n > 1, f(n) = n^2 + f(n - 1)$. If $f(4) = 23$, find $f(1)$.

Section 2—Sequence of Partial Sums Let us begin this day by looking back into the recent past and illustrate the two common methods of defining sequences and at some related problems.

Example 1 (the nonrecursive method). Let f be the sequence defined by $f(n) = \frac{2n + 1}{n}$. Now $f(1) = 3, f(2) = 5/2$, and so forth, and we have been cautioned that we might see this sequence shown in the following manner:

$$f = \left\{3, 5/2, 7/3, 9/4, \ldots, \frac{2n + 1}{n}, \ldots\right\}.$$

Even without the graph of this sequence, we can see that it is a decreasing and bounded function. Its least upper bound is 3,3 $\in$ rng f, while its greatest lower bound is 2, 2 $\notin$ rng f. Now consider the task of finding the smallest $m \in N$ such that $|f(m) - 2| < 0.001$. Since $|f(m) - 2| = \left|\frac{2m + 1}{m} - 2\right|$, we simplify this and write $\left|\frac{1}{m}\right| < 0.001$ or, since $m > 0$, as $1/m < 0.001$. Solving, we find $m > 1000$ so the smallest value that can be assigned m is 1001.

Question. So what?

Answer. The significance of this problem can be stated as follows: each term after the 1000th will differ from the GLB of f by less than 0.001.

Example 2 (the recursive method). Suppose that g is the sequence for which $g(1) = 1$ and, for $n > 1$, $g(n) = n \cdot g(n - 1)$. We see that $g(2) = 2$, $g(3) = 6$, and that $g(4) = 24$. Thus

$$g = \{1, 2, 6, 24, 120, \ldots\}$$

and in a few lucky cases, like this example, we can even see how to describe the sequence with a nonrecursive rule. For here $g(n) = n!$.

To further the cause, let's look at another example.

Example 3. Suppose that F is the sequence such that $F(n) = (-1)^n \cdot (n + 1)$ and that G is the sequence such that $G(n) = (-1)^{n+1} \cdot (n + 1)$. By making some rather simple computations we can see that

$$F = \{-2, 3, -4, 5, -6, \ldots\} \quad \text{and that}$$
$$G = \{2, -3, 4, -5, 6, \ldots\}.$$

It is surmised that, because the terms of these sequences alternate in "sign," they are called *alternating sequences.*

There are other things about sequences that we could discuss but it is best that we direct our attention to a major and important idea. Namely,

with each sequence F there is associated another sequence S, called *the sequence of partial sums,* defined as follows:

$S(1) = F(1)$ *and, for* $n > 1$, $S(n) = S(n - 1) + F(n)$.

Examine this definition and note that

$$S(1) = F(1)$$
$$S(2) = S(1) + F(2) = F(1) + F(2)$$
$$S(3) = S(2) + F(3) = F(1) + F(2) + F(3)$$

and so on, for then you can see why it is called what it is—the sequence of partial sums.

Let us rapidly return to the three examples given earlier for the purpose of looking at the sequence S that is now known to be associated with the given sequence.

Intending to leave example 1 for the last because of the onerous arithmetic, let us begin with the sequences given in example 2.

In example 2 we had the sequence $g = \{1, 2, 6, 24, \ldots\}$, so we can use the definition and construct the associated sequence of partial sums S as follows:

$$S(1) = g(1) = 1$$
$$S(2) = S(1) + g(2) = 1 + 2 = 3$$
$$S(3) = S(2) + g(3) = 3 + 6 = 9, \text{ and}$$
$$S(4) = S(3) + g(4) = 9 + 24 = 33$$

so we could display the sequence S in the following manner

$$S = \{1, 3, 9, 33, \ldots\}.$$

In example 3 we had the sequence $F = \{-2, 3, -4, 5, -6, \ldots\}$ so that the sequence S is indicated by $S = \{-2, 1, -3, 2, -4, \ldots\}$. Also had $G = \{2, -3, 4, -5, 6, \ldots\}$, hence, for this sequence,

$$S = \{2, -1, 3, -2, 4, \ldots\}.$$

Finally, for the sequence given in example 1,

$$f = 3, 5/2, 7/3, \ldots\}.$$

We can see that

$$S(1) = f(1) = 3$$
$$S(2) = S(1) + f(2) = 3 + 5/2 = 11/2$$

and

$$S(3) = S(2) + f(3) = 11/2 + 7/3 = 47/6.$$

Example 4. If $f(1) = 1, f(2) = 3$ and, for $n > 2, f(n) = f(n-1) - f(n-2)$, then $f(3) = f(2) - f(1) = 2$, $f(4) = f(3) - f(2) = -1$, $f(5) = -1 - 2 = -3$, and so forth. Thus

$$f = \{1, 3, 2, -1, -3, \ldots\}$$

and therefore

$$S = \{1, 4, 6, 5, 2, \ldots\}.$$

PROBLEM SET 9.2

1. Suppose that f is the sequence such that $f(n) = 2n - 3$, give the first 4 terms of f and of S.

2. Suppose that g is the sequence such that $g(n) = n^2 - 1$, give the first 4 terms of g and of S.

3. If $F(n) = (-1)^n \cdot n^2$, give the first 4 terms of F and of S.

4. If $f(n) = (-1)^{n+1} \cdot 2^n$, give the first 4 terms of f and of S.

5. If $F(1) = -1/2$ and, for $n > 1$, $F(n) = F(n-1) + \frac{3}{2^n}$, give the first 4 terms of F and of S.

6. If $F(1) = 2$, $F(2) = 3$ and, for $n > 2$, $F(n) = 2F(n-1) - F(n-2)$, give the first 5 terms of F and of S.

7. Give the first 5 terms of f and of S if $f(n) = \frac{1}{2}[1 + (-1)^{n-1}]$.

8. Suppose F is the arithmetic sequence (it is quite legal to look back at problem 3 in problem set 9.1), it can be shown that $F(n) = a + (n-1)d$. Give the first 4 terms of F and of S. Now then, try to conjecture a nonrecursive rule for $S(n)$. *Hint:* Compare the terms of the sequence $C(n) = n(n-1)$ and the coefficients of d in each term of S.

9. If $f(n) = n$, give the first 5 terms of f and of S. Now try to conjecture a nonrecursive rule for $S(n)$.

10. If $f(1) = 1$ and, for $n > 1$, $f(n) = (-1)^{n-1} \cdot \frac{(n-1)!}{f(n-1)}$, give the first 5 terms of f and of S.

11. Suppose that F is the sequence such that $F(1) = F(2) = 1$ and, for $n > 2$, $F(n) = F(n-1) + F(n-2)$. Give the first 7 terms of F and of S. (The sequence F is known as the *Fibonacci sequence* and its terms are called *Fibonacci numbers*. These numbers have some curious properties and have pleased some mathematicians).

12. Suppose that F is the Fibonacci sequence given in problem 11. Show that each of the following is true:

a. $F(4) = 2F(2) + F(1)$ **b.** $F(5) = 2F(3) + F(2)$
c. $F(6) = 2F(4) + F(3)$

13. If F is the Fibonacci sequence given in problem 11, then $F(k+2) = F(k+1) + F(k)$ for each $k \in N$, hence so does $F(k+3) = F(k+2) + F(k+1)$. Combine these two formulas to find an expression for $F(k+3)$ in terms of $F(k+1)$ and $F(k)$.

14. *Bonus problem.* The equation $F(n) = F(n-1) + F(n-2)$, which helps to define the terms of F, may be written as

$$F(k+2) - F(k+1) = F(k)$$

where $k \in N$. Starting with

$$S(n) = F(1) + F(2) + F(3) + \ldots + F(n-1) + F(n)$$

use the underlined equation using $k = 1, 2, 3, \ldots, (n-1), n$ and rewrite the above formula for $S(n)$. Take a hard look at this new formula for $S(n)$ and come up with a very brief formula for $S(n)$.

15. *Bonus problem.* If $g(1) = 1$ and, for $n > 1$, $g(n) = \dfrac{g(n-1)}{2 + g(n-1)}$, give the first few terms of g and on the basis of this evidence, conjecture a nonrecursive rule for $g(n)$. *Hint:* Evaluate 2^n for a few values of n.

Section 3—Sequence S The problem with which we now concern ourselves can be basically stated as follows: given the sequence S of partial sums to find the associated sequence f.

The solution of this problem begins with recalling that the sequence S was defined as follows: $S(1) = f(1)$ and, for $n > 1$, $S(n) = S(n-1) + f(n)$. For, from this definition we can readily see that for $n > 1$, $f(n) = S(n) - S(n-1)$ so that if we are given $S(n)$ we can produce a rule or formula that will describe the sequence f except for possibly the first term.

Example. Suppose S is the sequence for which $S(n) = \dfrac{2n}{n+1}$. It follows that $S(n-1) = \dfrac{2n-2}{n}$ thus, for $n > 1$, $f(n) = \dfrac{2n}{n+1} - \dfrac{2n-2}{n}$. Simplifying this we find that $f(n) = \dfrac{2}{n(n+1)}$ for $n > 1$.

Armed with the knowledge that this rule either gives $f(1)$ or it does not, we can easily verify which is the case for we know that $f(1) = S(1)$. In this example, the rule not only gives $f(n)$ for $n > 1$, but it also gives $f(1)$ and so we can describe the sequence f by the statement

$$\forall n, f(n) = \frac{2}{n(n+1)}.$$

Example. If $S(n) = n^2(n + 1)$ then, for $n > 1, f(n) = n^2(n + 1) - (n - 1)^2 n$. Thus, simplifying this, we find that (at least for $n > 1$) $f(n) = n(3n - 1)$. Again you are cautioned that this rule may not yield the first term of f, but, as in the preceding example, we are cheered to see that it does, since $S(1) = 2$. Hence f is the sequence such that

$$\forall n, f(n) = n(3n - 1).$$

This next example illustrates the worst that can happen in the solution of the basic problem of this section.

Example. Suppose $S(n) = 3^n$ then, since $S(n - 1) = 3^{n-1}$, we know that, for $n > 1, f(n) = 3^n - 3^{n-1}$. Thus, for $n > 1, f(n) = 2 \cdot 3^{n-1}$. We now examine the idea that this rule either gives $f(1)$ or it does not. If it does, then (by the rule) $f(1) = 2$ but $S(1) = 3$, and so the sequence f must be described by

$$f(1) = 3 \text{ and, for } n > 1, f(n) = 2 \cdot 3^{n-1}.$$

Now let us return to the definition of the sequence S and from it obtain some other ideas. We know that $S(1) = f(1)$ and that $S(2) = S(1) + f(2)$, so we know that $S(2) = f(1) + f(2)$. Further, $S(3) = S(2) + f(3)$, hence $S(3) = f(1) + f(2) + f(3)$. Armed with this evidence, it should seem reasonable that

$$S(n) = f(1) + f(2) + f(3) + \ldots + f(n).$$

This last statement can be compactly written as

$$S(n) = \sum_{k=1}^{n} f(k).$$

The notation on the right-hand side is known as the sigma notation and it can be read as "the sum of $f(k)$ from $k = 1$ to $k = n$," and when we want to know what it means, we do not argue about its meaning but look at the statement

$$S(n) = f(1) + f(2) + f(3) + \ldots + f(n).$$

Now, indicating $S(n)$ by this compact notation is not possible unless we know $f(k)$, but if we know $f(k)$, then it is easy. Thus for our first two examples respectively we can write

$$S(n) = \sum_{k=1}^{n} \frac{2}{k(k+1)} \quad \text{and} \quad S(n) = \sum_{k=1}^{n} k(3k-1).$$

For the third example, we can write

$$S(n) = 3 + \sum_{k=2}^{n} 2 \cdot 3^{k-1}.$$

PROBLEM SET 9.3

[1] $S(n) = xn$

[2] $S(n) = \dfrac{n(n+1)}{2}$

3. $S(n) = \dfrac{n+1}{n}$

4. $S(n) = \dfrac{2n-1}{n}$

5. $S(n) = n^2(n-1)$

6. $S(n) = n^2$

7. $S(n) = \dfrac{n}{2n+1}$

8. $S(n) = \dfrac{n}{n+1}$

9. $S(n) = \dfrac{n(3n+1)}{2}$

10. $S(n) = \dfrac{n(2n+1)}{2}$

11. $S(n) = (n+1)! - 1$

12. $S(n) = 1 - 1/n!$

13. $S(n) = 2^n$

14. $S(n) = 3^{n+1}$

15. $S(n) = \dfrac{n(4n^2-1)}{3}$

16. $S(n) = \dfrac{n(n+1)(n+2)}{6}$

17. $S(n) = \dfrac{a(1-x^n)}{1-x}$

18. $S(n) = \dfrac{a(1-r^n)}{1-r}$

Try to conjecture the n-th of a sequence F whose first four terms are:

19. 1, 2/3, 3/5, and 4/7

20. $1 \cdot 3$, $2 \cdot 4$, $3 \cdot 5$, and $4 \cdot 6$

21. $x, -x^2, x^3$, and $-x^4$

22. 1, -3, 5, and -7

Hint: $(-1)^{n+1}$ will take care of the alternating signs.

[23] Suppose that F is the sequence such that $F(1) = 4$, and for $n > 1$, $F(n) = \dfrac{-F(n-1)}{2}$. Give the first four terms of F *and* the first four terms of S.

Bonus Problems

24. If $F(n) = \dfrac{3n+1}{2n-1}$, find the smallest counting number m such that $|F(m - 3/2| < 0.01$.

25. If $S(n) = L(n!)$, give $F(n)$. *Hint:* Theorem L-2 helps.

Section 4—Sigma Notation Let us begin by reviewing the ideas of the previous section with the following.

If $S(n) = \dfrac{1 - x^n}{1 - x}$, $x \neq 1$, then for $n > 1$ we have

$$f(n) = \frac{1 - x^n}{1 - x} - \frac{1 - x^{n-1}}{1 - x} = \frac{x^{n-1}(1 - x)}{1 - x} = x^{n-1}.$$

Now, if we assume (for all that we have shown is that $x^0 = 1$ when $x > 0$) that $x^0 = 1$, we can describe the sequence f with the statement: for all n, $f(n) = x^{n-1}$, which we may write as $\forall n$, $f(n) = x^{n-1}$. Thus $S(n) = \sum_{k=1}^{n} x^{k-1}$.

In order to help develop our understanding of the sigma notation, we continue with some easy examples.

Example. To evaluate $\sum_{k=1}^{4} 2^{k-1}$ we must first realize that what is required is the sum $f(1) + f(2) + f(3) + f(4)$ for some sequence f. Question, what sequence f? Ah so, for the sequence f for which $f(n) = 2^{n-1}$. Thus $\sum_{k=1}^{4} 2^{k-1} = 1 + 2 + 4 + 8 = 15$.

Example. $\sum_{k=1}^{5} (k^2 - 2k) = -1 + 0 + 3 + 8 + 15 = 25$. In this example, $f(n) = n^2 - 2n$.

Example.

$$\sum_{k=1}^{3} 1/k! = \frac{1}{1!} + \frac{1}{2!} + \frac{1}{3!} = \frac{5}{3}.$$

Example.

$$\sum_{k=1}^{4} (k + 1)! - k!$$
$$= (2! - 1!) + (3! - 2!) + (4! - 3!) + (5! - 4!)$$
$$= 5! - 1!.$$

Example

$$\sum_{k=1}^{3} [2^{k+1} - 2^k] = (2^2 - 2) + (2^3 - 2^2) + (2^4 - 2^3)$$
$$= 2^4 - 2.$$

In this last pair of examples, I have deliberately left the results as indicated rather than the natural simplifications to 119 and 14 respectively.

We continue our study of the sigma notation with some examples that are more general in nature and from which we shall conjecture certain properties of sigma.

Example. Supposing that f is a sequence and that c is a number distinct from 0, then

$$\sum_{k=1}^{n} c \cdot f(k) = c \cdot f(1) + c \cdot f(2) + c \cdot f(3) + \cdots + c \cdot f(n)$$
$$= c \cdot [f(1) + f(2) + f(3) + \cdots + f(n)]$$

and we will allow this example to suggest the following property of sigma.

Property 1. $\sum_{k=1}^{n} c \cdot f(k) = c \cdot \sum_{k=1}^{n} f(k).$

Example. If each of f and g is a sequence, we are sure that $[f + g]$ is a sequence and so

$$\sum_{k=1}^{n} [f + g] = [f + g](1) + [f + g](2) + \cdots + [f + g](n)$$
$$= f(1) + g(1) + f(2) + g(2) + \cdots$$
$$+ f(n) + g(n)$$
$$= [f(1) + f(2) + \cdots + f(n)] + [g(1) + g(2)$$
$$+ \cdots + g(n)]$$

and we allow this example to suggest the following property.

Property 2. $\sum_{k=1}^{n} [f + g](k) = \sum_{k=1}^{n} f(k) + \sum_{k=1}^{n} g(k).$

This next example is a little more subtle, for it deals with sums of a special form known as *telescoping sums*. We hope you will see one reason why this name is well chosen.

Example. Suppose that f is a sequence, then by using property 1 with $c = -1$, and property 2 we have that

$$\sum_{k=1}^{n} [f(k + 1) - f(k)] = \sum_{k=1}^{n} f(k + 1) - \sum_{k=1}^{n} f(k).$$

But

$$\sum_{k=1}^{n} f(k + 1) = f(2) + f(3) + \cdots + f(n) + f(n + 1)$$

and

$$\sum_{k=1}^{n} f(k) = f(1) + f(2) + f(3) + \cdots + f(n)$$

and again we conjecture, for the above is not a proof by any stretch of the imagination, a property of sigma.

Property 3. $\sum_{k=1}^{n} [f(k + 1) - f(k)] = f(n + 1) - f(1).$

Telescoping sums are characterized (easily distinguished) by the appearance of $k + 1$ and k. To illustrate this remark, I invite your attention to the pair of examples immediately preceding property 1, for each of these examples is an example of a telescoping sum, that is, one to which property 3 applies. In the first of these examples, $f(k) = k!$ while in the second $f(k) = 2^k$. To further illustrate these comments, consider the following example.

Example. Find $\sum_{k=1}^{n} \left[\frac{1}{k + 1} - \frac{1}{k}\right]$. The appearance of $k + 1$ and k is the clue that this is a telescoping sum in which $f(k) = 1/k$. Using property 3, we can quickly write

$$\sum_{k=1}^{n} \left[\frac{1}{k + 1} - \frac{1}{k}\right] = \frac{1}{n + 1} - 1.$$

I now ask you to engage in some extracurricular activities related to this last example. Start with the simple idea that $-1 \cdot \sum_{k=1}^{n} \left[\frac{1}{k + 1} - \frac{1}{k}\right] = 1 - \frac{1}{n + 1}$. Now then, use property 1 and elementary algebra to simplify this equation.

Question. What is the result of this activity?

Partial Answer. Elementary algebra shows that

$$\frac{1}{k + 1} - \frac{1}{k} = \frac{-1}{k(k + 1)}$$

PROBLEM SET 9.4

1. Evaluate:

a. $\sum_{k=1}^{5} (2k - 1)$ **b.** $\sum_{k=1}^{6} 2^{k-1}$ **c.** $\sum_{k=1}^{3} \frac{k}{2k + 1}$

2. Evaluate:

a. $\sum_{k=1}^{6} (k/2)$ **b.** $\sum_{k=1}^{3} \frac{k^k}{k!}$ **c.** $\sum_{k=1}^{5} (2k - 2)$

3. Evaluate:

a. $\sum_{k=1}^{5} (3k - 2)$ **b.** $\sum_{k=1}^{5} ck$ **c.** $\sum_{k=1}^{3} \left[(-1)^{k+1} \cdot \frac{2^k}{k!}\right]$

4. Use property 3 of this section to evaluate the following:

a. $\sum_{k=1}^{8} [\sqrt{(k + 1)} - \sqrt{k}]$ **b.** $\sum_{k=1}^{4} [2^{k+1} - 2^k]$

c. $\sum_{k=1}^{6} \left[\frac{1}{k + 1} - \frac{1}{k}\right]$ **d.** $\sum_{k=1}^{3} [(k + 1)^3 - k^3]$

5. Evaluate each of the indicated sums and thus verify the given statements:

a. $\sum_{k=1}^{3} (k^2 + k) = \sum_{k=1}^{3} k^2 + \sum_{k=1}^{3} k$

b. $\sum_{k=1}^{5} 4k = 4 \cdot \sum_{k=1}^{5} k.$

6. Evaluate:

$$\sum_{k=1}^{3} k, \sum_{k=1}^{3} (k - 1), \text{ and } \sum_{k=1}^{3} (k^2 - k)$$

for this gives you a counterexample to the statement

$$\sum_{k=1}^{n} [f(k) \cdot g(k) = \sum_{k=1}^{n} f(k) \cdot \sum_{k=1}^{n} g(k)$$

and explains why we do not have this statement included in our properties of sigma.

In each of problems 7–15, write the indicated sum in the sigma notation. *Warning:* You must first be able to guess a rule for $f(n)$ or $f(k)$ before you can do this.

7. $1 + 4 + 9 + 16 + \cdots + f(n)$.

8. $\frac{1}{3} + \frac{1}{9} + \frac{1}{27} + \frac{1}{81} + \cdots + f(n)$.

9. $0 + \frac{1}{3} + \frac{2}{4} + \frac{3}{5} + \frac{4}{6} + \cdots + f(n)$.

10. $\frac{0}{1} + \frac{1}{1\cdot 2} + \frac{2}{1\cdot 2\cdot 3} + \frac{3}{1\cdot 2\cdot 3\cdot 4} + \frac{4}{1\cdot 2\cdot 3\cdot 4\cdot 5} + \cdots + f(n)$.

11. $1 + \frac{2^2}{1!} + \frac{3^3}{2!} + \frac{4^4}{3!} + \cdots + f(n)$.

12. $1\cdot 3 + 3\cdot 5 + 5\cdot 7 + 7\cdot 9 + \cdots + f(n)$.

13. $1\cdot 2 + 3\cdot 4 + 5\cdot 6 + 7\cdot 8 + \cdots + f(n)$.

14. $1 - \frac{1}{2} + \frac{1}{3} - \frac{1}{4} + \cdots + f(n)$.

15. $1 - \frac{1}{3} + \frac{1}{5} - \frac{1}{7} + \cdots + f(n)$.

[16] In problems 1 and 2 of set 9.3 we found that, if x is a number, then $\sum_{k=1}^{n} x = nx$ and that $\sum_{k=1}^{n} k = \frac{n(n+1)}{2}$. Use these results and properties 1 and 2 of sigma to find:

a. $\sum_{k=1}^{n} (2k - 3)$ **b.** $\sum_{k=1}^{n} (k + 1)$ **c.** $\sum_{k=1}^{n} \left(\frac{k}{2} - \frac{1}{4}\right)$

[17] Look at the last example of this section and answer the question.

18. *Bonus problem.* Find the values of 4!, 5!, and 6!. Now then, calculate the following sums:

a. $\sum_{k=1}^{3} k!\cdot k = 1!\cdot 1 + 2!\cdot 2 + 3!\cdot 3.$

b. $\sum_{k=1}^{4} k!\cdot k = 1!\cdot 1 + 2!\cdot 2 + 3!\cdot 3 + 4!\cdot 4.$

c. $\sum_{k=1}^{5} k!\cdot k = 1!\cdot 1 + 2!\cdot 2 + 3!\cdot 3 + 4!\cdot 4 + 5!\cdot 5.$

Study what you have just accomplished and then *conjecture* a formula for the sum

d. $\sum_{k=1}^{k} k!\cdot k = 1!\cdot 1 + 2!\cdot 2 + 3!\cdot 3 + \cdots + n!\cdot n.$

Section 5—More About Sigma We shall now try to show how we can use the properties of sigma given in the last section to determine $S(n)$ when $f(k)$ is a polynomial in k. I must first point out, however, that the first two properties of sigma are useful to us only if there is at least one sequence f (preferably two) for which $S(n)$, that is, $\sum_{k=1}^{n} f(k)$, is known. Let me cheerfully point out that in a previous problem set we obtained the following results:

$$\sum_{k=1}^{n} x = nx \quad \text{and} \quad \sum_{k=1}^{n} k = \frac{n(n+1)}{2}$$

so that we can use these conjectured properties to find certain sums as indicated in the examples that follow:

Example. To find $\sum_{k=1}^{n} (4k - 1)$, we can begin with property 2 and find that $\sum_{k=1}^{n} (4k - 1) = \sum_{k=1}^{n} 4k + \sum_{k=1}^{n} (-1)$ and then use property 1 to show that $\sum_{k=1}^{n} (4k - 1) = 4 \cdot \sum_{k=1}^{n} k + \sum_{k=1}^{n} (-1)$. Using the sums obtained in problem set S-2, we have that $\sum_{k=1}^{n} (4k - 1) = 4 \cdot \frac{n(n+1)}{2} - n$, which we simplify and obtain $\sum_{k=1}^{n} (4k - 1) = n(2n + 1)$.

Example. To find $\sum_{k=1}^{n} (2k^2 - k)$, we must first observe that $\sum_{k=1}^{n} (2k^2 - k) = \sum_{k=1}^{n} 2k^2 - \sum_{k=1}^{n} k = 2 \cdot \sum_{k=1}^{n} k^2 - \sum_{k=1}^{n} k$. We have already shown that $\sum_{k=1}^{n} k = \frac{n(n+1)}{2}$ and, if you will permit me to anticipate a problem in the exercise to follow, we will learn that $\sum_{k=1}^{n} k^2 = \frac{n(n+1)(2n+1)}{6}$. So, if we put all the pieces together correctly, we find that

$$\begin{aligned} \sum_{k=1}^{n} (2k^2 - k) &= \frac{n(n+1)(2n+1)}{3} - \frac{n(n+1)}{2} \\ &= \frac{n(n+1)}{6} \cdot [4n + 2 - 3] \\ &= \frac{n(n+1)(4n-1)}{6}. \end{aligned}$$

Through the use of a problem whose solution we have already found, we shall try to indicate a method by which other sums may be found.

Hoping to make the basic ideas behind this method clear before it is even used, I tell you that if our problem is to find $\sum_{k=1}^{n} k^2$, we would begin with the easily established fact that

$$(k + 1)^3 - k^3 \text{ is } 3k^2 + 3k + 1, \text{ hence}$$
$$\sum_{k=1}^{n} [(k + 1)^3 - k^3] \text{ is } \sum_{k=1}^{n} (3k^2 + 3k + 1).$$

More generally, if our problem is to find $\sum_{k=1}^{n} k^m$ where m is a counting number, we would start with the notion that $\sum_{k=1}^{n} [(k + 1)^{m+1} - k^{m+1}]$ is the place to start.

The example that follows should clear up the details. Note the use made of property 3, the so called telescoping sum. In this example, do not let the fact that you already know the solution deter you from understanding the method by which we obtain the solution.

Example. Find $\sum_{k=1}^{n} k$.

Solution. Begin by noting that $(k + 1)^2 - k^2 = 2k + 1$ so that $\sum_{k=1}^{n} [(k + 1)^2 - k^2] = \sum_{k=1}^{n} (2k + 1)$. Now, the sum on the left-hand side of this statement is a telescoping sum in which $f(k) = k^2$, and so by property 3 we know that $\sum_{k=1}^{n} [(k + 1)^2 - k^2] = (n + 1)^2 - 1 = n^2 + 2n$. Therefore $\sum_{k=1}^{n} (2k + 1) = n^2 + 2n$. But by properties 1 and 2, we can show that $\sum_{k=1}^{n} (2k + 1) = 2 \cdot \sum_{k=1}^{n} k + \sum_{k=1}^{n} 1$. Since $\sum_{k=1}^{n} 1$ is n, we can show that $n^2 + 2n = 2 \cdot \sum_{k=1}^{n} k + n$ and conclude our problem with the statement that $\sum_{k=1}^{n} k = \frac{n^2 + n}{2} = \frac{n(n + 1)}{2}$.

PROBLEM SET 9.5

1. Use property 3 to find each of the following:

a. $\sum_{k=1}^{n} [x^{k+1} - x^k]$

b. $\sum_{k=1}^{n} [(k + 1)! - k!]$

2. Use property 3 to find each of the following:

a. $\sum_{k=1}^{n} \left\{\frac{1}{(k+1)!} - \frac{1}{k!}\right\}$ **b.** $\sum_{k=1}^{n} \left[\frac{k+1}{2^{k+1}} - \frac{k}{2^k}\right]$

[3] **a.** The problem is to find $\sum_{k=1}^{n} k^2$ and, to help you grasp the procedure outlined in the text example, begin the solution with the idea that

$$\sum_{k=1}^{n} [(k+1)^3 - k^3] = \sum_{k=1}^{n} (3k^2 + 3k + 1)$$

and note that the sum on the left-hand side is a telescoping sum.

Answer. $\sum_{k=1}^{n} k^2 = \frac{n(n+1)(2n+1)}{6}$.

b. Use the same ideas to find $\sum_{k=1}^{n} k^3$.

In problems 4–12, use properties 1 and 2 of sigma and the results of previous problems to find *and* simplify each of the following sums.

4. $\sum_{k=1}^{n} (2k - 1)$ **5.** $\sum_{k=1}^{n} (k^2 + k)$ **6.** $\sum_{k=1}^{n} (3k^2 - k)$

7. $\sum_{k=1}^{n} (2k^3 - 3k)$ **8.** $\sum_{k=1}^{n} (k^3 - 2k^2)$ **9.** $\sum_{k=1}^{n} (k + 1/2)$

10. $\sum_{k=1}^{n} (k^2 + 2k)$ **11.** $\sum_{k=1}^{n} (k^3 + k/2)$ **12.** $\sum_{k=1}^{n} (4k + 3)$

In each of problems 13–16 do the following: (a) give and simplify $S(n)$, and (b) conjecture the answer to the question "as n increases, what number does $S(n)$ approach?" The language used here is pathetic, but you will learn how to improve it in a calculus course.

13. $\sum_{k=1}^{n} [(2k - 1)/n^2] = S(n)$ **14.** $S(n) = \sum_{k=1}^{n} [(k + 3)/n^2]$

15. $S(n) = \sum_{k=1}^{n} \left[\frac{1}{n^3}(2k^2 - k)\right]$ **16.** $S(n) = \sum_{k=1}^{n} \left[\frac{1}{n^2}(3k + 2)\right]$

Und jetzt, einige gescheite Aufgaben.

17. Assuming that x is not 1, make use of property 1 of sigma and problem 1a of this exercise to find

$$(1 - x)\cdot \sum_{k=1}^{n} x^k.$$

Additional hint: $(1 - x)x^k$ is $x^k - x^{k+1}$.

Use the above results to find and simplify $\sum_{k=1}^{n} x^k$.

18. Use problem 1b to find and simplify $\sum_{k=1}^{n} k \cdot k!$.

19. Use problem 2a to find and simplify $\sum_{k=1}^{n} \frac{k}{(k+1)!}$.

20. Use problem 2b to find and simplify $\sum_{k=1}^{n} \frac{k-1}{2^{k+1}}$.

21. *Bonus problem.* Suppose that $f(x) = 1 + x$. Find $f\left(1 + \frac{k}{n}\right)$ and then find a formula for $S(n) = \sum_{k=1}^{n} f\left(1 + \frac{k}{n}\right) \cdot 1/n$.

Section 6—Geometric Sequence One type of sequence is of such classical importance that it merits special attention. We have already introduced this sequence in problem 16 of problem set 9-1 and elsewhere, so all that remains is to formalize its introduction with the following definition.

> DEFINITION. *If each of a and x is a number, a is not* 0, *and if G is a sequence such that* $G(1) = a$ *and, for* $n > 1$, $G(n) = x \cdot G(n-1)$, *then G is called a* geometric sequence.

The number x is often called the *common ratio.* We shall assume when necessary that $x^0 = 1$ even when $x < 0$, but we shall avoid like the plague saying anything about 0^0. In what follows, then, we shall dismiss for one reason or another the cases where $x = 0$ and $x = 1$. One should also be able to guess that the geometric sequence G could be defined by the following rule (at least it is not unreasonable to do so).

6.1 $G(n) = ax^{n-1}$.

Because of the importance of this type sequence, the student should develop the ability to pick out from any given sequences those that are geometric. Thus each of the following is a geometric sequence* and, for clarity, we have given the common ratio x.

* This statement is false for there is no reason why the sequence indicated by 1, 1/2, 1/4, . . . cannot have 4 as its fourth term, in which case it would certainly not be a geometric sequence. What is meant, of course, is that if all we know about the sequence is that its first few terms are as indicated, then it would at least be an intelligent guess to surmise that it is a geometric sequence.

$1, 1/2, 1/4, \ldots$	$x = 1/2$
$2, -1, 1/2, -1/4, \ldots$	$x = -1/2$
$2, 6, 18, \ldots$	$x = 3$
$\sqrt{2}, 2, 2\sqrt{2}, \ldots$	$x = \sqrt{2}$

Example. Suppose that F is a geometric sequence whose third term is 9 and whose fifth term is 81/4, give $F(n)$.

Solution. From the definition we know that $F(3) = ax^2$ and $F(5) = x^2 \cdot F(3)$. From this last equation we find that $x^2 = F(5)/F(3) = 9/4$, so from the first equation we can deduce that $a = F(3)/x^2 = 4$. However, $x^2 = 9/4$ so that $x = 3/2$ or $x = -3/2$ so that there are two geometric sequences satisfying the given conditions, and thus either is a solution to our example. For clarity I shall give the two sequences,

$$F(n) = 4 \cdot (3/2)^{n-1} \quad \text{and} \quad F(n) = 4 \cdot (-3/2)^{n-1}.$$

Example. Find a number t so that $t - 1$, $t + 8$, and 48 are the first three terms of a geometric sequence G and give $G(n)$.

Solution. Letting x denote the common ratio, we know that $t + 8 = x \cdot (t - 1)$ and that $48 = x \cdot (t + 8)$. Solving the first of these two equations for x and substituting in the second, we find that $t^2 - 32t + 112 = 0$. Solving this quadratic equation, we find that $t = 4$ or $t = 28$. If $t = 4$, then $x = 4$ and $a = 3$ so that sequence indicated by $3, 12, 48, \ldots, 3 \cdot 4^{n-1}$ is a solution to our example.

To complete our ideas on this topic, we are urged to recall that with each sequence there is associated another sequence, which we consistently denote by S, called the *sequence of partial sums.* Now, if G is a geometric sequence, then this associated sequence S is called a *geometric series.* With the objective of obtaining a formula or nonrecursive rule for $S(n)$, we give the following "plausible" argument: If S is a geometric series, then

$$S(n) = a + ax + ax^2 + \cdots + ax^{n-1}, \text{ hence}$$
$$x \cdot S(n) = ax + ax^2 + ax^3 + \cdots + ax^{n-1} + ax^n$$

so that it should seem reasonable to write that

$$S(n) - x \cdot S(n) = a - ax^n.$$

Factoring and simplifying this equation produces the rule:

$$(6.2) \quad S(n) = \frac{a(1 - x^n)}{1 - x} \quad \text{if} \quad x \neq 1.$$

So we may add to our knowledge by observing that if G is the sequence indicated by $3, 2, 4/3, \ldots, 3\cdot(2/3)^{n-1}$, then S is the sequence (geometric series) indicated by $3, 5, 19/3, \cdots$ and more precisely indicated by the rule

$$S(n) = \frac{3\cdot[1 - (2/3)^n]}{1 - 2/3} = 9\cdot[1 - (2/3)^n].$$

Example. If G is a geometric sequence in which $G(1) = a = 1/8$ and $G(7) = 8$, find x and $S(n)$.

Solution. Since $G(7) = ax^6$, we can substitute and find that $x^6 = 64$, thus $x = 2$ or $x = -2$. Now, if x is 2, then $S(n) = \dfrac{2^n - 1}{8}$.

This next example presents the same ideas that were used earlier in arriving at equation 6.2.

Example. $(x - 1)\cdot \sum_{k=1}^{n} x^{k-1} = \sum_{k=1}^{n} [x^k - x^{k-1}]$ by property 1 of sigma. Examine closely the right-hand side of this equation and imagine that $f(k) = x^{k-1}$. If you do this, you will see that the sum on the right-hand side is a telescoping sum so that property 3 of sigma assures us that

$$(x - 1)\cdot \sum_{k=1}^{n} x^{k-1} = x^n - 1.$$

From this we can deduce that, if x is not 1, then

$$\sum_{k=1}^{n} x^{k-1} = \frac{x^n - 1}{x - 1} = \frac{1 - x^n}{1 - x}.$$

The rest is easy, for by property 1 of sigma

$$\sum_{k=1}^{n} ax^{k-1} = a\cdot \sum_{k=1}^{n} x^{k-1}.$$

Therefore, if G is a geometric sequence, $S(n) = \dfrac{a(1 - x^n)}{1 - x}$.

PROBLEM SET 9.6

1. Assuming that the given numbers are the first three terms of a geometric sequence G, find $G(6)$ and $S(6)$.

a. $96, 48, 24, \ldots$ **b.** $3, -6, 12, \ldots$

2. Same as problem 1 for

   a. 3, 3/2, 3/4, . . . b. −1/5, 3/5, −9/5, . . .

3. Find $G(n)$ and $S(n)$ if $a = 1/3$ and $x = 3$.

4. Find $G(n)$ and $S(n)$ if $a = 27$ and $G(4) = 1$.

5. Find $G(n)$ and $S(n)$ if $G(5) = 1$ and $G(7) = 1/4$.

6. If $a = 3$ and $x = -3/4$, find $G(4)$ and $S(4)$.

7. If $a = 2$ and $x = -2/3$, find $G(4)$ and $S(4)$.

8. If $a = 5/8$ and, for some counting number m, $G(m) = 40$ while $S(m) = \frac{215}{8}$, find m and x.

9. If $a = 3\sqrt{2}$ and $x = \sqrt{2}/3$, find $G(6)$ and $S(6)$.

10. Suppose that t^3, t^2s, ts^2, and s^3 are the first four terms of a geometric sequence G, find $G(n)$ and $S(n)$.

11. Find a number t so that $-3/2$, t, and $-8/27$ are the first three terms of a geometric sequence.

12. Find a number t and a number z such that 3/4, t, z, and 2/9 are the first four terms of a geometric sequence.

13. If t and z are positive numbers, then the statement that "u is the *geometric mean of t and z*" means that u is a positive number such that t, u, and z are the first three terms of a geometric sequence. Find the geometric mean to each of the following pairs of positive numbers.

    (a) 4 and 25 (b) 12 and 3 (c) 6 and 18 (d) t and z.

14. Suppose that p and q are positive numbers and $p < q$. The *arithmetic mean to p and q* is the positive number $\frac{p+q}{2}$.

    a. Give the geometric mean to p and q.
    b. If we denote the arithmetic mean to p and q by M_1 and the geometric mean by M_2, find $M_1 - M_2$.
    c. Which of the following is true? $M_1 < M_2$, $M_1 = M_2$, or $M_1 > M_2$. Give reason for your answer.

15. If $S(n) = \sum_{k=1}^{n} (1/2)^{k-1}$, find the smallest counting number m such that $|S(m) - 2| < 0.01$.

16. If $S(n) = \sum_{k=1}^{n} 2\cdot(1/3)^{k-1}$, find the smallest counting number m such that $|S(m) - 3| < 0.01$.

17. Which of the following sequences are *not* geometric?

$$f_1 = \{2\sqrt{2}, 4, 4\sqrt{2}, 8, \ldots\} \qquad F_1 = \{1/2, 1, 3/2, 2, \ldots\}$$
$$f_2 = \{1/2, 1/6, 1/12, \ldots\} \qquad F_2 = \{1/2, 2, 8, 32, \ldots\}.$$

18. If $a = 16$ and $x = -1/4$, give $G(5)$ and $S(5)$.

19. If $a = 3$ and $x = -2/3$, give $G(4)$ and $S(4)$.

20. If $n > 3$, examine and explain whether the following statement is true or false:

$$\sum_{k=1}^{n} (1/2)^{k-1} < \sum_{k=1}^{n} \frac{1}{(k-1)!} < 1 + \sum_{k=1}^{n} (1/2)^{k-1}.$$

Section 7—Convergence If S is the geometric series, that is,

$$S(n) = \sum_{k=1}^{n} ax^{k-1}$$

we can observe that $S(n) = \dfrac{a}{1-x} - \dfrac{ax^n}{1-x}$ and we now devote our attention to those geometric series in which the absolute value of the common ratio is less than 1.

We begin by considering some examples.

Example 1. Suppose that $S(n) = \sum_{k=1}^{n} (1/2)^{k-1} = 2(1 - (1/2)^n)$. It is easy to see that $S(1) = 1$, $S(2) = 3/2$, $S(3) = 7/4$, $S(4) = 15/8$, and so forth. One should also note that the sequence S is bounded above and that the LUB of $S(n)$ is 2. Figure 9.6 may help you to understand

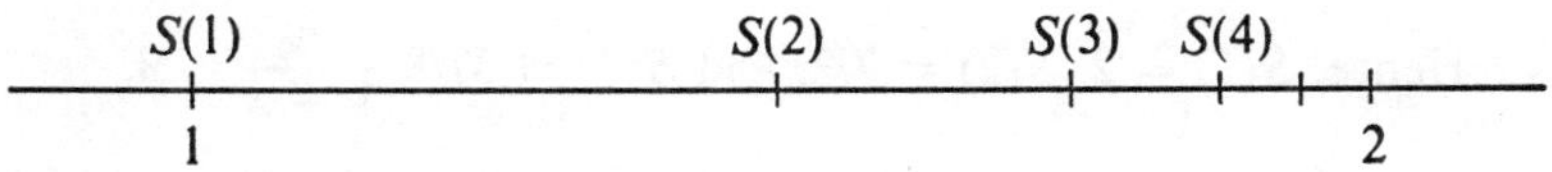

Fig. 9.6

the idea that while no term of S is 2, each term of S is closer to 2 than the preceding term.

The student may well wonder what all the fuss is about, for this example seems to be a rehash of our previous study of the axiom of completeness.

Let him be warned, however, that there is more to our madness than this. Let him be temporarily satisfied with the ideas that (a) in the example above, $\frac{a}{1-x} = 2$ and that (b) each open interval (no matter how small the interval) containing 2 also contains at least one term of S. In the following examples, keep these ideas in mind as we grope our way toward understanding a major concept.

Example 2. Suppose that

$$S(n) = \sum_{k=1}^{n} (-2/3)^{k-1} = 3/5 \cdot [1 - (-2/3)^n].$$

We can easily show that $S(1) = 1$, $S(2) = 1/3$, $S(3) = 7/9$, $S(4) = 13/27$, and so forth. Again we hope that figure 9.7 may help.

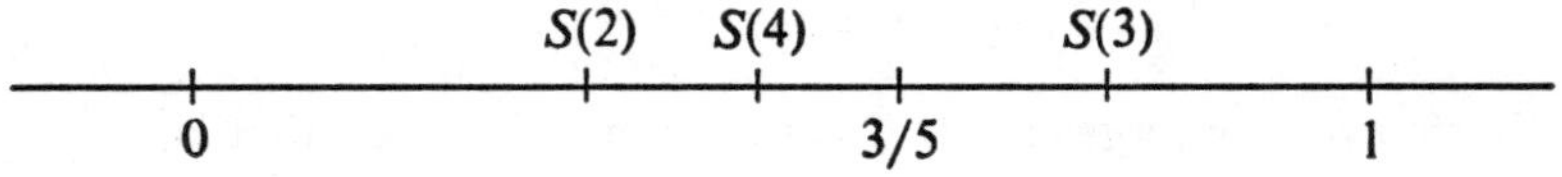

Fig. 9.7

Now note that 1 is the LUB of $S(n)$ and also note that $S(2)$ is the GLB of $S(n)$, but observe further that $\frac{a}{1-x} = 3/5$ and that the terms of S hop to and fro about this number. However, each term of S is closer to 3/5 than the preceding term and each open interval containing 3/5 also contains at least one term of S.

Example 2 gives some evidence that the concept we are struggling with is not merely a rehash of the axiom of completeness. This next example is presented without either comment or diagrams.

Example 3. If

$$S(n) = \sum_{k=1}^{n} 2 \cdot (3/4)^{k-1} = 8 \cdot [1 - (3/4)^n].$$

Hence, $S(1) = 2$, $S(2) = 7/2$, and $S(3) = 37/8 \cdot \frac{a}{1-x} = 8$.

The central theme we are trying to see and understand in these examples is often expressed as follows:

for example 1, "S converges to 2,"
for example 2, "S converges to 3/5,"
for example 3, "S converges to 8," and, in general,

"S converges to $\frac{a}{1-x}$."

For a more formal presentation, let us begin by noting that

$$\frac{a}{1-x} - S(n) = \frac{ax^n}{1-x}$$

and advance to the following argument.

If $0 < |x| < 1$, then $0 < |x|^2$, and if we assume that for all n, $0 < |x|^n$, then $0 < \frac{|a|\cdot|x|^n}{|1-x|}$. Hence, if $0 < p$, there is a counting number m such that whenever $n > m$ we have $\frac{|a|\cdot|x|^n}{|1-x|} < p$. For, if this were not true, then it must be true that for every counting number m,

$$\frac{|a|\cdot|x|^m}{|1-x|} \geq p, \text{ hence } |x|^m \geq \frac{p\cdot|1-x|}{|a|}.$$

[In the interest of brevity, let the positive number $\frac{p\cdot|1-x|}{|a|}$ be denoted by z in the rest of this argument.] Since L is an increasing function, it follows that $L(|x|^m) \geq L(z)$. However, $L(|x|^m)$ is $m\cdot L(|x|)$. Thus we have now established that for each counting number m, $m\cdot L(|x|) \geq L(z)$. Since $|x| < 1$, we can deduce that $L(|x|) < 0$ and so $m \leq \frac{L(z)}{L(|x|)}$. This last statement means that the counting numbers are bounded above which is absurd. QED.

The application of this argument to the geometric series is expressed in the following statement: "If $0 < |x| < 1$ and $0 < p$, then there is a counting number m such that whenever $n > m$, then $\left|S(n) - \frac{a}{x-1}\right| < p$." That is,

$$\text{if } n > m, \text{ then } S(n) = \frac{a}{1-x} + [\textit{some number between } -p \textit{ and } p].$$

This statement, we shall say, means that if $0 < |x| < 1$, then the geometric series converges to the number $\frac{a}{1-x}$, and we will denote this by writing either

$$a + ax^2 + ax^3 + \cdots + ax^n + \cdots = \frac{a}{1-x}$$

or, using the compact sigma notation,

$$\sum_{k=1}^{\infty} ax^{k-1} = \frac{a}{1-x}.$$

So, for example 2, we could write either

$$1 - 2/3 + 4/9 - \cdots + (-2/3)^n + \cdots = 3/5$$

or

$$\sum_{k=1}^{\infty} (-2/3)^{k-1} = 3/5$$

and hope that you will prefer the latter.

Example 4. Find the number to which the series

$$3 - 1 + 1/3 - 1/9 + \cdots \text{ converges.}$$

Solution. $G(n) = 3(-1/3)^{n-1}$ and, since $|-1/3| < 1$, we can write

$$\sum_{k=1}^{\infty} 3(-1/3)^{k-1} = 9/4.$$

PROBLEM SET 9.7

Find the number to which the following geometric series converge and give your answer in one of the two notations described in this section.

1. $S(n) = \sum_{k=1}^{n} (3/4)^{k-1}$

2. $S(n) = \sum_{k=1}^{n} (2/3)^{k-1}$

3. $S(n) = \sum_{k=1}^{n} (-1/2)^{k-1}$

4. $S(n) = \sum_{k=1}^{n} (-3/4)^{k-1}$

5. $S(n) = \sum_{k=1}^{n} 4\cdot(-2/3)^{k-1}$

6. $S(n) = \sum_{k=1}^{n} 6\cdot(-1/2)^{k-1}$

7. $S(n) = \sum_{k=1}^{n} 3\cdot(0.1)^{k-1}$

8. $S(n) = \sum_{k=1}^{n} 2\cdot(0.1)^{k-1}$

9. $S(n) = \sum_{k=1}^{n} 6\cdot(-1/3)^{k-1}$

10. $S(n) = \sum_{k=1}^{n} 4\cdot(-2/3)^{k-1}$

11. $8 + 2 + 1/2 + \cdots$

12. $3 + 6/5 + 12/25 + \cdots$

13. $S(n) = \sum_{k=1}^{n} -3\cdot(3/5)^{k-1}$

14. $S(n) = \sum_{k=1}^{n} -2\cdot(4/7)^{k-1}$

15. $-3/2 + 1 - 2/3 + \cdots$

16. $-5/4 + 1 - 4/5 + \cdots$

17. $2 - 0.2 + 0.02 - 0.002 + \cdots$

18. $3 - 0.3 + 0.03 - 0.003 + \cdots$

The following statement is the basis of problems 19 and 20: The length of each side of a square is 4 centimeters. A second square is inscribed by connecting the midpoints of the sides of the first square, a third by connecting the midpoints of the second, and so on.

19. Show that the sum of the areas of the squares thus formed is a geometric series *and* find the number to which this series converges.

20. Show that the sum of the perimeters of the squares thus formed is a geometric series *and* find the number to which this series converges.

21. If we start with an equilateral triangle (instead of a square) whose sides are 4 cm. long and form other equilateral triangles in the same way, show that the sum of the areas of the triangles thus formed is a geometric series *and* find the number to which this series converges.

SUMMARY

In this chapter we learned about functions whose domain is the set of counting numbers. Such functions are called *sequences.* Attention was called to the fact that with each sequence there is associated another sequence called *the sequence of partial sums.*

The sigma notation, a very important notation, was introduced as a compact notation for the sequence of partial sums. Some properties of this notation were investigated and formulas developed for finding certain sums.

The chapter concluded with a look at the geometric sequence and its associated series. The convergence of this series was dealt with.

Review Problems

1. Give the first 4 terms of F and S if $F(n) = n^2 - 3b$.

2. Give the first 4 terms of the sequences F and S if $F(1) = 2/3$ and, for $n > 1$, $F(n) = F(n - 1) - (2/3)^n$.

3. If $S(n) = n^2 + n$, find $F(n)$ and write $S(n)$ in sigma notation.

4. If $S(n) = \dfrac{n+1}{n}$, find $F(n)$ and write $S(n)$ in sigma notation.

5. If $G(n) = 8 \cdot (1/3)^{2(n-1)}$, show that G is a geometric sequence and then find the number to which the geometric series converges.

6. Find:

a. $\sum_{k=1}^{n} \left[\frac{3^k}{k+1} - \frac{3^{k-1}}{k}\right]$

b. $\sum_{k=1}^{n} (2k^3 - 3k)$

c. Show that $\sum_{k=1}^{n} f(k) = \sum_{j=0}^{n-1} f(j+1)$.

CHAPTER 10
Mathematical Induction

Section 1—Theorem MI If we examine the statements that have been proven, two methods of proof may be distinguished and labeled "direct" and "indirect." In the direct method of proof, we proceed from the hypothesis to a conclusion through a series of statements, each of which is justified by some axiom, definition, or previously proven theorem. In the indirect proofs, we relied upon the simple idea that a statement was either true or false so that, by assuming the theorem false, we were able to obtain, again through a series of statements justified as in the direct method, a statement that was a contradiction.

If a statement (theorem) is to be true for all of the counting numbers, then it is often possible to construct a proof based upon an idea called *mathematical induction.* The idea mentioned and the method of proof are embodied in the following theorem.

> THEOREM MI. *If T is a subset of N such that* $1 \in T$ *and, whenever* $k \in T$, $k + 1 \in T$, *then T is N.*

It should seem perfectly natural to argue that since 1 is in T, then $1 + 1 = 2$ is in T. Again, since 2 is in T, then $2 + 1 = 3$ is in T. Continuing in this manner, one should be able to show that every counting number is in T, that is, T is N. It may come as a surprise to learn that this argument lacks precision the moment we say "continuing in this manner" and cannot, therefore, be considered as a proof of the theorem. In order that you may have time to understand why this is so and at the same time learn how this theorem can be used to establish many important results, we shall temporarily assume that theorem MI is true. In the examples that follow, we shall attempt to see how the assumption of theorem MI can help us prove other useful statements.

Example. If n is a counting number, prove that

$$(-1)^{2n-1} = -1.$$

Proof. Let T denote the number set to which m belongs only if m is a connecting number and

$$(-1)^{2m-1} = -1.$$

Since $(-1)^{2.1-1} = -1$, 1 belongs to T. So T is not empty.

If $k \in T$, then k is a counting number and $(-1)^{2k-1}$ is -1 by definition of T given above.

From this it follows that $(-1)^2 \cdot (-1)^{2k-1} = (-1)^2 \cdot (-1)$, that is, by adding exponents on the left-hand side, that

$$(-1)^{2k+1} = -1.$$

But this means that $k + 1$ belongs to T. Therefore we have shown that $1 \in T$ and that, whenever $k \in T$, $k + 1 \in T$ so by theorem MI, T is N and we have established that

$$(-1)^{2n-1} = -1 \text{ for all } n \in N.$$

Example. If $x > 1$, show that $x^n > 1$ for all $n \in N$.

Proof. Let T be the number set to which m belongs only if m is a counting number and $x^m > 1$. It should be clear (from the hypothesis) $1 \in T$. If $k \in T$, then, by definition, k is a counting number and $x^k > 1$.

From this it follows that $x \cdot x^k > 1 \cdot x$ by order axiom 4, thus $x^{k+1} > x$ and, since $x > 1$ by hypothesis,

$$x^{k+1} > 1.$$

But this means that whenever $k \in T$, $k + 1 \in T$, so by theorem MI, T is N and the desired result is proven.

The proof of formulas for finding certain sums is another illustration of the use of proof by mathematical induction. If we have such a formula that seems to hold for all of the counting numbers—we may have just guessed this formula because it seemed to work—then theorem MI provides us with the procedure for showing that the formula is valid or is false. If false, then this fact should be enough reason for us to discontinue the use of the formula. As an example, we earlier conjectured the formula

$$\sum_{k=1}^{n} k = \frac{n(n+1)}{2}.$$

So, as another example of proof by mathematical induction, let us establish this formula.

Proof. Let T be the number set to which m belongs only if $m \in N$ and $\sum_{k=1}^{m} k = \frac{m(m+1)}{2}$. $1 \in T$ since $\sum_{k=1}^{1} k = 1 = \frac{1(1+1)}{2}$ if $j \notin T$, then j is a counting number (k is being used here as the index of summation) and $\sum_{k=1}^{j} k = \frac{j(j+1)}{2}$. Add the next term of this sum to both sides of this last equation and get $(j+1) + \sum_{k=1}^{j} k = \frac{j(j+1)}{2} + (j+1)$. If we simplify this result, we obtain the statement

$$\sum_{k=1}^{j+1} k = \frac{j(j+1) + 2(j+1)}{2} = \frac{(j+1)(j+2)}{2}.$$

However, this means (check this) that $j + 1 \in T$. Thus we have shown that $1 \in T$ and that, whenever $j \in T$, $j + 1 \in T$, hence T is N by theorem MI.

Example. If $n \in N$, show that $\sum_{k=1}^{n} k \cdot k! = (n+1)! - 1$.

Remark. This first appeared as a *bonus problem* in chapter 9, section 3.

Proof. Define T as before. Since $1! = 1$, it is easy to see that $1 \in T$. If $j \in T$, then j is a counting number and

$$\sum_{k=1}^{j} k \cdot k! = (j+1)! - 1$$

Adding $(j+1) \cdot [(j+1)!]$ to both sides, we can see that

$$(j+1) \cdot [(j+1)!] + \sum_{k=1}^{j} k \cdot k! \text{ is } \sum_{k=1}^{j+1} k \cdot k!$$

and that

$$\begin{aligned}(j+1) \cdot [(j+1)!] + (j+1)! - 1 &= ([(j+1)!] \cdot [j+2]) - 1 \\ &= (j+2)! - 1.\end{aligned}$$

All this activity means that $j + 1 \in T$.

Thus, by thoerem MI, we have proven our formula.

PROBLEM SET 10.1

If $n \in N$, show that each of the following is true.

1. If $0 < x < 1$, then $0 < x^n < 1$.

2. If $1 < x$, then $x^{n-1} < x^n$.

3. $2n \leq 2^n$.

4. $n + 1 \leq 2n$.

5. If $0 < a < b$, then $a^n < b^n$.

6. $\sum_{k=1}^{n} c = cn$.

7. $\sum_{k=1}^{n} (2k - 1) = n^2$.

8. $\sum_{k=1}^{n} (k^2 + k) = \frac{n(n + 1)(n + 2)}{3}$.

9. $\sum_{k=1}^{n} \frac{1}{k(k + 1)} = \frac{n}{n + 1}$.

10. Just for review, find the sum indicated in problem 8 in the same manner that we found such sums in chapter 9. This has very little (nothing as a matter of fact) to do with mathematical induction.

Are the following statements true for all $n \in N$? If not, why not? If yes, justify.

11. $\sum_{k=1}^{n} 1/k < 1 + \frac{10}{n}$.

12. $\sum_{k=1}^{n} k! = 3^{n-1}$.

Section 2—Proof of Conjectures The last set of problems gave the student some practice in the art of proof by methematical induction, that is, an opportunity to learn something about the principles involved. It is hoped that any questions he may have will be answered before he goes on.

On many occasions in mathematics, one gathers some information about a given problem and a pattern seems to emerge. In other words, the computations that are made are taken as evidence that a general rule could be formulated. The mathematician uses this evidence to guess the general rule or formula and then tries to verify it. For example, the calculus student has the problem of finding a function f' which is derived from a given function f. Let us suppose that table 10.1 represents the results of his investigations and efforts. Let us further assume that he sees some pattern in all of this. For, on the basis of what he sees, he conjectures a general formula and sets out to verify it. If, like Manfred v. Dumbkopf, he does not see a pattern, then there is nothing else that can be done for him. We cannot verify the rule even if we are able to guess a formula

TABLE 10.1

$f(x)$	$f'(x)$
x	1
x^2	$2x$
x^3	$3x^2$
x^4	$4x^3$

for $f'(x)$, for we do not know how to find the function f'. In our own field of studies, many similar situations have arisen that we do know how to handle. For example, the properties of sigma or nonrecursive rules for sequences. Certainly, then, we can and should verify some of our earlier conjectures. This is the primary objective of the examples that follow and of the subsequent problem set.

Example. If F is the sequence for which $F(1) = 1$ and, for $n > 1$, $F(n) = n \cdot F(n - 1)$, then show that $F(n) = n!$.

Remark. The underlined rule is the conjecture that originated in Example 2 on page 356.

Proof. Let T be the number set to which m belongs only if $m \in N$ and $F(m) = m!$.

Since $F(1) = 1$ and $1! = 1$, 1 belongs to T. If $k \in T$, then $k \in N$ and $F(k) = k!$. Further, $k + 1$ is a counting number and $k + 1 > 1$, hence

$$(k + 1) \cdot F(k) = (k + 1) \cdot k! \text{ or } F(k + 1) = (k + 1)!.$$

This last equation means that $k + 1 \in T$, and we are finished by virtue of theorem MI.

Many of our problems of finding sums of certain sequences were based upon the three conjectured "properties of sigma." It would seem to me, then, that we can give some substance to our previous efforts along this line by proving these properties. Let us do the decent thing, and prove these three conjectures. Let us then at least look at this next example.

Example. If F is a sequence and c is a number, then $\sum_{k=1}^{n} c \cdot F(k) = c \cdot \sum_{k=1}^{n} F(k)$. (This is property 1 for sigma.)

Since I wish to leave to the student the pleasure of justifying this statement, I shall merely talk it to death.

If we define a number set T in a manner that should by now be clear, we should have no trouble in seeing that $1 \in T$. If $m \in T$, then $\sum_{k=1}^{m} c \cdot F(k) = c \cdot \sum_{k=1}^{m} F(k)$. It should seem reasonable that we can add the same number to both sides, but what number? The answer to this question is based upon knowing what we are trying to show and upon knowing that the sigma notation is just a compact way of writing a sum. Again the question, what number? Patience. Try to see that $c \cdot F(m + 1)$ is the answer.

Let us conclude this day's work by pointing out some ideas that I really hoped you would discover yourself. The manner in which we defined the number set T made the following obvious: T is a subset of N and we knew that T existed (was not empty) the moment we were able to show that $1 \in T$. Further, a number is placed in, or belongs to, T only if the statement to be proven is true for that counting number.

PROBLEM SET 10.2

Prove each of the following statements.

1. If $|x| < 1$, then $|x|^n < 1$.

2. $2^{n-1} \leq n!$.

3. If d is a number and f is a sequence such that $f(1) = a$ and, for $n > 1$, $f(n) = d + f(n - 1)$, then $f(n) = a + (n - 1)d$ for all n.

4. If $r \neq 0$ and if f is a sequence such that $f(1) = a$ and, for $n > 1$, $f(n) = r \cdot f(n - 1)$, then for all n $f(n) = ar^{n-1}$.

5. $\sum_{k=1}^{n} c \cdot F(k) = c \cdot \sum_{k=1}^{n} F(k)$.

6. $\sum_{k=1}^{n} [F + G](k) = \sum_{k=1}^{n} F(k) + \sum_{k=1}^{n} G(k)$.

7. $\sum_{k=1}^{n} [F(k + 1) - F(k)] = F(n + 1) - F(1)$.

8. $\sum_{k=1}^{n} x^{k-1} = \frac{1 - x^n}{1 - x}$. Assume that $x \neq 1$ and that $x^0 = 1$.

9. If $x > -1$, then $(1 + x)^n \nless 1 + nx$.

10. $(x + y)^n \geq x^{n-1} \cdot (x + ny)$.

11. If $0 < x$, then $Ln(x^n) = n \cdot Ln(x)$.

12. $\sum_{k=1}^{n} \frac{1}{(2k - 1)(2k + 1)} = \frac{n}{2n + 1}$.

Section 3—Binomial Theorem The student is probably familiar with the products listed below, but, if not, he should verify them at once.

$$(a + b)^2 = a^2 + 2ab + b^2$$
$$(a + b)^3 = a^3 + 3a^2b + 3ab^2 + b^3$$
$$(a + b)^4 = a^4 + 4a^3b + 6a^2b^2 + 4ab^3 + b^4$$
$$(a + b)^5 = a^5 + 5a^4b + 10a^3b^2 + 10a^2b^3 + 5ab^4 + b^5$$

The coefficients in the above products form an interesting pattern when written as illustrated in the following scheme known as Pascal's triangle.

$$\begin{array}{ccccccccccc} & & & & 1 & & 1 & & & & \\ & & & 1 & & 2 & & 1 & & & \\ & & 1 & & 3 & & 3 & & 1 & & \\ & 1 & & 4 & & 6 & & 4 & & 1 & \\ 1 & & 5 & & 10 & & 10 & & 5 & & 1 \end{array}$$

In this array, each horizontal row begins and ends with the number 1. One may also note (surely Pascal must have) that in each row after the first, each other entry is the sum of the two adjacent numbers in the preceding row.

Question. What are the coefficients in the expansion of $(a + b)^6$? That is, what is the sixth row of Pascal's triangle?

Answer. 1, 6, 15, 20, 15, 6, 1.

Whenever a pattern like this is discovered, we should suspect that there might be some general way of describing it, so we are led to ask, Is there some general formula for expanding $(a + b)^n$ for any counting number n? Indeed, there is such a formula, and it is known as the *binomial theorem*. We shall now attempt to conjecture this theorem. It seems quite likely that this theorem was born of conjectures very similar to what we shall now present.

Surely we can begin by conjecturing (with some evidence to support us) that

$$(a + b)^n = c_0a^n + c_1a^{n-1}b + c_2a^{n-2}b^2 + \cdots + c_nb^n$$

where each of $c_0, c_1, c_2, \ldots, c_n$ is a number. You can also note that the sum of the exponents of a and b is n and you can be told that the numbers $c_0, c_1, c_2, \ldots$ are called the *binomial coefficients.*

In our attempt to develop the formula referred to, we can see that one problem is to find a pattern by which the coefficients can be determined. We have already noted that $c_0 = c_n = 1$ and that Pascal's triangle is a pattern. However, there is one small drawback to this pattern, namely, it is not easy to determine the n-th row of the triangle, that is, we have some difficulty in generalizing this pattern. Let us persist. Look at the triangle and note that the ratio of each coefficient, after the first, to the previous one in the same row, for example, the fourth row, is the array

$$1 \quad 4 \quad 6 \quad 4 \quad 1$$

and the ratios mentioned above are

$$4/1 \quad 6/4 \quad 4/6 \quad \text{and} \quad 1/4.$$

For the fifth row

$$1 \quad 5 \quad 10 \quad 10 \quad 5 \quad 1$$

we can form the ratios

$$5/1 \quad 10/5 \quad 10/10 \quad 5/10 \quad \text{and} \quad 1/5.$$

If we assume that this same pattern prevails, then for the sixth row of Pascal's triangle we could write the ratios

$$6/1 \quad 5/2 \quad 4/3 \quad 3/4 \quad 2/5 \quad \text{and} \quad 1/6.$$

Thus the sixth row could be written in the form

$$1 \quad \frac{6}{1} \quad \frac{6\cdot 5}{1\cdot 2} \quad \underline{\frac{6\cdot 5\cdot 4}{1\cdot 2\cdot 3}} \quad \underline{\frac{6\cdot 5\cdot 4\cdot 3}{1\cdot 2\cdot 3\cdot 4}} \quad \frac{6\cdot 5\cdot 4\cdot 3\cdot 2}{1\cdot 2\cdot 3\cdot 4\cdot 5} \quad 1.$$

Take another look at this and then focus your attention on the underlined (for less eye strain) coefficients. Long before your vision fails you, you will note that these coefficients could be written as follows:

$$\frac{6\cdot 5\cdot 4}{1\cdot 2\cdot 3} = \frac{(1\cdot 2\cdot 3)\cdot 4\cdot 5\cdot 6}{(1\cdot 2\cdot 3)\cdot 1\cdot 2\cdot 3} = \frac{6!}{3!\cdot 3!}$$

$$\frac{6\cdot 5\cdot 4\cdot 3}{1\cdot 2\cdot 3\cdot 4} = \frac{(1\cdot 2)\cdot 3\cdot 4\cdot 5\cdot 6}{(1\cdot 2)\cdot 1\cdot 2\cdot 3\cdot 4} = \frac{6!}{2!\cdot 4!}.$$

Now note that the first of these is the fourth coefficient c_3 in the sixth row of Pascal's triangle while the other is the fifth coefficient c_4 in this same row. In a similar way you can see that the third coefficient c_2 in row 6 could be written in the form

$$c_2 = \frac{6!}{4!\cdot 2!}.$$

At the risk of insulting your intelligence, I ask you to note that the first and second coefficients c_0 and c_1 in the fifth row could be written in the form

$$c_0 = \frac{5!}{5!\cdot 0!} \qquad c_1 = \frac{5!}{4!\cdot 1!}$$

and, by using our knowledge of factorials, we can show that the above gives $c_0 = 1$ and $c_1 = 5$, which is good since that is what they are.

Thus we are happy to report that we think we have found a pattern for writing the coefficients for the expansion of $(a + b)^n$, to wit

$$c_k = \frac{n!}{(n - k)!\cdot k!} \text{ for } k = 0, 1, 2, 3, \ldots, n$$

and as a result of this activity we can conjecture that the binomial theorem should be written as follows:

$$(a + b)^n = \frac{n!}{n!\cdot 0!} a^n + \frac{n!}{(n - 1)!\cdot 1!} a^{n-1}b + \frac{n!}{(n - 2)!\cdot 2!} a^{n-2}b^2 + \cdots + \frac{n!}{(n - n)!\cdot n!} b^n.$$

After some years had passed, an enterprising student noted the relationship between the factorials in the denominator of the binomial coefficient and the exponents of a and b. As an example, he noted that in the expansion of $(a + b)^9$ the terms containing a^3b^6 and a^5b^4 had the coefficients

$$\frac{9!}{3!\cdot 6!} \text{ and } \frac{9!}{5!\cdot 4!} \text{ respectively.}$$

Perhaps this same student, with an eye for simplicity of notation, wrote the above coefficients in the form

$$\binom{9}{6} \text{ and } \binom{9}{4} \text{ respectively.}$$

That is, he made the following definition

Binomial Coefficients

$$c_k = \frac{n!}{(n-k)!\cdot k!} = \binom{n}{k}$$

where $k = 0, 1, 2, 3, \ldots, n$.

He then rewrote the binomial formula as follows:

$$(a+b)^n = \binom{n}{0} a^n + \binom{n}{1} a^{n-1}b + \binom{n}{2} a^{n-2}b^2 + \cdots + \binom{n}{n} b^n.$$

Much later, someone who had studied the sigma notation and who had an eye not only for simplicity but also for beauty, wrote the final form of the

BINOMIAL THEOREM. *If each of a and b is a number and n is a counting number, then* $(a+b)^n = \sum_{k=1}^{n} \binom{n}{k} a^{n-k}\cdot b^k$.

Examples.

$$(x+2)^4 = \sum_{k=0}^{4} \binom{4}{k} x^{4-k}\cdot 2^k$$

$$= x^4 + \binom{4}{2} x^3 2 + \binom{4}{2} x^2 2^2 + \binom{4}{3} x 2^3 + 2^4.$$

Moral. Do not evaluate the coefficients other than the first and last unless you must. If we must, because we are told to, then we consult the definition of the binomial coefficients and write

$$(x+2)^4 = x^4 + 8x^3 + 24x^2 + 32x + 16$$

and not only are we irked with the arithmetic, but we have also destroyed the beauty in the symmetry of the coefficients. But alas, we had to.

Example. Give the term involving x^6t^2 in $(x+t)^8$.

Solution (after only a glance at what we have done).

$$\binom{8}{2} x^6t^2$$

or, if we have to simplify, $28x^6t^2$.

Example. Give the first three terms of $(a+1)^{10}$.

Solution. First term is a^{10}

Second term is $\binom{10}{1}a^9$

Third term is $\binom{10}{2}a^8$

PROBLEM SET 10.3

1. This problem deals with the observation made by Pascal in his triangle of coefficients. Verify each of the following:

a. $\binom{6}{2} + \binom{6}{3} = \binom{7}{4}$ **b.** $\binom{n}{k-1} + \binom{n}{k} = \binom{n+1}{k}$

2. Find the following sums:

a. $1 - \binom{4}{1} + \binom{4}{2} - \binom{4}{3} + \binom{4}{4}$

b. $1 + \binom{4}{1} + \binom{4}{2} + \binom{4}{3} + \binom{4}{4}$

3. Show that $2\cdot\binom{n}{2} + \binom{n}{1} = n^2$.

In problems 4 and 5, leave coefficients in the form $\binom{n}{k}$.

4. Expand $(a + b)^5$. **5.** Expand $(a + b)^6$.

6. Expand *and simplify* $(x + 2)^5$. *Hint:* problem 4 with $a = x$ and $b = 2$.

7. Expand *and simplify* $(2t - 1)^5$. *Hint:* problem 4 with $a = 2t$ and $b = -1$.

In each of the following, leave the coefficients in the form $\binom{n}{k}$.

8. Give first four terms in $(x + t)^{10}$.

9. Give first four terms in $(z + 5)^{12}$.

10. Give first four terms of $(3s - 1)^8$.

11. a. Give the term involving x^4y^{10} in $(x + y)^{14}$.
b. Give the term involving x^3y^8 in $(2x + y)^{11}$.

12. a. Give the fifth term of $(x - 2)^{10}$.
b. Give the third term of $(m + 3)^7$.

13. Use the binomial theorem to show that

$$1 + \sum_{k=1}^{n} \binom{n}{k} = 2^n. \textit{ Hint: } (1 + 1)^n = 2^n.$$

14. *Bonus problem.* Alas and alack, in spite of all that has been said about the binomial theorem, we have not yet proven it. Remedy this by using mathematical induction to prove this ancient theorem.

Section 4—Proof of Theorem MI (optional) In some classes, the statement which we have labeled theorem MI is added to the list of assumptions (axioms) and called *the axiom of induction.* There is nothing wrong with this, and it may very well be the best thing to do at this stage in the student's development. For those students who demand more than this, however, this section will concern itself with indicating how the theorem may be deduced from the axioms already given in this text. The student who reads this section any further should do so only after (not before) he has reviewed axiom N and the axiom of completeness, for they will surely be needed. We pause for you to either do this or to decide that you will read no further.

We begin with a statement which is sometimes said to be an example of the "well-ordering principle," but is the statement that we shall label as theorem 14.

THEOREM 14. *If K is a number set each element of which is a counting number, then K contains a least element.*

Proof (the reason for considering this theorem is simple; we need it to prove theorem MI):

K is bounded below by 1 for if $m \in K$ and $m < 1$,

then m is not a counting number contrary to the hypothesis. Hence, there is a GLB of K, say b, that is, GLB of $K = b$. Now either b is a counting number or not. If b is a counting number, then $b \in K$, for if not, then (since $b + 1$ is a counting number whenever b is) $b + 1$ would be a lower bound of K which is absurd. Thus, if $b \in N$, the theorem is proven. Let us suppose, therefore, that b is not a counting number.

If $b \notin N$, then there is an element of K, call it c, between b and $b + 1$. Now, either c is the least element in K or there is a counting number c_1 such that $b < c_1 < c < b + 1$. But $b + 1 < c_1 + 1$, thus $c_1 < c < c_1 + 1$. Therefore c is not a counting number which is a contradiction establishing that c is the least element of K and so the theorem is proven.

In addition to theorem 14, we need to know that the following statement is true.

> THEOREM 15. *If n is a counting number and $n > 1$, then $n - 1$ is a counting number.*

We shall rely upon our intuition about theorem 15 and assume that it is true, at least for the present. We conclude this section with the following argument.

> *Theorem MI: proof of.* Let us suppose that the statement is false and let K be the set of counting numbers not in T. By theorem 14, K contains a least element, call it c. Now $c \neq 1$ since 1 belongs to T by hypothesis, hence $c > 1$. By theorem 15, $c - 1$ is a counting number. Since $c - 1 < c$, $c - 1$ belongs to T and from this it follows, by hypothesis, that $(c - 1) + 1$ belongs to T, that is, c belongs to T. This contradiction establishes the theorem.

SUMMARY

This chapter introduced two concepts that will be of some importance in any study of analysis: proof by mathematical induction and the binomial theorem.

The principal of mathematical induction provides a method for proving certain statements that hold for all positive integers. The proof of formulas for certain sums is an excellent example of the use of this principle.

In this chapter, we again used the sigma notation so that the student could become familiar with this important notation.

Review Problems

1. Show that, for every counting number n, $\dfrac{4^n - 1}{3}$ is an integer.

2. Show that $\sum_{k=1}^{n} \dfrac{k}{2^k} = 2 - \dfrac{n+2}{2^n}$ for $\forall n\ N$.

3. Use the binomial theorem to expand $(2x + 3)^4$.

4. Give and simplify the fourth term in the expansion of $(x + 4)^7$.

CHAPTER 11
Slope Function

Section 0—Prelude, the Slope Quotient This section will deal mainly with the number $\dfrac{F(x) - F(a)}{x - a}$ where F is a continuous function. Our mission is to do the dirty work preliminary to, and vital to, the main assault on the slope function which follows. Furthermore, it is almost everywhere agreed that frequent review of our skill in factoring and simplifying quotients, though not revered, is essential.

If F is a continuous function and if x and a are numbers ($x \neq a$) in the domain of F, then we shall call the number $\dfrac{F(x) - F(a)}{x - a}$ the *slope quotient* since it is obviously a quotient and because the student can see

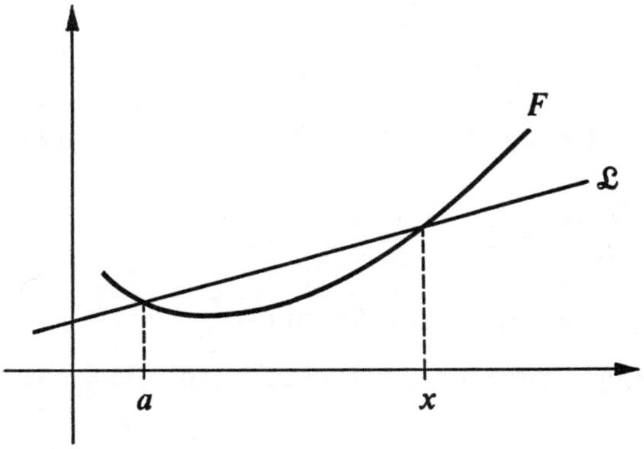

Fig. 11.1

that this number is the slope of the line $\mathscr{L}$ shown in figure 11.1. Prior to any examples, you are urged to recall that you know how to factor a difference of two squares and thus are able to understand the following:

1. If z and t are positive numbers, then $(\sqrt{z} + \sqrt{t})(\sqrt{z} - \sqrt{t}) = z - t$ from which it follows that $\sqrt{z} - \sqrt{t} = \dfrac{z - t}{\sqrt{z} + \sqrt{t}}$.
2. If z and t are numbers, then $(|z| + |t|)(|z| - |t|) = z^2 - t^2$ hence $|z| - |t| = \dfrac{(z - t)(z + t)}{|z| + |t|}$.

Example. If $F(x) = x^2 - 2x$, then $\dfrac{F(5) - F(1)}{5 - 1} = \dfrac{16}{4} = 4$. Further,

$$\begin{aligned} F(x) - F(a) &= (x^2 - 2x) - (a^2 - 2a) \\ &= x^2 - a^2 - 2x + 2a \\ &= (x - a)(x + a) - 2(x - a) \\ &= (x - a)(x + a - 2) \end{aligned}$$

thus,

$$\frac{F(x) - F(a)}{x - a} = x + a - 2.$$

Note that, with $x = 5$ and $a = 1$, the last result agrees with the first computation made in this example. As a matter of fact, it would be amazing if it did not.

Example. If $G(x) = \sqrt{(x - 1)}$, then $G(x) - G(a) = \sqrt{(x - 1)} - \sqrt{(a - 1)}$. If we now think about statement (1) on the preceding page, we see that

$$\begin{aligned} G(x) - G(a) &= \frac{(x - 1) - (a - 1)}{\sqrt{(x - 1)} + \sqrt{(a - 1)}} \\ &= \frac{x - a}{\sqrt{(x - 1)} + \sqrt{(a - 1)}} \end{aligned}$$

hence,

$$\frac{G(x) - G(a)}{x - a} = \frac{1}{\sqrt{(x - 1)} + \sqrt{(a - 1)}}.$$

Example. If $U(x) = |x + 3|$, then $U(x) - U(a) = |x + 3| - |a + 3|$ and a glance at statement (2) will explain why

$$U(x) - U(a) = \frac{(x - a)(x + a + 6)}{|x + 3| + |a + 3|}$$

therefore,

$$\frac{U(x) - U(a)}{x - a} = \frac{x + a + 6}{|x + 3| + |a + 3|}.$$

Example. If $Q(x) = \frac{x + 1}{2x}$, then we are able to see that

$$Q(x) - Q(a) \text{ is } \frac{x + 1}{2x} - \frac{a + 1}{2a} = \frac{a - x}{2ax} = \frac{-1(x - a)}{2ax}$$

therefore

$$\frac{Q(x) - Q(a)}{x - a} \text{ is } -\frac{1}{2ax}.$$

In this next example, we wish to remind the student that the quotient we are now concentrating upon is the slope of the line containing the points $[x, f(x)]$ and $[a, f(a)]$. It is also time for us to review the point-slope equation of a line.

Problem. If $f(x) = x^2 - 4$, give the point-slope equation of the line containing $[2, f(2)]$ and $[7, f(7)]$.

Solution. The slope m of this line is $\frac{f(7) - f(2)}{7 - 2}$, thus $m = 9$. Since we have two points given, we certainly have a point on this line. Selecting the point $[2, f(2)] = (2,0)$, we then write the desired equation as

$$y = 9(x - 2).$$

PROBLEM SET 11.0

1. Recalling the definitions of the special functions I and H, give **(a)** $\frac{I(x) - I(a)}{x - a}$ and **(b)** $\frac{H(x) - H(a)}{x - a}$.

2. If $p(x) = x^2$, give $\frac{p(x) - p(a)}{x - a}$ and $\frac{p(0.5) - p(0.25)}{0.25}$.

3. If $P(x) = x^3$, give $\frac{P(x) - P(a)}{x - a}$ and $\frac{P(3) - P(1)}{2}$.

4. If $Z(x) = \sqrt{x}$, give $\frac{Z(x) - Z(a)}{x - a}$.

5. If $V(x) = |x|$, give $\dfrac{V(x) - V(a)}{x - a}$.

6. If $F(x) = x^2 + x$, give $\dfrac{F(x) - F(a)}{x - a}$.

7. If $g(x) = \sqrt{(x + 2)}$, give $\dfrac{g(x) - g(a)}{x - a}$.

8. If $G(x) = |x - 2|$, give $\dfrac{G(x) - G(a)}{x - a}$.

9. If $Q(x) = \dfrac{x}{x + 1}$, give $\dfrac{Q(x) - Q(a)}{x - a}$.

10. If $W(x) = \dfrac{1}{x - 1}$, give $\dfrac{W(x) - W(a)}{x - a}$.

11. If $s(x) = mx + b$, give $\dfrac{s(x) - s(a)}{x - a}$.

12. If $F(x) = 2x - 1$ and $G(x) = x^2 + 1$, give $\dfrac{[F \circ G](x) - [F \circ G](a)}{x - a}$.

13. a. Give the point-slope equation of the line which contains the point $(a,1/a)$ and has slope $-1/a^2$.
b. On one set of coordinate axis, *carefully* sketch the graphs of the function H and the line mentioned above using $a = 2$.

14. a. If $S(x) = x^2 - x$, give $\dfrac{S(x) - S(1)}{x - 1}$.
b. Give the point-slope equation of the line which contains $[1,S(1)]$ and has slope 1.
c. On one set of coordinate axis, carefully sketch the graphs of the function S and the line in part *b*.

15. a. If $G(x) = 2x - x^2$, give $\dfrac{G(x) - G(2)}{x - 2}$.
b. On one set of coordinate axis, carefully sketch the graphs of the function G and of the line which contains (2,0) and has slope -2.

16. a. If $Q(x) = \dfrac{x + 1}{x}$, dom $Q = \langle 0,4]$, give $\dfrac{Q(x) - Q(1)}{x - 1}$.
b. On one set of coordinate axis, carefully sketch the graphs of the

function Q and of the line which contains $[1,Q(1)]$ and has slope -1.

17. a. If $Z(x) = \sqrt{x}$, give $\dfrac{Z(x) - Z(1/2)}{x - 1/2}$.

b. Give the point-slope equation of the line which contains $[1/2,Z(1/2)]$ and has slope 1.

c. On one set of coordinate axis, carefully sketch the graphs of the function Z and of the line in part *b*.

Introduction In this chapter, we shall try to follow a trail blazed long ago by Gottfried Wilhelm von Leibniz. Though this great mathematician clearly marked the path for anyone to follow, you will note that the guide (the text) is prone to stagger from the trail all too frequently and to lose itself in the surrounding jungle.

In fairness, I suppose we ought to mention that Isaac Newton also marked this same trail. While Leibniz considered the problem of a line tangent to a graph, Newton started with the concept of velocity. He imagined a point moving along a straight line. At some time t_0, the point will be at A. What shall we mean by the velocity of this moving point at the instant t_0? (See figure 11.2.) If, at another instant t, the point is at B,

Fig. 11.2

we can determine the distance AB that is traversed in the elapsed time $t - t_0$. Divide this distance by the elapsed time and get the so-called average velocity. If t is very close to t_0, then this mean or average velocity will be an approximation to the velocity at the instant t_0. The closer t is to t_0, the better the approximation. So, if t is chosen closer and closer to t_0, the average velocity approaches closer and closer to some definite value which is called the *velocity of the point* at the instant t_0.

Section 1—Tangent Line In plane geometry, the tangent to a circle at a point P of the circle is defined as the line that intersects the circle at P and only at P (see figure 11.3). Such a definition would not work for the parabola. For, at each point on the parabola, there are two lines that meet the parabola at this point only: a line parallel to the axis and one other line. (See figure 11.4.) We feel that this "other line" is the "true" tangent; for it "just touches" the parabola. For a parabola, then, the tangent is a line having exactly one point in common with the parabola and not

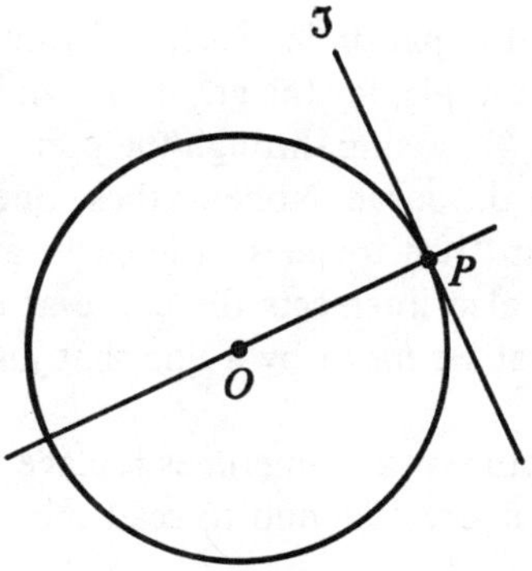

Fig. 11.3

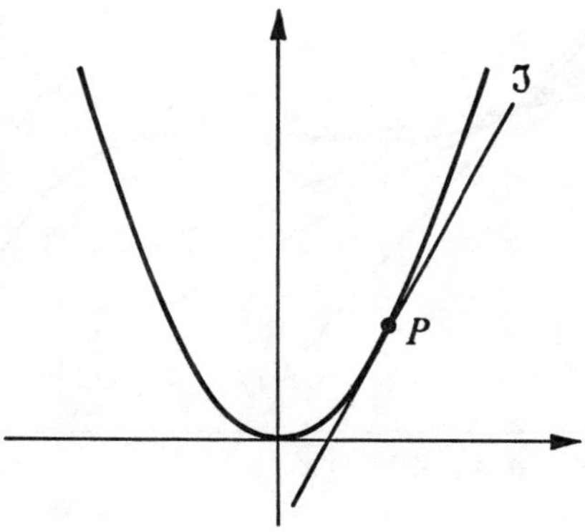

Fig. 11.4

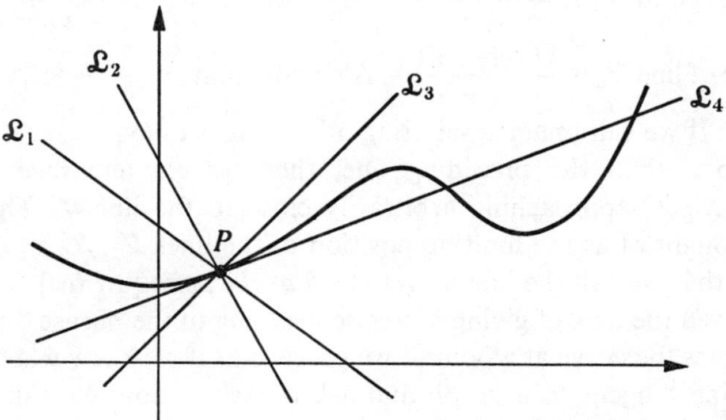

Fig. 11.5

parallel to the axis of the parabola. Such a definition, however, will not work in general. For example, for the graph shown in figure 11.5, there are several lines ($\mathscr{L}_1$, $\mathscr{L}_2$, $\mathscr{L}_3$) passing through the point P and having only this point in common with the curve. None of these lines fits the notion of the tangent as the line that "just touches" the curve at the point P. The line $\mathscr{L}_4$ does, but this line also intersects the curve at other points. Our task then is to describe what we mean by a line that just touches a curve at a point P of the curve.

Because a single picture may sometimes replace a thousand words, you are invited to look at figure 11.6 and to read the following comments:

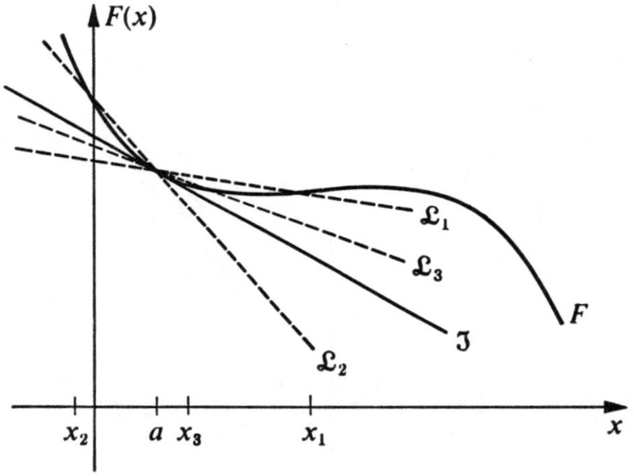

Fig. 11.6

The slope of $\mathscr{L}_1$ is $\dfrac{F(x_1) - F(a)}{x_1 - a}$, the slope of $\mathscr{L}_2$ is $\dfrac{F(x_2) - F(a)}{x_2 - a}$, and the slope of line $\mathscr{L}_3$ is $\dfrac{F(x_3) - F(a)}{x_3 - a}$. Also note that $|x_3 - a| < |x_2 - a| < |x_1 - a|$. If we can imagine selecting other values of $(x_4, x_5, \ldots, x_n)$, each closer to a than the preceding one, then we can envisage the lines $\mathscr{L}_4, \mathscr{L}_5, \ldots, \mathscr{L}_n$ approaching arbitrarily close to the line $\mathscr{T}$. The line $\mathscr{T}$ being thought of as the limiting position of the lines $\mathscr{L}_1, \mathscr{L}_2, \ldots, \mathscr{L}_n$. It is this line that we call the line *tangent to F at the point* $[a,F(a)]$.

So, given the task of giving a precise meaning to the phrase "a line that just touches the curve at a point," we proceed as if we knew what is meant by the line tangent to a graph and ask ourselves how we can compute the slope of this line. We shall, then, turn the method of computing this slope into the desired definition.

If we are given a function F and a point in F, say $[a,F(a)]$, we want to compute the slope m of the tangent line—that is, the line tangent to the graph of F at the given point. The trouble is, to compute the slope of any line you need another point. *The key observation is that, if we pick another point* $Q = [x_1,F(x_1)]$ *which is close to P, then the slope* m_1 *of the line* $\mathscr{L}_1$ (see figure 11.7) *will be close to m.* But we are pleased to report, m_1 can be computed for

$$m_1 = \frac{F(x_1) - F(a)}{x_1 - a}.$$

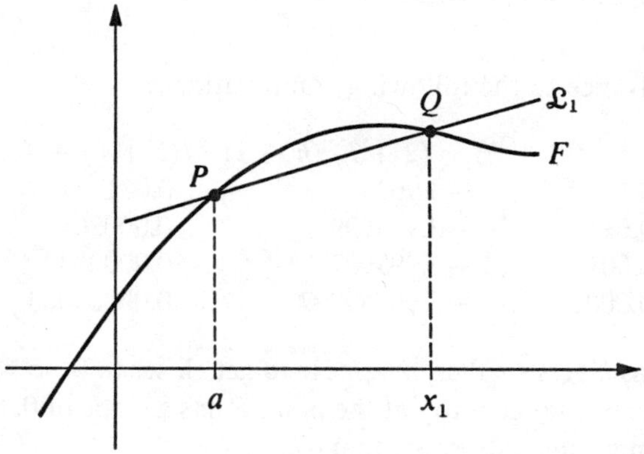

Fig. 11.7

The point Q being close to P (the pathetic language used above) means that x_1 is close to a, that is, $|x_1 - a|$ is rather small. While this is not much better, we expect that the smaller $|x_1 - a|$ is, the closer m_1 is to m.

Example 1. Let us test these remarks using $F(x) = x^2$ and $P = [1,F(1)] = (1,1)$.

Selecting $Q = [x,F(x)]$, we write $m = \dfrac{F(x) - F(1)}{x - 1}$.
In making the necessary computations, it is expedient to let $x = 1 + h$ and to write $m = \dfrac{F(1 + h) - F(1)}{h}$.

h	$F(1 + h) = (1 + h)^2$	$F(1 + h) - F(1)$	m
0.1	1.21	0.21	2.1
−0.01	0.9801	−0.0199	1.99
0.001	1.002001	0.002001	2.001
0.0001	1.00020001	0.00020001	2.0001

From the above, it seems reasonable to assert that as h gets closer to 0, x gets closer to 1, and m appears to get closer to 2. So we conjecture that the slope of the line tangent to F at the point P in F has a slope of 2. We should also note that $\frac{F(x) - F(1)}{x - 1} = \frac{x^2 - 1}{x - 1} = x + 1$ and thus, as x approaches 1, the slope quotient approaches 2.

Example 2. $f(x) = x^3 - 3x^2$ and $P = [2,f(2)] = (2,-4)$. Again, for expedience, let $x = 2 + h$ and write

$$m = \frac{f(x) - f(2)}{x - 2} = \frac{f(2 + h) - f(2)}{h}$$

and advance to the following computations:

h	$f(2 + h) = (2 + h)^2(h - 1)$	$f(2 + h) - f(2)$	m
0.1	-3.969	0.031	0.31
0.01	-3.999699	0.000301	0.0301
0.001	-3.999997	0.000003	0.003
0.0001	-3.99999997	0.00000003	0.0003

Thus, as h gets smaller, m appear to get closer to 0, so we conjecture that the line tangent to f at the point P has a slope of 0, that is, it is a horizontal line. Also note that

$$\frac{f(x) - f(2)}{x - 2} = \frac{x^3 - 3x^2 + 4}{x - 2} = x^2 - x - 2$$

which approaches 0 as x approaches 2.

One more time!

Example 3. a. $G(x) = \sqrt{(x + 1)}$; $P = [3,G(3)] = (3,2)$.

Let $x = 3 + h$, so then $m = \frac{G(x) - G(3)}{x - 3} = \frac{G(3 + h) - G(3)}{h}$

h	$G(3 + h) - G(3) = \sqrt{(4 + h)} - 2$	m
0.2	0.0493901	0.24695
-0.1	0.0246951	0.251582
0.05	0.0124612	0.249224
-0.01	0.000250156	0.250156
0.001	0.0002499984	0.2499984
10^{-5}	$0.2499998 \cdot 10^{-5}$	0.2499998

so it seems that as h gets smaller, m gets closer to 0.25. Also note that $\frac{G(x) - G(3)}{x - 3} = \frac{1}{\sqrt{(x + 1)} + 2}$ for this allows us to conjecture that as x approaches 3, the slope quotient approaches 1/4.

Example 3. b. For contrast and to make another point, use the same function G and let $P = (-1,0)$. Again we let $x = -1 + h$ but note that $h > 0$ is necessary in order that $-1 + h \in \text{dom } G$.

h	$G(-1 + h) - G(-1) = \sqrt{h}$	m
0.1	0.3162278	3.16228
0.01	0.1	10.0
0.0001	0.01	100.0
10^{-6}	0.001	1000.0

We should also note that

$$\frac{G(x) - G(-1)}{x + 1} = \frac{1}{\sqrt{(x + 1)}}.$$

so that as x gets closer to -1, the slope quotient starts giving us trouble. To be specific, the slope quotient increases beyond all bound. The implications of this can be seen by sketching the graph of G and by noting that the lines $\mathscr{L}_1, \mathscr{L}_2, \mathscr{L}_3, \ldots$ seem to approach the vertical line. (See figure 11.8.) The implication of this can be expressed by

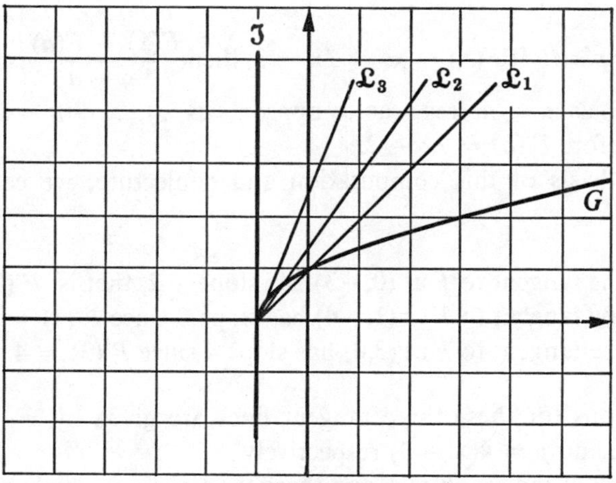

Fig. 11.8

the statement, "the line tangent to G at the point $(-1,0)$ is a vertical line which has no slope."

It should now be reasonably clear that, though the graph of a function is not always a straight line and therefore may not have a fixed slope, at each point $[a,F(a)]$ on the graph there may be a line tangent to the graph that either has a fixed slope or is vertical. You have also noted that we can find the slope of this tangent line by:

1. simplifying the slope quotient $\dfrac{F(x) - F(a)}{x - a}$ and
2. conjecturing an answer to the question, "As x approaches a, does the slope quotient approach some number?"

When the answer to this question is "yes," then this number is the slope of the tangent line at the point $[a,F(a)]$ and we shall denote this number by $F'(a)$.

Now we are able to define the line tangent to the graph of F at the point $[a,F(a)]$, completing our mission, by using ye olde point-slope equation of a line.

DEFINITION. *If* $a \in \operatorname{dom} F$ *and* $F'(a)$ *exists, then the line* $\mathcal{T}$ *tangent to the graph of* F *at the point* $[a,F(a)]$ *is given by*

$$\mathcal{T} = \{(x,y) \mid y - F(a) = F'(a)\cdot(x - a)\}.$$

Example 4. If $F(x) = x^2 - 2x - 3$, then $\dfrac{F(x) - F(a)}{x - a}$ is $x + a - 2$. Thus, as $x \to a$ (read as x approaches a), $x + a - 2 \to 2a - 2$. Therefore $F'(a) = 2a - 2$.

On the basis of this computation and conjecture, we can conclude that:

the line tangent to F at $(0,-3)$ has slope -2, that is, $F'(0) = -2$.
the line tangent to F at $(1,-4)$ has slope 0 since $F'(1) = 0$.
the line tangent to F at $(3,0)$ has slope 4 since $F'(3) = 4$.

The equations for these three tangent lines are given by $y + 3 = -2x$, $y + 4 = 0$, and $y = 4(x - 3)$ respectively.

Now look at the graph of F and these tangent lines which are shown in figure 11.9.

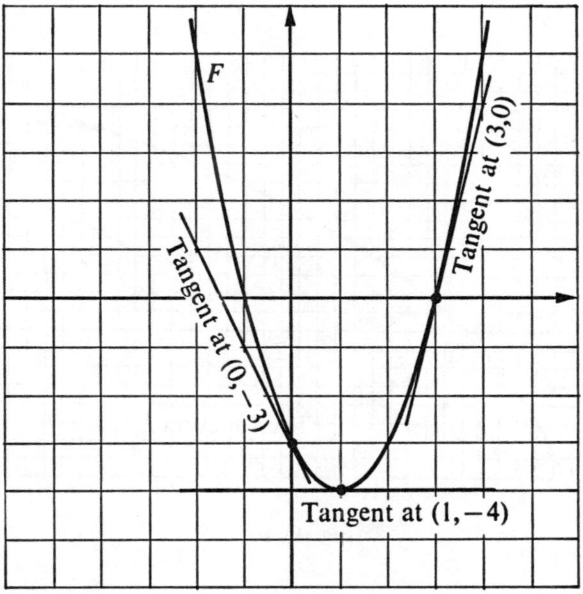

Fig. 11.9

PROBLEM 11.1

1. Figure 11.10 is the graph of the function L. Draw the lines tangent to the graph of L at the points $[1,L(1)]$, $[2,L(2)]$, and $[3,L(3)]$. Now, estimate the slope of each of these tangent lines.
2. On the basis of what you have done in problem 1, can you make a guess at what is the slope of the line tangent to L at any point $[a,L(a)]$? If you can, do so. If you cannot make a guess, you are probably one of the great silent majority of the class.
3. Figure 11.11 is the graph of the function E. Draw the lines tangent to the graph of E at the points $[-1/2,E(-1/2)]$, $[0,E(0)]$, and $[1,E(1)]$. Now, estimate the slope of each of these tangents.
4. Using the above figure, estimate the ordinates of the points of tangency. Compare with the slopes obtained in the above problem. Can you now make a guess about the slope of the line tangent to E at any point $[a,E(a)]$? See remarks in problem 2.

In each of problems 5–9, find the slope of the line tangent to the graph of F at the indicated point. On the same set of axis, sketch the graph of F and of the tangent line.

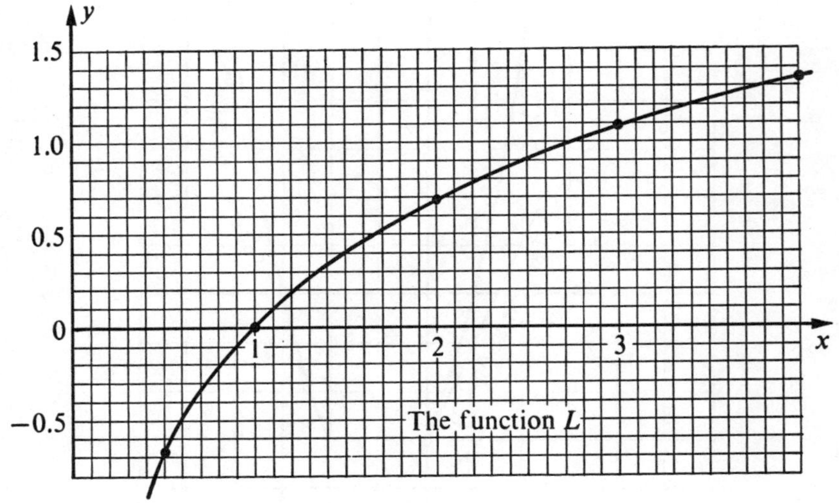

Fig. 11.10

5. $F(x) = x^2 + 2x$; $[-1, F(-1)]$.

6. $F(x) = x^3$; $[1, F(1)]$.

7. $F(x) = \frac{1}{x}$; $[1, F(1)]$.

8. $F(x) = \sqrt{x}$; $[1, F(1)]$.

9. $F(x) = \frac{x}{x+1}$; $[1, F(1)]$.

10. Suppose $G(x) = x^3 - 3x^2$. Give the coordinates of the point(s) $[a, G(a)]$ where the tangent line is horizontal, that is, where $G'(a) = 0$.

11. Same as problem 10 for $x^2 - 3x$.

12. Suppose that $f(x) = x^2 + 3x - 4$. Find $f'(a)$ and use it to find a number c between -2 and 2 such that $f'(c) = \frac{f(2) - f(-2)}{4}$. Now, on one set of axis, sketch the graph of f, the line containing the points $[-2, f(-2)]$ and $[2, f(2)]$, *and* the line containing the point $[c, f(c)]$ whose slope is $f'(c)$.

13. Suppose that $g(x) = \sqrt{x}$. Find $g'(a)$ and use it to find a number c between 0 and 4 such that $g'(c) = \frac{g(4) - g(0)}{4}$. Now, on one set of

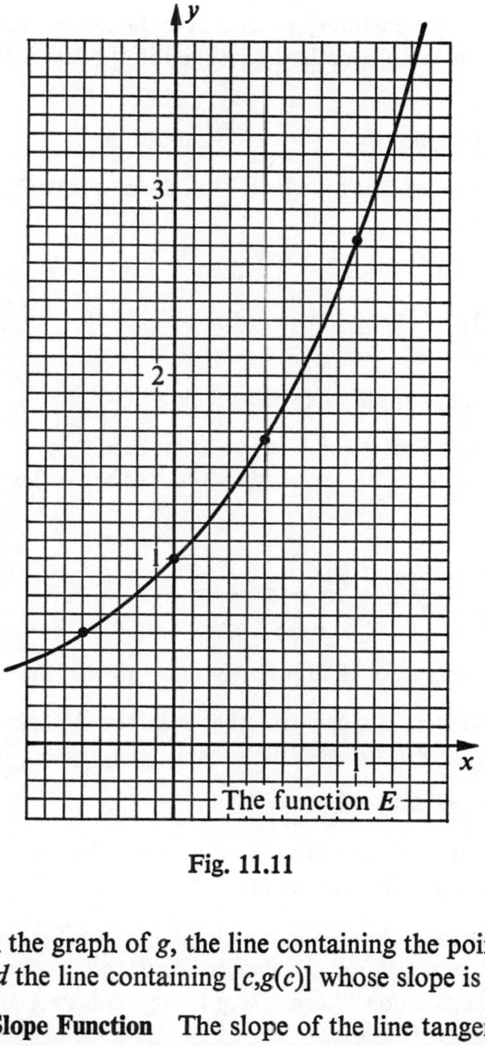

Fig. 11.11

axis, sketch the graph of g, the line containing the points $[0,g(0)]$ and $[4,g(4)]$, *and* the line containing $[c,g(c)]$ whose slope is $g'(c)$.

Section 2—Slope Function The slope of the line tangent to the graph of a function F at the point $[a,F(a)]$ is the number denoted by $F'(a)$. We shall call this number *the slope of F at a.*

A knowledge of the slope of the graph of a function can be useful. For example, the fact that the line tangent to the function $F(x) = x^2 - 2x - 3$ (see last example of the previous section) at the point $(1,-4)$ is horizontal makes it clear that the point $(1,-4)$ is a minimum point of the function. The fact that the slope of F at 0 is a negative number indicates that the function is decreasing in the vicinity of the point $(0,-3)$. The fact that the slope of F at 3 is 4 indicates that the function is increasing in the vicinity of the point $(3,0)$.

The slope of any elementary function can be determined at every point in its graph by the simple methods outlined in the last section:

1. Find and simplify $\frac{F(x) - F(a)}{x - a}$ and then
2. find the number to which the slope quotient converges as x approaches a.

Example. If $F(x) = x^3 - 3x$, then

$$\frac{F(x) - F(a)}{x - a} = \frac{(x^3 - 3x) - (a^3 - 3a)}{x - a}$$

which we can simplify as shown below

$$\begin{aligned} F(x) - F(a) &= x^3 - a^3 - 3x + 3a \\ &= (x - a)(x^2 + ax + a^2) - 3(x - a) \end{aligned}$$

hence

$$\frac{F(x) - F(a)}{x - a} = x^2 + ax + a^2 - 3.$$

So as $x \to a$ (i.e., x is sufficiently close to (a), then $\frac{F(x) - F(a)}{x - a} \to 3a^2 - 3$. Hence, for each number a in the domain of F there is a unique number $F'(a)$ and, as we have just argued,

$$F'(a) = 3a^2 - 3.$$

We might just as well have written that $F'(x) = 3x^2 - 3$ for each number x in the domain of F.

In other words, the slope of F at $[X, F(X)]$ is itself a function of X. We will denote this function by F' and we will call it the slope function. *The domain of the function F' consists of those numbers X in the domain of F for which $F'(X)$ is a number.*

In figure 11.12 we have sketched the graphs of the function $F(x) = x^3 - 3x$ and the slope function F' where $F'(X) = 3x^2 - 3$, dom $F' = R$.

Example 2. If $G(x) = \frac{x + 1}{x}$, $x \neq 0$, then, for $a \neq 0$, $G(x) - G(a) = \frac{x + 1}{x} - \frac{a + 1}{a} = -\frac{x - a}{ax}$, hence as $x \to a$, $\frac{G(x) - G(a)}{x - a} \to -1/a^2$ and so $G'(a) = -1/a^2$ or, making a slight notational change, the slope function G' has the rule of pairing $G'(x) = -1/x^2$ and dom G' $= \{x \mid x \neq 0\}$.

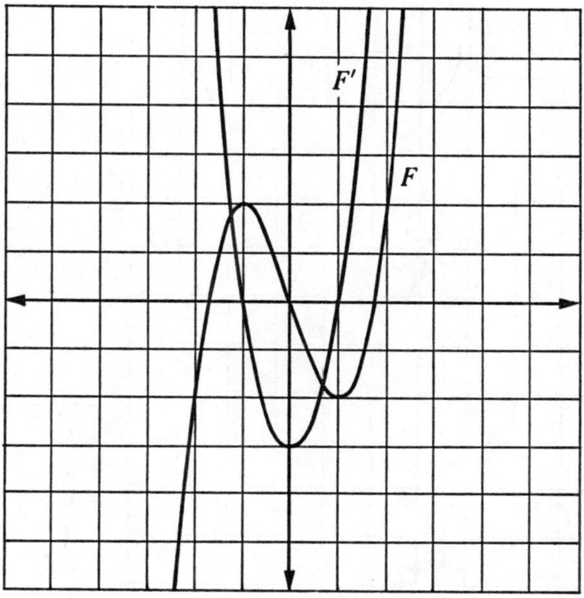

Fig. 11.12

The slope of a function is important in the study of the function. The elementary functions, as a matter of fact, can be defined in terms of their slope. The function E, for example, could be defined by simply requiring that $E(1) = 0$ and that its slope at each point be equal to the value of the function itself at that point. We don't intend doing this because we have no desire to rob you of the pleasure (in calculus) of seeing how this is done. We do intend, however, to indicate how the slope function can be used to help sketch the graph of a given function.

In making the graph of a function, it can be helpful to know the slope of the curve at each point and to indicate this slope by drawing a short segment of the tangent line at each point.

Example 3. If $f(x) = x^3 + 2x^2 - 4x + 1$, then $f'(x) = 3x^2 + 4x - 4$. Dom $f' = R$. You should verify this. We tabulate values of $f(x)$ and $f'(x)$ for certain values of x. A little factoring helps. Note that $f'(x) = (3x - 2)(x - 2)$. (See figure 11.13.)

To be certain that you have not been misled by the previous examples, let me stress the idea that the dom F' need not be the same as the dom F.

Example 4. Suppose that $G(x) = \sqrt{(x + 1)}$. (This is the same function used in example 3b on p. 383.) We can show that

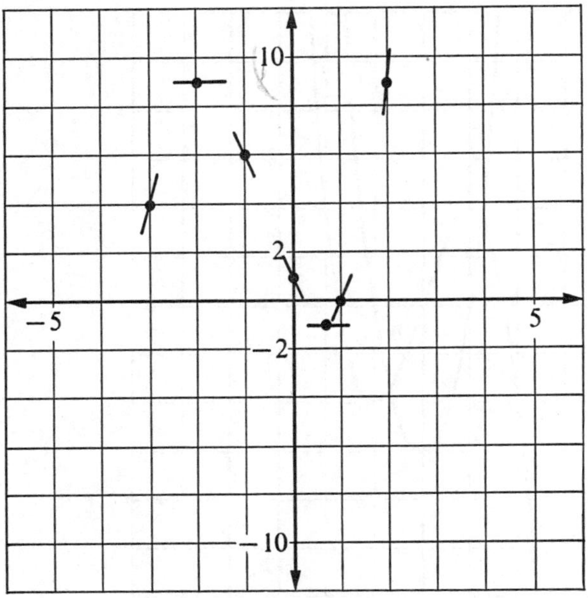

Fig. 11.13

x	$f(x)$	$f'(x)$
−3	4	11
−2	9	0
−1	6	−5
0	1	−4
2/3	−13/27	0
1	0	3
2	9	16

$\dfrac{G(x) - (Ga)}{x - a} = \dfrac{1}{\sqrt{(x + 1)} + \sqrt{(a + 1)}}$. So it seems reasonable to assert that as x approaches a, the slope quotient approaches $\dfrac{1}{2\sqrt{(a + 1)}}$ and to write $G'(a) = \dfrac{1}{2\sqrt{(a + 1)}}$. Hence, $G'(x) = \dfrac{1}{2\sqrt{(x + 1)}}$. Note, however, that in spite of the fact that $G(-1) = 0$, $G'(-1)$ does not exist. *Conclusion?* Dom $G' = \{x \mid x > -1\}$.

Example 5. Suppose that $V(x) = |x|$, dom $V = R$.

$$V(x) - V(a) = |x| - |a| = \frac{(x - a)(x + a)}{|x| + |a|}, \text{ hence}$$

$$\frac{V(x) - V(a)}{x - a} = \frac{x + a}{|x| + |a|}.$$

$$\text{As } x \to a, \frac{V(x) - V(a)}{x - a} \to \frac{2a}{2|a|} = \frac{a}{|a|}$$

that is, $V'(a) = \dfrac{a}{|a|}$. Thus $V'(x) = \dfrac{x}{|x|}$.

This should make us suspicious about the existence of $V'(0)$. To confirm this suspicion, let $a = 0$ and note that

$$\text{for } x > 0, \frac{V(x) - V(0)}{x - 0} = \frac{x}{|x|} = \frac{x}{x} = 1 \text{ but}$$

$$\text{for } x < 0, \frac{V(x) - V(0)}{x - 0} = \frac{x}{|x|} = \frac{x}{-x} = -1.$$

Therefore there is no number to which the slope quotient converges or approaches as x approaches 0. *Conclusion:* $V'(0)$ does not exist.

Summary If $V(x) = |x|$, then $V'(x) = \dfrac{x}{|x|}$; dom $V' = \{x \mid x \neq 0\}$.

Let us conclude this section with an example which will serve as a guide to the form we expect you to use in many of the exercises or problems to follow.

Example. $F(x) = x^3 - 2x$, give $F'(x)$.

Solution.
$$\begin{aligned} F(x) - F(a) &= (x^3 - 2x) - (a^3 - 2a) \\ &= x^3 - a^3 - 2x + 2a \\ &= (x - a)(x^2 + ax + a^2) - 2(x - a) \\ &= (x - a)(x^2 + ax + a^2 - 2) \end{aligned}$$

$$\text{thus } \frac{F(x) - F(a)}{x - a} = x^2 + ax + a^2 - 2.$$

$$\text{As } x \to a, \frac{F(x) - F(a)}{x - a} \to a^2 + a^2 + a^2 - 2 = 3a^2 - 2, \text{ hence}$$

$$F'(a) = 3a^2 - 2$$

and therefore

$$F'(x) = 3x^2 - 2; \text{ dom } F' = R.$$

PROBLEM SET 11.2

1. Before you become immersed in the algebra involved in the problems that follow, stop a moment and consider this question. If F is a function, what geometric meaning can be attached to the number $F'(x)$?

2. If $F(x) = mx + b$, give $F'(x)$ and dom F'.

3. If $F(x) = x^2$, give $F'(x)$ and dom F'.

4. If $F(x) = x^3$, give $F'(x)$ and dom F'.

5. If $F(x) = x^4$, give $F'(x)$ and dom F'.

6. If $P(x) = x^2 + 2x$, give $P'(x)$ and dom P'.

7. If $P(x) = 2x^2 - x$, give $P'(x)$ and dom P'.

8. If $G(x) = x^3 + 2x - 3$, give $G'(x)$ and dom P'.

9. If $p(x) = x^3 - x^2 + 1$, give $p'(x)$ and dom p'.

10. If $U(x) = x^3 + 2x$, give $U'(x)$ and dom U'.

11. If $F(x) = \sqrt{(x + 2)}$, give $F'(x)$ and dom F'. *Hint:* $\sqrt{z} - \sqrt{t} = \dfrac{z - t}{\sqrt{z} + \sqrt{t}}$.

12. If $F(x) = \sqrt{(2x)}$, give $F'(x)$ and dom F'. See hint to problem 11.

13. If $V(x) = |3x|$, give $V'(x)$ and dom V'. *Hint:* $|z| - |t| = \dfrac{(z - t)(z + t)}{|z| + |t|}$.

14. If $V(x) = |x - 2|$, give $V'(x)$ and dom V'. See hint to problem 13.

15. If $Q(x) = \dfrac{x + 2}{x}$, give $Q'(x)$ and dom Q'.

16. If $q(x) = \dfrac{x^2}{x - 1}$, give $q'(x)$ and dom q'.

In each of problems 17–20, give the values of $x \in$ dom F for which $F'(x) = 0$ or for which $F'(x)$ does not exist (if there are any) *and* then sketch the graphs of F and F'.

17. $F(x) = x^2 + 2x - 3$.

18. $F(x) = x^3 - x^2$.

19. $F(x) = \dfrac{x^2}{x - 1}$.

20. $F(x) = \sqrt{(x + 2)}$.

21. *Bonus problem.* If $Z(x) = |x^2 - 1|$, give: **(a)** $Z'(x)$, **(b)** the values of x for which $Z'(x)$ does not exist, and **(c)** sketch the graph of Z.

Section 3—Rate of Change At this time and in this place we shall merely "skim the surface" of an application of the slope function that is quite useful. It is hoped that this brief encounter with the notion of *rate of change* (alias "related rate" in some texts) will arouse your interest in the topic to the point where you will want to read more about it in any calculus textbook.

If f is a linear function, $f(x) = mx + b$, then the slope m gives the rate of change of $f(x)$ with respect to x:

$$m = \frac{\text{change in } f(x)}{\text{change in } x} = \frac{f(x) - f(x_1)}{x - x_1}.$$

In other words, on a line whose slope is m, the ordinates change m times as fast as the abscissa. Also note that if $f(x) = mx + b$, then $f'(x) = m$.

For the more general case, if f is a function that has slope at each of its points, the graph of f is some curve. The slope $f'(x)$ still gives the rate of change of $f(x)$ with respect to x, but this rate of change varies from point to point. For example (see figure 11.14), at x_1 the rate of change is $f'(x_1)$;

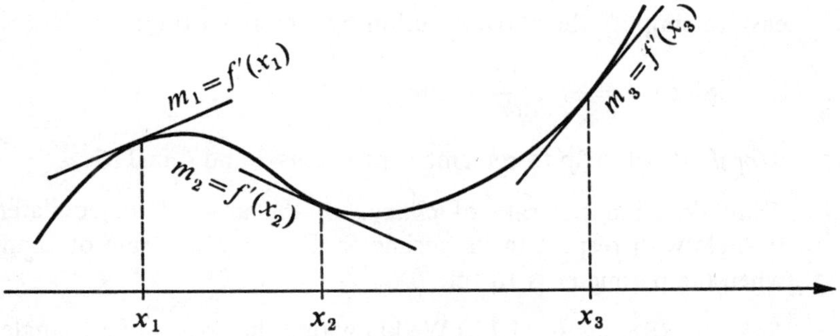

Fig. 11.14

at x_2 the rate of change is $f'(x_2)$; at x_3 the rate of change is $f'(x_3)$. Thus $f'(x)$ gives us the rate of change of $f(x)$ with respect to x for each x in the domain of f.

There is a notation that is very attractive to persons who apply the notion of rate of change to physical problems. This notation is not only suggestive of a rate or ratio but it is also widely used. I am therefore morally bound to call it to your attention.

If $y = f(x)$, then $\dfrac{dy}{dx} = f'(x)$.

So, whenever you see the notation $\dfrac{dy}{dx}$, don't argue or worry over what it means. Just look at what we have said it means.

Example. Find the rate of change of the area of a circle with respect to its radius. Evaluate this rate of change when the radius is 2 cm.

Solution. Since $A = \pi r^2$, that is $A(r) = \pi r^2$, then

$$A'(r_0) = \lim_{r \to r_0} \frac{\pi r^2 - \pi r_0^2}{r - r_0} = \lim_{r \to r_0} \pi(r + r_0) = 2\pi r_0$$

thus the rate of change of the area of a circle with respect to its radius is $2\pi r$ and we write

$$\frac{dA}{dr} = 2\pi r.$$

When $r = 2$, $\dfrac{dA}{dr} = 4\pi$.

Remark. If we had recalled the solution to problem 3 of problem set 11.2, ($f(x) = x^2$ implies that $f'(x) = 2x$), it would have been easy to simplify the above solution by merely writing:

$$\text{Since } A = \pi r^2, \frac{dA}{dr} = 2\pi r.$$

Moral. It will help to remember problems 3 and 4 in 11.2.

Example. Find the rate of change of the area of an equilateral triangle with respect to its perimeter. Evaluate this rate of change when the perimeter is 18 cm.

Solution. (See figure 11.15.) We know that the area A of a triangle is

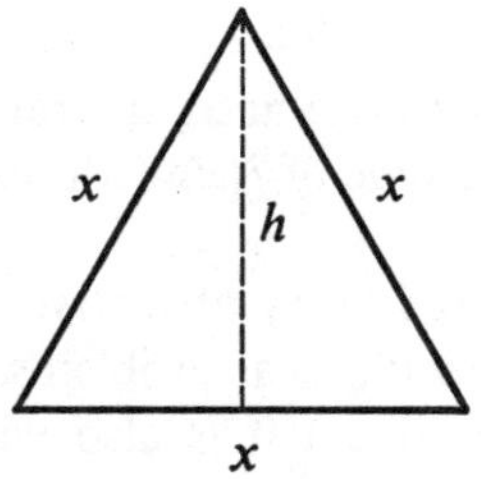

Fig. 11.15

given by the formula $A = \frac{1}{2}$ (base)(altitude). If we let x denote the length of each side and h the altitude, then $h^2 = \frac{3}{4}x^2$ so that $h = \frac{\sqrt{3}}{2}x$. Substituting in the formula above we have $A = \frac{\sqrt{3}}{4}x^2$, hence $\frac{dA}{dx} = \frac{\sqrt{3}}{2}x$. But this is not what we wished to find. (What does $\frac{dA}{dx}$

denote?) After we have stifled our irritation, we realize that if p is the perimeter, $p = 3x$. This implies that $x^2 = \frac{p^2}{9}$ so that $A = \frac{\sqrt{3}}{36} p^2$. From this it easily follows that

$$\frac{dA}{dp} = \frac{\sqrt{3}}{18} p \text{ and, for } p = 18, \frac{dA}{dp} = \sqrt{3}.$$

Example. The dimensions of a rectangle are changing in such a way that the area of the rectangle remains constant. Find the rate of change of the height h with respect to the base b.

Solution. Now $A = bh$ so if $A = k$ where k is a constant, $h = \frac{k}{b}$, that is, $h(b) = \frac{k}{b}$. Hence

$$h'(b_0) = \lim_{b \to b_0} \frac{-k(b - b_0)}{bb_0(b - b_0)} = \lim_{b \to b_0} \frac{-k}{bb_0} = -\frac{k}{b_0^2}$$

and so

$$\frac{dh}{db} = -\frac{k}{b^2}.$$

The negative sign means that the height decreases as the base increases.

Velocity One of the most useful (maybe the most interesting) rate-of-change problems is that of a particle moving along a straight line (rectilinear motion). We restrict ourselves to motion along a straight line for the simple reason that we are not prepared to handle curvilinear motion.*

We choose a point O on the straight line along which our particle moves and we designate one direction as positive. We also choose a unit of length, a unit of time, and an instant from which we count time. Then the particle moves along a number line and the motion defines a function s which associates with each number t the coordinate $s(t)$ of the particle (point) at that time. This function s is often called the "position function of the particle" and the graph of this function can be thought of as the picture of the motion.

For example, if $s(t) = t^2 - 3t$, then at the time $t = 0$, the position is $s(0) = 0$ and the particle P is at the origin. At the time $t = 1$, $s(1) = -2$

* The problem of tangents to a curve was only one of two problems that led to the development of the derivative calculus. The other was the description of motion. This latter problem dominated the work of Newton.

and P is two units to the left of the origin. Since $s(3) = 0$, then P returns to the origin after three units of time. When $t = 5$, the position is given by $s(5) = 10$ or 10 units to the right of the origin.

Uniform motion The motion of the particle is called *uniform motion* if the distance traversed by the particle is proportional to the elapsed time. Assume that a particle has uniform motion and let δ be a number defined as follows: $|\delta|$ is the distance traversed during one time unit; δ is positive or negative according to whether the particle moves in the positive or negative direction. Thus our particle has the velocity

$$\delta \frac{\text{(length units)}}{\text{(time units)}}$$

for example, δ ft./min., δ km./hr., and so forth. If, at $t = 0$, our particle is at a distance b from the origin, then after t time units, its distance from b will be δt. Hence its distance from O will be $b + \delta t$. This motion, then, is described by the function

$$s(t) = \delta t + b$$

and it is easy to see that the graph of a uniform motion is a straight line and that the slope of this line is the velocity.

Nonuniform Motion (See figure 11.16.) Let us next consider a particle

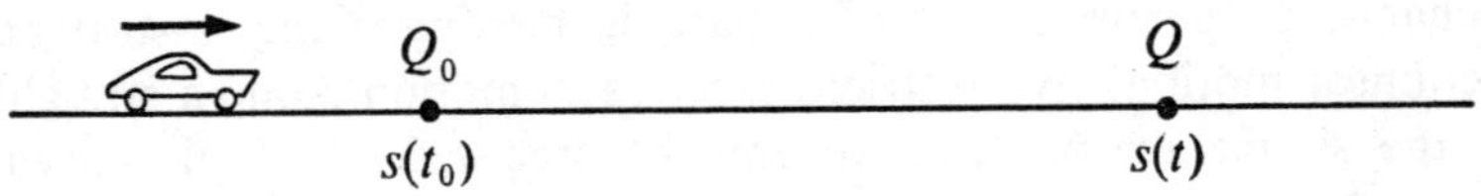

Fig. 11.16

whose motion is not uniform. For instance, let us imagine an automobile traveling along a straight road so that at some time t_0 it is at the point Q_0 whose coordinate is $s(t_0)$ and, that at some later time t, it is at the point Q whose coordinate is $s(t)$. The important question we put to the driver of this automobile is, "What is the velocity at Q_0?" Since his answer is likely to be the number read on the speedometer," let us suppose (just to be difficult) that this automobile has an odometer but no speedometer. Now, we are all familiar with the notion that the *average velocity* for the journey from Q_0 to Q is given by

$$\text{avg vel} = \frac{s(t) - s(t_0)}{t - t_0}.$$

But our driver (ill equipped, perhaps, but not stupid) realizes that this average velocity is quite likely to be a poor approximation to the velocity v at the instant t_0, especially so if the distance from Q_0 to Q is rather large. Sooner or later he will deduce that the closer t is to t_0, the better the agreement between the "average velocity" and the "true" or instantaneous velocity. In other words, he will define the velocity v as follows:

$$v = \lim_{t \to t_0} \frac{s(t) - s(t_0)}{t - t_0} = s'(t_0)$$

or, using the rate of change notation, as

$$v = \frac{ds}{dt}.$$

Applying this definition to the uniform motion

$$s(t) = \delta t + b$$

we obtain

$$v = \frac{ds}{dt} = \delta$$

as we should.

Free Fall A simple example of nonuniform rectilinear motion is that of a body released at a certain height and permitted to fall to the ground. Aristotle (384–322 B.C.) taught that the heavier the body, the faster it falls. Galileo (1564–1642), however, discovered that this motion is not uniform. He asserted that:

> GALILEO'S LAW. *The distance traversed by a freely falling body released from rest is proportional to the square of the time it has fallen, the constant of proportionality being the same for all bodies.*

Accordingly, if we consider the vertical direction of fall as negative, choose the origin O as the point of release, and count the time from the moment of release, then the motion of the falling body is described by the equation

$$s = -Kt^2.$$

The value of K is the same for all bodies, but it does depend on the units used. If we use feet and seconds, then K is approximately 16; for centimeters and seconds, K is approximately 490.

The velocity of the body at time t is

$$v = \frac{ds}{dt} = -2Kt$$

thus the *speed* (absolute value of the velocity) is proportional to the time of fall.

A particle released from rest and sliding down an inclined plane obeys essentially the same law (Galileo's). (See figure 11.17.) If s now denotes the

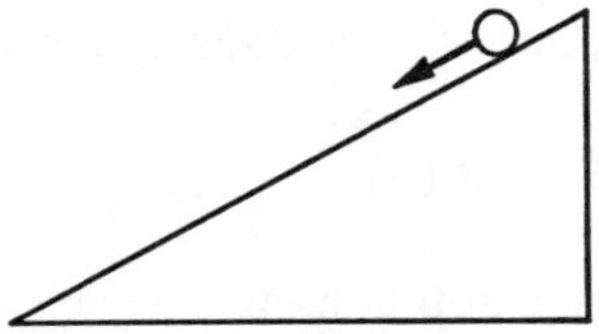

Fig. 11.17

length measured along the inclined plane from the moment of release, and if the upward direction is defined as positive, then Galileo's law reads

$$s = -\frac{Km}{\sqrt{(1 + m^2)}}\, t^2$$

where m is the slope of the inclined plane. In this case the velocity is given by

$$v = \frac{ds}{dt} = -\frac{2Km}{\sqrt{(1 + m^2)}}\, t.$$

If m is rather small, the speed increases rather slowly. This made it rather nice for Galileo since he had no clock and was unable to measure fast motions accurately.

PROBLEM SET 11.3

1. Find the rate of change of $y = \sqrt{x}$ with respect to x. Evaluate this rate of change when $x = 4$; when $x = \frac{1}{4}$.

2. Find the rate of change of the area A of a square with respect to the length x of a side. Evaluate this rate of change when $x = 3$.

3. Find the rate of change of the volume V of a cube with respect to the length x of a side. Evaluate this rate of change when $x = 3$.

4. Find the rate of change of the volume V of a sphere with respect to its radius r. *Hint:* $V = \frac{4}{3}\pi r^3$.

5. Find a number x_0 so that the rate of change of $y = x^2 + 2x$ with respect to x is 4 when $x = x_0$.

6. Given: $y = 2x^2 + 3x$ and $z = 3x^2 + 2x$.
Find: a number x such that $\frac{dy}{dx} = \frac{dz}{dx}$.

7. If the temperature of a confined gas is kept constant, its pressure p and its volume V are related by the equation $pV = k$ where k is a constant. Find the rate of change of the pressure with respect to the volume when the volume is 100 $cm.^3$.

8. A right circular cone has a radius of 3 cm. Find the rate of change of its volume V with respect to its altitude h.

9. The base of an isosceles triangle is 4 cm. Find the rate of change of its altitude h with respect to the length x of either of the two equal legs.

10. Water is pouring into an empty cylindrical tank at a rate of 10 $ft.^3$/min. If the radius of this tank is 3 ft., find the rate of change of the height of the water h with respect to the time t. *Hint:* Express h as a function of t and find $\frac{dh}{dt}$.

Velocity and Free Fall

11. If $s(t) = \frac{1}{2}t^2$ where we measure distance in meters and time in seconds, find the average velocity on the time interval from: **(a)** $t = 4$ to $t = 5$; **(b)** $t = 4$ to $t = 4.5$; **(c)** $t = 4$ to $t = 4.1$; and **(d)** find the velocity at $t = 4$. (See table 11.1.)

TABLE 11.1

Given: distance s in meters at time t seconds.	*Find:* (a) avg. velocity on the time interval from:	(b) velocity at
12. $s = t^2 + 96t$	$t = 0$ to $t = 2$	$t = 2$
13. $s = 4t - 2t^2$	$t = 0$ to $t = \frac{3}{2}$	$t = \frac{3}{2}$
14. $s = 3t + 6$	$t = 0$ to $t = 6$	$t = 3$
15. $s = t^2 + 2t + 1$	$t = 0$ to $t = 3$	$t = 3$
16. $s = \sqrt{(t + 1)}$	$t = 0$ to $t = 8$	$t = 8$

17. A body is dropped from a height of 100 ft. Find its velocity after 2 sec. How far has it fallen during these 2 sec.?

18. A body is dropped from a height of 10 m. (1000 cm.). Find its velocity after 0.1 sec.; after 0.2 sec.

19. A balloon 1600 ft. altitude drops a sandbag ballast. How long will it take the sandbag to hit the ground? What is its velocity upon impact?

20. A brick is dropped from a building. How far has it fallen when its velocity is -1176 cm./sec.?

21. A body slides along a smooth inclined plane that has a slope of $\frac{1}{2}$. If distance is measured in centimeters, what is the velocity after 2 sec.?

22. A particle sliding down an inclined plane achieves a velocity of -16 ft./sec. after 2 sec. What is the slope of the plane?

APPENDIX

PROOFS OF STATEMENTS RELEGATED TO THE APPENDIX

Proofs of statements from chapter 1, section 6

THEOREM. *If* $0 < p$, *then there is a positive number* x *such that* x^2 *is* p.

Proof. Let S be the set of all positive x such that $x^2 \leq p$. Since $\frac{p}{1 + p}$ is in S, S is not empty.

Now $(1 + p)^2 > p$ so $(1 + p)$ is an upper bound of S. Let LUB of $S = b$. Note that $b \geq \frac{p}{1 + p}$, hence $b > 0$. There are three possibilities: $b^2 > p$, $b^2 < p$, $b^2 = p$.

Suppose that $b^2 > p$ and let $c = b - \frac{(b^2 - p)}{2b} = \frac{1}{2}\left(b + \frac{p}{b}\right)$.
It follows that $0 < c < b$ and that

$$c^2 = \frac{1}{4}\left[b^2 + 2p + \left(\frac{p}{b}\right)^2\right] = \frac{1}{4}\left[4p + b^2 - 2p + \left(\frac{p}{b}\right)^2\right] \text{ so}$$

$$c^2 = p + \frac{1}{4}\left[b - \frac{p}{b}\right]^2 > p.$$

Therefore $c^2 > x^2$ for each x in S, hence $c > x$ for each x in S. But this means that c is an upper bound of S smaller than the least upper bound which is absurd. Thus $b^2 \ngtr p$.

Suppose that $b^2 < p$. Since $b > 0$, choose a positive number δ such that $\delta < b$ and $\delta < \frac{p - b^2}{3b}$. Then we have

$$(b + \delta)^2 = b^2 + \delta(2b + \delta) < b^2 + \delta(2b + b) \text{ so that}$$
$$(b + \delta)^2 < b^2 + 3b\delta < b^2 + (p - b^2) = p.$$

Hence $(b + \delta)$ is in S and, since $b + \delta > b$, b is not an upper bound of S. This contradiction establishes that $b^2 \nless p$. Therefore the only alternative is $b^2 = p$.

We now give a proof of the Archimedian property.

> *Proof.* Suppose that for every counting number n, $nt \leq x$. Let $S = \{nt \mid n \in N\}$. By our supposition, S is bounded above by x so there is a LUB of S.

Let b = LUB of S. Now $b - t < b$, hence $b - t$ cannot be an upper bound of S and so there is some element of S, call it mt, such that $mt > b - t$. From this, we deduce that $mt + t = (m + 1)t > b$. However, m is a counting number (since mt is in S), so $m + 1$ is a counting number. Hence $(m + 1)t$ is in S. It now follows that b is not even an upper bound of S, and this contradiction shows that our original assumption is false, thus proving the theorem.

Proof of a statement from chapter 6, section 5

Proof of the Statement

"If h is a number, then h contains a point of L."

The case $h = 0$ is trivial. We shall consider the case $h > 0$ only, assuming that the case $h < 0$ would follow because $L(1/x) = -L(x)$. Let us begin.

We know that there is a point of L below h and one above h. Label them

$[x_1,L(x_1)]$ and $[x_2,L(x_2)]$ respectively. Let $B = \{x \mid 1 < x \text{ and } L(x) < h\}$. Note that $x_1 \in B$. Let $A = \{x \mid 1 < x \text{ and } L(x) > h\}$. Note that $x_2 \in A$. B is bounded above by x_2 and A is bounded below by x_1. So, we let $b =$ LUB of B and $c =$ GLB of A. Note that $1 < b$. It follows that $c \nless b$. *Why?* Let us *assume* that $b < c$ since this appears to be a reasonable assumption. If $b < c$, then there is a number z between them (see figure A.1) and hence

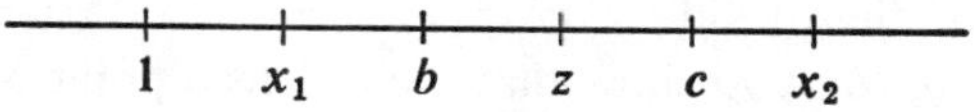

Fig. A.1.

$L(b) < L(z) < L(c)$. *Why?* If $L(z) < h$, then b is not an upper bound of B which is absurd. If $L(z) > h$, then c is not a lower bound of A. Also absurd. Hence, if $b < c$, then for each number z between b and c—and there are quite a few—$[z,L(z)] \in$ h. But you will be asked to argue that this is a contradiction. Therefore, not only is our assumption that $b < c$ unreasonable, it is false. Conclusion? $b = c$. (See figure A.2.) One detail remains to be settled. Does $L(c) = h$?

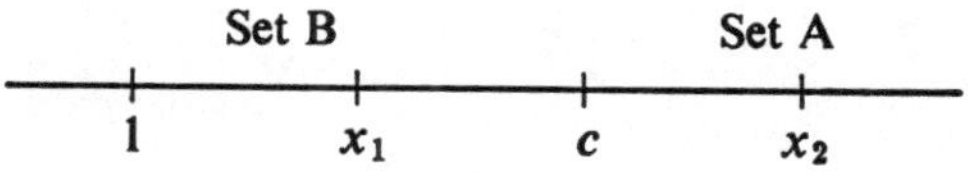

Fig. A.2.

Suppose that $L(c) < h$, then there is a positive number δ such that $L(c) = h - \delta$. We can show that $\int_c^{c+\delta} H < \delta/c \leq \delta$ so that $\int_1^c H + \int_c^{c+\delta} H < h - \delta + \delta$. Hence $L(c + \delta) < h$ and c is not an upper bound of B. Therefore $L(c) \nless h$.

Suppose that $L(c) > h$. If this is so, then there is a positive number ε such that $L(c) = h + \varepsilon$. Now, $L(c) = \int_1^c H$, so $L(c) < c - 1$. Hence $h + \varepsilon < c - 1$ and, with $h > 0$, it follows that $\varepsilon < c - 1$, that is, $1 < c - \varepsilon$. Equipped with this information, we find that

$$\int_1^{c-\varepsilon} H + \int_{c-\varepsilon}^{c} H = \int_1^c H \text{ and} \int_{c-\varepsilon}^{c} H < \frac{\varepsilon}{c - \varepsilon} < \varepsilon.$$

But $(c - \varepsilon) \in B$ and so $L(c - \varepsilon) < h$. Thus

$$L(c) = \int_1^{c-\varepsilon} H + \int_{c-\varepsilon}^{c} H < h + \varepsilon$$

and this contradiction shows that $L(c) \not> h$. Therefore, all things considered, $L(c) = h$ and the statement that h contains a point of L can be proven. This means that the *range of L is the set of real numbers.*

Bonus question for the student:

1. Why is the statement $c < b$ false? *Hint:* Suppose it were true.
2. Can you show that h does not contain two points of L? *Hint:* Suppose that it did.
3. If $h < 0$, can you show that h contains a point of L?

TABLES

These tables are presented on the following pages:

Table A.1—Squares, Cubes, Roots
Table A.2—Trigonometric Functions of Real Numbers
Table A.3—Trigonometric Functions of Angles

TABLE A.1

SQUARES, CUBES, ROOTS

n	n^2	$\sqrt{n}$	n^3	$\sqrt[3]{n}$	n	n^2	$\sqrt{n}$	n^3	$\sqrt[3]{n}$
1	1	1.000	1	1.000	51	2,601	7.141	132,651	3.708
2	4	1.414	8	1.260	52	2,704	7.211	140,608	3.733
3	9	1.732	27	1.442	53	2,809	7.280	148,877	3.756
4	16	2.000	64	1.587	54	2,916	7.348	157,464	3.780
5	25	2.236	125	1.710	55	3,025	7.416	166,375	3.803
6	36	2.449	216	1.817	56	3,136	7.483	175,616	3.826
7	49	2.646	343	1.913	57	3,249	7.550	185,193	3.849
8	64	2.828	512	2.000	58	3,364	7.616	195,112	3.871
9	81	3.000	729	2.080	59	3,481	7.681	205,379	3.893
10	100	3.162	1,000	2.154	60	3,600	7.746	216,000	3.915
11	121	3.317	1,331	2.224	61	3,721	7.810	226,981	3.936
12	144	3.464	1,728	2.289	62	3,844	7.874	238,328	3.958
13	169	3.606	2,197	2.351	63	3,969	7.937	250,047	3.979
14	196	3.742	2,744	2.410	64	4,096	8.000	262,144	4.000
15	225	3.873	3,375	2.466	65	4,225	8.062	274,625	4.021
16	256	4.000	4,096	2.520	66	4,356	8.124	287,496	4.041
17	289	4.123	4,913	2.571	67	4,489	8.185	300,763	4.062
18	324	4.243	5,832	2.621	68	4,624	8.246	314,432	4.082
19	361	4.359	6,859	2.668	69	4,761	8.307	328,509	4.102
20	400	4.472	8,000	2.714	70	4,900	8.367	343,000	4.121
21	441	4.583	9,261	2.759	71	5,041	8.426	357,911	4.141
22	484	4.690	10,648	2.802	72	5,184	8.485	373,248	4.160
23	529	4.796	12,167	2.844	73	5,329	8.544	389,017	4.179
24	576	4.899	13,824	2.884	74	5,476	8.602	405,224	4.198
25	625	5.000	15,625	2.924	75	5,625	8.660	421,875	4.217
26	676	5.099	17,576	2.962	76	5,776	8.718	438,976	4.236
27	729	5.196	19,683	3.000	77	5,929	8.775	456,533	4.254
28	784	5.292	21,952	3.037	78	6,084	8.832	474,552	4.273
29	841	5.385	24,389	3.072	79	6,241	8.888	493,039	4.291
30	900	5.477	27,000	3.107	80	6,400	8.944	512,000	4.309
31	961	5.568	29,791	3.141	81	6,561	9.000	531,441	4.327
32	1,024	5.657	32,768	3.175	82	6,724	9.055	551,368	4.344
33	1,089	5.745	35,937	3.208	83	6,889	9.110	571.787	4.362
34	1,156	5.831	39,304	3.240	84	7,056	9.165	592,704	4.380
35	1,225	5.916	42,875	3.271	85	7,225	9.220	614,125	4.397
36	1,296	6.000	46,656	3.302	86	7,396	9.274	636,056	4.414
37	1,369	6.083	50,653	3.332	87	7,569	9.327	658,503	4.431
38	1,444	6.164	54,872	3.362	88	7,744	9.381	681,472	4.448
39	1,521	6.245	59,319	3.391	89	7,921	9.434	704,969	4.465
40	1,600	6.325	64,000	3.420	90	8,100	9.487	729,000	4.481
41	1,681	6.403	68,921	3.448	91	8,281	9.539	753,571	4.498
42	1,764	6.481	74,088	3.476	92	8,464	9,592	778,688	4.514
43	1,849	6.557	79,507	3.503	93	8,649	9.644	804,357	4.531
44	1,936	6.633	85,184	3.530	94	8,836	9.695	830,584	4.547
45	2,025	6.708	91,125	3.557	95	9,025	9.747	857,375	4.563
46	2,116	6.782	97,336	3.583	96	9,216	9.798	884,736	4.579
47	2,209	6.856	103,823	3.609	97	9,409	9.849	912,673	4.595
48	2,304	6.928	110,592	3.634	98	9,604	9.899	941,192	4.610
49	2,401	7.000	117,649	3.659	99	9,801	9.950	970,299	4.626
50	2,500	7.071	125,000	3.684	100	10,000	10.000	1,000,000	4.642

n n^2 $\sqrt{n}$ n^3 $\sqrt[3]{n}$ n n^2 $\sqrt{n}$ n^3 $\sqrt[3]{n}$

TABLE A.2

TRIGONOMETRIC FUNCTIONS OF REAL NUMBERS

x	$\sin x$	$\tan x$	$\cot x$	$\cos x$
.00	.00000	.00000	∞	1.00000
.01	.01000	.01000	99.997	.99995
.02	.02000	.02000	49.993	.99980
.03	.03000	.03000	33.323	.99955
.04	.03999	.04002	24.987	.99920
.05	.04998	.05004	19.983	.99875
.06	.95996	.06007	16.647	.99820
.07	.06994	.07011	14.262	.99755
.08	.07991	.08017	12.473	.99680
.09	.08988	.09024	11.081	.99595
.10	.09983	.10033	9.9666	.99500
.11	.10978	.11045	9.0542	.99396
.12	.11971	.12058	8.2933	.99281
.13	.12963	.13074	7.6489	.99156
.14	.13954	.14092	7.0961	.99022
.15	.14944	.15144	6.6166	.98877
.16	.15932	.16138	6.1966	.98723
.17	.16918	.17166	5.8256	.98558
.18	.17903	.18197	5.4954	.98384
.19	.18886	.19232	5.1997	.98200
.20	.19867	.20271	4.9332	.98007
.21	.20846	.21314	4.6917	.97803
.22	.21823	.22362	4.4719	.97590
.23	.22798	.23414	4.2709	.97367
.24	.23770	.24472	4.0864	.97134
.25	.24740	.25534	3.9163	.96891
.26	.25708	.26602	3.7591	.96639
.27	.26673	.27676	3.6133	.96377
.28	.27636	.28755	3.4776	.96106
.29	.28595	.29841	3.3511	.95824
.30	.29552	.30934	3.2327	.95534
.31	.30506	.32033	3.1218	.95233
.32	.31457	.33139	3.0176	.94924
.33	.32404	.34252	2.9195	.94604
.34	.33349	.35374	2.8270	.94275
.35	.34290	.36503	2.7395	.93937
.36	.35227	.37640	2.6567	.93590
.37	.36162	.38786	2.5782	.93233
.38	.37092	.39941	2.5037	.92866
.39	.38019	.41105	2.4328	.92491
.40	.38942	.42279	2.3652	.92106
.41	.39861	.43463	2.3008	.91712
.42	.40776	.44657	2.2393	.91309
.43	.41687	.45862	2.1804	.90897
.44	.42594	.47078	2.1241	.90475
.45	.43497	.48306	2.0702	.90045
.46	.44395	.49545	2.1084	.89605
.47	.45289	.50797	1.9686	.89157
.48	.46178	.52061	1.9208	.88699
.49	.47063	.53339	1.8748	.88233
.50	.47943	.54630	1.8305	.87758

x	$\sin x$	$\tan x$	$\cot x$	$\cos x$
.50	.47943	.54630	1.8305	.87758
.51	.48818	.55936	1.7878	.87274
.52	.49688	.57256	1.7465	.86782
.53	.50553	.58592	1.7067	.86281
.54	.51414	.59943	1.6683	.85771
.55	.52269	.61311	1.6310	.85252
.56	.53119	.62695	1.5950	.84726
.57	.53963	.64097	1.5601	.84190
.58	.54802	.65517	1.5263	.83646
.59	.55636	.66956	1.4935	.83094
.60	.56464	.68414	1.4617	.82534
.61	.57287	.69892	1.4308	.81965
.62	.58104	.71391	1.4007	.81338
.63	.58914	.72911	1.3715	.80803
.64	.59720	.74454	1.3431	.80210
.65	.60519	.76020	1.3154	.79608
.66	.61312	.77610	1.2885	.78999
.67	.62099	.79225	1.2622	.78382
.68	.62879	.80866	1.2366	.77757
.69	.63654	.82534	1.2116	.77125
.70	.64422	.84229	1.1872	.76484
.71	.65183	.85953	1.1634	.75836
.72	.65938	.87707	1.1042	.75181
.73	.66687	.89492	1.1174	.74517
.74	.67429	.91309	1.0952	.73847
.75	.68164	.93160	1.0734	.73169
.76	.68892	.95045	1.0521	.72484
.77	.69614	.96967	1.0313	.71791
.78	.70328	.98926	1.0109	.71091
.79	.71035	1.0092	.99084	.70385
.80	.71736	1.0296	.97121	.69671
.81	.72429	1.0505	.95197	.68950
.82	.73115	1.0717	.93309	.68222
.83	.73793	1.0934	.91455	.67488
.84	.74464	1.1156	.89635	.66746
.85	.75128	1.1383	.87848	.65998
.86	.75784	1.1616	.86091	.65244
.87	.76433	1.1853	.84365	.64483
.88	.77074	1.2097	.82668	.63715
.89	.77707	1.2346	.80998	.62941
.90	.78333	1.2602	.79355	.62161
.91	.78950	1.2864	.77738	.61375
.92	.79560	1.3133	.76146	.60582
.93	.80162	1.3409	.75478	.59783
.94	.80756	1.3692	.73034	.58979
.95	.81342	1.3984	.71511	.58168
.96	.81919	1.4284	.70010	.57352
.97	.82489	1.4592	.68531	.56530
.98	.83050	1.4910	.67071	.55702
.99	.83603	1.5237	.65631	.54869
1.00	.84147	1.5574	.64209	.54030

x $\sin x$ $\tan x$ $\cot x$ $\cos x$ x $\sin x$ $\tan x$ $\cot x$ $\cos x$

TABLE A.2 (*continued*)

x	$\sin x$	$\tan x$	$\cot x$	$\cos x$
1.00	.84147	1.5574	.64209	.54030
1.01	.84683	1.5922	.62806	.53186
1.02	.85211	1.6281	.61420	.52337
1.03	.85730	1.6652	.60051	.51482
1.04	.86240	1.7036	.58699	.50622
1.05	.86742	1.7433	.57362	.49757
1.06	.87236	1.7844	.56040	.48887
1.07	.87720	1.8270	.54734	.48012
1.08	.88196	1.8712	.53441	.47133
1.09	.88663	1.9171	.52162	.46249
1.10	.89121	1.9648	.50897	.45360
1.11	.89570	2.0143	.49644	.44466
1.12	.90010	2.0660	.48404	.43568
1.13	.90441	2.1198	.47175	.42666
1.14	.90863	2.1759	.45959	.41759
1.15	.91276	2.2345	.44753	.40849
1.16	.91680	2.2958	.43558	.39934
1.17	.92075	2.3600	.42373	.39015
1.18	.92461	2.4273	.41199	.38092
1.19	.92837	2.4979	.40034	.37166
1.20	.93204	2.5722	.38878	.36236
1.21	.93562	2.6503	.37731	.35032
1.22	.93910	2.7328	.36593	.34365
1.23	.94249	2.8198	.35463	.33424
1.24	.94578	2.9119	.34341	.32480
1.25	.94898	3.0096	.33227	.31532
1.26	.95209	3.1133	.32121	.30582
1.27	.95510	3.2236	.31021	.29628
1.28	.95802	3.3413	.29928	.28672
1.29	.96084	3.4672	.28842	.27712
1.30	.96356	3.6021	.27762	.26750
1.31	.96618	3.7471	.26687	.25785
1.32	.96872	3.9033	.25619	.24818
1.33	.97115	4.0723	.24556	.23848
1.34	.97348	4.2556	.23498	.22875
1.35	.97572	4.4552	.22446	.21901
1.36	.97786	4.6734	.21398	.20924
1.37	.97991	4.9131	.20354	.19945
1.38	.98185	5.1774	.19315	.18964
1.39	.98370	5.4707	.18279	.17981
1.40	.98545	5.7979	.17248	.16997
1.41	.98710	6.1654	.16220	.16010
1.42	.98865	6.5811	.15195	.15023
1.43	.99010	7.0555	.14173	.14033
1.44	.99146	7.6018	.13155	.13042
1.45	.99271	8.2381	.12139	.12050
1.46	.99387	8.9886	.11125	.11057
1.47	.99492	9.8874	.10114	.10063
1.48	.99588	10.983	.09105	.09067
1.49	.99674	12.350	.08097	.08071
1.50	.99749	14.101	.07081	.07074
x	$\sin x$	$\tan x$	$\cot x$	$\cos x$

x	$\sin x$	$\tan x$	$\cot x$	$\cos x$
1.50	.99749	14.101	.07091	.07074
1.51	.99815	16.428	.06087	.06076
1.52	.99871	19.670	.05084	.05077
1.53	.99917	24.498	.04082	.04079
1.54	.99953	32.461	.03081	.03079
1.55	.99978	48.078	.02080	.02079
1.56	.99994	92.621	.01080	.01080
1.57	1.00000	1255.8	.00080	.00080
1.58	.99996	−108.65	−.00920	−.00920
1.59	.99982	−52.067	−.01921	−.01920
1.60	.99957	−34.233	−.02921	−.02920
1.61	.99923	−25.495	−.03922	−.03919
1.62	.99879	−20.307	−.04924	−.04918
1.63	.99825	−16.871	−.05927	−.05917
1.64	.99761	−14.427	−.06931	−.06915
1.65	.99687	−12.599	−.07937	−.07912
1.66	.99602	−11.181	−.08944	−.08909
1.67	.99508	−10.047	−.09953	−.09904
1.68	.99404	−9.1208	−.10964	−.10899
1.69	.99290	−8.3492	−.11977	−.11892
1.70	.99166	−7.6966	−.12993	−.12884
1.71	.99033	−7.1373	−.14011	−.13875
1.72	.98889	−6.6524	−.15032	−.14865
1.73	.98735	−6.2281	−.16056	−.15853
1.74	.98572	−5.8535	−.17084	−.16840
1.75	.98399	−5.5204	−.18115	−.17825
1.76	.98215	−5.2221	−.19149	−.18808
1.77	.98022	−4.9534	−.20188	−.19789
1.78	.97820	−4.7101	−.21231	−.20768
1.79	.97607	−4.4887	−.22278	−.21745
1.80	.97385	−4.2863	−.23330	−.22720
1.81	.97153	−4.1005	−.24387	−.23693
1.82	.96911	−3.9294	−.25449	−.24663
1.83	.96659	−3.7712	−.26517	−.25631
1.84	.96398	−3.6245	−.27590	−.26596
1.85	.96128	−3.4881	−.28669	−.27559
1.86	.95847	−3.3608	−.29755	−.28519
1.87	.95557	−3.2419	−.30846	−.29476
1.88	.95258	−3.1304	−.31945	−.30430
1.89	.94949	−3.0257	−.33051	−.31381
1.90	.94630	−2.9271	−.34164	−.32329
1.91	.94302	−2.8341	−.35284	−.33274
1.92	.93965	−2.7463	−.36413	−.34215
1.93	.93618	−2.6632	−.37549	−.35153
1.94	.93262	−2.5843	−.38695	−.36087
1.95	.92896	−2.5095	−.39849	−.37018
1.96	.92521	−2.4383	−.41012	−.37945
1.97	.92137	−2.3705	−.42185	−.38868
1.98	.91744	−2.3058	−.43368	−.39788
1.99	.91341	−2.2441	−.44562	−.40703
2.00	.90930	−2.1850	−.45766	−.41615
x	$\sin x$	$\tan x$	$\cot x$	$\cos x$

TABLE A.3

TRIGONOMETRIC FUNCTIONS OF ANGLES

Degrees	Radians	sin	cos	tan	cot	sec	csc		
0	.0000	.0000	1.0000	.0000	∞	1.0000	∞	1.5708	90
1	.0175	.0175	.9998	.0175	57.2900	1.0002	57.299	1.5533	89
2	.0349	.0349	.9994	.0349	28.6363	1.0006	28.654	1.5359	88
3	.0524	.0523	.9986	.0524	19.0811	1.0014	19.107	1.5184	87
4	.0698	.0698	.9976	.0699	14.3007	1.0024	14.336	1.5010	86
5	.0873	.0872	.9962	.0875	11.4301	1.0038	11.474	1.4835	85
6	.1047	.1045	.9945	.1051	9.5144	1.0055	9.5668	1.4661	84
7	.1222	.1219	.9925	.1228	8.1443	1.0075	8.2055	1.4486	83
8	.1396	.1392	.9903	.1405	7.1154	1.0098	7.1853	1.4312	82
9	.1571	.1564	.9877	.1584	6.3138	1.0125	6.3925	1.4137	81
10	.1745	.1736	.9848	.1763	5.6713	1.0154	5.7588	1.3963	80
11	.1920	.1908	.9816	.1944	5.1446	1.0187	5.2408	1.3788	79
12	.2094	.2079	.9781	.2126	4.7046	1.0223	4.8097	1.3614	78
13	.2269	.2250	.9744	.2309	4.3315	1.0263	4.4454	1.3439	77
14	.2443	.2419	.9703	.2493	4.0108	1.0306	4.1336	1.3265	76
15	.2618	.2588	.9659	.2679	3.7321	1.0353	3.8637	1.3090	75
16	.2793	.2756	.9613	.2867	3.4874	1.0403	3.6280	1.2915	74
17	.2967	.2924	.9563	.3057	3.2709	1.0457	3.4203	1.2741	73
18	.3142	.3090	.9511	.3249	3.0777	1.0515	3.2361	1.2566	72
19	.3316	.3256	.9455	.3443	2.9042	1.0576	3.0716	1.2392	71
20	.3491	.3420	.9397	.3640	2.7475	1.0642	2.9238	1.2217	70
21	.3665	.3584	.9336	.3839	2.6051	1.0711	2.7904	1.2043	69
22	.3840	.3746	.9272	.4040	2.4751	1.0785	2.6695	1.1868	68
23	.4014	.3907	.9205	.4245	2.3559	1.0864	2.5593	1.1694	67
24	.4189	.4067	.9135	.4452	2.2460	1.0946	2.4586	1.1519	66
25	.4363	.4226	.9063	.4663	2.1445	1.1034	2.3662	1.1345	65
26	.4538	.4384	.8988	.4877	2.0503	1.1126	2.2812	1.1170	64
27	.4712	.4540	.8910	.5095	1.9626	1.1223	2.2027	1.0996	63
28	.4887	.4695	.8829	.5317	1.8807	1.1326	2.1301	1.0821	62
29	.5061	.4848	.8746	.5543	1.8040	1.1434	2.0627	1.0647	61
30	.5236	.5000	.8660	.5774	1.7321	1.1547	2.0000	1.0472	60
31	.5411	.5150	.8572	.6009	1.6643	1.1666	1.9416	1.0297	59
32	.5585	.5299	.8480	.6249	1.6003	1.1792	1.8871	1.0123	58
33	.5760	.5446	.8387	.6494	1.5399	1.1924	1.8361	.9948	57
34	.5934	.5592	.8290	.6745	1.4826	1.2062	1.7883	.9774	56
35	.6109	.5736	.8192	.7002	1.4281	1.2208	1.7434	.9599	55
36	.6283	.5878	.8090	.7265	1.3764	1.2361	1.7013	.9425	54
37	.6458	.6018	.7986	.7536	1.3270	1.2521	1.6616	.9250	53
38	.6632	.6157	.7880	.7813	1.2799	1.2690	1.6243	.9076	52
39	.6807	.6293	.7771	.8098	1.2349	1.2868	1.5890	.8901	51
40	.6981	.6428	.7660	.8391	1.1918	1.3054	1.5557	.8727	50
41	.7156	.6561	.7547	.8693	1.1504	1.3250	1.5243	.8552	49
42	.7330	.6691	.7431	.9004	1.1106	1.3456	1.4945	.8378	48
43	.7505	.6820	.7314	.9325	1.0724	1.3673	1.4663	.8203	47
44	.7679	.6947	.7193	.9657	1.0355	1.3902	1.4396	.8029	46
45	.7845	.7071	.7071	1.0000	1.0000	1.4142	1.4142	.7854	45
		cos	sin	cot	tan	csc	sec	Radians	Degrees

Answers to Selected Problems

PROBLEMS FOR REVIEW

Group I: Sets

1. $\{3, 5, 7, 9, \ldots\}$

2. $\{0, 3, 8, 15, \ldots\}$

3. $\{1, 1/2, 1/3, 1/4, \ldots\}$

4. $\{2, 3/2, 4/3, 5/4, \ldots\}$

5. $\{1, 4, 7, 10, \ldots\}$

6. $\{1/2, 1/6, 1/12, 1/20, \ldots\}$

7. $\{x \mid x = n^2, n \in N\}$

8. $\left\{x \,\middle|\, x = \frac{1}{2n+1}, n \in N\right\}$

9. $\{x \mid x = 2n, n \in N\}$

10. $\{x \mid x = 9 - 2n, n \in N\}$

11. $K \cup L = \{-2, 0, 1, 2, 3, 4, 5\}$

12. $K \cap L = \{0, 2, 4\}$

13. $L \cup M = \{0, 1, 2, 3, 4, 5, 7\}$

14. $L \cap M = \{1, 4\}$

15. $K \cap M = \{4\}$

16. $\{3, 5, 7\}$

17. $\left\{x \,\middle|\, x = \frac{1}{2n+1}, n \in N\right\}$, that is, the set given in problem 8.

PROBLEMS FOR REVIEW

Group II: Special Products

1. a. $x^2 + (r + s)x + rs$ **b.** $x^2 + 2sx + s^2$
c. $x^3 + p^3$ **d.** $x^4 - r^4$

2. a. $z^2 - m$ **b.** $kx^2 + (ks + r)x + rs$
c. $x^3 - p^3$ **d.** $u^{2m} - a^2$

3. a. $x^2 - x - 6$ **b.** $x^2 + 3x - 10$
c. $z^2 - 15z + 56$ **d.** $t^2 + 13t + 36$

4. a. $2x^2 + x - 6$ **b.** $3w^2 + w - 2$
c. $2s^2 + 3s + 1$ **d.** $4x^2 + 5x - 6$

5. a. $m^2 + 9m + 20$ **b.** $x^2 + 7x - 8$
c. $4t^2 + 4t - 3$ **d.** $s^2 - s - 72$

6. a. $4t^2 + 8t + 3$ **b.** $2t^2 - 7t - 15$
c. $3x^2 + 4x - 15$ **d.** $6n^2 - n - 1$

7. **a.** $x^2 + 5/2x + 1$ **b.** $x^2 + 10/3x + 1$
c. $x^2 + 1/2x - 3$ **d.** $x^2 + 5/12x - 1/4$

8. **a.** $y^2 - 3$ **b.** $4x^2 - 1$
c. $x^2 - 5$ **d.** $9s^2 - 4t^2$

9. **a.** $x^{2n} + x^{2n-1}$ **b.** $p^{m+1} - p^m$
c. $z^{m+1} + z^m$ **d.** $z^m - z^{m-1}$

10. **a.** $x^2 + 2x - 1$ **b.** $s^2 + 2s - 2$
c. $t^2 - 6t + 4$ **d.** $z^2 + 4z - 3$

11. **a.** $m^3 - 27$ **b.** $t^3 - 8$
c. $z^3 - 2\sqrt{2}$ **d.** $8x^3 - 1$

12. **a.** $t^3 + 1$ **b.** $x^3 + 27$
c. $8m^3 + 27$ **d.** $s^3 + 3\sqrt{3}$

13. **a.** $x^2 + 24x + 144$ **b.** $x^2 - 16x + 64$
c. $s^2 - 10s + 25$ **d.** $4t^2 + 28t + 49$

14.
$$[x + r/2]^2 - \frac{r^2 - 4s}{4} = x^2 + rx + r^2/4 - r^2/4 + s$$
$$= x^2 + rx + s$$

15.
$$\frac{r^2 + 4s}{4} - [x - r/2]^2 = r^2/4 + s - [x^2 - rx + r^2/4]$$
$$= s + rx - x^2$$

16. **a.** $(x - a)(x + 3)$ **b.** $(x - t)(x + 3)$
c. $(x - a)(x + a + 2)$ **d.** $(m - t)(m + t - 7)$

17. **a.** $(x - a)(x - 3)$ **b.** $(x - a)(4 - x)$
c. $(x - a)(x + a - 5)$ **d.** $(x - 1)(x - m)$

18. **a.** $z^{m-1}(z + 1)$ **b.** $t^m(t - 1)$
c. $x(x^{m-1} - 1)$ **d.** $x^{m-1}(x - 1)$

19. **a.** $(x + 4)(x + 2)$ **b.** $(y + 6)(y + 5)$
c. $(s + 7)(s + 4)$ **d.** $(x + 6)(x + 7)$

20. **a.** $(w - 8)(w - 5)$ **b.** $(x - 6)(x - 4)$
c. $(m - 3)(m - 2)$ **d.** $(x - 2)(x - 6)$

21. **a.** $(x - 7)^2$ **b.** $(x - 1)^2$
c. $(w - 5)^2$ **d.** $(t - 4)^2$

22. **a.** $(m - 7)(m + 4)$ **b.** $(x - 1)(x + 3)$
c. $(x - 2)(x + 7)$ **d.** $(z - 8)(z + 7)$

23. **a.** $(3 + x)(1 - x)$ **b.** $(3 + x)(2 - x)$
c. $(5 - t)(1 + t)$ **d.** $(4 - t)(3 + t)$

24. **a.** $(t + 6)^2$ **b.** $(m + 9)^2$
c. $(x + 3)^2$ **d.** $(x + 1)^2$

25. **a.** $(2x - a)(2x + a)$ **b.** $(w - 6)(w + 6)$
c. $(x - 8)(x + 8)$ **d.** $(3m - 2)(3m + 2)$

26. **a.** $(z^n - 2)(z^n + 2)$ **b.** $(x^m - 1)(x^m + 1)$
c. $(x^m - 12)(x^m + 12)$ **d.** $(z^n - p)(z^n + p)$

27. **a.** $(x - 1)(x + 5)$ **b.** $(s - 3)(s + 1)$
c. $(m + 2)(m + 4)$ **d.** $(s - 3)(s + 5)$

28. **a.** $(2x + 1)(x + 3)$ **b.** $(3x - 1)(x - 2)$
c. $(4w - 5)(w - 1)$ **d.** $(3z + 1)(z - 4)$

29. **a.** $(3x - 1)(x + 1)$ **b.** $(2x - 3)(x - 3)$
c. $(4m - 1)(m - 2)$ **d.** $(5s + 1)(s + 2)$

30. **a.** $(2t + 5)(2t - 3)$ **b.** $(3x + 1)(2x + 1)$
c. $(3s + 1)(2s - 1)$ **d.** $(4x - 1)(2x + 3)$

31. **a.** $(m - 1)(m^2 + m + 1)$ **b.** $(x - 2)(x^2 + 2x + 4)$
c. $(w - 1)(w^2 + w + 1)$ **d.** $(x - 3)(x^2 + 3x + 9)$

32. **a.** $(x + 2)(x^2 - 2x + 4)$ **b.** $(m + 1)(m^2 - m + 1)$
c. $(w + 4)(w^2 - 4w + 16)$ **d.** $(z + 1)(z^2 - z + 1)$

33. **a.** $(2z - 3)(4z^2 + 6z + 9)$ **b.** $(3x - 1)(9x^2 + 3x + 1)$
c. $(x - 2a)(x^2 + 2ax + 4a^2)$ **d.** $(x + 2t)(x^2 - 2tx + 4t^2)$

34. **a.** $(x - a)(x + a + 2)$ **b.** $(x - 2)(x + 2)$
c. $(x - a)(x + a + 3)$ **d.** $4(x - 2)(x^2 + 2x + 4)$

35. **a.** $(z + 3)(z + 2)$ **b.** $2(s + 2)(s + 4)$
c. $t^2(s - 9t)(s + 7t)$ **d.** $(w - 3)(w + 3)(w - 1)(w + 1)$

36. **a.** $(x + 3)^2 - 5$ **b.** $(x - 4)^2 - 13$
c. $(x - 1)^2 - 6$ **d.** $(x + 5)^2 - 20$

37. **a.** $5 - (x - 2)^2$ **b.** $7 - (x + 1)^2$
c. $9 - (x - 3)^2$ **d.** $19 - (x - 4)^2$

38. **a.** $(x - 3/2)^2 - 5/4$ **b.** $(x + 5/2)^2 - 41/4$
c. $(x + 1/2)^2 - 9/4$ **d.** $(x - 5/2)^2 - 21/4$

39. **a.** $2(x + 5/4)^2 - 33/8$ **b.** $2(x + 1)^2 - 5$
c. $3(x - 1/6)^2 - 49/12$ **d.** $2(x + 1/4)^2 - 49/8$

40. **a.** $(x + 1/2)^2 + 7/4$ **b.** $2(x - 3/4)^2 + 23/8$
c. $(x + 2)^2 - 1$ **d.** $3(x - 1/2)^2 + 1/4$

PROBLEMS FOR REVIEW

Group III: Quotients

1. $\dfrac{x + y}{xy}$ **2.** $\dfrac{5t + 7}{30}$ **3.** $\dfrac{5(x + 2)}{x + 3}$

4. $\dfrac{7(2 - x)}{30}$ **5.** $\dfrac{10}{y^2 - 4}$ **6.** $\dfrac{4a^2 + 9a + 29}{a^3 - 27}$

7. $\dfrac{x(y - x)}{x^3 + y^3}$ **8.** $\dfrac{2(3x - 7)}{(x - 5)(x + 1)(x - 1)}$

9. 1 **10.** x/z **11.** $x + 1$

12. $x + y$ **13.** xy **14.** $y + 2$

15. $\dfrac{5 - y}{4}$ **16.** $\dfrac{(s - 1)(s + 3)}{(s + 7)(2 - s)}$ **17.** $\dfrac{xy}{x - y}$

18. $t - 1$ **19.** $x + w$ **20.** $\dfrac{x - 1}{x + 1}$

21. $x + 4$ **22.** $x^2 + 2x + 4$ **23.** $2(x^2 - 3x + 9)$

24. $\dfrac{-1}{5(2x + 1)}$ **25.** $\dfrac{-1}{ax}$ **26.** $\dfrac{-(x + a)}{a^2x^2}$

27. $\dfrac{-1}{2\sqrt{x}(\sqrt{x} + 2)}$ **28.** $\dfrac{1}{\sqrt{x} + \sqrt{a}}$

29. $\dfrac{1}{\sqrt{(x + 1)} + \sqrt{2}}$ **30.** $\dfrac{x + 2}{\sqrt{(x^2 - 1)} + \sqrt{3}}$

PROBLEMS FOR REVIEW

Group IV: Equations

1. **a.** $\{1/4\}$ **b.** $\{-2/3\}$ **c.** $\{2\}$

2. **a.** $\left\{\dfrac{2a}{a - 1}\right\}$ **b.** $\left\{\dfrac{4a + 1}{a - 1}\right\}$ **c.** $\left\{\dfrac{3a + 2}{1 - a}\right\}$

3. **a.** $\{-2,2\}$ **b.** $\{-7,1\}$ **c.** $\{-7,1\}$

4. a. $\{5\}$ **b.** $\{17\}$ **c.** $\{8\}$

5. a. $\{-2,7\}$ **b.** $\{-4,5\}$ **c.** $\{-5,-3\}$

6. a. $\{-3,-1/2\}$ **b.** $\{1/2,4\}$ **c.** $\{3/2,4\}$

7. a. $\{3 \pm 2\sqrt{2}\}$ **b.** $\{2 \pm \sqrt{7}\}$ **c.** $\{-1 \pm \sqrt{3}\}$

8. a. $\left\{\frac{1 \pm \sqrt{6}}{2}\right\}$ **b.** $\{1 \pm \sqrt{13}\}$ **c.** $\{3 \pm \sqrt{26}\}$

9. $\{1/2,3/4\}$ **b.** $\{-2/3,-1/2\}$ **c.** $\{1/4,3/2\}$

10. a. no soln. **b.** no soln. **c.** no soln.

11. a. $\{2,3\}$ **b.** $\{3\}$ **c.** $\{1,6\}$

12. a. $\{1,9\}$ **b.** $\{-5\}$ **c.** $\{4,8\}$

13. a. 4 **b.** 2 **c.** 2

14. a. $\{-4a,2a\}$ **b.** $\{-s,r/2\}$ **c.** $\{0,1/2,2\}$

15. a. 3 **b.** 2/3 **c.** -4

ANSWERS CHAPTER 1

Problem Set 1.1

1. a. $0 < x$ and $x < 1$, given $xx < 1x$ by axiom 4, $x^2 < x$ and $x < 1$ implies $x^2 < 1$ by axiom 2
b. $0 < t, 0 < x$ and $t < x$ given $tt < tx$ and $tx < xx$ by axiom 4, thus $t^2 < x^2$ by axiom 2

3. a. $1 < 1 + 1$ by theorem 3 **b.** $2 < 2 + 1$ by theorem 3
c. $0 < 1$ by theorem 2 and $1 < 2$ by (a) above, $0 < 2$ by axiom 2, also $2 < 3$ by (b) so $0 < 3$ by axiom 2
d. $2 < 3$ by (b) thus $-3 < -1$ by theorem 1

5. a. $x < 7/3$ **b.** $x \mid x < 1/5$

7. a. $x > 4$ **b.** $x > 3/2$ **c.** $x < 2$

9. $m = 1/4, M = 1$

15. a. $\frac{p+q}{2}$ is between p and q **b.** $(p + q) > p$ and $p + q > q$.
c. from part b it follows that $1/(p + q) < 1/p$ and $1/(p + q) < 1/q$ thus multiplying by pq we have $pq/(p + q) < q$ and $pq/(p + q) < p$

Problem Set 1.2

1. $\langle -2,1\rangle$ **3.** $x > 1$ or $x < -2$ **5.** $[-4,0]$

7. $\langle -1,1/4\rangle$ **9.** $x < 3$ **11.** $X \geq 3$ or $x < -5$

13. $x < 4$ or $x > 9$ **15.** $\langle -2,1/3\rangle$

17. $-2 < x < 1$ or $x \geq 2$ **19.** $x < -5$ or $-2 < x < 4$

21. $[0,5/2\rangle$ **23.** $\langle 0,3/2\rangle$

Problem Set 1.3

1. $\dfrac{b-a}{2}$ **3.** $\{1,2\}$ **5.** $\{1/2, 5/2\}$

7. $\{-3,-2\}$ **9.** $\langle -1,11/3\rangle$ **11.** $x > 5$ or $x < 1$

13. $[-1/5,1]$ **15.** $\langle -1,5/3\rangle$ **17.** $\langle -1,4\rangle$

19. $[-3/4,-1/4]$ **21.** $\langle 2,6\rangle$ **23.** $[a,b]$

25. $1/7 < \dfrac{1}{x+4} < 1/3$

Problem Set 1.4

3. Case 1: $r - s < 0$, $|r - s| = -(r - s) = s - r$ since $s - r > 0$, $|s - r| = s - r$, hence $|r - s| = |s - r|$.

5. $|x - a| + |x - b| = |x - a| + |b - x|$ and by theorem 10 $|x - a + b - x| \leq |x - a| + |x - b|$, hence $|b - a| \leq |x - a| + |x - b|$.

7. $|x - t| + |y - t| < p/2 + p/2$, further $|x - t| + |y - t| = |x - t| + |t - y|$, by theorem 10 $|x - t + t - y| \leq |x - t| + |y - t| < p$, hence $|x - y| < p$.

9. Given: $-|x| \leq x \leq |x|$ and $-|t| \leq t < |t|$, by theorem 4 $-(|x| + |t|) \leq x + t \leq |x| + |t|$ so, by theorem 8 $|x + t| \leq |x| + |t|$.

11. $M = 6$ **13.** $M = 6$ **15.** $M = 7$

Problem Set 1.5

1.

Set:	A	C	F	W	M	T	V
LUB:	$\nexists$	1	3	2	1,	3	$\nexists$
GLB:	1	−1/3	3/2	0	0	1	$\nexists$

4.

Set:	B	D	G	Z
LUB:	0	1/2	$\nexists$	1
GLB:	−3	−1	0	−1

Problem Set 1.6

1. $6/5 < 5/4$ **3.** $27/100 < 3/11$ **5.** $13/7 < 13/6 < 7/3$

7. $21/50 < 19/45 < 14/33$

9. $0 < p^2$ so $pq < p^2 + pq$ by axiom 3 since $0 < p + q$, $0 < 1(p + q)$ by theorem 5 so $pq \cdot \dfrac{1}{p+q} < (p^2 + pq) \cdot \dfrac{1}{p+q}$ by axiom 4, hence $\dfrac{pq}{p+q} < p$. The same sort of argument will show that $\dfrac{pq}{p+q} < q$.

15. $\sqrt{2} < 1.5$ and $1.7 < \sqrt{3}$. Also $1.5 < 1.6 < 1.7$ so $\sqrt{2} < 1.6 < \sqrt{3}$.

ANSWERS CHAPTER 2

Problem Set 2.1

5. a. $\sqrt{53}$ **b.** $5\sqrt{2}$ **c.** $\sqrt{10}$ **d.** $2\sqrt{10}$

7. a. neither (isosceles) **b.** right, hyp. is $\sqrt{82}$

9. a. neither (isosceles) **b.** equilateral, side is 4

11. a. right, hyp. is 10 **b.** neither

13. a. (5,11/2) **b.** (1,1)

15. $r = 20, s = -7$ **17.** Area $= 25\sqrt{3}$ **19.** $d(M_1M_2) = a/2$

Problem Set 2.2

1. a. $(x,y) \mid (x - 3)^2 + (y - 2)^2 = 9$
b. $(x,y) \mid (x + 3)^2 + (y + 1)^2 = 36$

3. a. $(x,y) \mid (x + 1)^2 + (y - 1)^2 = 25/4$
b. $(x,y) \mid (x + 3)^2 + y^2 = 25$

5. **a.** $(x,y) \mid (x - 2)^2 + (y + 2)^2 = 2$
b. $(x,y) \mid (x + 3)^2 + (y + 4)^2 = 4/9$

7. $(x,y) \mid (x - 3)^2 + (y + 1)^2 = 29$

9. $(x,y) \mid x^2 + y^2 = 1$

11. $(x,y) \mid x^2 + (y - 3)^2 = 37$

13. **a.**

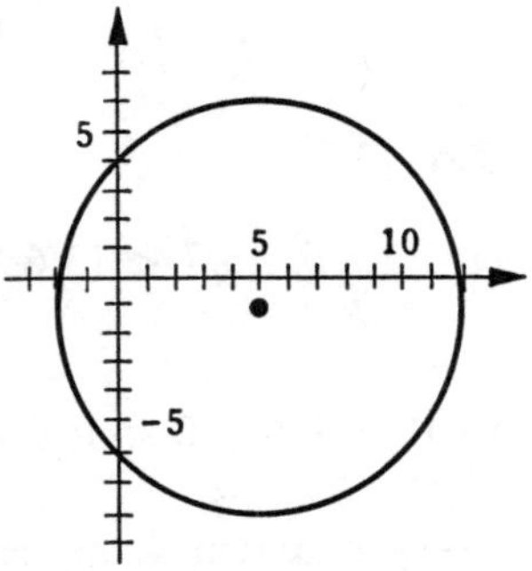

Fig. A.3a

b.

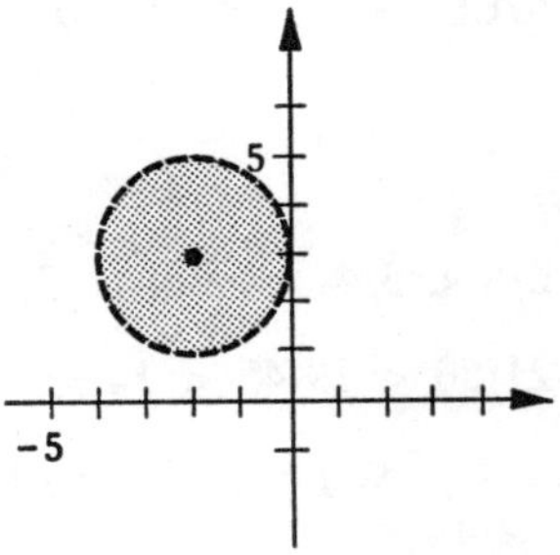

Fig. A.3b. Center Radius.

15. **a.**

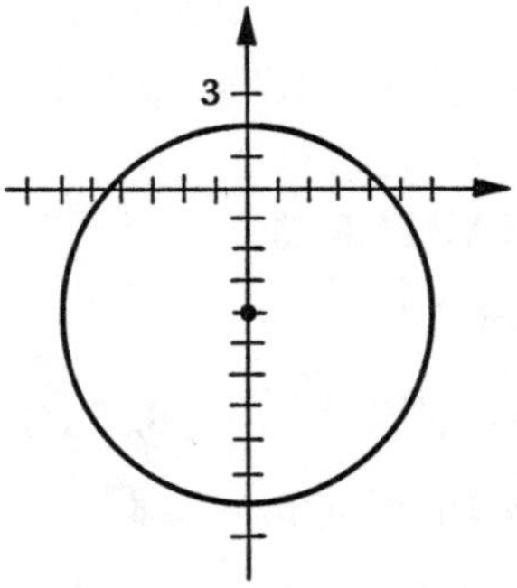

Fig. A.4a

b.

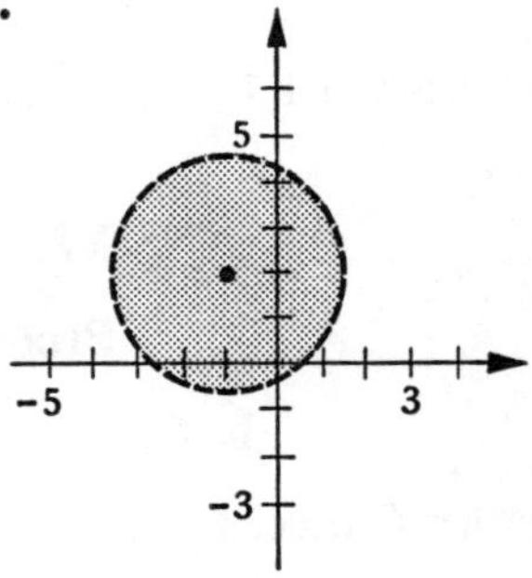

Fig. A.4b

17. **a.**

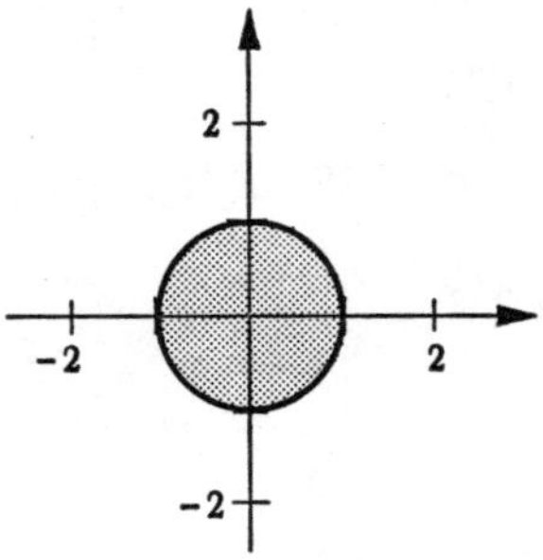

Fig. A.5a

b.

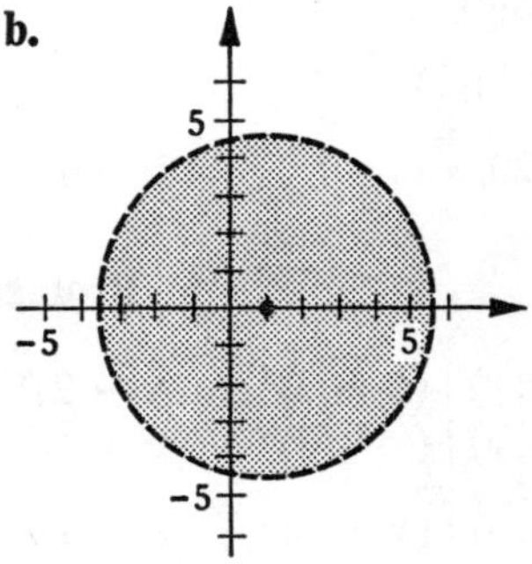

Fig. A.5b

19. $(-1,2)$	4
21. $(4,0)$	3
23. $(5,12)$	13
25. $(0,-5)$	10
27. $(-1,3/2)$	5/2

Problem Set 2.3

1. a. Since $h < p, 0 < p - h$ by axiom 3. For $m > 0, \frac{q-k}{p-h} \cdot (p-h) > 0$, so $q - k > 0$, therefore $q > k$.

3.

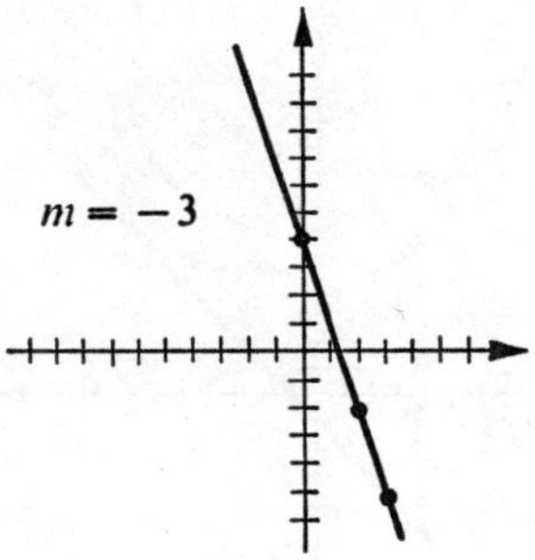

Fig. A.6

5.

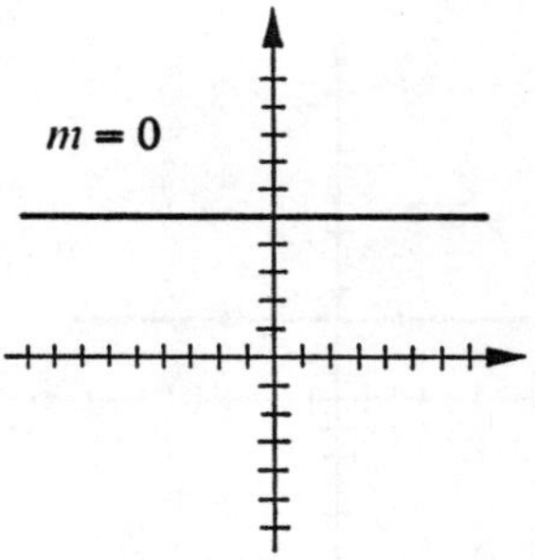

Fig. A.7

7.

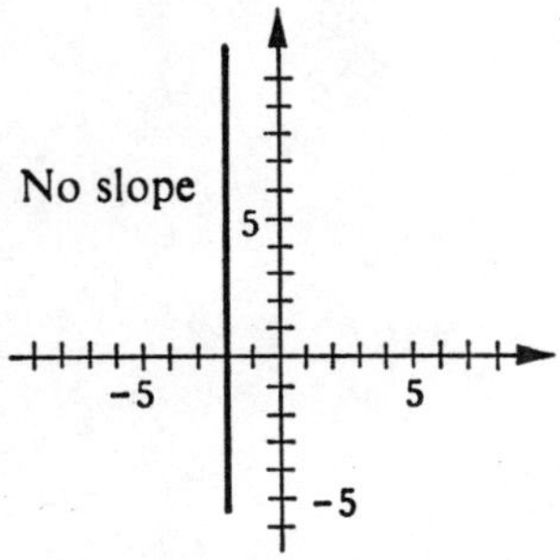

Fig. A.8

9.

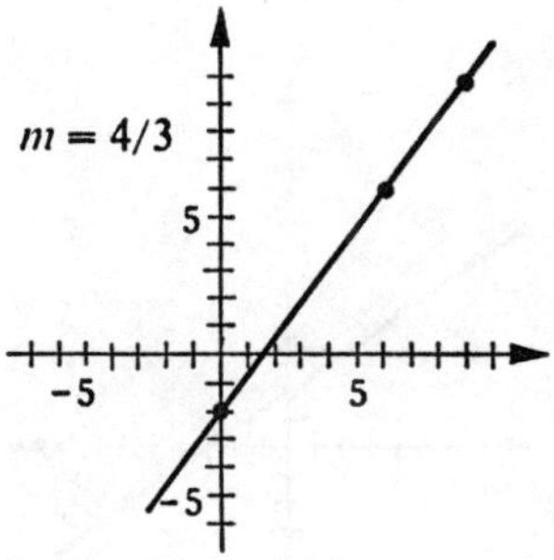

Fig. A.9

11.

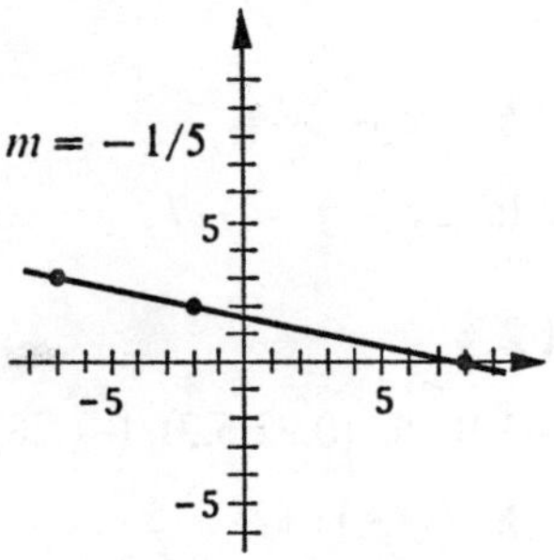

Fig. A.10

13.

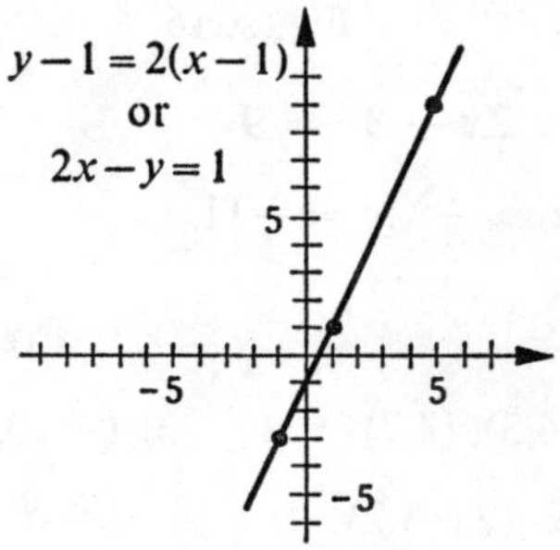

Fig. A.11

15.

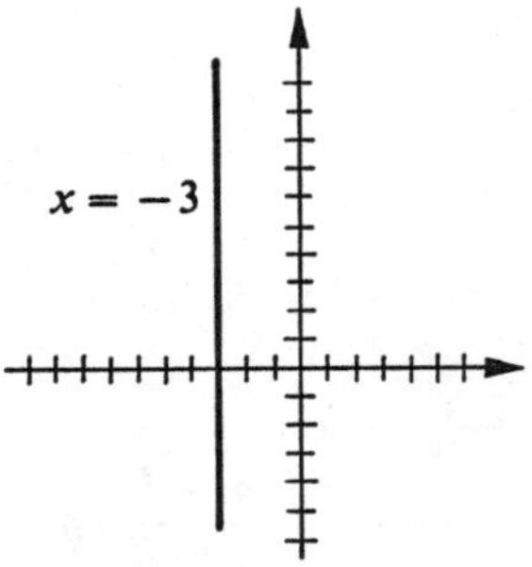

Fig. A.12

17.

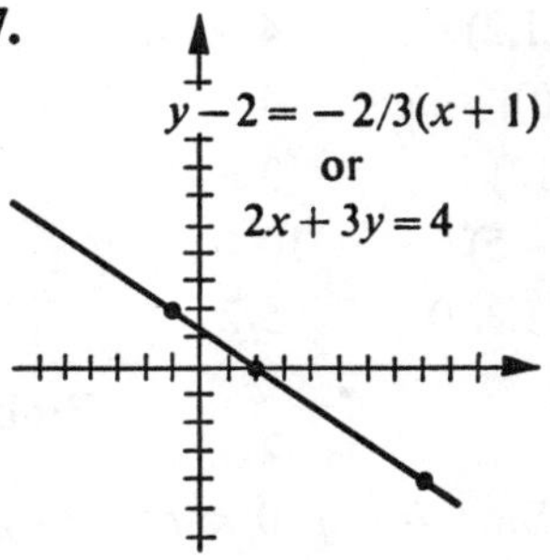

Fig. A.13

19.

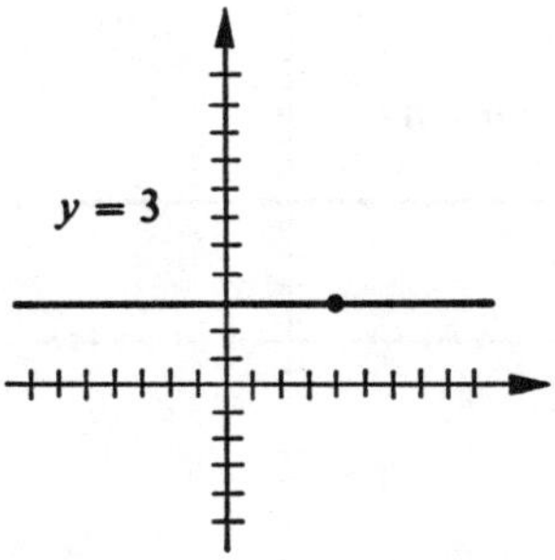

Fig. A.14

21.

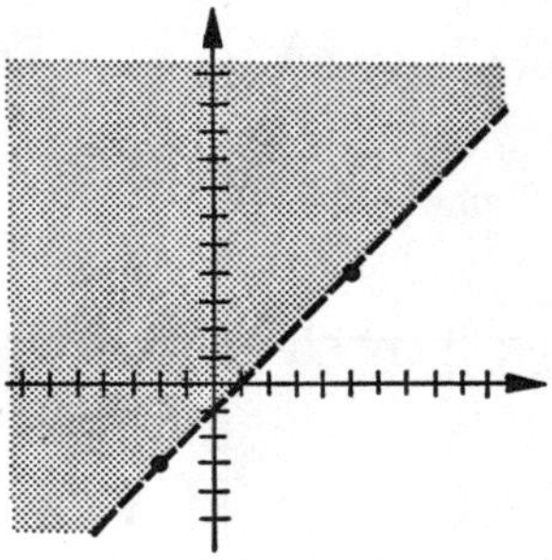

Fig. A.15

23.

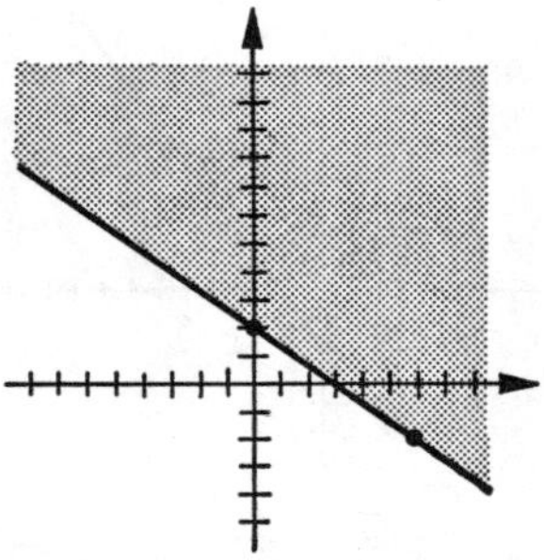

Fig. A.16

25. a. $2x + 3y = 9$ **b.** $3x - 2y = 7$

27. a. $x - 2y = -11$ **b.** $2x + y = -7$

Problem Set 2.4

1. (4,3), (2,7) **3.** $(-10,-4)$, $(6,-10)$ **5.** (0.92,5.3), $(-.28,2.7)$

7. a. $(2\sqrt{3},2)$ **b.** $(\sqrt{3},1)$
c. $(-3\sqrt{2}, -\sqrt{6})$ **d.** $(\sqrt{6} + 3\sqrt{2}, \sqrt{2} + \sqrt{6})$

11. $\mathbf{z} = (-2.5 + t, 4 + 2t)$

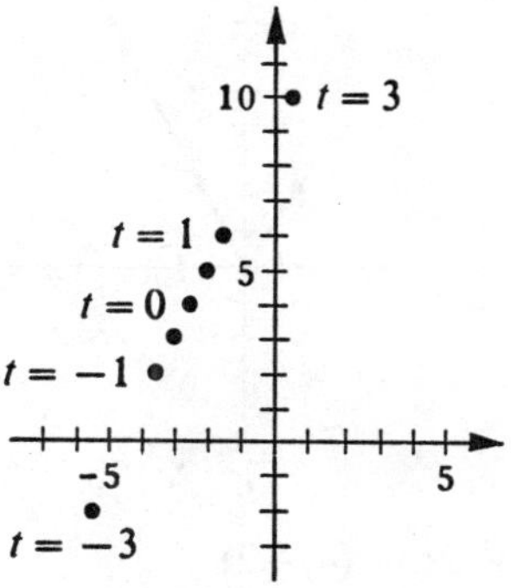

Fig. A.17

Problem Set 2.5

1. a. $\mathbf{u} = (4,-3)$ **b.** $\mathbf{v} = (-3,8)$ **c.** $\mathbf{w} = (-5,-3)$
d. $\mathbf{x} = (5,5)$ **e.** $\mathbf{z} = (3,2)$

3. a. $|\mathbf{u}| = 5$ **b.** $|\mathbf{v}| = \sqrt{73}$ **c.** $|\mathbf{w}| = \sqrt{34}$
d. $|\mathbf{x}| = 5\sqrt{2}$ **e.** $|\mathbf{z}| = \sqrt{13}$

5.

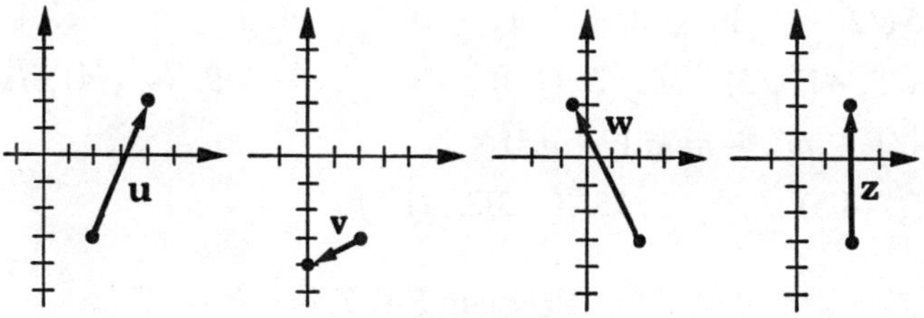

Fig. A.18

7. a. $\mathbf{u} + \mathbf{v} = (-4,2)$;
$\mathbf{u} - \mathbf{v} = (-2,10)$

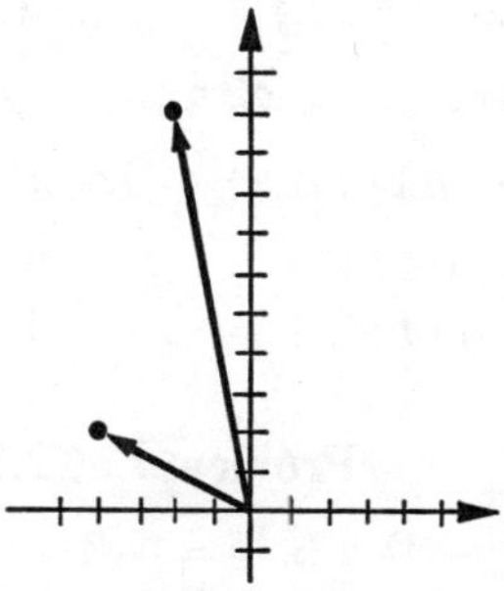

Fig. A.19a

b. $\mathbf{u} + \mathbf{v} = (4,7)$;
$\mathbf{u} - \mathbf{v} = (-2,3)$

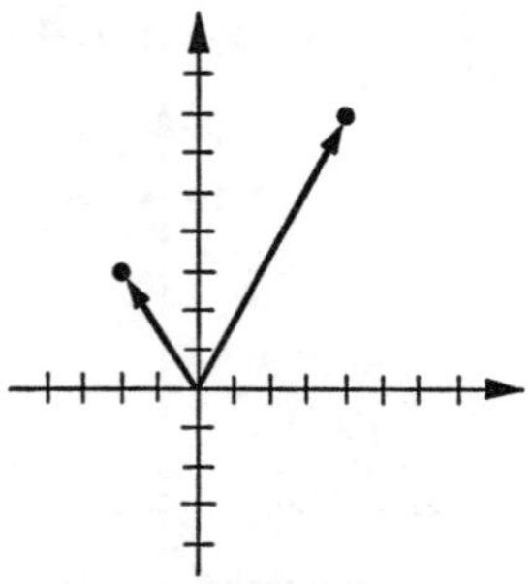

Fig. A.19b

9. $2\cdot\mathbf{z} = (2,-6)$; $\mathbf{w} + 2\mathbf{z} = (4,-2)$

11. $2\cdot\mathbf{w} = (4,8)$; $3\cdot\mathbf{z} = (3,-9)$; $2\mathbf{w} + 3\mathbf{z} = (7,-1)$

13. $|\mathbf{u} - \mathbf{v}| = 2\sqrt{14}$; $|\mathbf{u} + \mathbf{v}| = 5$

15. $|\mathbf{u} - \mathbf{v}| = \dfrac{5\sqrt{2}}{6}$; $|\mathbf{u} + \mathbf{v}| = \dfrac{\sqrt{2}}{6}$

17. a. $(1,1)$ **b.** $(3,1)$ **c.** $(2,-3)$
d. $(2\sqrt{2},2)$ **e.** $(1,-1)$ **f.** $(-1,1)$

Problem Set 2.6

1. a. parallel **b.** parallel **c.** not parallel **d.** parallel

3. a. $5\sqrt{2}$ **b.** $\sqrt{10}$ **c.** $\sqrt{10}$ **d.** 1 **e.** 1

5. $(2/\sqrt{5},-1/\sqrt{5})$ **7.** $(1,0)$ **9.** $(4/\sqrt{41},5/\sqrt{41})$

11. $(1/\sqrt{1 + m^2}, -m/\sqrt{1 + m^2})$

13. $(3/5,-4/5)$ **15.** $(\sqrt{3}/2,-1/2)$

Problem Set 2.7

1. $\mathscr{L} = \{(5 + t, t) \mid t \in R\}$ **3.** $\mathscr{L} = \{(1 + 3t, 3 - t) \mid t \in R\}$

5. $\mathscr{L} = \{(-3t, 1 + 2t) \mid t \in R\}$ **7.** $\mathscr{L} = \{(3 + 2t, 4 - 6t) \mid t \in R\}$

9. a. no **b.** yes **c.** no **d.** yes **e.** yes

11. $\mathscr{L} = \{(h + t, k + mt) \mid t \in R\}$ **13.** $\mathscr{L} = \{(2 + t, 4t - 1) \mid t \in R\}$

15. a. $\{(1 + t, t - 6) \mid t \in R\}$ **b.** $\{(1, t - 6) \mid t \in R\}$
c. $\{(1 + 4t, 2t - 6) \mid t \in R\}$ **d.** $\{(1 + t, t - 6) \mid t \in R\}$

Problem Set 2.8

3. a. $y - 2 = -3(x - 1)$ **b.** $y = -3 + 5$ **c.** $3x + y = 5$

5. a. $y + 3 = 1/2(x - 0)$ **b.** $y = 1/2\cdot x - 3$ **c.** $x - 2y = 6$

7. a. $x - 2y = 9$ **b.** $3x - y = 4$ **c.** $4x - 3y = 4$

9. $y = x + 3$ **11.** $5x + 2y = 22$ **13.** $3x - 4y = 0$

Problem Set 2.9$\mathscr{E}$

1. $F_1 = (0,-c)$, $F_2 = (0,c)$, $P = (x,y)$; $d(PF_1) + d(PF_2) = 2a$; $\sqrt{[x^2 + (y + c)^2]} + \sqrt{[x^2 + (y - c)^2 = 2a]}$; $a\sqrt{[x^2 + (y - c)^2} = a^2 - cy$; $a^2x^2 + (a^2 - c^2)y^2 = a^2(a^2 - c^2)$; let $b^2 = a^2 - c^2$; $\frac{x^2}{b^2} + \frac{y^2}{a^2} = 1$

3.

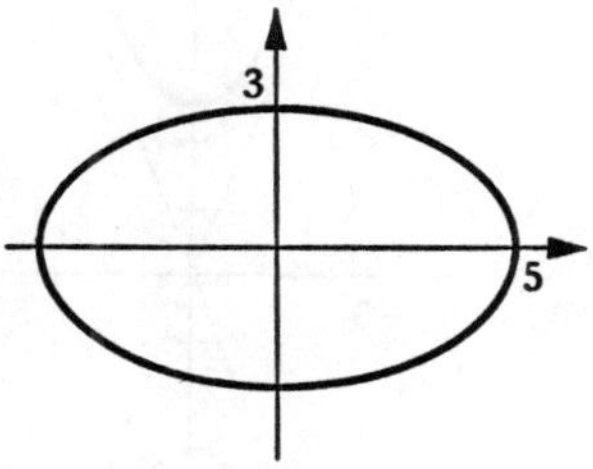

Fig. A.20

5.

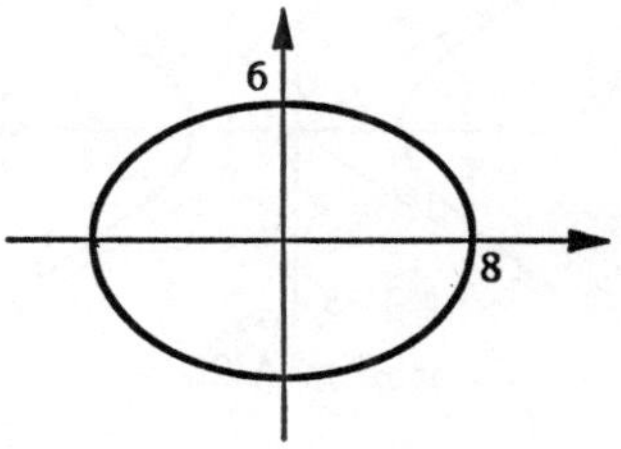

Fig. A.21

7.

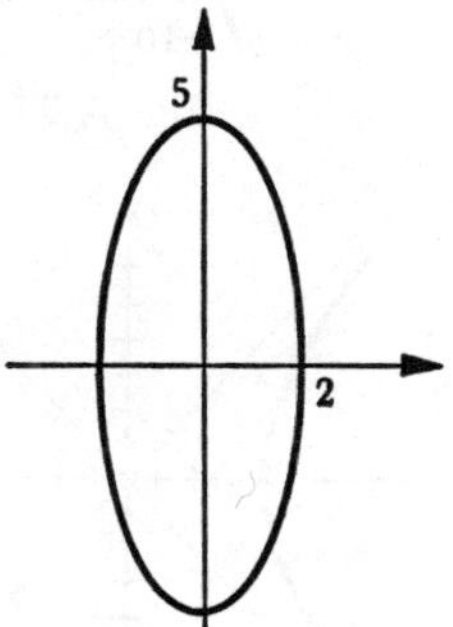

Fig. A.22

9.

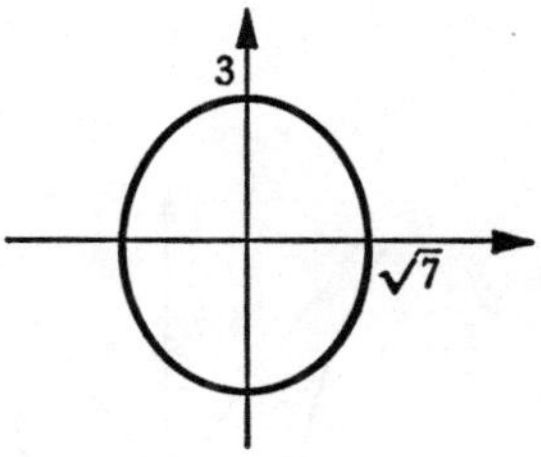

Fig. A.23

11.

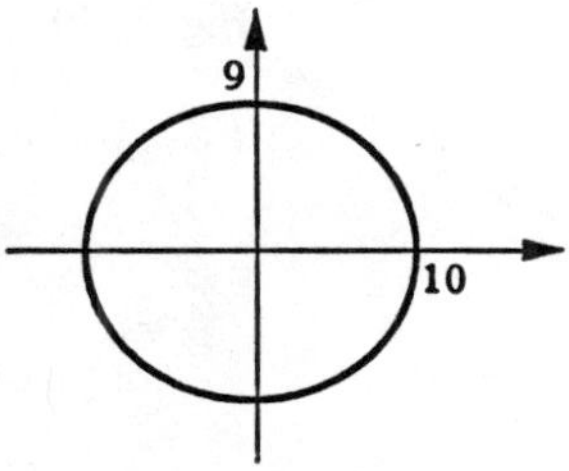

Fig. A.24

13.

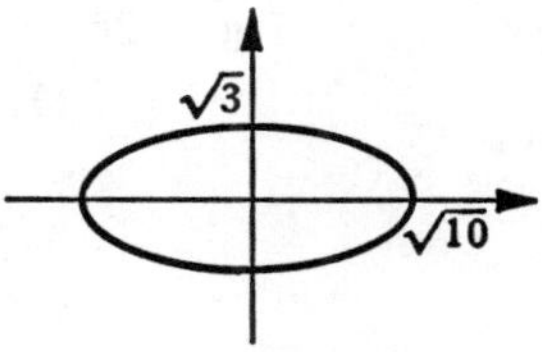

Fig. A.25

15. $\frac{x^2}{25} + \frac{y^2}{16} = 1$

17. $\frac{x^2}{12} + \frac{y^2}{16} = 1$

19. $\dfrac{x^2}{20} + \dfrac{y^2}{4} = 1$

21. $\dfrac{x^2}{9} + \dfrac{y^2}{25} = 1$

23. $\dfrac{x^2}{\frac{25}{4}} + \dfrac{y^2}{\frac{45}{4}} = 1$

25. $a = \sqrt{14}/2$; $b = \sqrt{5}/2$; $c = 3/2$

Problem Set 2.9 $\mathscr{H}$

3.

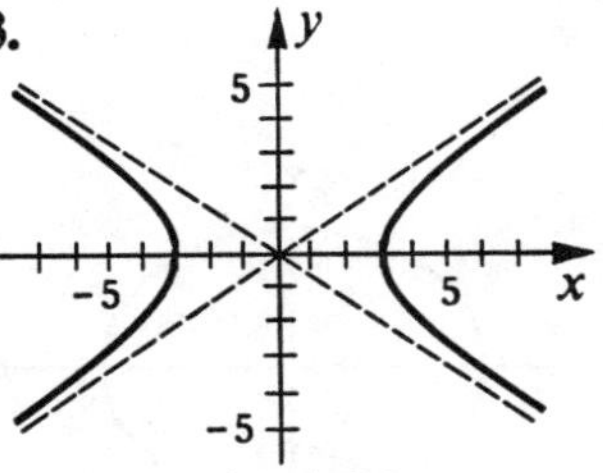

Fig. A.26

5.

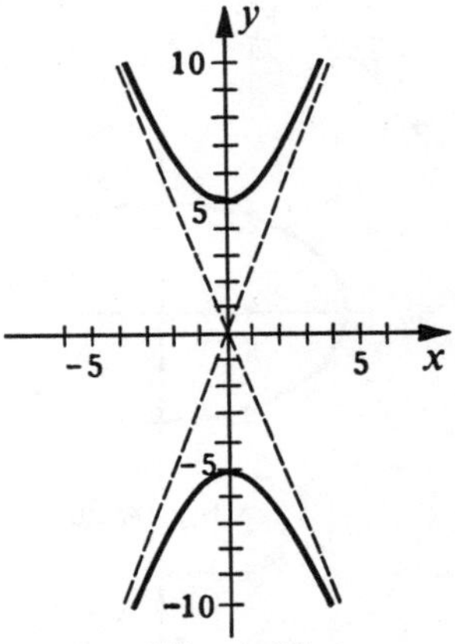

Fig. A.27

7.

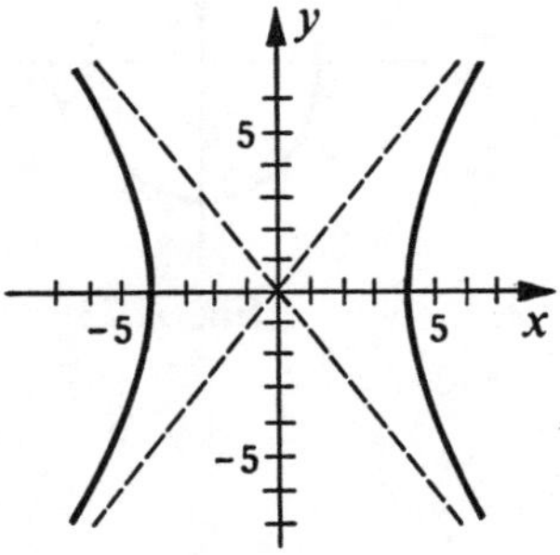

Fig. A.28

9.

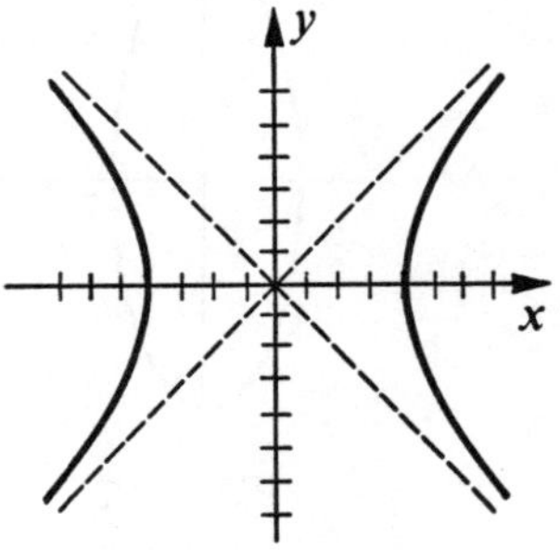

Fig. A.29

11.

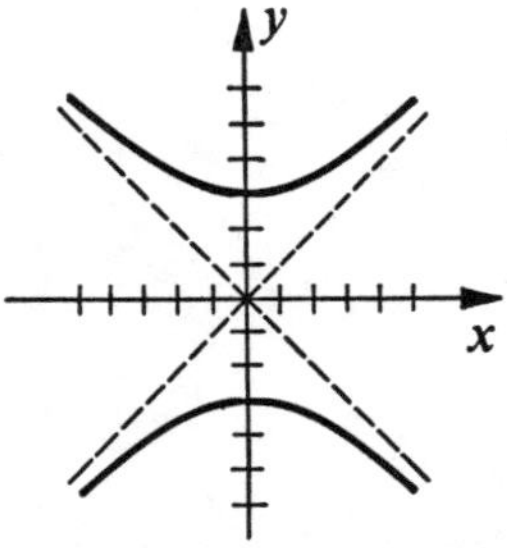

Fig. A.30

13. Transverse and conjugate axes have lengths respectively: (3) 6 and 4; (4) 8 and 6; (5) 10 and 2; (6) 12 and 12; (7) 8 and 10; (8) 6 and 8; (9) 8 and 8; (10) 4 and 6; (11) 6 and 6; (12) 6 and 12.

15. $\dfrac{x^2}{15} - \dfrac{y^2}{1} = 1$

17. $\dfrac{x^2}{1} - \dfrac{y^2}{1} = 1$

19. $\dfrac{x^2}{16} - \dfrac{y^2}{25} = 1$

21. $\dfrac{x^2}{4} - \dfrac{y^2}{36} = 1$

23. In $(1 - e^2)x^2 + y^2 = a^2(1 - e^2)$ divide by the term on the right and get

$$\frac{x^2}{a^2} + \frac{y^2}{a^2(1 - e^2)} = 1.$$

Subst. $b^2 = -a^2(1 - e^2)$ and get

$$\frac{x^2}{a^2} - \frac{y^2}{b^2} = 1.$$

Problem Set 2.9$\mathscr{P}$

1. $d(PF) = d(PQ)$; $\sqrt{[x^2 + (y - c)^2]} = y + c$; $x^2 = 4cy$

3. $y^2 = -2x$

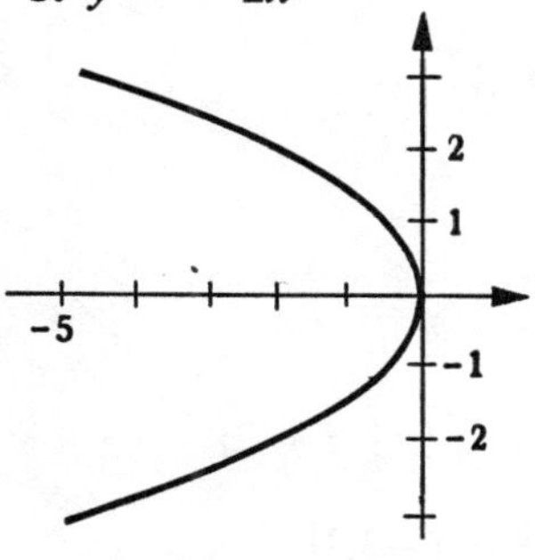

focus: $(-1/2, 0)$
directrix: $x = 1/2$
Fig. A.31

5. $x^2 = -y$

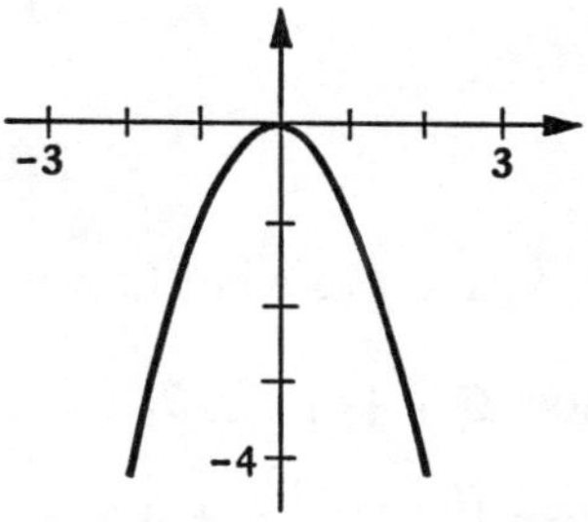

focus: $(0, -1/4)$
directrix: $y = 1/4$
Fig. A.32

7. $y^2 = 4x$

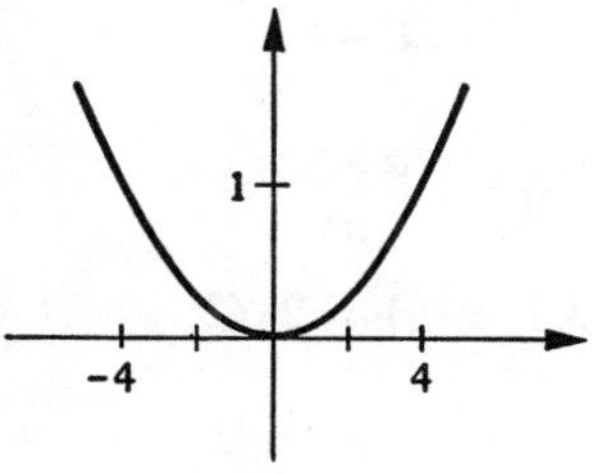

focus: $(1,0)$
directrix: $x = -1$
Fig. A.33

9. $x^2 = 16y$

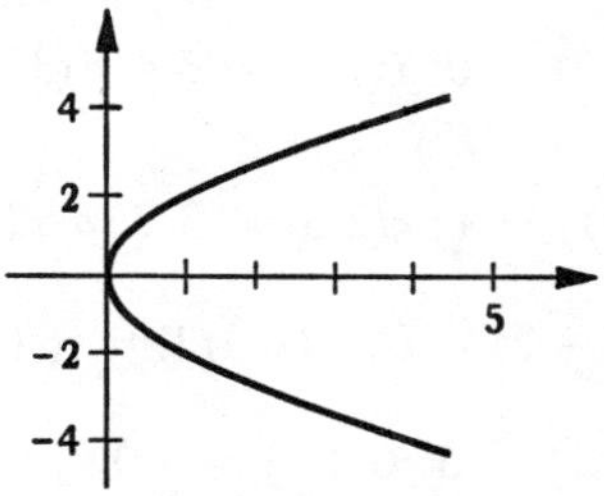

focus: $(0,4)$
directrix: $y = -4$
Fig. A.34

11. $x^2 = -8y$ **13.** $y^2 = -16x$ **15.** $4c$

Problem Set 2.10

3. $(-1,1)$

5. ellipse, major axis horizontal; center: $(-2,3)$; $\frac{(x+2)^2}{4} + \frac{(y-3)^2}{1} = 1$

7. parabola, opens right; vertex: $(2,-3)$; $(y+3)^2 = 4(x-2)$

9. ellipse, major axis vertical; center; $(0,-2)$; $\frac{x^2}{4} + \frac{(y+2)^2}{9} = 1$

11. parabola, opens up; vertex: $(-3/2,-25/4)$; $(x+3/2)^2 = y + 25/4$

13. $\frac{(y-3)^2}{4} + \frac{(x-2)^2}{12} = 1$ **15.** $\frac{(y-5)^2}{36} + \frac{(x+3)^2}{27} = 1$

ANSWERS CHAPTER 3

Problem Set 3.1

1. f, V, and C are not functions. **3.** $\operatorname{dom} g = R$

5. $\operatorname{dom} G = \{x \mid x \leq 2\}$ **7.** $\operatorname{dom} U = \{x \mid x \neq 2\}$

9. $\operatorname{dom} W = \{x \mid x \neq 0, x \neq 2\}$ **11.** $\operatorname{dom} F = [-3,3]$

13. $g(2) = 3$; $g(-2) = -1$; $g(1/2) = 3/2$; $1/g(2) = 3/2$; $g(a+b) = a + b + 1$; $g(a) + g(b) = a + b + 2$

15. $f(2) = 0$; $f(1/2) = 3/4$; $f(3/2) = 3/4$; $\frac{f(t) - f(2)}{t-2} = -t$

17. $h(1) = 1$; $h(2/3) = 3/2$; $a \cdot h(a) = 1$; $h(x) - h(a) = \frac{a-x}{ax}$

19. $F(-2) = 0$; $F(-1/2) = \sqrt{(3/2)}$; $F(1/4) = 3/2$; $F(6) = 2\sqrt{2}$

21. $G(1) = 0$; $G(1.9) = -0.9$; $G(2) = -1$; $G(2.1) = -0.8$; $G(4) = 3$

23. $F(t) = F(-t)$; $F(2t) = 4F(t)$; $F(t^2) = [F(t)]^2$; if $t \geq 0$, $\sqrt{[F(t)} = t$; $\sqrt{[F(t)]} = t$ for all t

Problem Set 3.2

5.

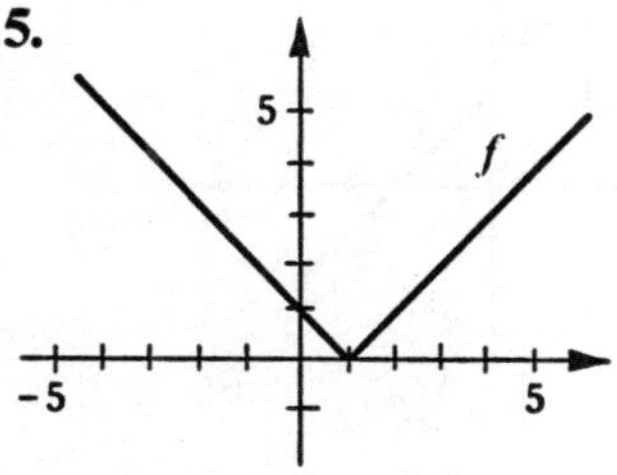

dom $F = R$
rng $f = \{y: y \geq 0\}$

Fig. A.35

7.

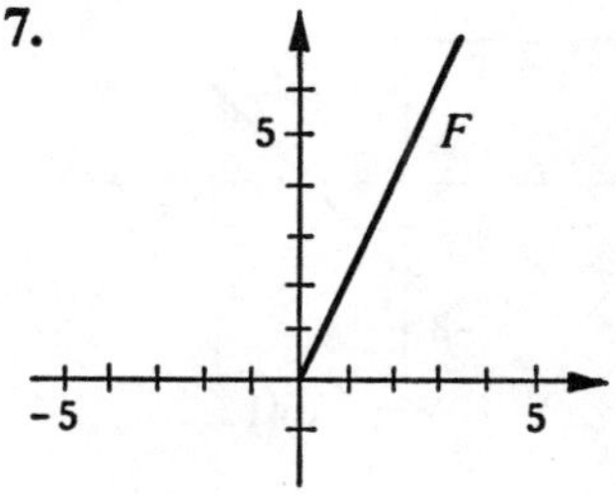

dom $F = R$
rng $F = \{y: y \geq 0\}$

Fig. A.36

9.

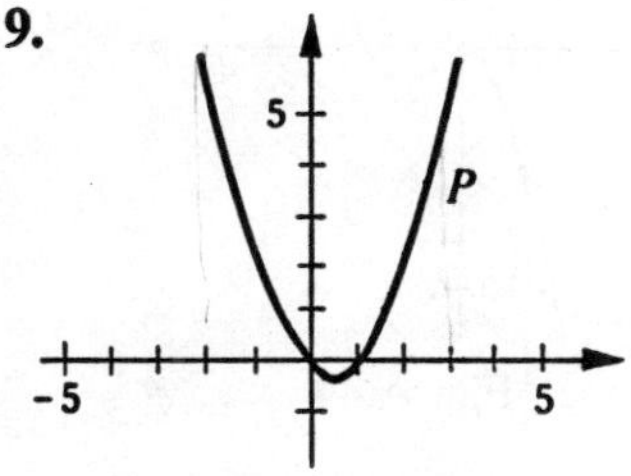

dom $P = R$
rng $P = \{y: y \geq -1/4\}$

Fig. A.37

11.

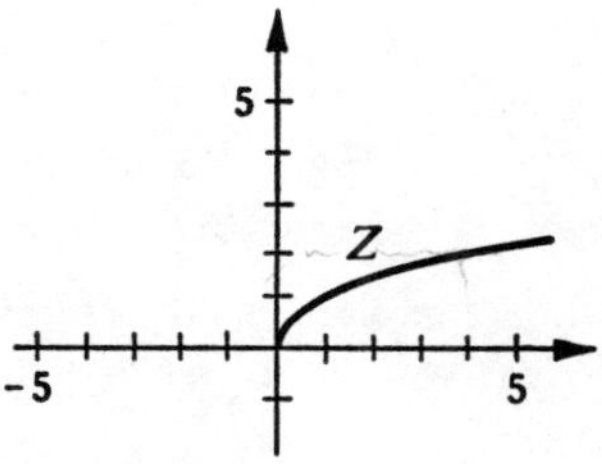

dom $Z = \{x: x \geq 0\}$
rng $Z = \{y: y \geq 0\}$

Fig. A.38

13.

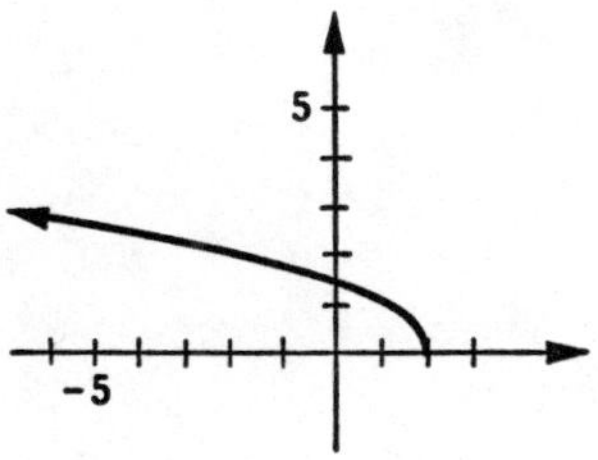

dom $G = \{x \mid x \leq 2\}$
rng $G = \{y \mid y \geq 0\}$

Fig. A.39

15.

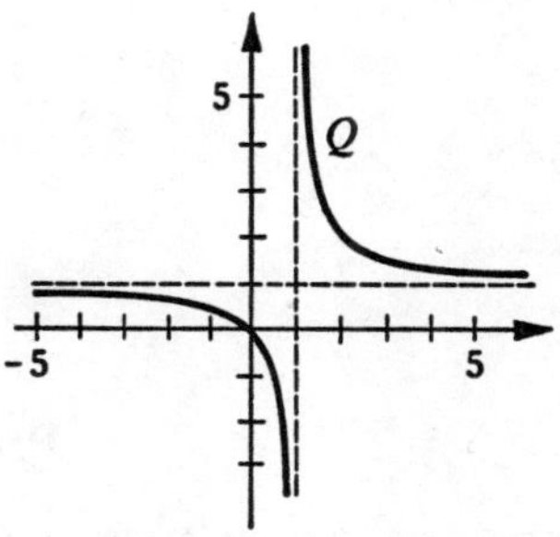

dom $Q = R$
rng $Q = \{y: y \neq 1\}$

Fig. A.40

17.

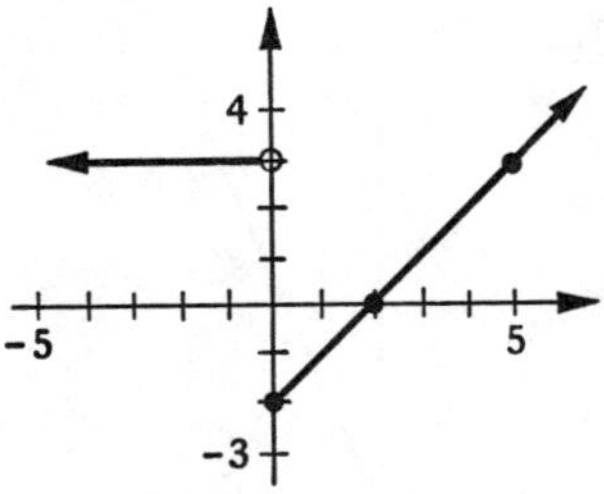

Fig. A.41

19.

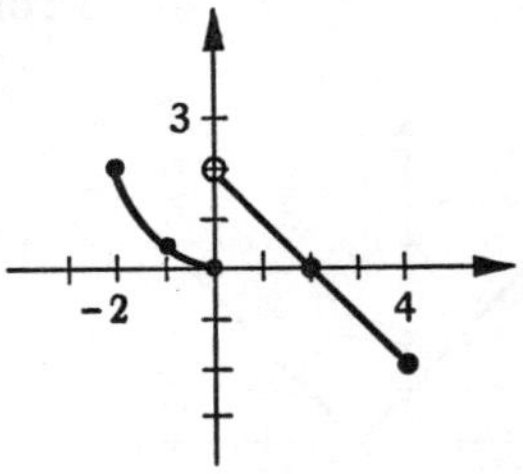

Fig. A.42

21.

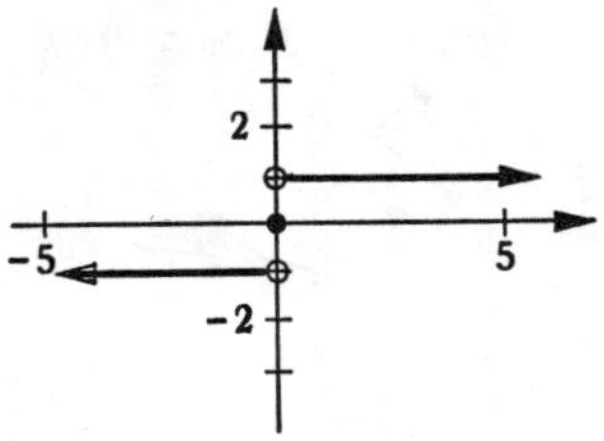

Fig. A.43

23.

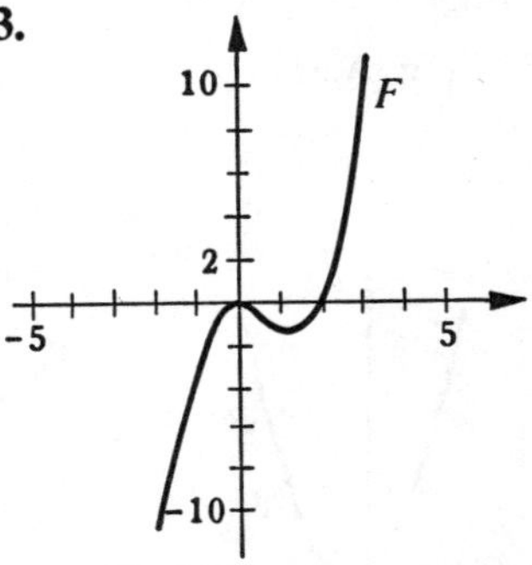

Fig. A.44

25.

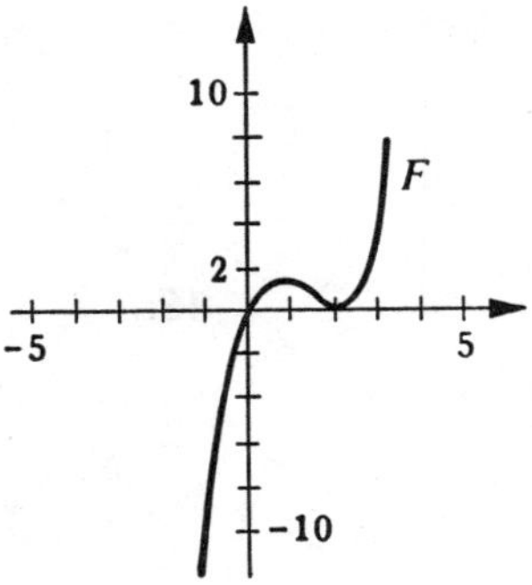

Fig. A.45

Problem Set 3.3

1. $[F + G] = \{(-1,3), (0,3), (1,2), (2,3)\}$
$[FG] = \{(-1,2), (0,-4), (1,0), (2,-10\}$
$[FG] = \{(-1,\ 1/2), (0,-4), (2,-5/2)\}$

3. **a.** $[F + G](x) = x + 1$ **b.** $[FG](x) = -2x^2 + 5x - 2$

c. $[F/G](x) = \dfrac{2x - 1}{2 - x}$

d. dom $[F + G] = R$; dom $[FG] = R$; dom $[F/G] = \{x \mid x \neq 2\}$

5. **a.** $[F + G](x) = x^2 + 4x - 12$

b. $[FG](x) = x^3 + x^2 - 16x + 20$

c. $[F/G](x) = x + 5$

d. dom $[F + G] = R$; dom $[FG] = R$; dom $[F/G] = \{x \mid x \neq 2\}$

7. $[H + I](x) = \dfrac{x^2 + 1}{x}$; dom $[H + I] = \{x \mid x > 0\}$;
rng $[H + I] = \{y \mid y \geq 2\}$

9. **a.** $[F + G](x) = \dfrac{x^2 - x + 1}{x + 2}$ **b.** $[F/G](x) = x^2 - x$

11. **a.** $[fg](x) = 1 - x$ **b.** $[f + g](x) = 2$

13.

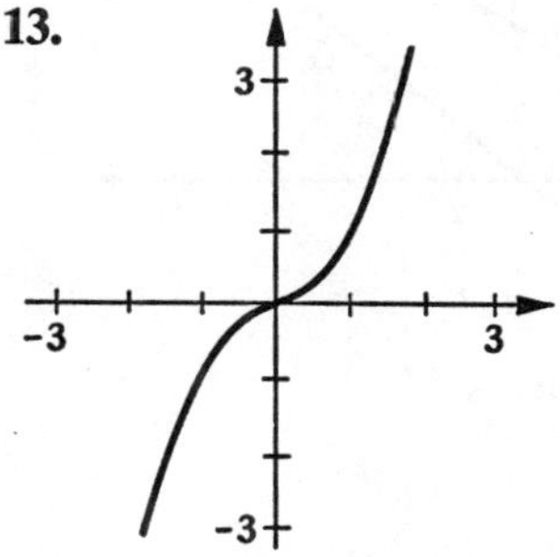

Fig. A.46

15. **a.** The statement that "G is a function" means that G is a point set that no two points in G have the same abscissa.

b. If G is a function, then $G(a)$ denotes the ordinate of that point in G whose abscissa is a.

17. For $x = -1, 0$, and $1/2$, $[F \circ G](x) \nexists$,
$[F \circ G](1) = 0$, $[F \circ G](2) = 1$.

19.

x:	-1	0	$1/2$	1	2
$[F \circ G](x)$:	1	0	$-1/2$	$\nexists$	-2

21. $[F \circ G](\delta) = -\delta$

Problem Set 3.4

1. $[F \circ G] = \{(2,2), (4,2), (5,4)\}$

3. $\sqrt{(x-1)}$; $\sqrt{x} - 1$

5. $-x$; $\dfrac{x}{2x-1}$

7. $x^2 - 1$; $(x-1)^2$

9. $1/\sqrt{x}$

11. $x^3 + x$

Problem Set 3.5

3.

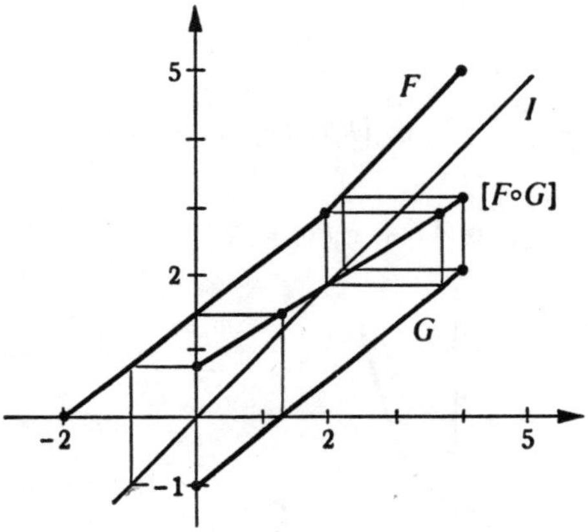

Fig. A.47

5.

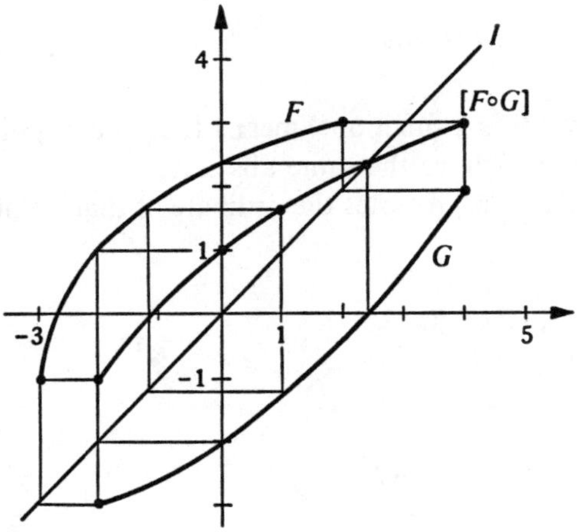

Fig. A.48

7. $[F \circ G](x) = x^2 + 11x + 24$; $[G \circ F](x) = x^2 + 3x$

9. $[F \circ G](x) = \dfrac{x-2}{2(x-4)}$; $[G \circ F](x) = \dfrac{1-7x}{4x}$

11. $[F \circ G](x) = \dfrac{4x+1}{x+2}$; $[G \circ F](x) = \dfrac{3x+2}{x+3}$

13. $[F \circ G](x) = x$; $[G \circ F](x) = \sqrt{x^2}$

15. $y - 4 = -1/2 \cdot (x - 0)$ or $y - 2 = -1/2 \cdot (x - 4)$

17. $y + 3 = \dfrac{15}{16} \cdot (x + 2)$ or $y + 1/2 = \dfrac{15}{16} \cdot (x - 2/3)$

19. $y - k = -\dfrac{1}{2a}(x + a)$ or $y - k + 1 = -\dfrac{1}{2a}(x - a)$

Problem Set 3.6

1. $[F \circ G](x) = 5 - 2x$;
dom $[F \circ G] = R$;
rng $[F \circ G] = R$

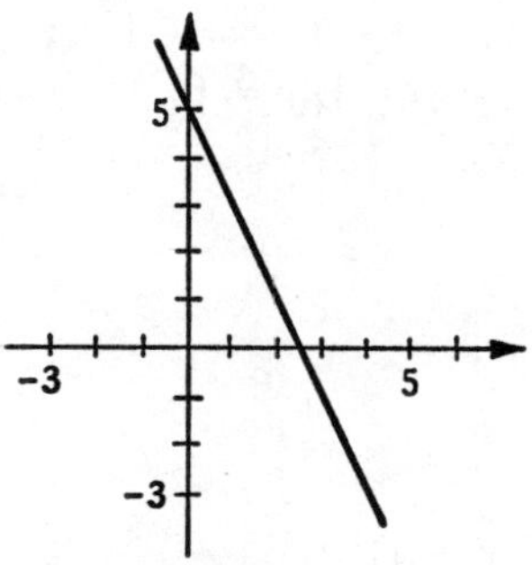

Fig. A.49

3. $[F \circ G](x) = x^2 - 3x + 2$;
dom $[F \circ G] = R$;
rng $[F \circ G] = \{y : y \geq -1/4\}$

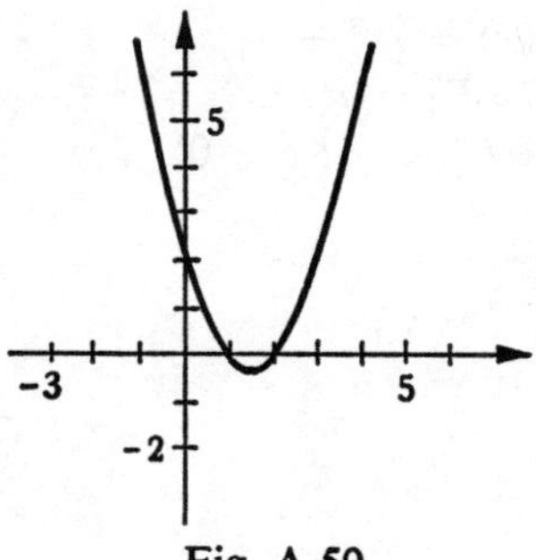

Fig. A.50

5. $[F \circ G](x) = x - 2\sqrt{x}$;
dom $[F \circ G] = \{x: x \geq 0\}$;
rng $[F \circ G] = \{y: y \geq -1\}$

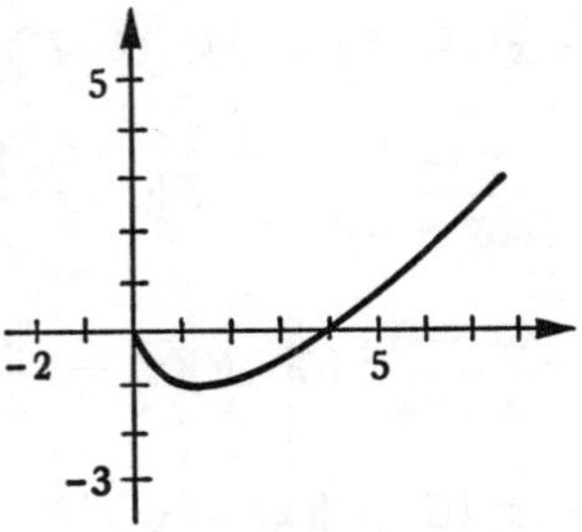

Fig. A.51

7. $[F \circ G](x) = x$;
dom $[F \circ G] = \{x: x \geq -1\}$; rng $[F \circ G] = \{y: y \geq -1\}$

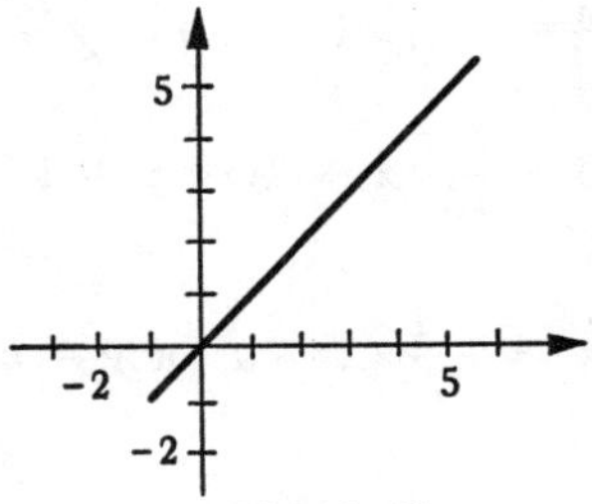

Fig. A.52

9. $[F \circ G](x) = \dfrac{1}{2x + 1}$;
dom $[F \circ G] = \{x: x \neq -1, x \neq -1/2\}$;
rng $[F \circ G] = \{y: y \neq -1, y \neq 0\}$

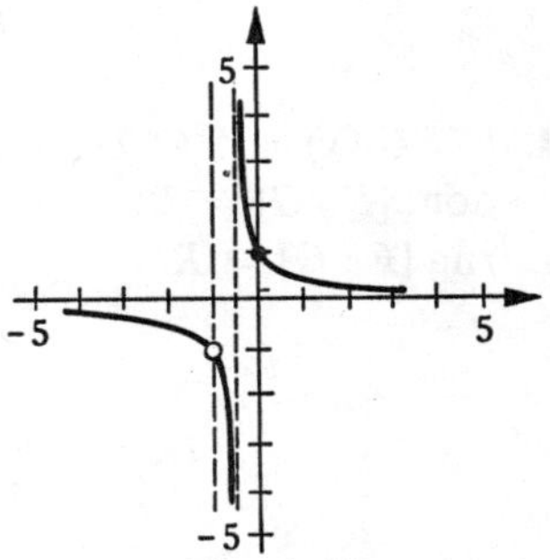

Fig. A.53

11. $[P \circ I](x) = \sqrt{(2x + 1)}$;
dom $[P \circ I] = \{x: x \geq -1/2\}$;
rng $[P \circ I] = \{y: y \geq 0\}$

13. $[G \circ F](x) = |x^2 - x|$;
dom $[G \circ F] = R$;
rng $[G \circ F] = \{y: y \geq 0\}$

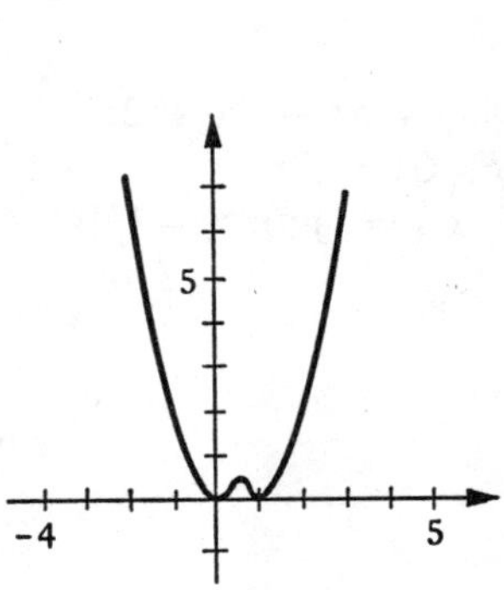

Fig. A.54

15. $[G \circ F](x) = x^2 + x - 2;$
dom $[G \circ F] = R;$
rng $[G \circ F] = \{y : y \geq -9/4\}$

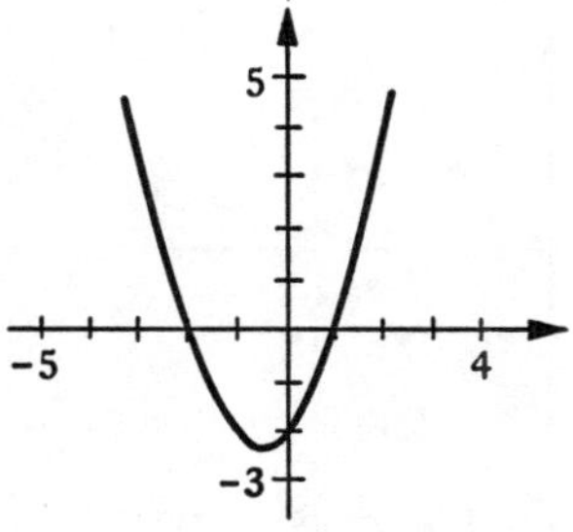

Fig. A.55

17.

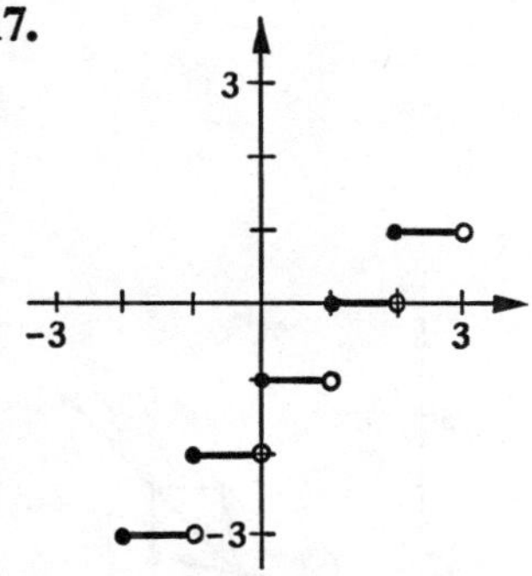

Fig. A.56

ANSWERS CHAPTER 4

Problem Set 4.1

1. **a.** incr. over domain
c. incr. over $\langle 2/3,\infty\rangle$; decr. over $[0,2/3\rangle$
e. incr. over $\langle 1,\infty\rangle$
g. incr. over $\langle 1/2,\infty\rangle$; decr. over $\langle -\infty,1/2\rangle$

3. 1 *a*, *b*, and *f* are 1-to-1.

5. **a.**

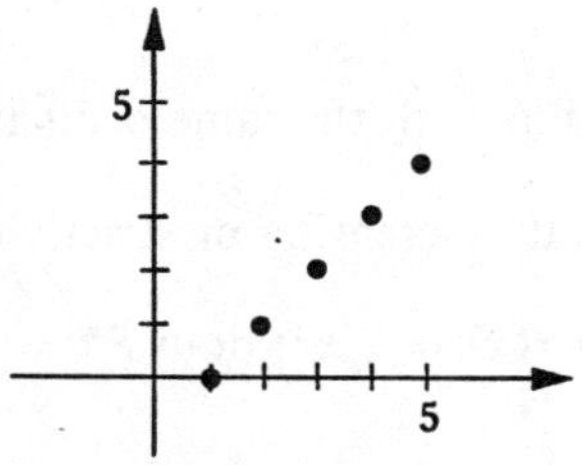

Fig. A.57

b. incr. over domain
c. Yes, is 1-to-1

7. a.

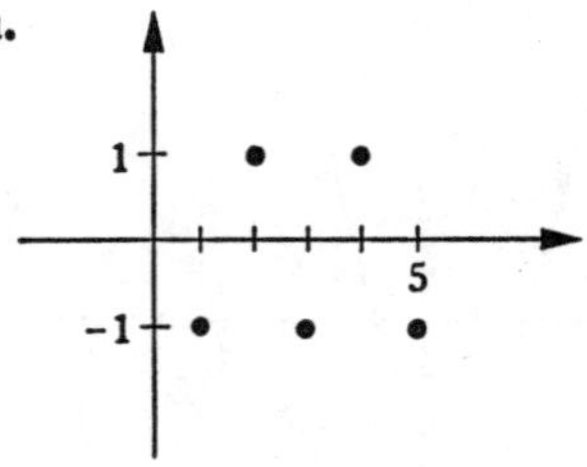

Fig. A.58

b. No
c. No

Problem Set 4.2

1. $F = \{(0,1), (3,2), (1,3), (2,4)\}$; Yes

3.

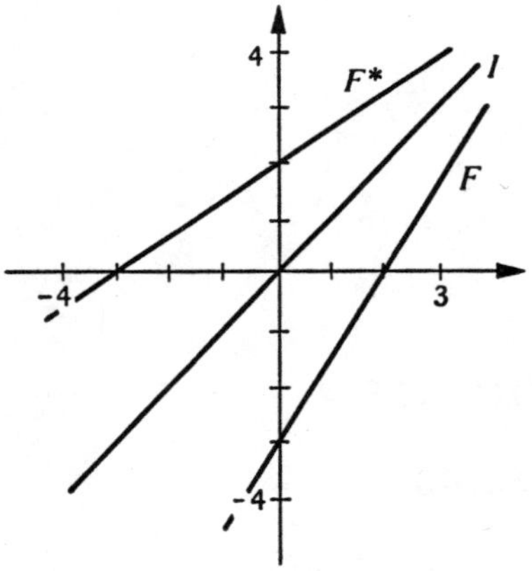

Fig. A.59

5.

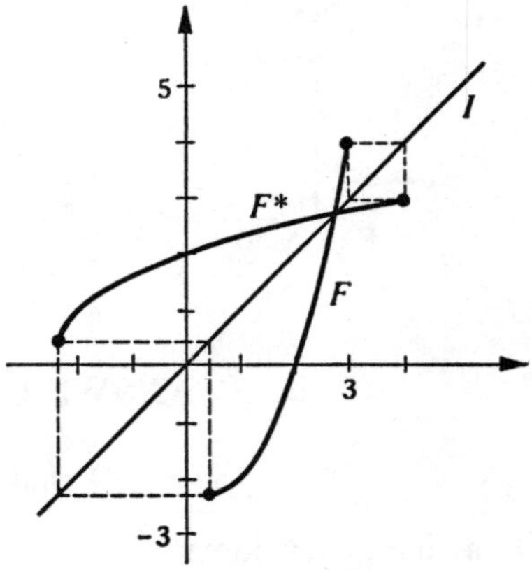

Fig. A.60

7. F^* and G^* are *not* functions, for each contains two points with the same abscissa.
Complete the statement: "If F is a function, then F^* is a function if ________."

Answer: F does not have two points with the same ordinate.

Alternate answer: F is either strictly increasing or strictly decreasing.

9. $F^* = \{(x,y): y = 1/x^2\}$ that is, $F^*(x) = 1/x^2$; dom $F^* = \{x: x > 0\}$; rng $F^* = \{y: y > 0\}$

11. $F^* = \{(y,x): y = x^2\}$ or $F^* = \{(x,y): y^2 = x\}$; not a function

13. $F^* = \left\{(x,y): y = \dfrac{2x}{1-x}\right\}$ that is, $F^*(x) = \dfrac{2x}{1-x}$;
dom $F^* = \{x: x \neq 1\}$; rng $F^* = \{y: y \neq -2\}$

15. **a.** $F^* = \{(y,x): y = x^2 - 1\}$ and F^* is not a function
b. G^* is a function and $G^*(x) = \sqrt{(x+1)}$; dom $G^* = \{x: x \geq -1\}$

Problem Set 4.3

1.

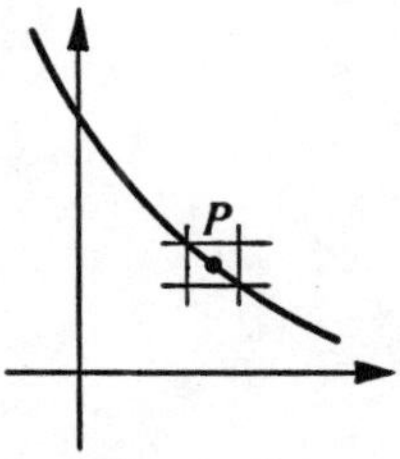

Fig. A.61a

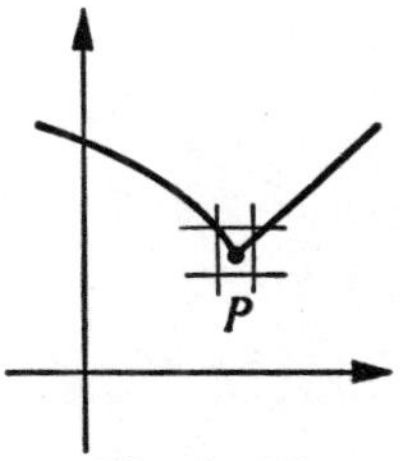

Fig. A.61b

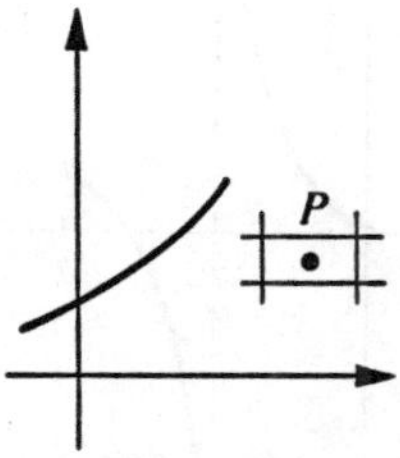

Fig. A.61c

3.

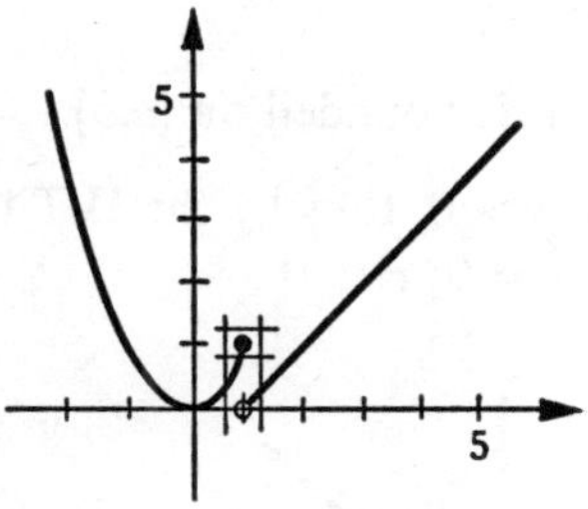

Fig. A.62
Not continuous.

5.

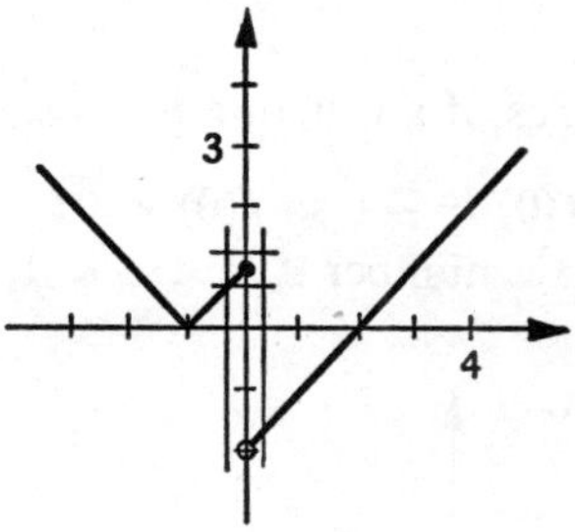

Fig. A.63
Not continuous.

7.

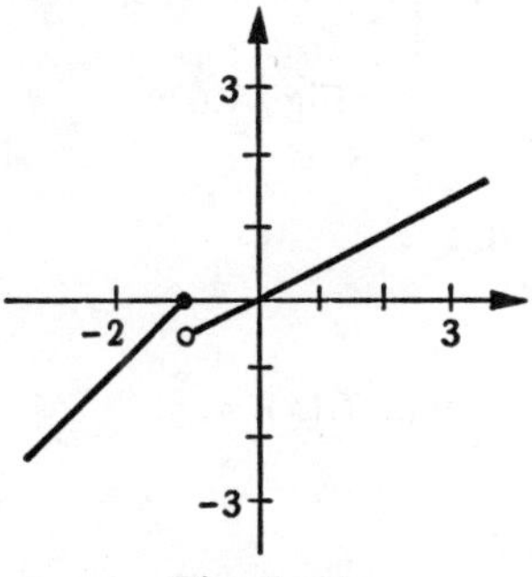

Fig. A.64
Not continuous.

9.

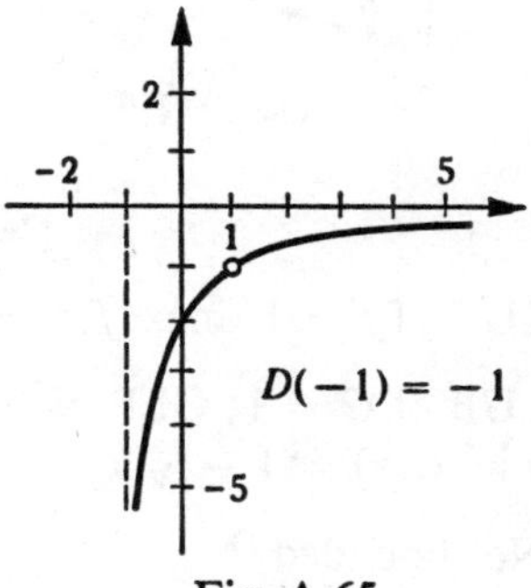

Fig. A.65

11.

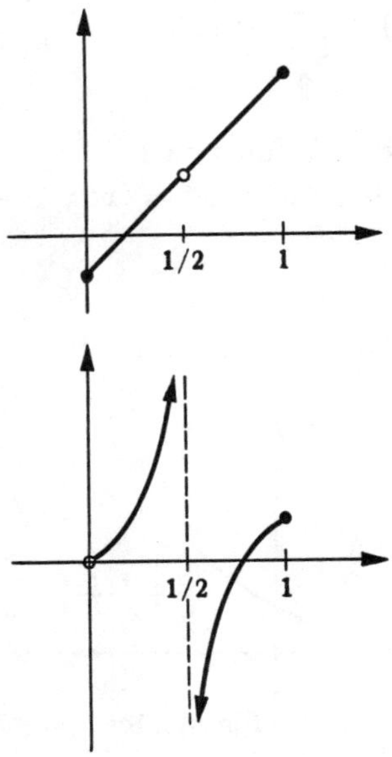

Fig. A.66

13. Yes, if a function is cont. on $[a,b]$, then it is bounded on $[a,b]$.

15. $f(0) = -1$ so $f(0) < 0$; $f(1) = 1$, so $f(1) > 0$, thus by the IVT there is a number c, $0 < c < 1$, such that $f(c) = 0$; $c \approx 0.6$

17. a.

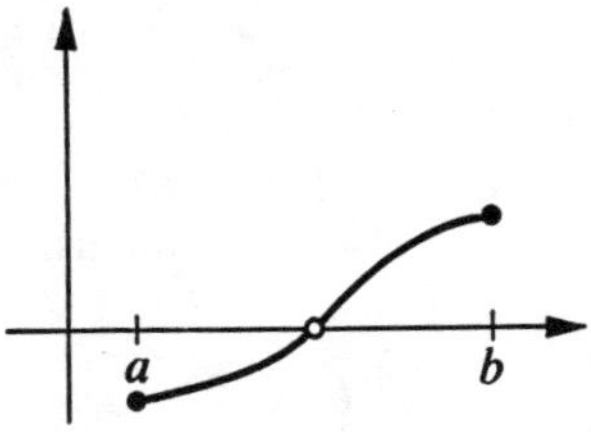

Fig. A.67a

b.

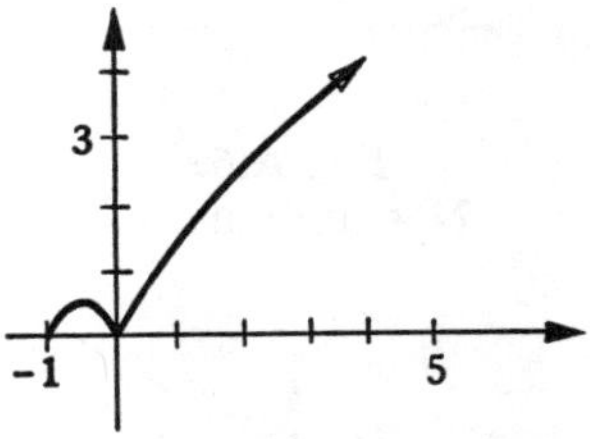

Fig. A.67b

Problem Set 4.4

1. LUB of $f = 1$; max. $f(x) = 1$

3. LUB of $G = 1$; GLB of $G = 1 - \sqrt{5}$; max. $G(x) = 1$; min. $G(x) = 1 - \sqrt{5}$

5. Not bounded

7. GLB of $w = -4$; min. $w(x) = -4$

9. GLB of $Z = 0$; no min.

11. GLB of $F = 0$; min. $F(x) = 0$

13. LUB of $p = 4$; max. $p(x) = 4$

15. GLB of $g = -4$; no min.

17. LUB of $p = 0$; GLB of $p = -2$; max. $p(x) = 0$, no min.

19. $A(x) = 100x - 2x^2$; rectangle 25 m. by 50 m. has max. area of 1250 m.^{2}

ANSWERS CHAPTER 5

Problem Set 5.1

1. a. $P(x) = 2x - x^2$; degree is 2
c. $P(x) = 3x^3 + x^2 - 1$; degree is 3

3. zeros of $P = \{1,3\}$

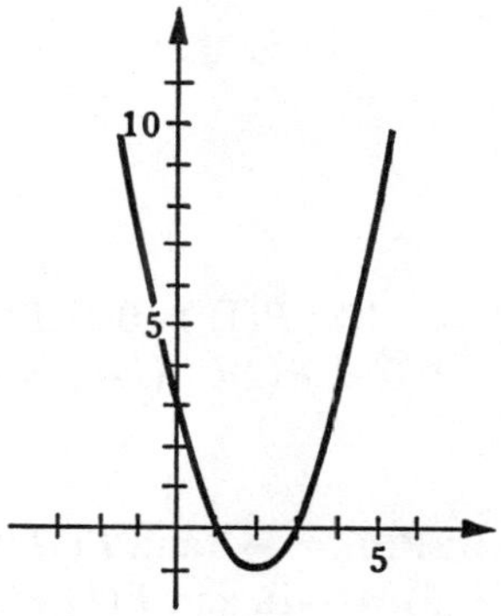

Fig. A.68

5. zeros of $P = \{1\}$

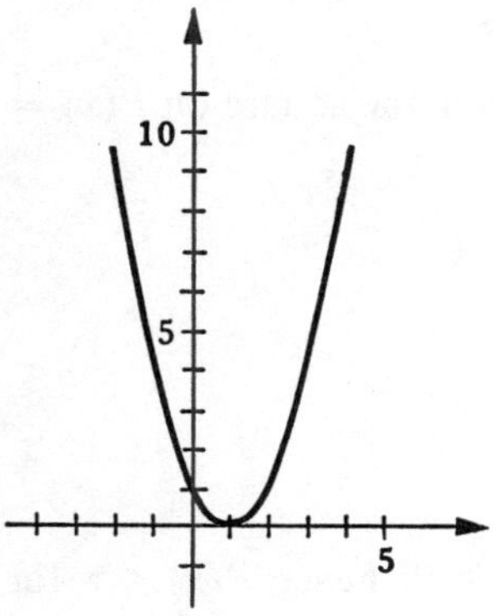

Fig. A.69

7. zeros of $P = \{-2, -1, 1, 2\}$

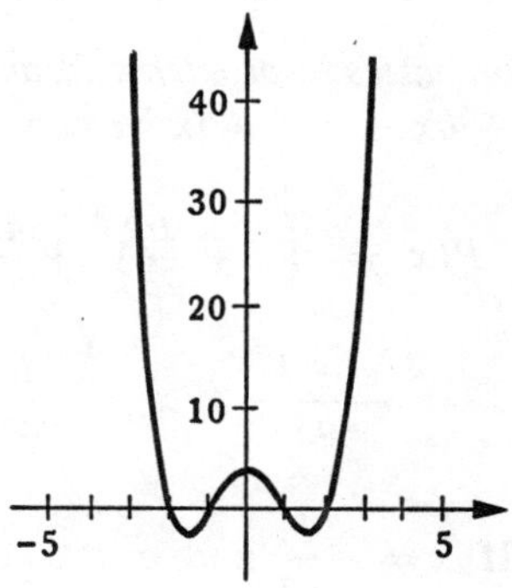

Fig. A.70

9. zeros of $P = -1, 1, 3$

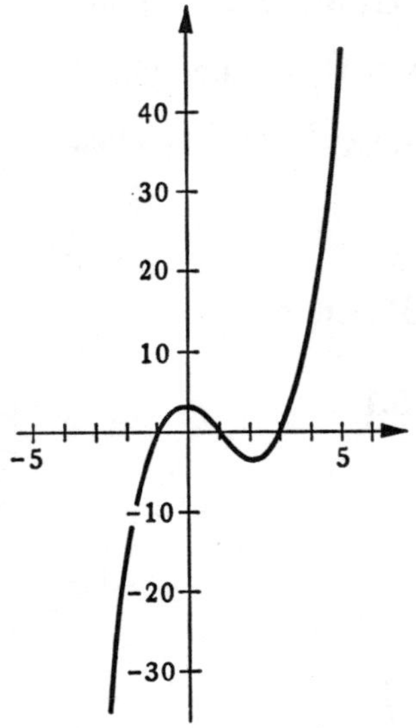

Fig. A.71

11. zeros of $P = -5, 0, 3$

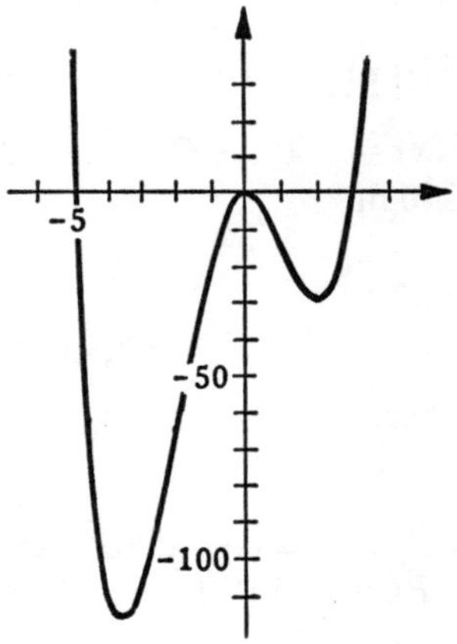

Fig. A.72

13. $P(1) = 1$ and $P(3/2) = -1/8$; $P(1) > 0 > P(3/2)$; $P(3/2) = -1/8$ and $P(2) = 1$; $P(3/2) < 0 < P(2)$; $P(-4) = -29$ and $P(-3) = 1$; $P(-4) < 0 < P(-3)$

15. $P(-1) = 0$; $P(1) = 0$; $P(0) = -1$ and $P(1/2) = 7/16$; $P(0) < 0 < P(1/2)$; $P(3) = -16$ and $P(4) = 15$; $P(3) < 0 < P(4)$

17. **a.** If $b^2 - 4ac > 0$, then P has two zeros.
b. If $b^2 - 4ac = 0$, then P has only one zero.
c. If $b^2 - 4ac < 0$, then P has no real zeros.

Preliminaries to problems 18 *and* 19*:* By completing the square on $P(x) = ax^2 + bx + c$, $a \neq 0$, we can show that

$$P(x) = a\left(x + \frac{b}{2a}\right)^2 + \frac{4ac - b^2}{4a}.$$

Let $z = \dfrac{4ac - b^2}{4a}$. If $x = -\dfrac{b}{2a}$, then $P(x) = z$.

19. If $x \neq -\dfrac{b}{2a}$ and $a < 0$, then $a\left(x + \dfrac{b}{2a}\right)^2 < 0$, hence $P(x) \leq z$ for all x.

Problem Set 5.2

1. a. $P(2/3) = 8/3$

b. $P(x) = 3x^2 + 3x + 1 + \dfrac{8/3}{x - 2/3}$

c.
$$\begin{array}{r|rrrr} 2/3 & 3 & 1 & -1 & 1 \\ & & 2 & 2 & 2/3 \\ \hline & 3 & 3 & 1 & 8/3 \end{array}$$

3.
$$\begin{array}{r|rrrr} -3 & 1 & -2 & 4 & -12 \\ & & -3 & 15 & -57 \\ \hline & 1 & -5 & 19 & -69 \end{array}$$

5.
$$\begin{array}{r|rrrrrr} 1 & 1 & 0 & -1 & 2 & 0 & 1 \\ & & 1 & 1 & 0 & 2 & 2 \\ \hline & 1 & 1 & 0 & 2 & 2 & 3 \end{array}$$

7.
$$\begin{array}{r|rrrrr} -3 & 4 & 0 & 3 & -2 & 5 \\ & & -12 & 36 & -117 & 357 \\ \hline & 4 & -12 & 39 & -119 & 362 \end{array}$$

9.
$$\begin{array}{r|rrrr} 3/4 & 4 & -7 & 15 & -9 \\ & & 3 & -3 & 9 \\ \hline & 4 & -4 & 12 & 0 \end{array}$$

3/4 is a zero of P

11. a.
$$\begin{array}{r|rrrr} 2 & 1 & -4 & -8 & -50 \\ & & 2 & -4 & -24 \\ \hline & 1 & -2 & -12 & -74 \end{array}$$

$P(2) = -74$

b. $P(3) = -83$

c. $P(-3) = -89$

13. 4 is an upper bound

15. a. $P(10) = 213$ **b.** $P(10) = 1{,}301$ **c.** $P(10) = 2{,}154$
d. $P(10) = 10{,}205$ **e.** $P(x) = x^3 + 2x + 7$

Problem Set 5.3

1. at most 3

3. at most 5

5. -4 mult. of 3; 2 mult. of 1; 3 mult. of 2

7. -2 mult. of 2

9. 4 is an upper bound; -4 is a lower bound

11. 2 is an upper bound; -2 is a lower bound

13. 3 is an upper bound; -4 is a lower bound

15.
$$\begin{array}{r|rrrr} 3/2 & 2 & -2 & 1 & 6 \\ & & 3 & 3/2 & 15/4 \\ \hline & 2 & 1 & 5/2 & 39/4 \end{array}$$

thus $P(3/2) = 39/4$.

Problem Set 5.4 Part A

1. no rational zeros **3.** $\{-3, -2, 1/2, 2\}$ **5.** $\{1, 2, 3\}$

7. $\{-2,1\}$ **9.** no rational zeros **11.** $\{4/3\}$

13. $\{-3,1/2\}$

ANSWERS CHAPTER 6

Problem Set 6.1

1. The sum of the area of the two parts equals the area of the whole.

3. $\frac{p+q}{2} - r = \frac{p+q-2r}{2} < \frac{p+q-2p}{2} = \frac{q-p}{2ab}$

5. $\frac{b^2-a^2}{2ab}$ **7.** $\frac{b-a}{b} < x < \frac{b-a}{a}$

Problem Set 6.4

1. $\int_{5/6}^{1} H = \int_{1}^{6/5} H = \int_{1}^{1.2} H = L(1.2) = 0.182$

3. $L(2.2) = L(2) + L(1.1) = 0.788$; $L(3.6) = L(3) + L(1.2) = 1.280$; $L(4) = L(2) + L(2) = 1.386$; $L(4/3) = L(4) - L(3) = 0.288$

Problem Set 6.5

3. If $L(s) = L(t)$, then L would have two points with the same ordinate contrary to theorem L-4, that is, contrary to the fact that L is an increasing function.

5. $L(p) = q$; $L^*(q) = p$; $[L \circ L^*](q) = q$; $[L^* \circ L](p) = p$

7. a. $\int_{1/27}^{27} H = L(27) - L(1/27)$ [by theorem L-1] $= L(27) + L(27)$ [by theorem L-2ii] $= 6 \cdot L(3) = 6.594$

11. **a.** $P = (1,1)$; $Q = \left(1, \dfrac{2c-1}{c^2}\right)$; $R = (1,1/c)$; $T = (c,1/c)$

b. $\dfrac{c^2-1}{2c}$ **c.** $\dfrac{(3c-1)(c-1)}{2c^2}$ **d.** $\dfrac{c-1}{c}$

Problem Set 6.6

3. Since $(1,0) \in L$ $(0,1) \in E$, thus $E(0) = 1$. In like manner $E(1) = e$.

5. **a.** 0.693 **b.** 1.609 **c.** −1.098 **d.** −0.693

7. GLB of $E = 0$; no min.; E not bounded above

Problem Set 6.7

1. 5.6 **3.** 3.5 **5.** 3.05 **7.** 7.23 **9.** 1.43 **11.** 0.45

Problem Set 6.8

1. $\int_{7.3}^{7.4} H = \dfrac{1.47}{108.04} = 0.014$. Thus $L(7.4) = 2.002$. Since $L(e^2) = 2$ and L is an increasing function, we have $7.3 < e^2 < 7.4$.

2. **c.** $b^x c^x = E[xL(b)] \cdot E[xL(c)]$ [by defn. of b^x] $= E[xL(b) + xL(c)]$ [by theorem E-3] $= E\{x[L(b) + L(c)]\}$ [factoring] $= E[xL(bc)]$ [by theorem L-2] $= (bc)^x$ [by defn.]

3.

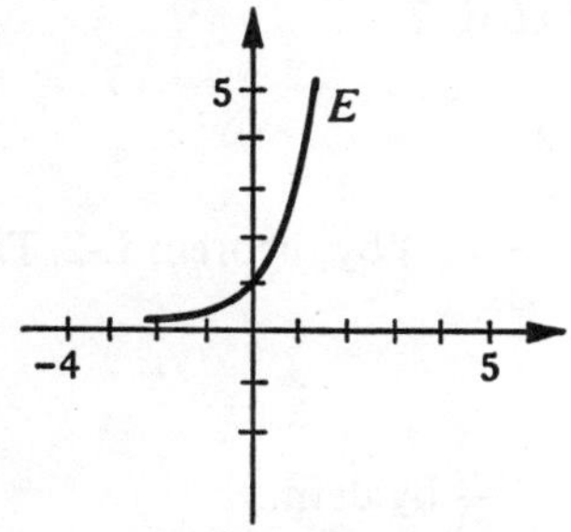

Fig. A.73

5.

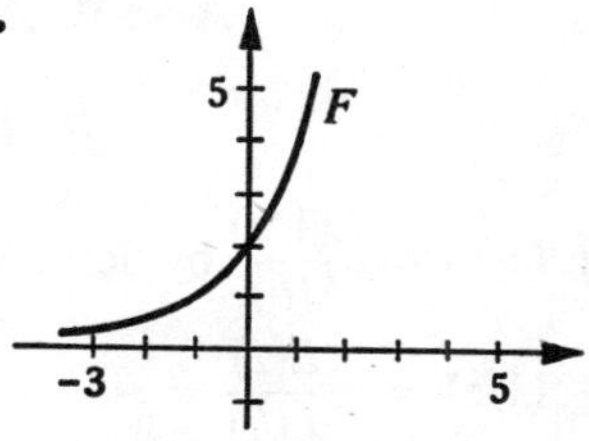

Fig. A.74

7.

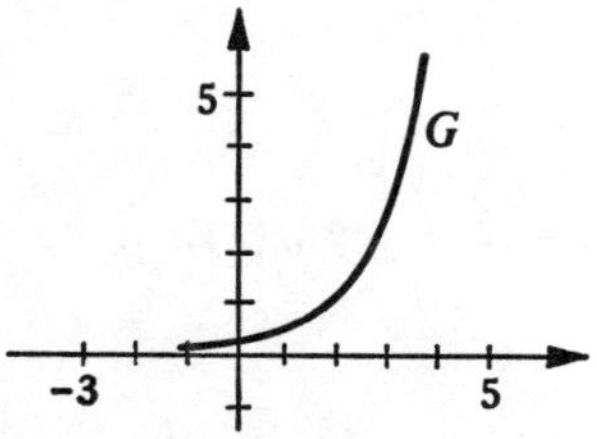

Fig. A.75

9.

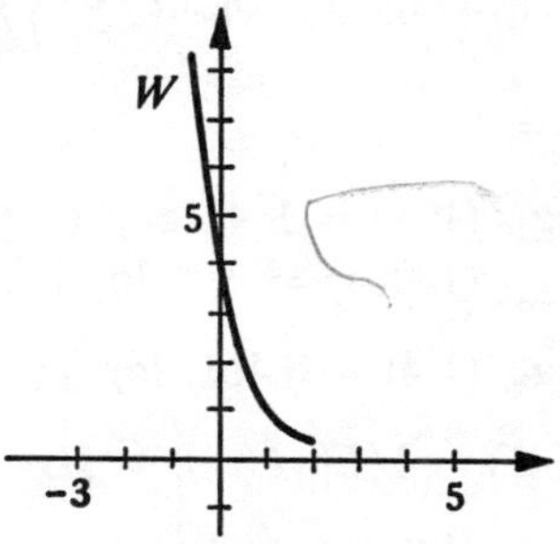

Fig. A.76

11.

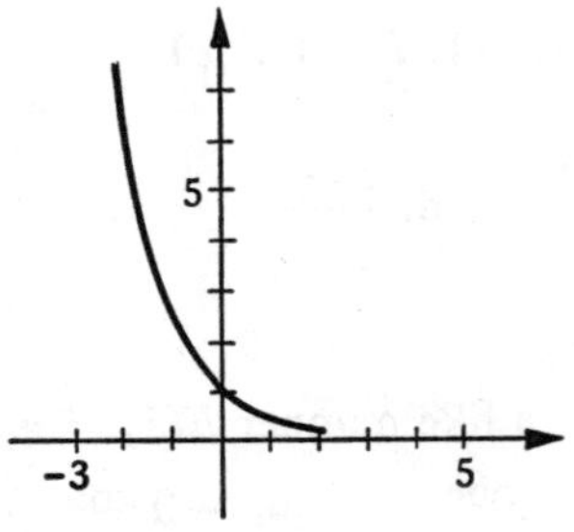

Fig. A.77

13.

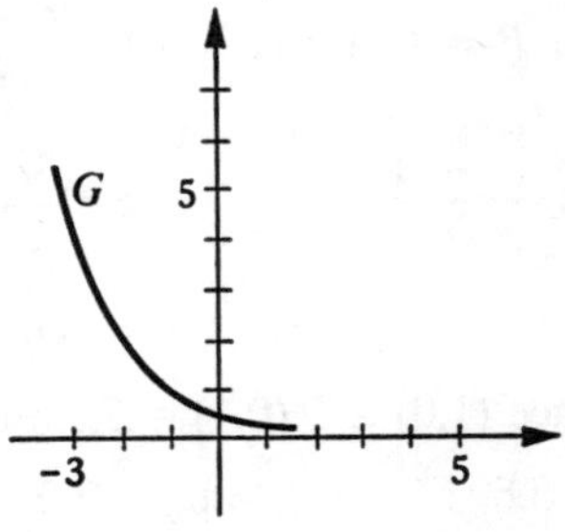

Fig. A.78

15.

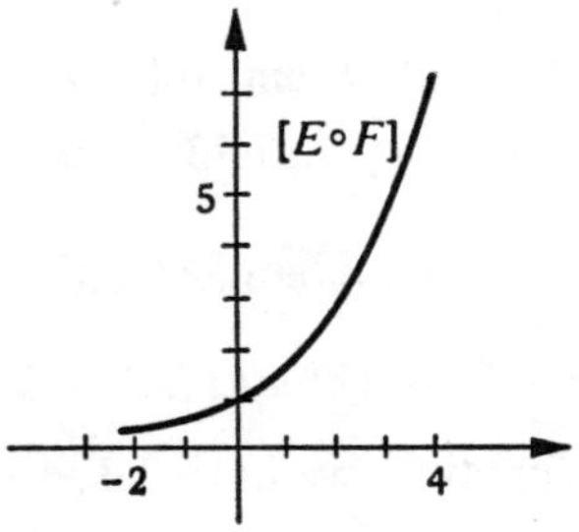

Fig. A.79

ANSWERS CHAPTER 7

Problem Set 7.1

1. $\log_b(1/z) = \dfrac{L(1/z)}{L(b)}$ by defn. But $L(1/z) = -L(z)$ by theorem L-2. Thus

$\log_b(1/z) = -\dfrac{L(z)}{L(b)} = -\log_b(z)$.

3. $\log_b(c) = \dfrac{L(c)}{L(b)}$ by defn. Also, $\log_c(b) = \dfrac{L(b)}{L(c)}$ by defn.

Hence $\log_b(c)\cdot\log_c(b) = \dfrac{L(c)}{L(b)}\cdot\dfrac{L(b)}{L(c)} = 1$.

5. $x = 1$

7. $\log_{10}(10x) = 1 + \log_{10}(x)$; $\log_{10}(100x) = 2 + \log_{10}(x)$;
$\log_{10}(1000x) = 3 + \log_{10}(x)$

11. $\log_{10}(1.4) = 0.461$; $\log_{10}(14) = 1.461$

13. $L(a^x) = L\{E[xL(a)]\} = [L \circ E][xL(a)] = x\cdot L(a)$

15. **a.** $\ln(10)\cdot\log_{10}(e) = 1$ **b.** $\log_{10}(e) = 0.434$

Problem Set 7.2

1. **a.** $\log_b(b) = \frac{L(b)}{L(b)} = 1$ **b.** $\log_b(1) = \frac{L(1)}{L(b)} = \frac{0}{L(b)} = 0$

c. $\log_b(1/b) = -1$ **d.** $\log_b(b^n) = n$

3. $x = 29$
5. $x = 6$
7. $x = 2$ or $x = 9$
9. $x = 1/2e$
11. $x = \frac{e+3}{2e-1}$
13. $x = 1 + \frac{L(5)}{L(3)} \approx 2.46$
15. $x = 1 + L(5) \approx 2.61$
17. $x = 1 + L(2) \approx 1.69$
19. $x = \frac{3L(3)}{2L(4)\text{-}L3} \approx 1.97$

Problem Set 7.3

1. $ph = 3 - 0.204 < 7$; Solution is acidic
3. $r = 0.020$
5. $x = \frac{k}{r}\ln\left(1 - \frac{rA}{V}\right)$
7. 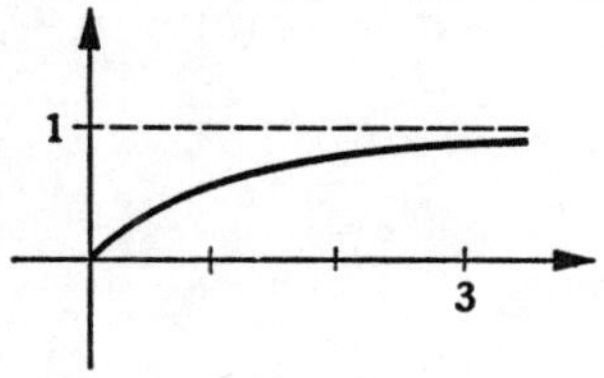

Fig. A.80

9. $h = \frac{\ln(2)}{0.144} = 4.81$ km.

Problem Set 7.4

1. 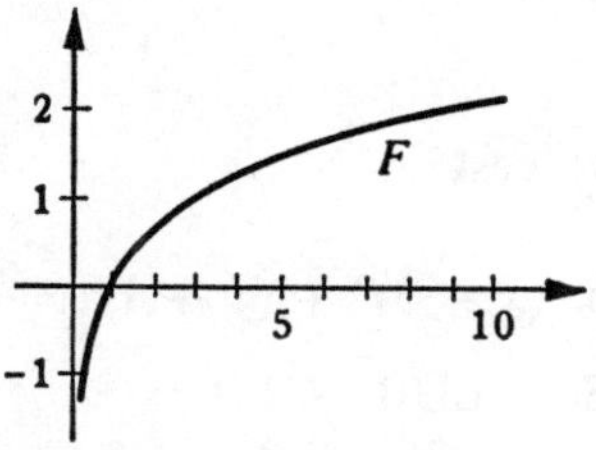

dom $F = \{x: x > 0\}$
Fig. A.81

3. 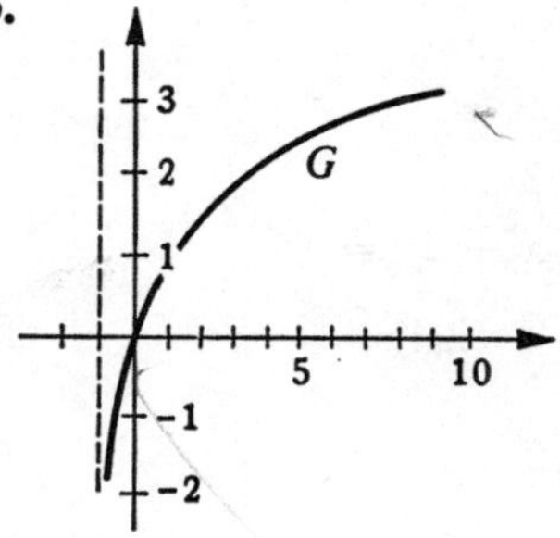

dom $G = \{x: x > -1\}$
Fig. A.82

5.

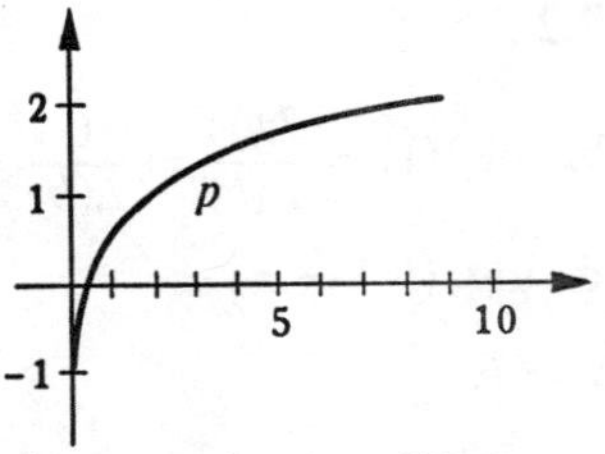

dom $p = \{x: x > 0\}$

Fig. A.83

7.

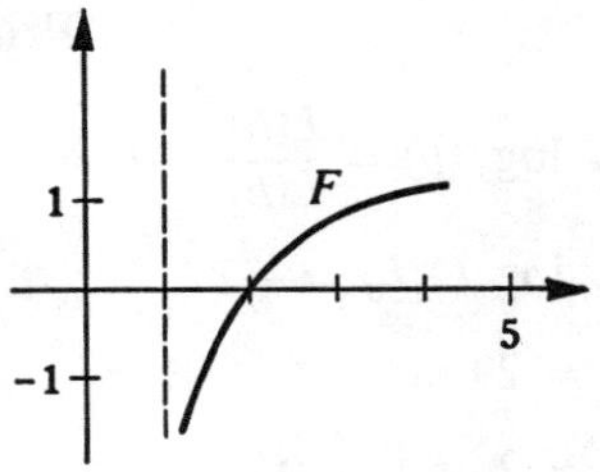

dom $F = \{x: x > 1\}$

Fig. A.84

9.

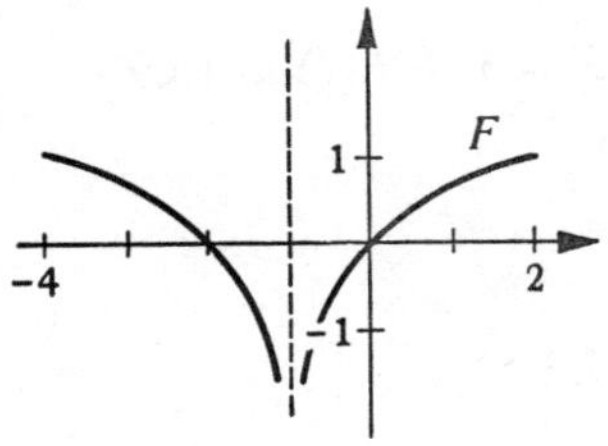

dom $F = \{x: x \neq -1\}$

Fig. A. 85

11.

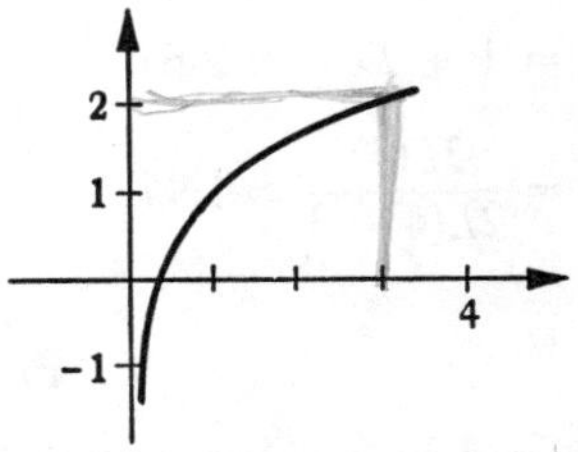

dom $F = x: x > 0$

Fig. A.86

13.

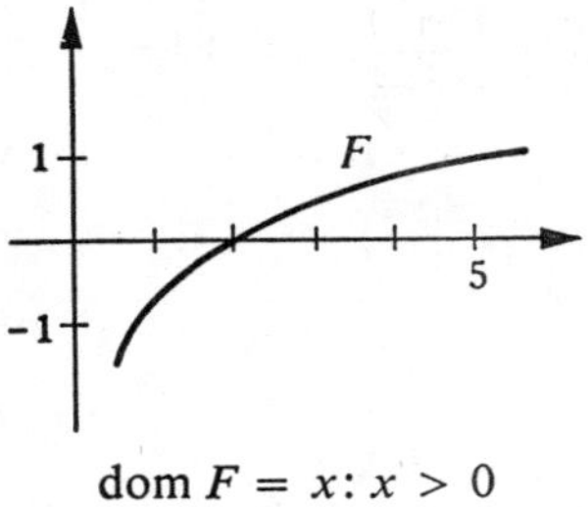

dom $F = x: x > 0$

Fig. A.87

15. $$\frac{F(x) - F(a)}{x - a} = \frac{1}{L(b)} \cdot \frac{L(x) - L(a)}{x - a}$$

Answers to Review Test

1. **a.** $x + 2y = 1$ **b.** $(x - 3)^2 + (y + 1)^2 = 8$

2. $U = \langle 2,4 \rangle$

3.

Set	LUB	GLB
A	2	0
B	2/3	1/3

4. **a.** "F is a function" means that F is a set of ordered pairs such that no two pairs in F have the same first element.

b. "F is a 1-to-1 function" means that F is a function such that no two pairs if F have the same second element.

5.

Function:	I	H	L	E	f	g	p
Domain:	R	$x > 0$	$x > 0$	R	R	$x \neq -2$	R
Range:	R	$y > 0$	R	$y > 0$	$y \geq 0$	$y \neq 1$	$y \geq -1$

6. $F(1) = 0$; $F(-1) = 2$; $F(a + 1) = \dfrac{a}{a + 1}$; $\dfrac{F(x) - F(a)}{x - a} = \dfrac{1}{ax}$

7. $[F \circ G](x) = \dfrac{2x + 1}{2x}$; $[G \circ F](x) = \dfrac{3x - 1}{x - 1}$

9. **a.** F is 1-to-1

b. $F(x) = x^2 - 1$; dom $F = \{x : x \geq 0\}$; rng $F = \{y : y \geq -1\}$

10. **a.** $\{\pm 1/3, \pm 2/3, \pm 1, \pm 2, \pm 3, \pm 6\}$

b.
$$\begin{array}{r|rrrr} 2/3 & 3 & -2 & -9 & 6 \\ & & 2 & 0 & -6 \\ \hline & 3 & 0 & -9 & 0 = \text{rem} \end{array}$$

$P(2/3)$ is 0

ANSWERS CHAPTER 8

Problem Set 8.1

1. a, c, e, and f are positive; b and d are negative.

3. $\cos(-\pi/4) = 1/\sqrt{2}$; $\cos(-\pi/2) = 0$; $\cos(3\pi/4) = -1/\sqrt{2}$;
$\sin(-\pi/4) = -1/\sqrt{2}$; $\sin(-\pi/2) = -1$; $\sin(3\pi/4) = 1/\sqrt{2}$;
$\cos(7\pi/4) = 1/\sqrt{2}$; $\sin(7\pi/4) = -1\sqrt{2}$

5. $\sin\theta = -3/5$

7. $\cos\theta = -\sqrt{11}/6$

9. $\cos\theta = 4/5$; $\sin\theta = 3/5$

11. $\cos(\pi/6) = \sqrt{3}/2$; $\sin(\pi/6) = 1/2$

13. $\cos(7\pi/6) = -\sqrt{3}/2$; $\sin(7\pi/6) = -1/2$

15. $\cos\theta = -4/5$; $\sin\theta = 3/5$

17. $\cos\theta = -1/\sqrt{2}$; $\sin\theta = -1/\sqrt{2}$

19. $\cos\theta = 1/\sqrt{3}$; $\sin\theta = 2/\sqrt{6}$

21. a.

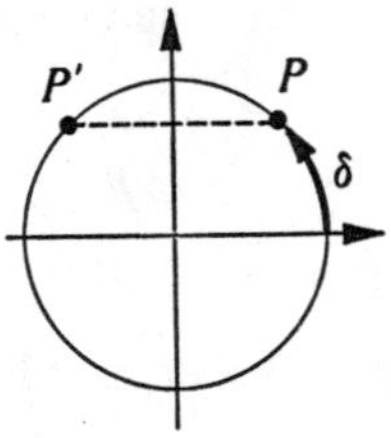

$\cos\theta = -\cos\delta$

$\sin\theta = \sin\delta$

Fig. A.88a

b.

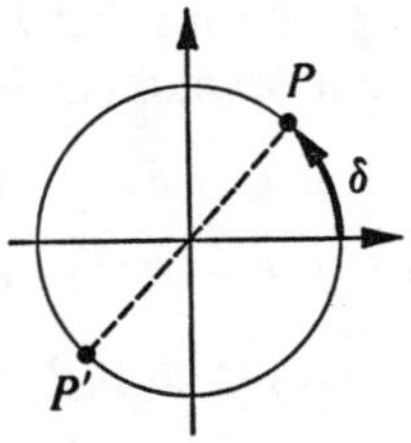

$\cos\theta = -\cos\delta$

$\sin\theta = -\sin\delta$

Fig. A.88b

c.

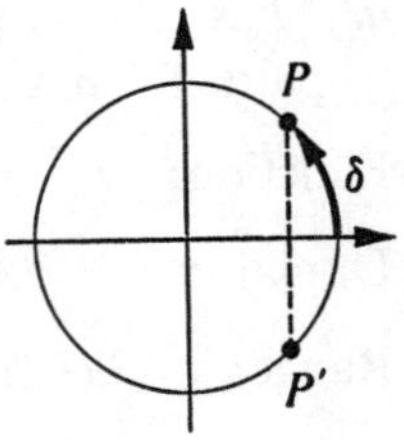

$\cos\theta = \cos\delta$

$\sin\theta = -\sin\delta$

Fig. A.88c

Problem Set 8.2

1. a. $\sin(5\pi/6) = \sin\pi/6 = 1/2$; $\cos(5\pi/6) = -\cos\pi/6 = -\sqrt{3}/2$
b. $\sin(4\pi/3) = -\sin\pi/3 = -\sqrt{3}/2$; $\cos(4\pi/3) = -\cos\pi/3 = -1/2$
c. $\sin(7\pi/4) = -\sin\pi/4 = -\sqrt{2}/2$; $\cos(7\pi/4) = \cos\pi/4 = \sqrt{2}/2$

3. $\sin(8\pi/5) = -\sin(2\pi/5)$

5. $\cos(7\pi/12) = -\cos(5\pi/12)$

7. $\sin(13\pi/10) = -\sin(3\pi/10)$

9. $\cos(4.24) = -\cos(1.10)$

11. $\sin(1.74) = \sin(1.40)$

13. $\sin u = -\sqrt{5}/5$ and $\cos u = -2\sqrt{5}/5$

Problem Set 8.3

1. $$\cos\left(u + \frac{\pi}{2}\right) = \cos(u)\cos\left(\frac{\pi}{2}\right) - \sin(u)\sin\left(\frac{\pi}{2}\right) = \cos(u)\cdot 0 - \sin(u)\cdot 1 = -\sin(u)$$

3. $$\sin(x - r) = \sin(x + [-r]) = \sin(x)\cos(-r) + \cos(x)\sin(-r) = \sin(x)\cos(r) - \cos(x)\sin(r)$$

5. $\cos(x + t) = \dfrac{16}{65}$ and $\sin(x + t) = \dfrac{63}{65}$

7. a. $$\cos(2x) = \cos(x + x) = \cos(x)\cos(x) - \sin(x)\sin(x) = \cos^2(x) - \sin^2(x)$$

b. $\cos(2x) = \cos^2(x) - [1 - \cos^2(x)] = 2\cos^2(x) - 1$

c. $\cos(2x) = [1 - \sin^2(x)] - \sin^2(x) = 1 - 2\sin^2(x)$

9. a. $\sin(x)$

b. $1 - \sin(x)$

11. a. 1

b. $\cos\left(\dfrac{\pi}{6}\right)$

c. $\sin\left(\dfrac{\pi}{6}\right)$

13. $\cos(\pi/12) = \dfrac{\sqrt{2} + \sqrt{6}}{4}$; $\sin(\pi/12) = \dfrac{\sqrt{6} - \sqrt{2}}{4}$

15. *Preliminaries:* $\cos(t) = 3/5$; $\sin(t) = 4/5$; $\cos(x) = -3/5$; $\sin(x) = 4/5$

a. $\cos(x + t) = -1$ **b.** $\sin(x + t) = 0$

c. $\cos(2t) = -7/25$ **d.** $\sin(2x) = -24/25$

Problem Set 8.4

1. $F(x) = 4 \sin x$; ampl. is 4; period 2π

3. $G(x) = \dfrac{3}{4} \cos x$; ampl. is 3/4; period 2π

5. $u(x) = \pi \sin x$; ampl. is π; period 2π

7. $v(x) = \cos(2x)$; ampl. is 1; period π

9. $W(x) = \sin(x/2)$; ampl. is 1; period 4π

11. $Z(x) = 2 \sin(\pi x/3)$; ampl. is 2; period 6

13. $f(x) = \dfrac{1}{2} \cos(2x)$; ampl. is 1/2; period π

15. $K(x) = -\sin(2x)$; ampl. is 1; period π

17. $Y(x) = \dfrac{3}{2} \cos(x/2)$; ampl. is 3/2; period 4π

19. By an addition formula, or by other considerations, $\cos\left(\dfrac{2x}{3} + 2\pi\right) = \cos\left(\dfrac{2x}{3}\right)$. Thus, if $F(x) = \cos\left(\dfrac{2x}{3}\right)$, then $F(t + 3\pi) = \cos\left(\dfrac{2t + 6\pi}{3}\right) = \cos\left(\dfrac{2t + 2\pi}{3}\right) = \cos\left(\dfrac{2t}{3}\right)$, hence $F(t + 3\pi) = F(t)$.

21. $F(x) = \sin x$

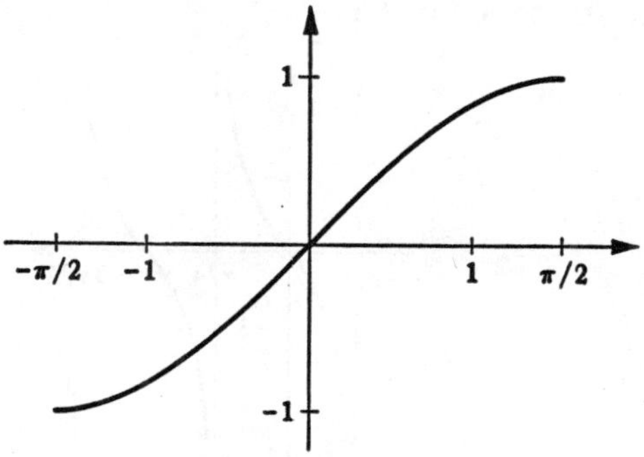

Fig. A.89

Problem Set 8.5

1. a. $\tan x > 0$ in qds I and III **b.** $\tan x < 0$ in qds II and IV

3. TABLE A.4

x	$\cos x$	$\sin x$	$\tan x$
0	1	0	0
$\frac{\pi}{6}$	$\frac{\sqrt{3}}{2}$	$\frac{1}{2}$	$\frac{\sqrt{3}}{3}$
$\frac{\pi}{4}$	$\frac{\sqrt{2}}{2}$	$\frac{\sqrt{2}}{2}$	1
$\frac{\pi}{3}$	$\frac{1}{2}$	$\frac{\sqrt{3}}{2}$	$\sqrt{3}$
$\frac{\pi}{2}$	0	1	not defn.

5. $\tan(5\pi/6) = -\sqrt{3}/3$

7. $\tan(-4\pi/3) = -\sqrt{3}$

9. $\tan(17\pi/6) = -\sqrt{3}/3$

11. $\tan x = -4/3$

13. $\tan x = 5/12$

15. $\tan(x + t) = 84/13$

17. $$\tan(x - t) = \tan(x + [-t]) = \frac{\tan(x) + \tan(-t)}{1 - \tan(x)\tan(-t)}$$
$$= \frac{\tan(x) - \tan(t)}{1 + \tan(x)\tan(t)}$$

19. $$\frac{2\tan(x)}{! + \tan^2(x)} = \frac{2\sin(x)}{\cos(x)}\cdot\cos^2(x) = 2\sin(x)\cos(x) = \sin(2x)$$

21. $\tan(2x + \pi) = \tan(2x) = \tan\left(2\left[x + \frac{\pi}{2}\right]\right)$, thus $F\left(x + \frac{\pi}{2}\right) = F(x)$ and so the period of F is $\frac{\pi}{2}$

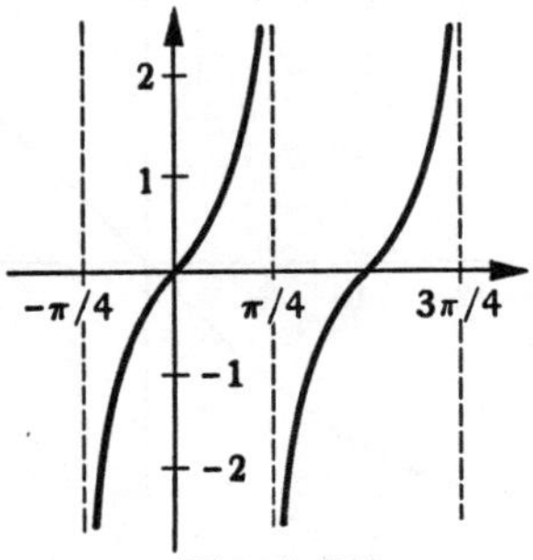

Fig. A.90

23. $\tan\left(\frac{x}{3}+\pi\right) = \tan\left(\frac{x}{3}\right) = \tan\left(\frac{1}{3}[x+3\pi]\right)$ thus $F(x+3\pi) = F(x)$ and so the period of F is 3π

25. $\tan\left(\frac{\pi x}{2}+\pi\right) = \tan\left(\frac{\pi x}{2}\right) = \tan\left(\frac{\pi}{2}[x+2]\right)$ thus $F(x+2) = F(x)$ and so the period of F is 2

27. dom $T = \langle -\pi/2, \pi/2 \rangle$

Problem Set 8.6

3. a. 1.18 **b.** 1.25 **c.** 0.35 **d.** 1.90
e. 1.44 **f.** 1.54 **g.** 0.90 **h.** 0.47

4. a. 33° **b.** 67° **c.** 56°

5. a. false **b.** true **c.** true

7. a. Let arc sin $x = z$ for then, $\sin z = x$ and $\cos^2 z = 1 - x^2$. If $x \geq 0$, $z \in [0,\pi/2]$, hence $\cos z \geq 0$. If $x \leq 0$, $z \in [-\pi/2,0]$ and again $\cos z \geq 0$. Therefore, in either case or for all $x \in [-1,1]$, $\cos z = \sqrt{(1-x^2)}$, i.e., $\cos(\text{arc sin } x) = \sqrt{(1-x^2)}$.

b. Let arc cos $x = z$ so that $\cos z = x$, hence $\sin^2 z = 1 - x^2$. As before, we can show that $\sin z \geq 0$, therefore $\sin z = \sqrt{(1-x^2)}$, that is,

$$\sin(\text{arc cos } x) = \sqrt{(1-x^2)}.$$

9. a. 0 **b.** 7/25

11. $\cos(2 \text{ arc cos } x) = 2x^2 - 1$

13. $\sin(2 \text{ arc cos } x) = 2x\sqrt{(1-x^2)}$

Problem Set 8.7

1. $f(x) = \sec x$

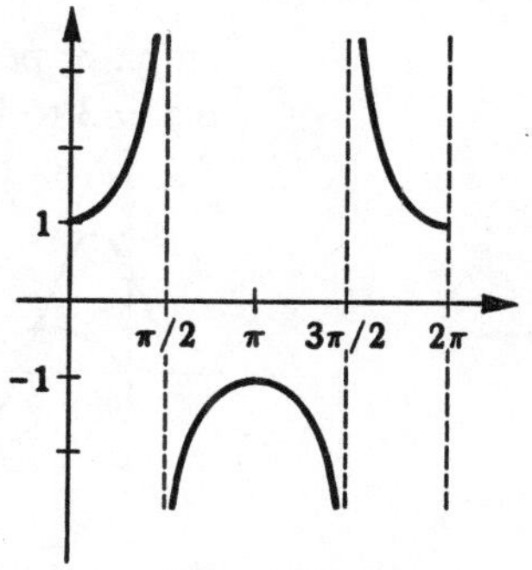

Fig. A.91

5. $\cot(x+\pi) = \dfrac{\cos(x+\pi)}{\sin(x+\pi)} = \dfrac{\cos(x)\cos(\pi) - \sin(x)\sin(\pi)}{\sin(x)\cos(\pi) + \cos(x)\sin(\pi)}$

$= \dfrac{-\cos(x)}{-\sin(x)} = \dfrac{\cos(x)}{\sin(x)} = \cot x$

7. $\csc(x+2\pi) = \dfrac{1}{\sin(x+2\pi)} = \dfrac{1}{\sin(x)} = \csc x$

9. **a.** 1 **b.** $2\sqrt{3}/3$ **c.** $\sqrt{2}$ **d.** 2 **e.** $2\sqrt{3}/3$ **f.** $2\sqrt{3}/3$

13. $\cos x = -4/5$; $\tan x = -3/4$; $\cot x = -4/3$; $\sec x = -5/4$; $\csc x = 5/3$

15. $4 + x^2 = 4\left(1 + \left[\dfrac{x}{2}\right]^2\right)$ and if $\dfrac{x}{2} = \tan z$, then $= 4(1 + \tan^2 z)$

$= 4\sec^2 z$

17. $x^2 - 9 = 9\left(\left[\dfrac{x}{3}\right]^2 - 1\right)$ and if we let $\dfrac{x}{3} = \sec z$, then

$= 9(\sec^2 z - 1) = 9\tan^2 z$

Problem Set 8.8

1. $S_p = \{\pi/3, 5\pi/3\}$ **3.** $S_p = \{\pi/6, 2\pi/3\}$ **5.** $S_p = \{3.922, 5.503\}$

7. $S_p = \{1.002, 2.140\}$ **9.** $S_p = \{0, 2\pi/3, 4\pi/3\}$

11. $S_p = \{\pi/2, 3\pi/2\}$ **13.** $S_p = \{\pi/6, 5\pi/6\}$

15. $S = \{\pi/2 + 2k\pi\} \cup \{3\pi/2 + 2k\pi\} \cup \{2\pi/3 + 2k\pi\} \cup \{4\pi/3 + 2k\pi\}$

17. $S = \{\pi/2 + 2k\pi\} \cup \{3\pi/2 + 2k\pi\} \cup \{7\pi/6 + 2k\pi\} \cup \{11\pi/6 + 2k\pi\}$

19. $S = \{x \mid x = (2k+1)\pi\}$

21. $S = \{\pi/6 + 2k\pi\} \cup \{5\pi/6 + 2k\pi\} \cup \{3\pi/2 + 2k\pi\}$

23. $S = \{5\pi/6 + 2k\pi\} \cup \{11\pi/6 + 2k\pi\} \cup \{\pi/2 + 2k\pi\}$

25. $x = 3\pi/8$ and $y = \pi/8$

Problem Set 8.9

1. ampl. 2; period 2π; p.s. $\pi/4$ to right

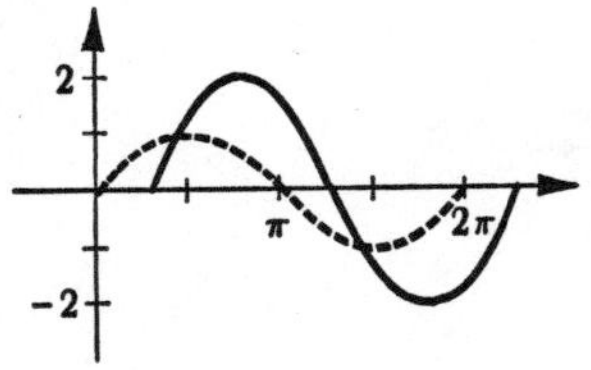

Fig. A.92

3. ampl. 3; period 2π; p.s. $\pi/3$ to left

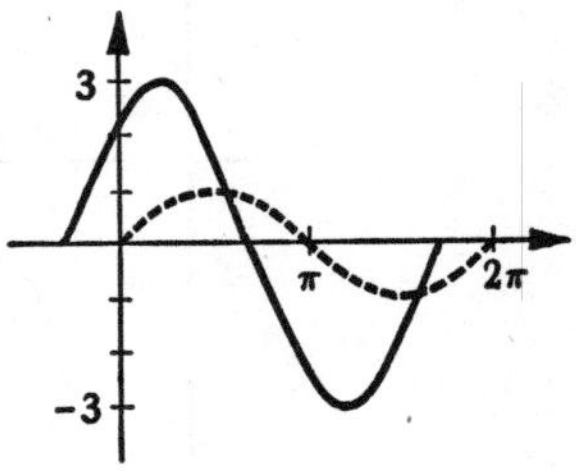

Fig. A.93

5. ampl. 1/2; period $2\pi/3$; p.s. 0

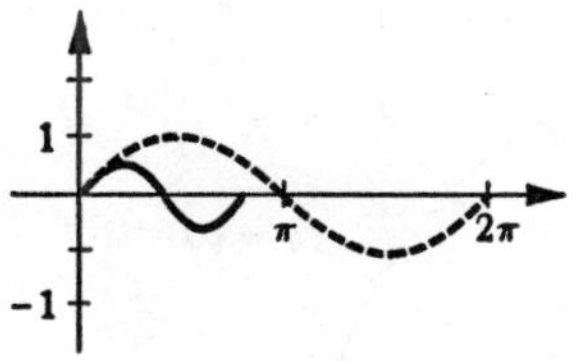

Fig. A.94

7. ampl. 2; period 2π; p.s. $\pi/2$ to left

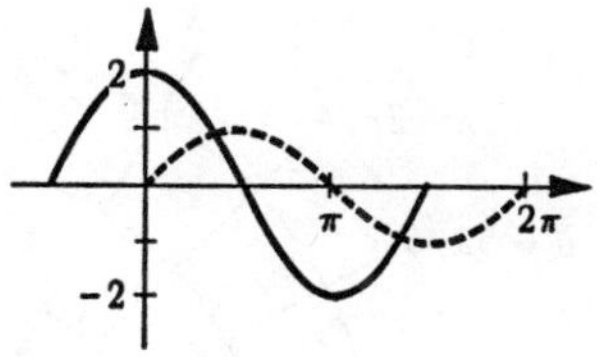

Fig. A.95

9. ampl. 3/2; period π; p.s. $\pi/2$ to left

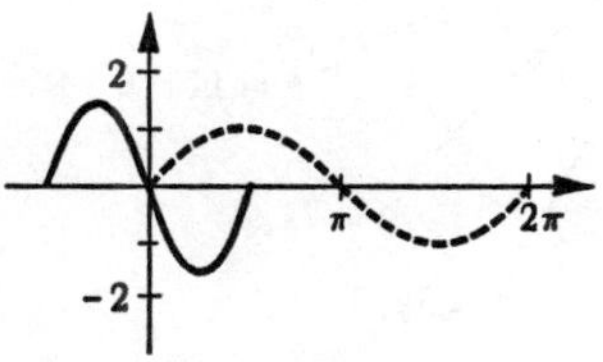

Fig. A.96

11. ampl. 2; period 2; p.s. 1/2 to left

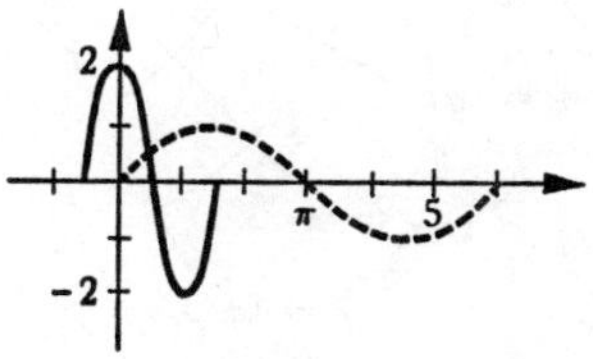

Fig. A.97

13. ampl. 1; period 4; p.s. 1 to left

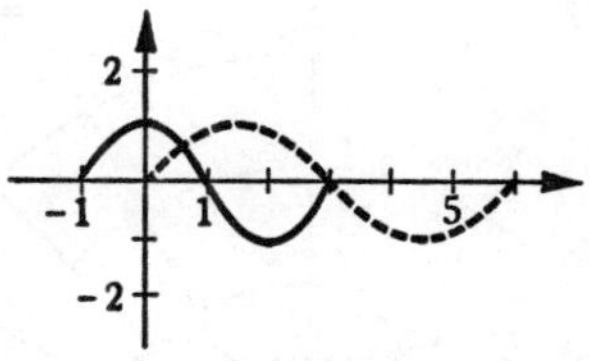

Fig. A.98

Problem Set 8.10

1. $(-1,0)$ **3.** $(-2,2\sqrt{3})$ **5.** $(3\sqrt{3},3)$ **7.** $(3,3)$

9. $(\sqrt{2},-\sqrt{2})$ **11.** $[3\sqrt{2},3\pi/4]$ **13.** $[5,31°]$ **15.** $[8,3\pi/4]$

17. $r = 6 \sin \theta$ **19.** $r = \tan \theta$ **21.** $r = \dfrac{2}{2 \cos \theta - \sin \theta}$

23. $r^2 = \cos(2\theta)$ **25.** $x^2 + y^2 - 3y = 0$ **27.** $x^2 - 2y - 1 = 0$

Problem Set 8.11

1.

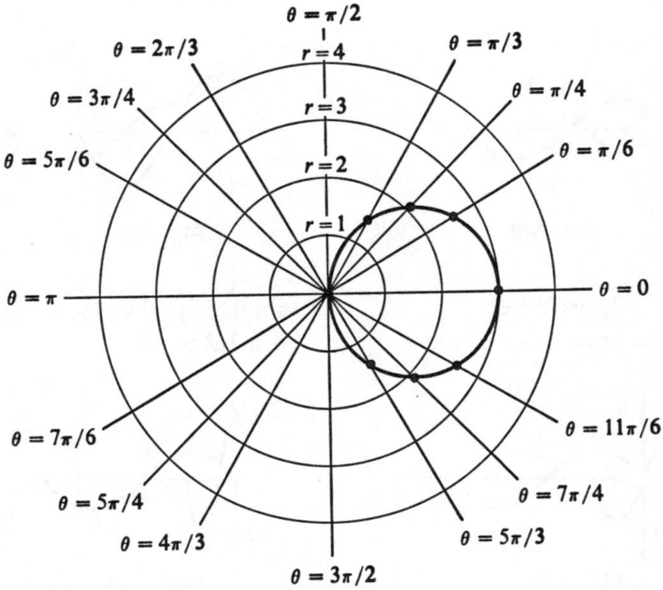

Fig. A.99

3.

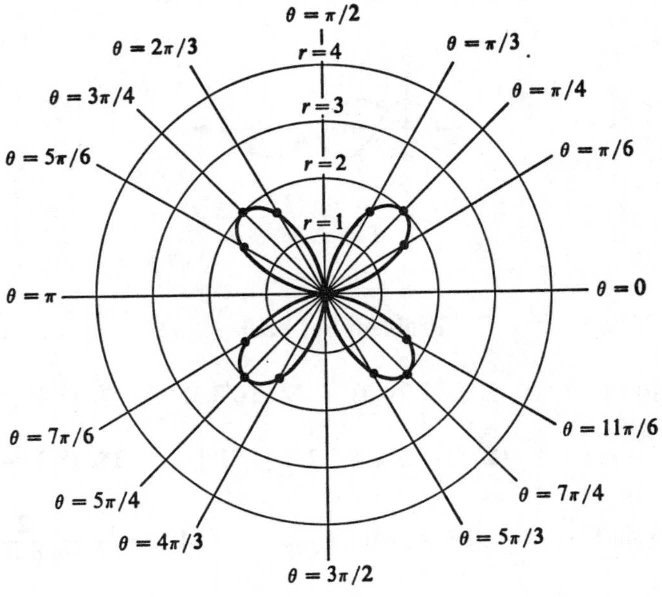

Fig. A.100

5.

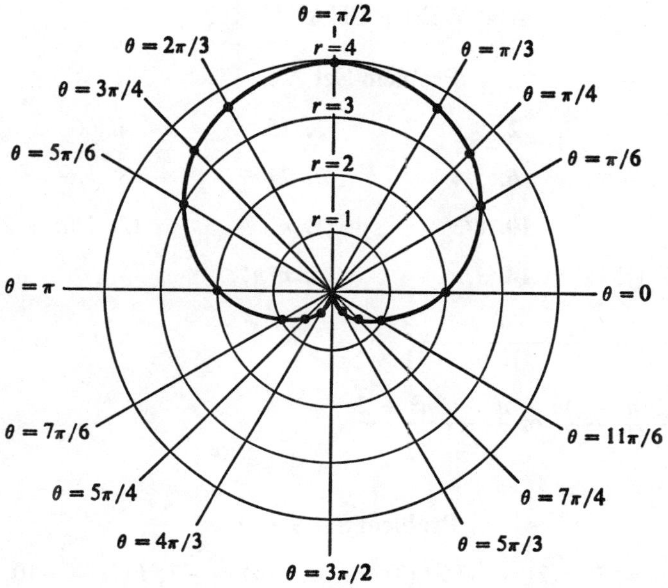

Fig. A.101

7.

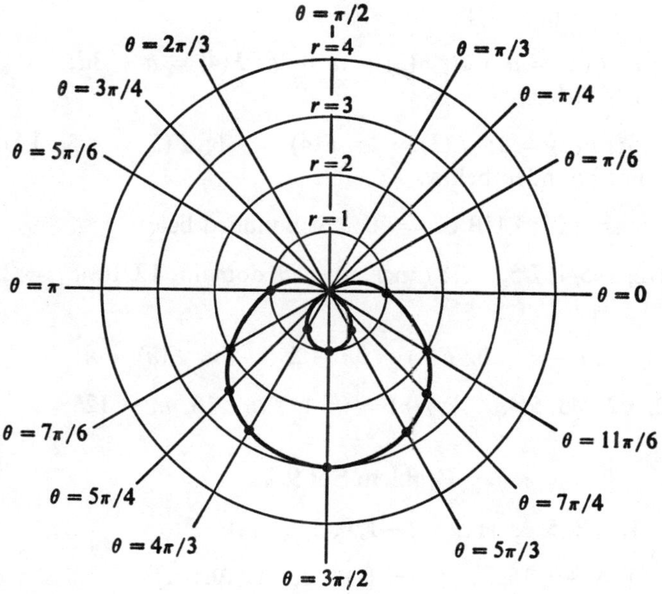

Fig. A.102

ANSWERS CHAPTER 9

Problem Set $N!$

1. 720 **2.** 9 **3.** 15 **4.** 66

5. 56 **6.** 5/4 **7.** 21 **8.** 1/6

9. 5/3 **10.** 8/3 **11.** $n!$ **12.** $1/(n+2)$

13. $1/(n-1)!$ **14.** $1/n$ **15.** $n \cdot n!$

16. $\dfrac{n}{(n+1)!}$ or $\dfrac{1}{(n-1)! \cdot (n+1)}$

17. $\dfrac{(n-2)(n-1)n}{6}$ or $\dfrac{n^3 - 3n^2 + 2n}{6}$

Problem Set 9.1

1. **a.** $F(1) = 2$; $F(2) = -1$; $F(3) = -4$; $F(4) = -7$; $F(5) = -10$; $F(n-1) = 8 - 3n$

b. $f(1) = 1/2$; $f(2) = 2/3$; $f(3) = 3/4$; $f(4) = 4/5$; $f(5) = 5/6$; $f(n-1) = (n-1)/n$

c. $g(1) = 0$; $g(2) = 0$; $g(3) = 3$; $g(4) = 20$; $g(5) = 115$; $g(n-1) = (n-1)! - (n-1)$

3. $F(1) = a$; $F(2) = a + d$; $F(3) = a + 2d$; $F(4) = a + 3d$; $F(5) = a + 4d$

5. $F(1) = 5$; $F(2) = 3$; $F(3) = 1$; $F(4) = -1$; $F(5) = -3$; LUB of $F = 5$, not bounded below.

7. f incr. over [1,3]; LUB of $f = 0$; not bounded below

9. $f = \{1/2, 1, 5/4, 7/5, \ldots\}$; f incr. over its domain; LUB of $f = 2$, GLB of $f = 1/2$

11. **a.** $F(n) = 2n - 1$ **b.** $F(n) = 3n - 2$ **c.** $F(n) = n!$

13. $f = \{2, 3/2, 4/3, 5/4, \ldots\}$; $f(n) = (n+1)/n$ **15.** $m = 126$

Problem Set 9.2

1. $f = \{-1, 1, 3, 5, \ldots\}$; $S = \{-1, 0, 3, 8, \ldots\}$

3. $F = \{-1, 4, -9, 16, \ldots\}$; $S = \{-1, 3, -6, 10, \ldots\}$

5. $F = \{-1/2, 1/4, 5/8, 13/16, \ldots\}$; $S = \{-1/2, -1/4, 3/8, 19/16, \ldots\}$

7. $f = \{1, 0, 1, 0, 1, \ldots\}$; $S = \{1, 1, 2, 2, 3, \ldots\}$

9. $f = \{1, 2, 3, 4, 5, \ldots\}$; $S = \{1, 3, 6, 10, 15, \ldots\}$; $S(n) = \dfrac{n(n+1)}{2}$

11. $F = \{1, 1, 2, 3, 5, 8, 13, \ldots\}$; $S = \{1, 2, 4, 7, 12, 20, 33, \ldots\}$

13. $F(k+3) = 2F(k+1) + F(k)$

Problem Set 9.3

1. $\forall n, f(n) = x$; $S(n) = \sum_{k=1}^{n} x$

3. $f(1) = 2$ and, for $n > 1, f(n) = \dfrac{-1}{n(n-1)}$;

   $S(n) = 2 - \sum_{k=2}^{n} \dfrac{1}{k(k-1)}$

5. $\forall n, f(n) = (n-1)(3n-2)$; $S(n) = \sum_{k=1}^{n} (3k^2 - 5k + 2)$

7. $\forall n, f(n) = \dfrac{1}{(2n+1)(2n-1)}$; $S(n) = \sum_{k=1}^{n} \dfrac{1}{4k^2 - 1}$

9. $\forall n, f(n) = 3n - 1$; $S(n) = \sum_{k=1}^{n} (3k - 1)$

11. $\forall n, f(n) = n \cdot n!$; $S(n) = \sum_{k=1}^{n} k \cdot k!$

13. $f(1) = 2$ and, for $n > 1, f(n) = 2^{n-1}$; $S(n) = 2 + \sum_{k=2}^{n} 2^{k-1}$

15. $\forall n, f(n) = (2n-1)^2$; $S(n) = \sum_{k=1}^{n} (2k-1)^2$

17. $\forall n, f(n) = ax^{n-1}$; $S(n) = \sum_{k=1}^{n} ax^{k-1}$

19. $F(n) = \dfrac{n}{2n-1}$

21. $F(n) = (-1)^{n+1} \cdot x^n$

23. $F(1) = 4, S(1) = 4$; $F(2) = -2, S(2) = 2$; $F(3) = 1, S(3) = 3$; $F(4) = -1/2, S(4) = 5/2$

Problem Set 9.4

1. a. 25 **b.** 63 **c.** $\dfrac{122}{105}$

3. a. 35 **b.** $15n$ **c.** 4/3

7. $\sum_{k=1}^{n} k^2$

9. $\sum_{k=1}^{n} \dfrac{k-1}{k+1}$

11. $\sum_{k=1}^{n} \dfrac{k^k}{(k-1)!}$

15. $\sum_{k=1}^{n} (-1)^{k+1} \cdot \dfrac{1}{2k-1}$

13. $\sum_{k=1}^{n} 2k(2k-1)$

17. $\sum_{k=1}^{n} \dfrac{1}{k(k+1)} = 1 - \dfrac{1}{n+1}$

Problem Set 9.5

1. a. $x(x^n - 1)$ **b.** $(n+1)! - 1$

3. b. $\sum_{k=1}^{n} k^3 = \dfrac{n^2(n+1)^2}{4}$

5. $\dfrac{n(n+1)(n+2)}{3}$

7. $\dfrac{n(n+1)(n^2+n-3)}{2}$

9. $\dfrac{n^2}{2}$

11. $\dfrac{n(n+1)(n^2+n+1)}{4}$

13. a. $S(n) = 1$ **b.** 1

15. a. $S(n) = \dfrac{(n+1)(4n-1)}{6n^2}$ **b.** 2/3

17. $\dfrac{x(1-x^n)}{1-x}$

19. $1 - \dfrac{1}{(n+1)!}$

Problem Set 9.6

1. $G(6) = 3$; $S(6) = 189$

3. $G(n) = 3^{n-2}$; $S(n) = 2/3(3^n - 1)$

5. $G(n) = 16(1/2)^{n-1}$; $S(n) = 32(1 - 2^{-n})$

7. $G(4) = \dfrac{-16}{27}$; $S(4) = \dfrac{26}{27}$

9. $G(6) = \dfrac{4}{81}$; $S(6) = \dfrac{721}{(3\sqrt{2}-2)81} = \dfrac{103(3\sqrt{2}+2)}{162}$

11. $t = -2/3$

13. a. 10 b. 6 c. $6\sqrt{3}$ d. $\sqrt{tz}$

15. $m = 7$

17. F_1 and F_2

19. $G(4) = -8/9$; $S(4) = 13/9$

Problem Set 9.7

1. $\sum_{k=1}^{\infty} (3/4)^{k-1} = 4$

3. $\sum_{k=1}^{\infty} (-1/2)^{k-1} = 2/3$

5. $\sum_{k=1}^{\infty} 4\cdot(-2/3)^{k-1} = \frac{12}{5}$

7. $\sum_{k=1}^{\infty} 3(0.1)^{k-1} = \frac{10}{3}$

9. $\sum_{k=1}^{\infty} 6(-1/3)^{k-1} = 9/2$

11. $\sum_{k=1}^{\infty} 8(1/4)^{k-1} = \frac{32}{3}$

13. $\sum_{k=1}^{\infty} -3(3/5)^{k-1} = \frac{-15}{2}$

15. $\sum_{k=1}^{\infty} -3/2(-2/3)^{k-1} = -9/10$

17. $\sum_{k=1}^{\infty} -2(-0.1)^{k-1} = \frac{20}{11}$

19. 32 cm^2

21. $\frac{16\sqrt{3}}{3}$ cm^2

ANSWERS CHAPTER 10

Problem Set 10.1

1. Let T be the number set to which m belongs only if $m \in N$ and $0 < x^m < 1$.

   $1 \in T$ since $0 < x < 1$ by hypothesis.

   If $k \in T$, then $k \in N$ and $0 < x^k < 1$. Since $0 < x$ by hypothesis, it follows that $0 < x \cdot x^k < x$ by axiom 4. But $x < 1$ by hypothesis, hence $x^{k+1} < 1$ by axiom 2. Therefore $0 < x^{k+1} < 1$, that is, if $k \in T$, then $(k + 1) \in T$. Thus, by theorem MI, T is N.

3. Let T be the number set to which m belongs only if $m \in N$ and $2m \leq 2^m$.

   $1 \in T$ since $2 \leq 2$.

   If $k \in T$, then $k \in N$ and $2k \leq 2^k$. From this it follows that $2 \cdot 2k \leq 2 \cdot 2^k$, that is, $4k \leq 2^{k+1}$. Keep this in mind while we consider the following: $1 \leq k$ since $k \in N$, so $2 \leq 2k$, hence $2 + 2k \leq 2k + 2k$. Thus $2 + 2k \leq 4k$. Now then, the above shows that $2 + 2k \leq 4k \leq 2^{k+1}$, hence $2(k + 1) \leq 2^{k+1}$. Therefore $(k + 1) \in T$ whenever $k \in T$. By theorem MI, T is N.

5. Let T be the number set to which m belongs only if $m \in N$ and $a^m < b^m$.

$1 \in T$ since $a < b$ by hypothesis.

If $k \in T$, then $k \in N$ and $a^k < b^k$. Since $0 < a$, it now follows that $a \cdot a^k < a \cdot b^k$ by axiom 4. Since $a < b$, it is also true that $0 < b$ by axiom 2. Thus $0 < b^k$ by problem 1 above. Thus $a < b$ implies that $a \cdot b^k < b \cdot b^k$ by axiom 4. We have now shown that whenever $k \in T$, then $a \cdot a^k < a \cdot b^k < b \cdot b^k$ so that $a^{k+1} < b^{k+1}$ which means that $(k + 1) \in T$. By theorem MI, therefore, T is N.

7. Let T be the number set to which m belongs only if $m \in N$ and $\sum_{k=1}^{m} (2k - 1) = m^2$.

$1 \in T$ since $\sum_{k=1}^{1} (2k - 1) = 1$ and $1^2 = 1$.

If $j \in T$, then $j \in N$ and $\sum_{k=1}^{j} (2k - 1) = j^2$. It follows that $(2j + 1) + \sum_{k=1}^{j} (2k - 1) = j^2 + (2j + 1)$. However, $(2j + 1) + \sum_{k=1}^{j} (2k - 1) = \sum_{k=1}^{j+1} (2k - 1)$ and $j^2 + 2j + 1 = (j + 1)^2$ hence $\sum_{k=1}^{j+1} (2k - 1) = (j + 1)^2$. This means that $(j + 1) \in T$ whenever $j \in T$ so T is N by theorem MI.

11. False. For $n = 8$, $1 + \dfrac{10}{n} = 2.25$ while

$$\sum_{k=1}^{8} 1/k = 1 + 1/2 + 1/3 + 1/4 + 1/5 + 1/6 + 1/7 + 1/8$$
$$> 1 + 0.5 + 0.3 + 0.2 + 0.2 + 0.1 + 0.1 + 0.1$$

and the sum on the right is 2.5.

Problem Set 10.2

1. Let T be the number set to which m belongs only if $m \in N$ and $|x|^m < 1$.

$1 \in T$ since $|x|^1 = |x| < 1$ by hypothesis.
If $k \in T$, then $k \in N$ and $|x|^k < 1$. From this it follows that

$|x| \cdot |x|^k < 1 \cdot |x|$ by axiom 4 so that $|x|^{k+1} < |x|$. But $|x| < 1$ so $|x|^{k+1} < 1$ by axiom 2. This last inequality means that $(k + 1) \in T$. Hence, by theorem MI, T is N.

3. Let T be the number set to which m belongs only if $m \in N$ and $f(m) = a + (m - 1)d$.

$1 \in T$ since $a + (1 - 1)d = a = f(1)$.
If $k \in T$, then $k \in N$ and $f(k) = a + (k - 1)d$. Now $f(k + 1) = d + f(k)$ by hypothesis, so $f(k + 1) = a + (k - 1)d + d = a + kd$. Thus $(k + 1) \in T$ whenever $k \in T$. So T is N by theorem MI.

7. Let T be the number set to which m belongs only if $m \in N$ and $\sum_{k=1}^{m} [F(k + 1) - F(k)] = F(m + 1) - F(1)$.

$1 \in T$ since $\sum_{k=1}^{1} [F(k + 1) - F(k)] = F(2) - F(1)$.

If $j \in T$, then $j \in N$ and $\sum_{k=1}^{j} [F(k + 1) - F(k)] = F(j + 1) - F(1)$. From this it follows that $F(j + 2) - F(j + 1) + \sum_{k=1}^{j} [F(k + 1) - F(k)] = F(j + 2) - F(j + 1) + F(j + 1) - F(1)$, that is, $\sum_{k=1}^{j+1} [F(k + 1) - F(k)] = F(j + 2) - F(1)$ and so $(j + 1) \in T$. By theorem MI, T is N.

9. Let T be the number set to which m belongs only if $m \in N$ and $(1 + x)^m \geq 1 + mx$.

$1 \in T$ since $(1 + x)^1 \geq 1 + x$.
If $k \in T$, then $k \in N$ and $(1 + x)^k \geq 1 + kx$. Now $x > -1$ so $1 + x > 0$, hence $(1 + x)^k(1 + x) \geq (1 + kx)(1 + x)$. Thus $(1 + x)^{k+1} \geq 1 + kx + x + kx^2 \geq 1 + (k + 1)x + kx^2 \geq 1 + (k + 1)x$. Therefore $(k + 1) \in T$ and so T is N.

Problem Set 10.3

1. a. $\binom{6}{2} + \binom{6}{3} = \frac{6!}{4!\,2!} + \frac{6!}{3!\,3!} = \frac{6!}{4!\,3!}(3 + 4) = \frac{7!}{3!\,4!} = \binom{7}{3}$

b. $\binom{n}{k-1} + \binom{n}{k} = \dfrac{n!}{(n-k+1)!\,(k-1)!} + \dfrac{n!}{(n-k)!\,k!}$

$$= \frac{n!\,k + n!\,(n-k+1)}{(n-k+1)!\,k!} = \frac{n!\,(n+1)}{(n-k+1)!\,k!}$$

$$= \binom{n+1}{k}$$

3. $2\cdot\binom{n}{2} + \binom{n}{1} = \dfrac{n!}{(n-2)!} + \dfrac{n!}{(n-1)!} = \dfrac{n\cdot n!}{(n-1)!} = n^2$

5. $(a+b)^6 = a^6 + \binom{6}{1}a^5b + \binom{6}{2}a^4b^2 + \binom{6}{3}a^3b^3 + \binom{6}{4}a^2b^4$

$$+ \binom{6}{5}ab^5 + b^6$$

7. $64t^6 - 192t^5 + 240t^4 - 160t^3 + 60t^2 - 12t + 1$

9. z^{12}; $\binom{12}{1}z^{11}5$; $\binom{12}{2}z^{10}5^2$; $\binom{12}{3}z^9 5^3$

11. a. 11th term: $\binom{14}{10}x^4y^{10}$ **b.** 9th term: $\binom{11}{8}(2x)^3y^8$

ANSWERS CHAPTER 11

Problem Set 11.0

1. a. $\dfrac{I(x) - I(a)}{x-a} = 1$ **b.** $\dfrac{H(x) - H(a)}{x-a} = -1/ax$

3. $\dfrac{P(x) - P(a)}{x-a} = x^2 + ax + a^2$

5. $\dfrac{V(x) - V(a)}{x-a} = \dfrac{x+a}{|x| + |a|}$

7. $\dfrac{g(x) - g(a)}{x-a} = \dfrac{1}{\sqrt{(x+2)} + \sqrt{(a+2)}}$

9. $\dfrac{Q(x) - Q(a)}{x - a} = \dfrac{1}{(x + 1)(a + 1)}$ **11.** $\dfrac{s(x) - s(a)}{x - a} = m$

13. a. $y - 1/a = -1/a^2 \cdot (x - a)$ **15. a.** $-x$

b.

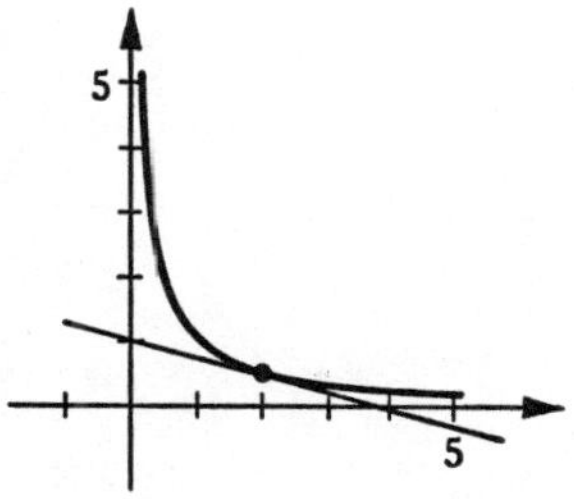

Fig. A.103

b.

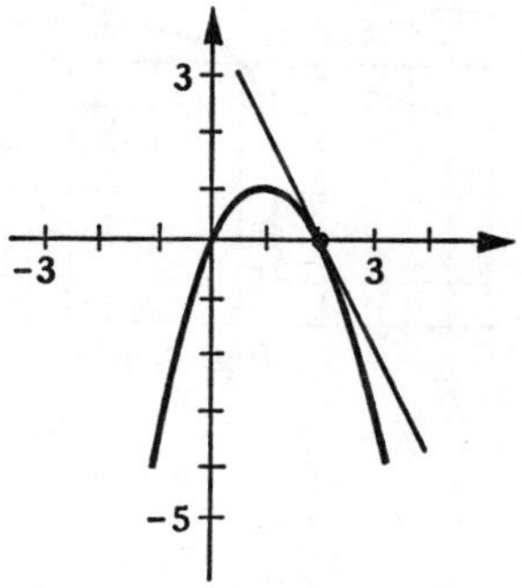

Fig. A.104

17. a. $\dfrac{1}{\sqrt{x} + 1/2}$ **b.** $y - 1/2 = 1 \cdot (x - 1/4)$

c.

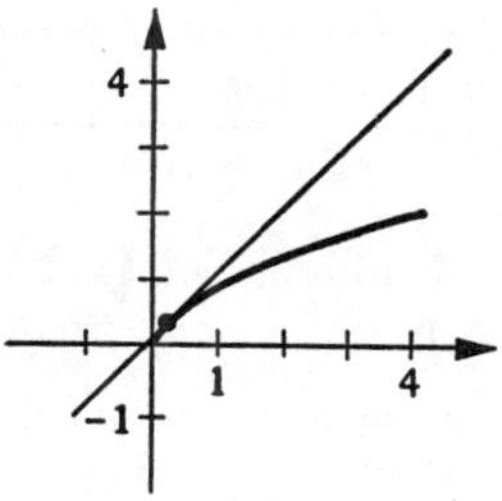

Fig. A.105

Problem Set 11.1

5. $F'(-1) = 0$

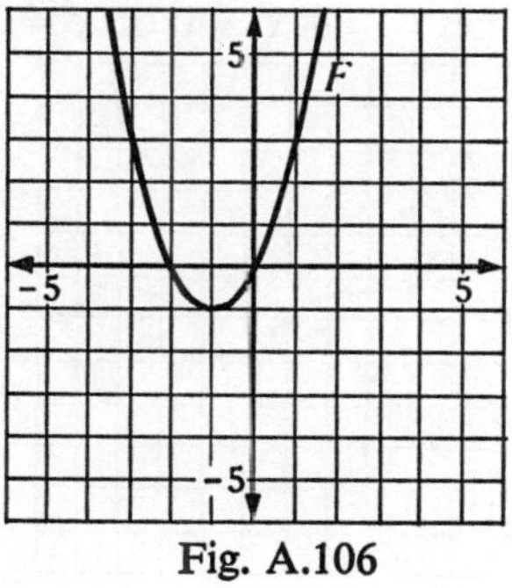

Fig. A.106

7. $F'(1) = 1$

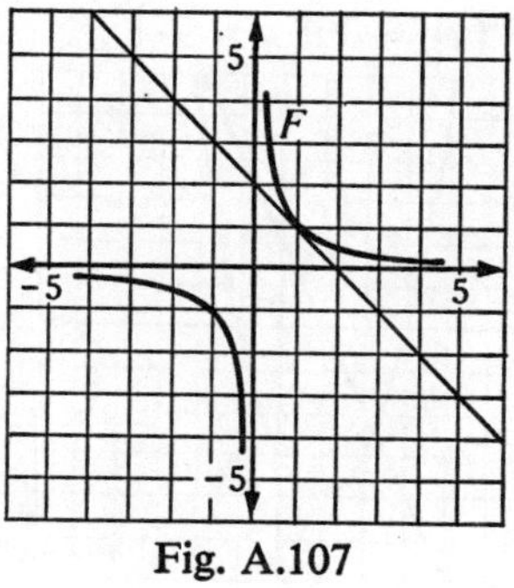

Fig. A.107

9. $F'(1) = 1/4$

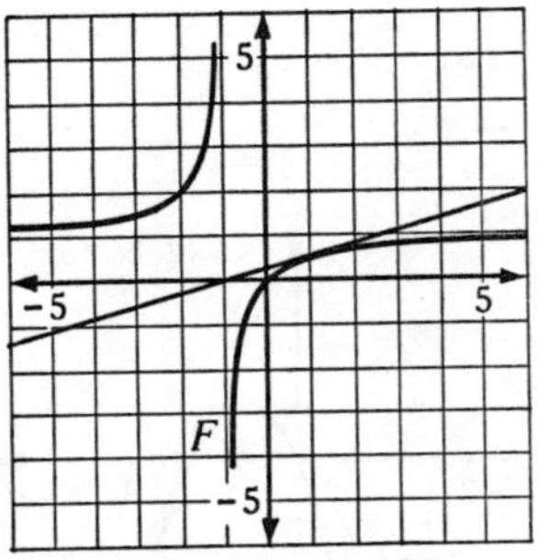

Fig. A.108

11. $(3/2, -9/4)$

13. $c = 1$

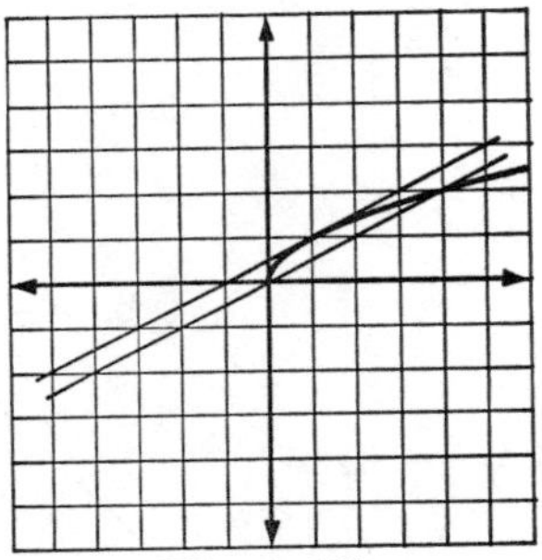

Fig. A.109

Problem Set 11.2

3. $F'(x) = 2x$; dom $F' = R$

5. $F'(x) = 4x^3$; dom $F' = R$

7. $P'(x) = 4x - 1$; dom $P' = R$

9. $p'(x) = 3x^2 - 2x$; dom $p' = R$

11. $F'(x) = \dfrac{1}{2\sqrt{x+2}}$; dom $F' = x \mid x \geq -2$

13. $V'(x) = \dfrac{3x}{|x|}$; dom $V' = x \mid x \neq 0$

15. $Q'(x) = -\dfrac{2}{x^2}$; dom $Q' = x \mid x \neq 0$

17. $F'(x) = 0$ if $x = -1$

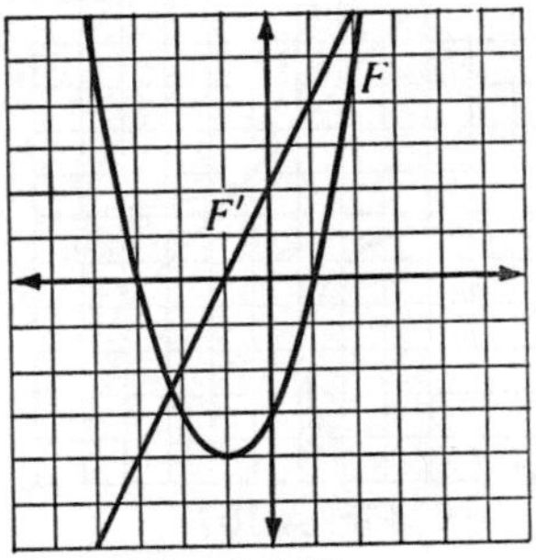

Fig. A.110

19. $F'(x) = 0$ if $x = 0$ or $x = 2$

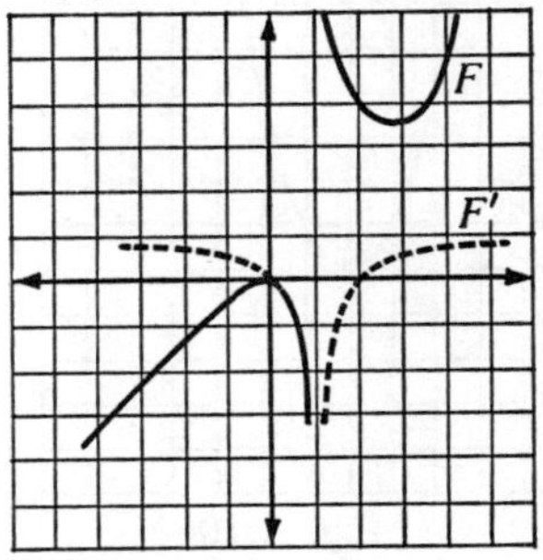

Fig. A.111

Problem Set 11.3

1. $\frac{dy}{dx} = \frac{1}{2\sqrt{x}}$; if $x = 4$, $\frac{dy}{dx} = \frac{1}{4}$; if $x = \frac{1}{4}$, $\frac{dy}{dx} = 1$.

3. $\frac{dV}{dx} = 3x^2$; if $x = 3$, $\frac{dV}{dx} = 27$. 5. $x_0 = 1$

7. $\frac{dp}{dV} = -\frac{k}{V^2}$; if $V = 100$ c.c., $\frac{dp}{dV} = -0.001 \cdot k$

9. $\frac{dh}{dx} = \frac{x}{\sqrt{x^2 - 4}}$

11. **a.** 4.5 m./sec. **b.** 4.25 m./sec. **c.** 4.05 m./sec. **d.** 4 m./sec.

13. **a.** 1 m./sec. **b.** −2 m./sec.

15. **a.** 5 m./sec. **b.** 8 m./sec.

17. **a.** 1960 cm./sec. **b.** 1960 cm.

19. 10 sec. −320 ft./sec. 21. approx. −880 cm./sec.

Answers to chapter review problems

CHAPTER 1 REVIEW PROBLEMS

1. Either $x < z$, $x > z$, or $x = z$. If $x < y$ and $y < z$, then $x < z$. If $x < y$, then $x + z < y + z$. If $x < y$ and $0 < z$, then $xz < yz$.

2. **a.** k is a counting number, the negative of a counting number, or zero.
   b. r is the quotient of an integer by a nonzero integer.

3. $x - 3 < 0$, that is, $x < 3$

4. **a.** $|x + t| \leq |x| + |t|$ **b.** $|x| \cdot |t| = |xt|$
   c. $|x - t| = |t - x|$ **d.** $|x| - |t| \leq |x - t|$
   e. $0 \leq |x|$

5. **a.** $x \in [-2,0\rangle$ **b.** $x < -1$ or $x > 4$
   c. $x \in \langle 0,5\rangle$ **d.** $x \in \langle -12,2\rangle$
   e. $x > 5$ or $x < 1$

6. **a.** $|x| < 4$ **b.** $|x| \leq 2$ **c.** $|x - 1| \leq 2$
   d. $|x - 6| < 4$ **e.** $|x + 7/2| < 5/2$

7. **a.** $\exists$ a number b such that $\forall x \in S$, $x \geq b$.
b. M is an upper bound of S and no number less than M is an upper bound of S.

8. LUB of $A = 2$, GLB of $A = 3/2$; no upper bound, GLB of $B = -4$; no bounds for C; LUB of $D = 1$, GLB of $D = -4$.

9. $p/q < m/n$ iff $np < mq$

10. **a.** 21 **b.** $2 \leq \sqrt{7} < 3$

CHAPTER 2 REVIEW PROBLEMS

1. **a.** 0 **b.** $\dfrac{n-b}{m-a}$

2. $y = -5/3 \cdot x - 1/3$

3. **a.** $y = 1/2 \cdot x + 1$ **b.** $y = -1/2 \cdot x + 3$

4. $(x - 2/3)^2 + (y + 10/3)^2 = \dfrac{260}{9}$, center $(2/3, -10/3)$; radius $\dfrac{2\sqrt{65}}{3}$

5. $(y - 1)^2 = 4(x - 1)$

6. $\dfrac{(x-6)^2}{25} + \dfrac{(y-1)^2}{9} = 1$

7. $\dfrac{(y+3)^2}{4} - \dfrac{(x-2)^2}{5} = 1$

8. **a.** hyperbola, transverse axis horizontal; center $(-3,1)$; foci $(-3 \pm 2\sqrt{2}, 1)$; transverse axis and conjugate axis have length 4
b. ellipse, major axis horizontal; center $(2,-1)$; foci $(-2,1)$, $(6,-1)$; length major axis 10; length minor axis 6
c. parabola, opens up; vertex $(0,-3/2)$; foci $(0,-11/8)$
d. line, slope $-1/2$; y intercept $-3/2$
e. hyperbola, transverse axis horizontal; $\dfrac{(x-2)^2}{4} - \dfrac{(y+1)^2}{12/5} = 1$
f. line, slope 3/[illegible] y intercept -3
g. ellipse, major axis vertical; $\dfrac{(x+2)^2}{6} + \dfrac{(y+2)^2}{12} = 1$; center $(-2,-2)$; length major axis $4\sqrt{3}$; length minor axis $2\sqrt{6}$

CHAPTER 3 REVIEW PROBLEMS

1. $f(x)$ denotes the ordinate of that point if f whose abscissa is x. The domain is the set of all first elements. The range is the set of all second elements.

2. $I(x) = x$, dom $I = R$; $H(x) = 1/x$, dom $H = \{x \mid x > 0\}$

3. $[F + G](x) = x^2 + 5x - 3$; dom $[F + G] = R$; $[F \cdot G](x) = x^3 + x^2 - 12x$, dom $[FG] = R$; $[F \circ G](x) = x^2 + 4x - 3$, dom $[F \circ G] = R$; $[G \circ F](x) = x^2 - 2x - 3$,dom $[G \circ F] = R$

4. $[f + g](1) = 8$, $[f + g](5/2) = 19/4$, $[f + g](3) = 0$; $[f \cdot g](1) = 7$, $[f \cdot g](5/2) = -75/8$, $[f \cdot g](3) = -4$; $[f/g](1) = 7$, $[f/g](5/2) = -6/25$, $[f/g](3) = -1$; $[f \circ g](1) = 7$, $[f \circ g](5/2) = -21/4$, $[f \circ g](3) = 10$

5.

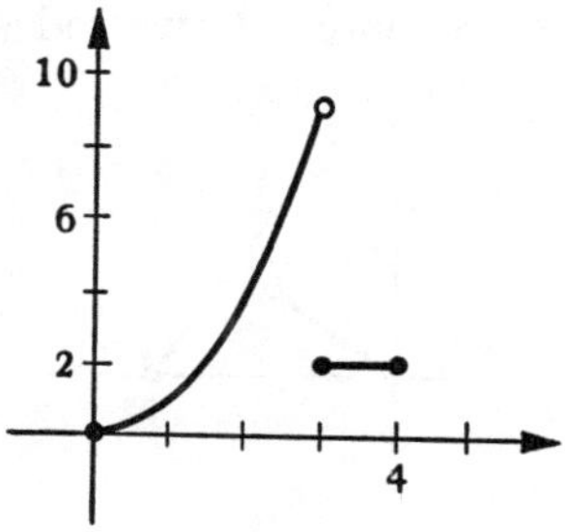

Fig. A.112

6. Suppose $n \le x < n + 1$, then $[x] = n$, $m \le y < m + 1$, then $[y] = m$, so that $n + m \le x + y < n + m + 2$. If $n + m \le x + y < n + m + 1$, then $[x] + [y] = [x + y]$, but if $n + m + 1 \le x + y < n + m + 2$, then $[x] + [y] < [x + y]$.

CHAPTER 4 REVIEW PROBLEMS

1. $f(x) =$ constant is neither incr. nor decr. over any subset of its domain.

2.

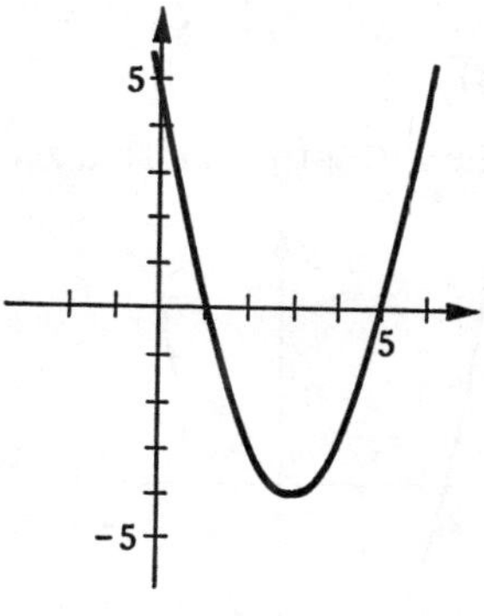

incr. over $\langle 3,\infty \rangle$
decr. over $\langle -\infty,2 \rangle$

Fig. A.113

3. **a.** $(a,b) \in F^*$ iff $(b,a) \in F$ **b.** F must be 1-to-1
c. dom F = rng F; rng F = dom F
d. If $x \in$ dom F, then $[x,F(x)] \in F$, thus $[F(x),x] \in F^*$. Therefore $[F^* \circ F](x) = F^*[F(x)] = x$ if $x \in$ dom F^*, then $[x,F^*(x)] \in F^*$, thus $[F^*(x),x] \in F$. Therefore $[F \circ F^*](x) = F[F^*(x)] = x$.

4. $F(x) = \dfrac{2x + 1}{1 - x}$ **5.** min. $f(x) = -4$; no max.

6. The following sketches are representative and are not the only solutions.

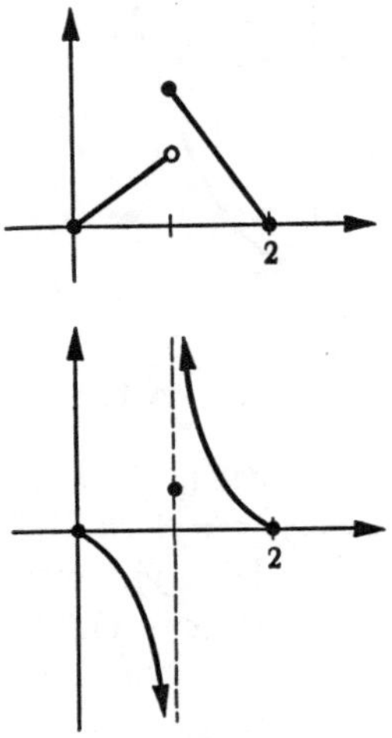

Fig. A.114

CHAPTER 5 REVIEW PROBLEMS

1. **a.** $P(x) = x^4 + 2x^3 + x^2 - 4$ **b.** $Q(x) = x^3 + x - 2$
c. $P(x) = (x + 2)\cdot Q(x)$

2. Zero of $Q = \{1\}$. Yes the zero of Q is also a zero of P.

3.

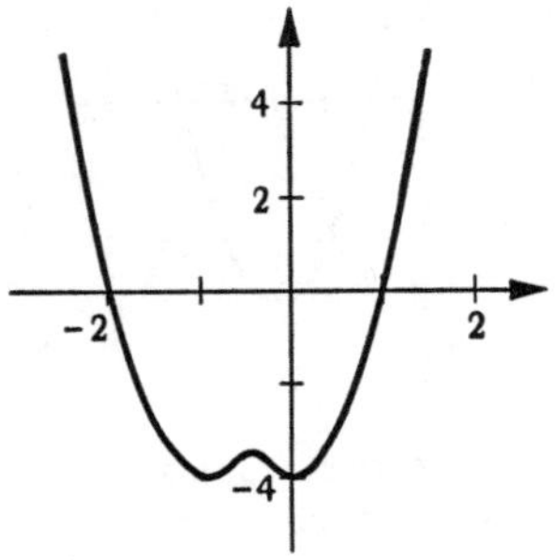

Fig. A.115

4. Zeros of $P = \left\{\dfrac{6 \pm \sqrt{39}}{2}\right\}$

5. Possible rational zeros of $P = \pm\{1, 2, 3, 6\}$

6. No.

$$\begin{array}{r|rrrrrr} 3 & 3 & -\,2\,+ & 0\,+ & 5\,+ & 1\,- & 11 \\ & & 9 & 21 & 63 & \ldots & \\ \hline & 3\,+ & 7\,+ & 21\,+ & 68\,+ & 205\,+ & 604 \end{array}$$

7. $c = 11$

$$\begin{array}{r|rrrrr} -2 & 1 & -\,2\,- & 1\,+ & c\,- & 6 \\ & & -2 & +8 & -14 & +(28 - 2c) \\ \hline & 1\,- & 4\,+ & 7\,+ & (c - 14)\,+ & (22 - 2c) \end{array}$$

and $(22 - 2c) = 0$ iff $c = 11$

8. $P(2) = 15$

$$\begin{array}{r|rrrrr} 2 & 1\,+ & 0\,- & 2\,+ & 0\,+ & 7 \\ & & 2 & 4 & 4 & 8 \\ \hline & 1\,+ & 2\,+ & 2\,+ & 4\,+ & 15 \end{array}$$

CHAPTER 6 REVIEW PROBLEMS

1. dom $L = \{x \mid x > 0\}$; rng $L = R$; dom $E = R$; rng $E = \{y \mid y > 0\}$

2.

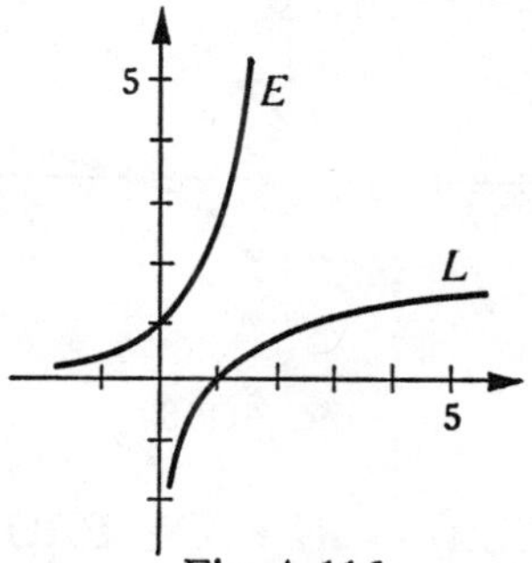

Fig. A.116

3. **a.** 0 **b.** -1 **c.** 0 **d.** 2
e. $1/e$ **f.** 1 **g.** 1 **h.** 3
i. $1/2$ **j.** $1/2$ **k.** 9

4. **a.** 1.82 **b.** -0.38 **c.** -0.72 **d.** 1.44
e. 1.72 **f.** 0.55 **g.** -1 **h.** 1.1
i. 0.90 **j.** $0 < 1/t < 1/e < 1 < s < e < t < s^2$

5. a. a b. k c. a d. k
e. c f. $1/c$ g. $-m$ h. ck

6. $b^x = E[xL(b)] = e^{xL(b)}$

7.

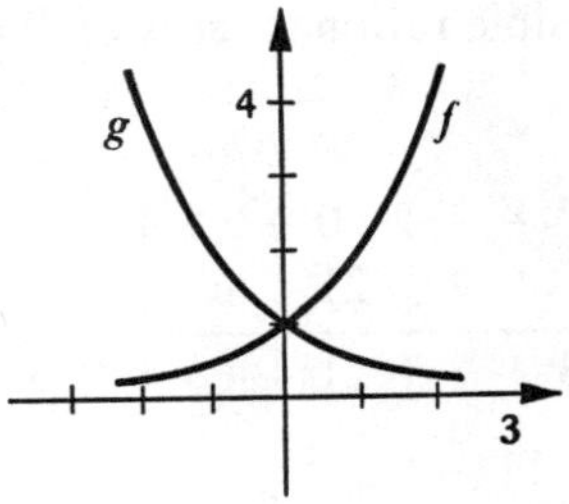

Fig. A.117

CHAPTER 7 REVIEW PROBLEMS

1. a. 1 b. 0 c. -1 d. 2 e. $1/L(b)$

2. a. $\log_b(x) + \log_b(z)$ b. $-\log_b(z)$
c. $\log_b(x) - \log_b(z)$ d. $p \log_b(x)$

3. a. $\log_b(6x)$ b. $\ln\left(\dfrac{2\pi}{\sqrt{(2\pi)} + 1}\right)$ c. $\log_{10}(6)$

4. a. $x = 3$ b. $x = \dfrac{e^2 + 2}{2}$ c. $x = 1$

5.

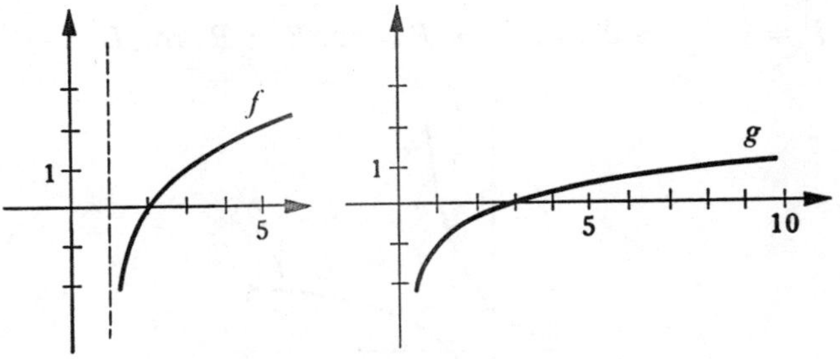

Fig. A.118

CHAPTER 8 REVIEW PROBLEMS

1. a. $\{k\pi\}$ b. $\{(2k - 1)\pi/2\}$
c. $\{2k\pi\}$ d. $\{\pi/6 + 2k\pi\} \cup \{5\pi/6 + 2k\pi\}$
e. $\{\pi/2 + 2k\pi\}$ f. $\{\pi/4 + k\pi\}$
g. $\{\pi/3 + 2k\pi\} \cup \{5\pi/3 + 2k\pi\}$ h. $\{\pi/6 + 2k\pi\} \cup \{11\pi/6 + 2k\pi\}$
i. $\{\pi/4 + 2k\pi\} \cup \{3\pi/4 + 2k\pi\}$

2. $\tan\theta = \dfrac{\sin\theta}{\cos\theta}$; $\cot\theta = \dfrac{\cos\theta}{\sin\theta}$; $\sec\theta = \dfrac{1}{\cos\theta}$; $\csc\theta = \dfrac{1}{\sin\theta}$

3. 2π and π

5. $S_p = \{\pi/3, 2\pi/3, 3\pi/2\}$

6. a. $[1,\pi]$ or $[1,180°]$
b. $[\surd(13),5.69]$ or $[\surd(13),326°]$
c. $[\surd 2,\pi/4]$ or $[\surd 2,45°]$

7. θ varies: $0 \to \pi/2$ $\quad \pi/2 \to \pi$ $\quad \pi \to 3\pi/2$ $\quad 3\pi/2 \to 2\pi$
r varies: $2 \to 1$ $\quad 1 \to 0$ $\quad 0 \to 1$ $\quad 1 \to 2$

CHAPTER 9 REVIEW PROBLEMS

1. $F = \{-2, -2, 3, 4, \ldots\}$; $S = \{-2, -4, -1, 3, \ldots\}$

2. $F = \{2/3, 2/9, -2/27, -22/81, \ldots\}$; $S = \{2/3, 8/9, 22/27, 44/81, \ldots\}$

3. $F(n) = 2n, \forall n$; $S(n) = \sum_{k=1}^{n} 2k$

4. $F(1) = 2$ and, for $n > 1$, $F(n) = -\dfrac{1}{n(n-1)}$;

$$S(n) = 2 - \sum_{k=2}^{n} \frac{1}{k(k-1)}$$

5. Sequence is geometric with $a = 8$ and $x = 1/9$. $\sum_{k=1}^{n} 8(1/3)^{2k-1} = 12$

6. a. $\dfrac{3^n}{n+1} - 1$

b. $n(n+1)(n^2 + n - 3)$

CHAPTER 10 REVIEW PROBLEMS

1. Let T be the number set to which m belongs only if $m \in N$ and $\dfrac{4^m - 1}{3} \in J$ where J is the set of integers. $1 \in T$ since $1 \in N$ and $\dfrac{4^1 - 1}{3} = 1$ and $1 \in J$. If $k \in T$, then $k \in N$ and $\dfrac{4^k - 1}{3} \in J$ by defn. of T.

Since the integers are closed under multiplication

$$4 \cdot \frac{4^k - 1}{3} J \quad \text{but} \quad 4 \cdot \frac{4^k - 1}{3} = \frac{4^{k+1} - 4}{3}$$

$$= \frac{4^{k+1} - 1}{3} - 1$$

and, since the integers are closed under addition we can add 1 and obtain that $\dfrac{4^{k+1} - 1}{3} \in J$. Thus $k + 1 \in T$. By Theorem MI,

T is N.

2. Let T be the number set to which m belongs only if $m \in N$ and $\sum_{k=1}^{m} \frac{k}{2^k} = 2 - \frac{m+2}{2^m}$. $1 \in T$ since $\sum_{k=1}^{1} \frac{k}{2^k} = 1/2$ and $2 - \frac{1+2}{2} = 1/2$.

If $j \in T$, then $j \in N$ and $\sum_{k=1}^{j} \frac{k}{2^k} = 2 - \frac{j+2}{2^j}$ by defn. of T. Hence

$$\frac{j+1}{2^{j+1}} + \sum_{k=1}^{j} \frac{k}{2^k} = 2 - \frac{j+2}{2^j} + \frac{j+1}{2^{j+1}}.$$

Simplifying,

$$\begin{aligned}\sum_{k=1}^{j+1} \frac{k}{2^k} &= 2 - \left[\frac{j+2}{2^j} - \frac{j+1}{2^{j+1}}\right]\\ &= 2 - \left[\frac{j+3}{2^{j+1}}\right]\\ &= 2 - \frac{(j+1)+2}{2^{j+1}}\end{aligned}$$

which means that $j + 1 \in T$. Therefore, by theorem MI,
T is N.

3. $$\begin{aligned}(2x+3)^4 &= \binom{4}{0}(2x)^4 + \binom{4}{1}(2x)^3 3 + \binom{4}{2}(2x)^2 3^2\\ &\quad + \binom{4}{3}(2x)3^3 + \binom{4}{4}3^4\\ &= (2x)^4 + \binom{4}{1}3(2x)^3 + \binom{4}{2}(2x)^2 9 + \binom{4}{3}27(2x) + 81\\ &= 16x^4 + 96x^3 + 216x^2 + 216x + 81\end{aligned}$$

4. Fourth term is $\binom{7}{3}x^3 4^4 = 8940x^3$.

Index